Lecture Notes in Computer Science   **8895**

Commenced Publication in 1973
Founding and Former Series Editors:
Gerhard Goos, Juris Hartmanis, and Jan van Leeuwen

More information about this series at http://www.springer.com/series/7410

Diego F. Aranha · Alfred Menezes (Eds.)

Progress in Cryptology – LATINCRYPT 2014

Third International Conference on Cryptology
and Information Security in Latin America
Florianópolis, Brazil, September 17–19, 2014
Revised Selected Papers

 Springer

Editors
Diego F. Aranha
University of Campinas
Campinas
Brazil

Alfred Menezes
University of Waterloo
Waterloo, ON
Canada

ISSN 0302-9743 ISSN 1611-3349 (electronic)
Lecture Notes in Computer Science
ISBN 978-3-319-16294-2 ISBN 978-3-319-16295-9 (eBook)
DOI 10.1007/978-3-319-16295-9

Library of Congress Control Number: 2015934898

Springer Cham Heidelberg New York Dordrecht London

Springer International Publishing AG Switzerland is part of Springer Science+Business Media
(www.springer.com)

Preface

Latincrypt 2014 was the Third International Conference on Cryptology and Information Security in Latin America and took place during September 17–19, 2014 in Florianópolis, Brazil. Latincrypt 2014 was organized by the Computer Security Laboratory (LabSEC) – Universidade Federal de Santa Catarina, in cooperation with The International Association for Cryptologic Research (IACR). The General Chairs of the conference were Ricardo Custódio and Daniel Panario.

The conference received 48 submissions, of which 5 were withdrawn under various circumstances (authors' request, being identified, randomly generated, or already published in another venue). Each submission was assigned to at least three committee members. Submissions co-authored by members of the Program Committee were assigned to at least five committee members. The reviewing process was challenging due to the large number of high-quality submissions, and we are deeply grateful to the committee members and external reviewers for their outstanding work. After careful deliberation, the Program Committee, which was chaired by Diego F. Aranha and Alfred Menezes, selected 19 submissions for presentation at the conference. In addition to these presentations, the program also included two sessions of talks by graduate students and four invited talks by Jacob Appelbaum, Claudia Diaz, J. Alex Halderman, and Kristin Lauter. The articles in this volume include the accepted submissions and an invited paper corresponding to Kristin Lauter's invited talk.

The reviewing process was run using the WebSubRev software, written by Shai Halevi from IBM Research and hosted on the IACR server. We are grateful to him for setting up the reviewing website and providing invaluable support during the entire process.

Finally, we would like to thank our sponsors CAPES, CNPq, FAPESC, CGI.br, and NIC.br for their financial support as well as all the people who contributed to the success of this conference. In particular, we are indebted to the members of the Latincrypt Steering Committee and the General Chairs for their diligent work and for making this conference possible. We would also like to thank Springer for accepting to publish the proceedings in the Lecture Notes in Computer Science series. It has been a great honor to be PC chairs for Latincrypt 2014 and we look forward to the next edition in the conference series.

October 2014

Diego F. Aranha
Alfred Menezes

LATINCRYPT 2014

Third International Conference on Cryptology and Information Security in Latin America

Florianópolis, Brazil
September 17–19, 2014

Organized by
Computer Security Laboratory (LabSEC), Departamento de Informática e
Estatística (INE) – Universidade Federal de Santa Catarina

In Cooperation with
The International Association for Cryptologic Research (IACR)

General Chairs

Ricardo Custódio   Universidade Federal de Santa Catarina, Brazil
Daniel Panario   Carleton University, Canada

Program Chairs

Diego F. Aranha   Universidade Estadual de Campinas, Brazil
Alfred Menezes   University of Waterloo, Canada

Steering Committee

Michel Abdalla	École Normale Supérieure, France
Paulo Barreto	Universidade de São Paulo, Brazil
Ricardo Dahab	Universidade Estadual de Campinas, Brazil
Alejandro Hevia	Universidad de Chile, Chile
Julio López	Universidade Estadual de Campinas, Brazil
Daniel Panario	Carleton University, Canada
Francisco Rodríguez-Henríquez	CINVESTAV-IPN, Mexico
Alfredo Viola	Universidad de la República, Uruguay

Program Commitee

Michel Abdalla	École Normale Supérieure, France
Jean-Philippe Aumasson	Kudelski Security, Switzerland
Paulo Barreto	Universidade de São Paulo, Brazil
Lejla Batina	Radboud University Nijmegen, The Netherlands
Joan Daemen	STMicroelectronics, Belgium
Ricardo Dahab	Universidade Estadual de Campinas, Brazil

Jeremie Detrey	Inria, France
Orr Dunkelman	University of Haifa, Israel
Joachim von zur Gathen	Universität Bonn, Germany
Jeroen van de Graaf	Universidade Federal de Minas Gerais, Brazil
Helena Handschuh	Cryptography Research, USA
	Katholieke Universiteit Leuven, Belgium
Nadia Heninger	University of Pennsylvania, USA
Alejandro Hevia	Universidad de Chile, Chile
Sorina Ionica	Microsoft Research, USA
Tanja Lange	Technische Universiteit Eindhoven, The Netherlands
Patrick Longa	Microsoft Research, USA
Julio López	Universidade Estadual de Campinas, Brazil
Anderson Nascimento	University of Brasília, Brazil
Gregory Neven	IBM Zurich Research Laboratory, Switzerland
Kenny Paterson	Royal Holloway, University of London, UK
Thomas Peyrin	Nanyang Technological University, Singapore
Francisco Rodríguez-Henríquez	CINVESTAV-IPN, Mexico
Peter Schwabe	Radboud University Nijmegen, The Netherlands
Nicolas Sendrier	Inria, France
Douglas Stebila	Queensland University of Technology, Australia
Damien Stehle	École Normale Supérieure de Lyon, France
Nicolas Thériault	Universidad del Bío-Bío, Chile
Emmanuel Thomé	Inria, France
Maribel Gonzalez Vasco	Universidad Rey Juan Carlos de Madrid, Spain
Alfredo Viola	Universidade de la República, Uruguay
Scott Yilek	University of St. Thomas, USA

External Reviewers

Rodrigo Abarzúa
Gora Adj
Janaka Alawatugoda
Shi Bai
Guido Bertoni
Gaetan Bisson
Olivier Blazy
Johannes Blömer
Joan Boyar
Angelo De Caro
Jung Hee Cheon
Baris Ege
Georg Fuchsbauer
Steven Galbraith

Conrado P.L. Gouvêa
Andreas Hülsing
Philipp Jovanovic
Carmen Kempka
Anna Krasnova
Adeline Langlois
Moonsung Lee
Loíck Lhote
Daniel Loebenberger
Karina M. Magalhães
Eduardo M. Morais
Michael Naehrig
Erick Nascimento
Ruben Niederhagen

Jesper Buus Nielsen
Michael Nüsken
Thomaz Oliveira
Louiza Papachristodolou
Kostas Papagiannopoulos
Luis J. Dominguez Perez
Kim Ramchen

Vanishree Hanumantha Rao
Boris Skoric
Antonio Vera
Hoeteck Wee
Erich Wenger
Konstantin Ziegler

Sponsoring Institutions

Coordination for the Improvement of Higher Education Personnel (CAPES)
National Council for Scientific and Technological Development (CNPq)
Universidade Federal de Santa Catarina (UFSC)
Computer Security Laboratory (LabSEC), UFSC
Departamento de Informática e Estatística (INE), UFSC
Fundação de Amparo à Pesquisa e Inovação do Estado de Santa Catarina (FAPESC)
The Brazilian Internet Steering Committee (CGI.br)
The Brazilian Networking Information Center (NIC.br)
Carleton University

Contents

Invited Talks

Private Computation on Encrypted Genomic Data

Kristin Lauter[1], Adriana López-Alt[2], and Michael Naehrig[1(✉)]

[1] Microsoft Research, Redmond, USA
{klauter,mnaehrig}@microsoft.com
[2] New York University, New York, USA
adrilopo@gmail.com

Abstract. A number of databases around the world currently host a wealth of genomic data that is invaluable to researchers conducting a variety of genomic studies. However, patients who volunteer their genomic data run the risk of privacy invasion. In this work, we give a cryptographic solution to this problem: to maintain patient privacy, we propose encrypting all genomic data in the database. To allow meaningful computation on the encrypted data, we propose using a homomorphic encryption scheme.

Specifically, we take basic genomic algorithms which are commonly used in genetic association studies and show how they can be made to work on encrypted genotype and phenotype data. In particular, we consider the Pearson Goodness-of-Fit test, the D' and r^2-measures of linkage disequilibrium, the Estimation Maximization (EM) algorithm for haplotyping, and the Cochran-Armitage Test for Trend. We also provide performance numbers for running these algorithms on encrypted data.

1 Introduction

As the cost of sequencing the human genome drops, more and more genomic data will become available for scientific study. At the same time, researchers are developing new methods for analyzing genomic data across populations to look for patterns and find correlations. Such research may help identify genetic risk factors for disease, suggest treatments, or find cures. But to make this data available for scientific study, patients expose themselves to risks from invasion of privacy [ADCHT13]. Even when the data is anonymized, individual patients' genomic data can be re-identified [GMG+13, WLW+09] and can furthermore expose close relatives to similar risks [HAHT13].

A number of databases to host genomic data for research have been created and currently house a wealth of genomic data, for example the 1,000 Genomes Project [TGP], the International Cancer Genome Consortium (ICGC) [ICG], and the International Rare Diseases Research Consortium (IRDiRC) [IRD]. There are also a number of shared research databases which house de-identified genomic

---

Adriana López-Alt—Research conducted while visiting Microsoft Research.

© Springer International Publishing Switzerland 2015
D.F. Aranha and A. Menezes (Eds.): LATINCRYPT 2014, LNCS 8895, pp. 3–27, 2015.
DOI: 10.1007/978-3-319-16295-9_1

sequence data such as the eMERGE Network [MCC+11], the Database of Geno-types and Phenotypes [dbG], the European Bioinformatics Institute [EBI], and the DNA Databank of Japan [Jap].

Various approaches to protecting genomic privacy while allowing research on the data include policy-based solutions, de-identification of data, approximate query-answering, and technological solutions based on cryptography. Emerging cryptographic solutions are quickly becoming more relevant. Encryption is a tool which essentially allows one to seal data in a metaphorical vault, which can only be opened by somebody holding the secret decryption key. *Homomorphic* Encryption (HE) allows other parties to operate on the data without possession of the secret key (metaphorically sticking their hands into the vault via a glove box and manipulating the data). *Fully* Homomorphic Encryption (FHE) allows the evaluation of *any* function on encrypted data but current implementations are widely inefficient. More practical variants of HE schemes allow for only a fixed amount of computation on encrypted data while still ensuring correctness and security. In particular, practical HE schemes allow for evaluation of polynomials of small degree.

In this work, we take basic genomic algorithms which are commonly used in genome wide association studies (GWAS) and show how they can be made to work on encrypted data using HE. We find a number of statistical algorithms which can be evaluated with polynomials of small degree, including the Pearson Goodness-of-Fit or Chi-Squared Test to test for deviation from Hardy-Weinberg equilibrium, the D' and r^2 measures of linkage disequilibrium to test for association in the genotypes at two loci in a genome, the Estimation Maximization (EM) Algorithm to estimate haplotype frequencies from genotype counts, and the Cochran-Armitage Test for Trend (CATT) to determine if a candidate allele is associated to a disease.

In our approach, these statistics are computed from encrypted genotype and phenotype counts in a population. Thus for a database containing encrypted phenotypes and genotypes, we consider two stages: in the first stage encrypted phenotype and genotype counts are computed using only simple additions. The parameters of the encryption scheme, as well as the running time of the computation in this stage depend on the size of the population sample being considered.[1] The second stage takes as input the encrypted genotype and phenotype counts obtained in the first stage and computes the output of the statistical algorithms mentioned above. In this stage the runtime of the statistical algorithms *does not depend* on the size of the population sample and only depends on the parameter set needed for the computation. Table 2 gives the timings to evaluate the statistical algorithms on encrypted genotype and phenotype counts. For example, the Cochran Armitage Test for Trend takes 0.94 s at the smaller parameter size and 3.63 s at the larger parameter size.

---

[1] The running time is linear in the population size for a fixed parameter set. For larger population sizes, parameters need to be increased and performance degrades, but not by a large factor (see Table 1 for a comparison of the running times for two typical parameter sets).

Genomic Databases: Hosted by a Trusted Party, Stored in an Untrusted Cloud. It is important to note that in this work we are considering single-key homomorphic encryption, which means that all data is encrypted under the same symmetric or asymmetric encryption key. To see how this can be used to protect privacy in genome databases as described above, consider the following scenario which captures one of the challenges facing government and research organizations currently deploying large-scale genomic databases for research.

A global alliance of government agencies, research institutes, and hospitals wants to pool all their patients' genomic data to make available for research. A common infrastructure is required to host all these data sets, and to handle the demands of distributed storage, providing a low cost solution which is scalable, elastic, efficient, and secure. These are the arguments for using commercial cloud computing infrastructure made by the Global Alliance [Glo13, p.17] in their proposal. Thus we arrive at the following requirement: data collected by a trusted host or hosts, such as a hospital or research facility, may need to be stored and processed in an untrusted cloud, to enable access and sharing across multiple types of boundaries. The mutually trusting data owners, i.e. the hospital or hospitals, can encrypt all data under a single key using homomorphic encryption. The cloud can then process queries on the encrypted data from arbitrary entities such as member organizations, registered individual researchers, clinicians etc. The cloud can return encrypted results in response to queries, and the trusted party can provide decryptions to registered parties according to some policy governing allowable queries. Note that the policy should not allow arbitrary queries, since this would expose the data to the same re-identification risks that an unencrypted public database faces. However, with a reasonable policy, this would allow researchers to study large data sets from multiple sources without making them publicly available to the researchers who query them.

Related Work. Much of the related work on genomic privacy focuses on the problem of pattern-matching for genomic sequences, which is quite different from the statistical algorithms we analyze here. Actually the circuits for pattern matching and edit distance are much deeper than those considered here, so less suitable as an efficient application of HE. On the other hand, De Cristofaro et al. [DCFT13] present an algorithm for private substring-matching which is extremely efficient. In another approach, Blanton et al. [BAFM12] efficiently carry out sequence comparisons using garbled circuits. Finally, Ayday et al. [ARH13] show how to use *additively* homomorphic encryption to predict disease susceptibility while preserving patient privacy.

Differential privacy techniques have also been investigated in several recent papers [FSU11, JS13]. Fienberg et al. [FSU11] propose releasing differentially private minor allele frequencies, chi-square statistics and p-values as well as a differentially-private approach to penalized logistic regression (see e.g. [PH08]). Johnson and Shmatikov [JS13] present a set of privacy-preserving data mining algorithms for GWAS datasets based on differential privacy techniques.

Finally a recent line of work investigating practical applications of HE to outsourcing computation on encrypted data has led to the present paper.

Lauter et al. [LNV11] introduce the idea of medical applications of HE, with optimizations, concrete parameters, and performance numbers. Graepel et al. [GLN13] apply HE to machine learning, both to train a model on encrypted data, and to give encrypted predictions based on an encrypted learned model. Bos et al. [BLN14] give optimized performance numbers for HE and a particular application in health care to predictive analysis, along with an algorithm for automatically selecting parameters. Yasuda et al. [YSK+13] give an application of HE to secure pattern matching.

2   Statistical Algorithms in Genetic Association Studies

In this section, we detail common statistical algorithms used in genetic association studies. We consider the Pearson Goodness-Of-Fit test, which is used to determine deviation from Hardy-Weinberg Equilibrium (HWE) (Sect. 2.1), the D' and r^2 measures of linkage disequilibrium, as well as the Estimation-Maximization (EM) algorithm for haplotyping (Sect. 2.2), and the Cochran-Armitage Test for Trend (CATT) used in case-control studies (Sect. 2.3).

2.1   Hardy-Weinberg Equilibrium and the Pearson Goodness-of-Fit Test

We begin by describing the Pearson Goodness-Of-Fit test, a test frequently used to determine whether a gene is in Hardy-Weinberg Equilibrium (HWE). We first review the notion of HWE and then describe the Pearson test.

Hardy-Weinberg Equilibrium (HWE). A gene is said to be in HWE if its allele frequencies are independent. More specifically, suppose A and a are two alleles of the gene being considered, and let N_{AA}, N_{Aa}, N_{aa} denote the observed population counts for genotypes AA, Aa, aa, respectively. Also let N be the total number of people in the sample population; that is, $N \overset{\text{def}}{=} N_{AA} + N_{Aa} + N_{aa}$. With this notation, the corresponding frequencies of the genotypes AA, Aa, aa are given by

$$p_{AA} \overset{\text{def}}{=} \frac{N_{AA}}{N}, \qquad p_{Aa} \overset{\text{def}}{=} \frac{N_{Aa}}{N}, \qquad p_{aa} \overset{\text{def}}{=} \frac{N_{aa}}{N}.$$

Moreover, the frequencies of the alleles A and a are given by

$$p_A \overset{\text{def}}{=} \frac{2N_{AA} + N_{Aa}}{2N}, \qquad p_a \overset{\text{def}}{=} \frac{2N_{aa} + N_{Aa}}{2N} = 1 - p_A,$$

since each count of genotype AA contributes two A alleles, each count of genotype aa contributes two a alleles, each count of genotype Aa contributes one A allele and one a allele, and the total number of alleles in a sample of N people is $2N$.

The gene is said to be in equilibrium if these frequencies are independent, or in other words, if

$$p_{AA} = p_A^2, \quad p_{Aa} = 2p_A p_a, \quad p_{aa} = p_a^2.$$

When a gene is in equilibrium, its allele frequencies stay the same from generation to generation unless perturbed by evolutionary influences. Researchers test for HWE as a way to test for data quality, and might discard loci that deviate significantly from equilibrium.

Pearson Goodness-of-Fit Test. The main observation made by the Pearson Goodness-of-Fit test is that if the alleles are independent (i.e. if the gene is in equilibrium) then we expect the observed counts to be

$$E_{AA} \stackrel{\text{def}}{=} Np_A^2, \quad E_{Aa} \stackrel{\text{def}}{=} 2Np_A p_a, \quad E_{aa} \stackrel{\text{def}}{=} Np_a^2.$$

Thus, deviation from equilibrium can be determined by comparing the X^2 test-statistic below to the χ^2-statistic with 1 degree of freedom[2]:

$$X^2 \stackrel{\text{def}}{=} \sum_{i \in \{AA, Aa, aa\}} \frac{(N_i - E_i)^2}{E_i}.$$

2.2 Linkage Disequilibrium

Another important notion in genetic association studies is linkage disequilibrium (LD). Linkage disequilibrium is an association in the genotypes at two loci in a genome. Suppose A, a are possible alleles at locus 1 and B, b are possible alleles at locus 2. In this case there are 9 possible genotypes: $AABB, AABb, AAbb,$ $AaBB, AaBb, Aabb, aaBB, aaBb, aabb$. For $i, i' \in \{A, a\}$ and $j, j' \in \{B, b\}$ we use $N_{ii'jj'}$ to denote the observed count of genotype $ii'jj'$. As before, let N be the total size of the population sample:

$$N \stackrel{\text{def}}{=} \sum_{\substack{i,i' \in \{A,a\} \\ j,j' \in \{B,b\}}} N_{ii'jj'}.$$

We consider the population frequencies of alleles A, a, B, b:

$$p_A \stackrel{\text{def}}{=} \frac{\sum_{j,j' \in \{B,b\}} 2N_{AAjj'} + N_{Aajj'}}{2N}, \quad p_a \stackrel{\text{def}}{=} \frac{\sum_{j,j' \in \{B,b\}} 2N_{aajj'} + N_{Aajj'}}{2N},$$

$$p_B \stackrel{\text{def}}{=} \frac{\sum_{i,i' \in \{A,a\}} 2N_{ii'BB} + N_{ii'Bb}}{2N}, \quad p_b \stackrel{\text{def}}{=} \frac{\sum_{i,i' \in \{A,a\}} 2N_{ii'bb} + N_{ii'Bb}}{2N}.$$

Moreover, there are exactly 4 haplotypes to consider: AB, Ab, aB, ab. For $i \in \{A, a\}$ and $j \in \{B, b\}$, we use N_{ij} to denote the observed count for the haplotype ij and consider the population frequencies

$$p_{AB} \stackrel{\text{def}}{=} \frac{N_{AB}}{2N}, \quad p_{Ab} \stackrel{\text{def}}{=} \frac{N_{Ab}}{2N}, \quad p_{aB} \stackrel{\text{def}}{=} \frac{N_{aB}}{2N}, \quad p_{ab} \stackrel{\text{def}}{=} \frac{N_{ab}}{2N}.$$

---

[2] 1 degree of freedom = 3 genotypes − 2 alleles.

Under *linkage equilibrium*, we expect the allele frequencies to be independent. In other words, we expect

$$p_{AB} = p_A p_B, \quad p_{Ab} = p_A p_b, \quad p_{aB} = p_a p_B, \quad p_{ab} = p_a p_b.$$

If the alleles are in *linkage disequilibrium*, the frequencies will deviate from the values above by a scalar D, so that

$$p_{AB} = p_A p_B + D, \quad p_{Ab} = p_A p_b - D, \quad p_{aB} = p_a p_B - D, \quad p_{ab} = p_a p_b + D.$$

The scalar D is easy to calculate: $D = p_{AB}p_{ab} - p_{Ab}p_{aB} = p_{AB} - p_A p_B$. However, the range of D depends on the frequencies, which makes it difficult to use it as a measure of disequilibrium. One of two scaled-down variants is used instead, the D'-measure or the r^2-measure.

D'-*Measure.* The D'-measure is defined as:

$$D' \overset{\text{def}}{=} \frac{|D|}{D_{\text{max}}} \quad \text{where} \quad D_{\text{max}} = \begin{cases} \min\{p_A p_b, p_a p_B\} & \text{if } D > 0, \\ \min\{p_A p_B, p_a p_b\} & \text{if } D < 0. \end{cases}$$

r^2- *Measure.* The r^2 measure is given by

$$r^2 \overset{\text{def}}{=} \frac{X^2}{N}, \quad \text{where} \quad X^2 \overset{\text{def}}{=} \sum_{\substack{i \in \{A,a\} \\ j \in \{B,b\}}} \frac{(N_{ij} - E_{ij})^2}{E_{ij}},$$

where N_{ij} is the observed count and $E_{ij} \overset{\text{def}}{=} N p_i p_j$ is the expected count. Using the fact that $|N_{ij} - E_{ij}| = ND$, it can be shown that

$$r^2 = \frac{D^2}{p_A p_B p_a p_b}.$$

The range of both D' and r^2 is $[0, 1]$. A value of 0 indicates perfect equilibrium and a value of 1 indicates perfect disequilibrium.

EM Algorithm for Haplotyping. Using the D' and r^2 LD measures described above requires knowing the observed haplotype counts or frequencies. However, haplotype counts (resp. frequencies) cannot be exactly determined from genotype counts (resp. frequencies). For example, consider 2 bi-allelic loci with alleles A, a and B, b. An observed genotype $AaBb$ can be one of two possible haplotypes: $(AB)(ab)$ or $(Ab)(aB)$. In practice, the Estimation Maximization (EM) algorithm can be used to *estimate* haplotype frequencies from genotype counts.

The EM algorithm starts with arbitrary initial values $p_{AB}^{(0)}, p_{Ab}^{(0)}, p_{aB}^{(0)}, p_{ab}^{(0)}$ for the haplotype frequencies, and iteratively updates them using the observed genotype counts. In each iteration, the current estimated haplotype frequencies are used in an *estimation step* to calculate the expected genotype frequencies (assuming the initial values are the true haplotype frequencies). Next, in a *maximization step*, these are used to estimate the haplotype frequencies for the next iteration. The algorithm stops when the haplotype frequencies have stabilized.

mTH ESTIMATION STEP

$$E_{AB/ab}^{(m)} \stackrel{\text{def}}{=} \mathbb{E}\left[N_{AB/ab} \mid N_{AaBb}, p_{AB}^{(0)}, p_{Ab}^{(0)}, p_{aB}^{(0)}, p_{ab}^{(0)}\right] = N_{AaBb} \cdot \frac{p_{AB}^{(m-1)} p_{ab}^{(m-1)}}{p_{AB}^{(m-1)} p_{ab}^{(m-1)} + p_{Ab}^{(m-1)} p_{aB}^{(m-1)}}$$

$$E_{Ab/aB}^{(m)} \stackrel{\text{def}}{=} \mathbb{E}\left[N_{Ab/aB} \mid N_{AaBb}, p_{AB}^{(0)}, p_{Ab}^{(0)}, p_{aB}^{(0)}, p_{ab}^{(0)}\right] = N_{AaBb} \cdot \frac{p_{Ab}^{(m-1)} p_{aB}^{(m-1)}}{p_{AB}^{(m-1)} p_{ab}^{(m-1)} + p_{Ab}^{(m-1)} p_{aB}^{(m-1)}}$$

mTH MAXIMIZATION STEP

$$N_{AB}^{(m)} = 2N_{AABB} + N_{AABb} + N_{AaBB} + E_{AB/ab}^{(m)}$$
$$N_{ab}^{(m)} = 2N_{aabb} + N_{aaBb} + N_{Aabb} + E_{AB/ab}^{(m)}$$
$$N_{Ab}^{(m)} = 2N_{AAbb} + N_{AABb} + N_{Aabb} + E_{Ab/aB}^{(m)}$$
$$N_{aB}^{(m)} = 2N_{aaBB} + N_{AaBB} + N_{aaBb} + E_{Ab/aB}^{(m)}$$

2.3 Cochran-Armitage Test for Trend (CATT)

Finally, we consider the Cochran-Armitage Test for Trend (CATT), which is used in case-control studies to determine if a candidate allele is associated to a disease. We first describe the basic structure of case-control studies, and then describe the CATT test.

Case-Control Studies. As mentioned above, a case-control study is used to determine if a candidate allele A associated to a specified disease. Such a study compares the genotypes of individuals who have the disease (cases) to the genotypes of individuals who do not (controls). A 2×3 contingency table of 3 genotypes vs. case/controls can be constructed with this information, as below, where the N_{ij} represent a number of individuals, R_i is the sum of the ith row, and C_j is the sum of the jth column. For example, N_{10} is the number of individuals with genotype AA who present the disease (affected phenotype), $R_0 = N_{00} + N_{01} + N_{02}$, $C_0 = N_{00} + N_{10}$, etc.

	AA	Aa	aa	Sum
Controls	N_{00}	N_{01}	N_{02}	R_0
Cases	N_{10}	N_{11}	N_{12}	R_1
Sum	C_0	C_1	C_2	N

Cochran-Armitage Test for Trend (CATT). Given a contingency table as above, the CATT computes the statistic

$$T \stackrel{\text{def}}{=} \sum_{i=0}^{2} w_i(N_{0i}R_1 - N_{1i}R_0),$$

where $\boldsymbol{w} \overset{\text{def}}{=} (w_0, w_1, w_2)$ is a vector of pre-determined weights[3], and the difference $(N_{0i}R_1 - N_{1i}R_0)$ can be thought of as the difference $N_{0i} - N_{1i}$ of controls and cases for a specific genotype, after reweighing the rows in the table to have the same sum.

The test statistic X^2 is defined to be

$$X^2 \overset{\text{def}}{=} \frac{T^2}{\text{Var}\,(T)},$$

where $\text{Var}\,(T)$ is the variance of T:

$$\text{Var}\,(T) = \frac{R_0 R_1}{N} \left(\sum_{i=0}^{2} w_i^2 C_i (N - C_i) - 2 \sum_{i=1}^{k-1} \sum_{j=i+1}^{k} w_i w_j C_i C_j \right).$$

To determine if a trend can be inferred, the CATT compares the test statistic X^2 to a χ^2-statistic with 1 degree of freedom.

2.4   Linear Regression

Linear regression is used in cases when the phenotype or trait is a continuous variable (e.g. tumor size) rather than a binary variable (e.g. whether a disease is present or not). It assumes a linear relationship between trait values and the genotype. The input data is a set of N pairs $(y_i, \boldsymbol{x}_i)$, where $y_i \in \{0, 1, 2\}$ is a genotype[4], and $\boldsymbol{x}_i \in \mathbb{R}^k$ is the vector of trait values corresponding to the individual with genotype x_i. Define

$$\boldsymbol{y} \overset{\text{def}}{=} \begin{pmatrix} y_1 \\ \vdots \\ y_N \end{pmatrix}, \quad \mathbf{X} \overset{\text{def}}{=} \begin{pmatrix} \boldsymbol{x}_1^\top \\ \vdots \\ \boldsymbol{x}_N^\top \end{pmatrix}$$

Linear regressing finds $\boldsymbol{\beta} \in \mathbb{R}^k$ and $\boldsymbol{\varepsilon} \in \mathbb{R}^N$ such that $\boldsymbol{y} = \mathbf{X}\boldsymbol{\beta} + \boldsymbol{\varepsilon}$.

Linear regression models can be found using the least squares approach, and a solution to approximating least squares on homomorphically encrypted data is considered by Graepel, et al. [GLN13, Sect. 3.1]. Their work focuses on Fisher's linear discriminant classifier, but as noted there, linear regression can be cast in a similar framework. We refer the reader to their work for more details.

3   Practical Homomorphic Encryption

Fully homomorphic encryption (FHE) enables one to perform arbitrary computations on encrypted data, without first decrypting the data and without any

---

[3] Common choices for the set of weights $\boldsymbol{w} = (w_0, w_1, w_2)$ are: $\boldsymbol{w} = (0, 1, 2)$ for the additive (co-dominant) model, $\boldsymbol{w} = (0, 1, 1)$ for the dominant model (A is dominant over a), and $\boldsymbol{w} = (0, 0, 1)$ for the recessive model (A is recessive to allele a).

[4] For a bi-allelic gene with alleles A and a, the value 0 corresponds to the genotype AA, the value 1 corresponds to the genotype Aa and the value 2 corresponds to the genotype aa.

knowledge of the secret decryption key. The result of the computation is given in encrypted form and can only be decrypted by a legitimate owner of the private decryption key. The first construction of FHE was shown by Gentry in 2009 [Gen09], and many improvements and new constructions have been presented in recent years [vDGHV10, BV11b, BV11a, BGV12, FV12, LTV12, Bra12, BLLN13, GSW13, BV14].

Model of Computation. In Gentry's initial work and many follow-up papers, computation is modeled as a boolean circuit with XOR and AND gates, or equivalently, as an arithmetic circuit over $\mathbb{F}_2$[5]. The data is encrypted bit-wise, which means that a separate ciphertext is produced for each bit in the message. Addition and multiplication operations are then performed on the encrypted bits. Unfortunately, breaking down a computation into bit operations can quickly lead to a large and complex circuit, thus making homomorphic computation very inefficient.

Luckily, most known constructions allow the computation to take place over a larger message space. In particular, if the desired computation only needs to compute additions and multiplications of integer values (as is (almost)[6] the case with all algorithms presented in Sect. 2), then the data does not necessarily need to be expressed in a bitwise manner. Indeed, most known constructions allow the integers (or appropriate encodings of the integers) to be encrypted and homomorphic additions and multiplications to be performed over these integer values. The advantage of this approach is clear: a ciphertext now contains much more information than a single bit of data, making the homomorphic computation much more efficient.

It is important to note that in the latter approach, the only possible homomorphic operations are addition (equivalently, subtraction) and multiplication. It is currently not known how to perform division of integer values without performing an inefficient bitwise computation, as described above. For practical reasons, in this work we limit homomorphic operations to include only addition, subtraction, and multiplication.

Levels of Homomorphism. In all known FHE schemes, ciphertexts inherently contain a certain amount of noise, which "pollutes" the ciphertext. This noise grows during homomorphic operations and if the noise becomes too large, the ciphertext cannot be decrypted even with the correct decryption key. A *somewhat homomorphic encryption scheme* is one that can evaluate a limited number of operations (both addition and multiplication) before the noise grows large enough to cause decryption failures. Somewhat homomorphic schemes are usually very practical.

In order to perform an unlimited number of operations (and achieve *fully* homomorphic encryption), ciphertexts need to be constantly refreshed in order to reduce their noise. This is done using a costly procedure called *bootstrapping*.

---

[5] An arithmetic circuit over $\mathbb{F}_t$ has addition and multiplication gates modulo t.

[6] The algorithms in Sect. 2 include divisions. In Sect. 4, we show how to get around this issue.

A *leveled homomorphic encryption scheme* is one that allows the setting of parameters so as to be able to evaluate a given computation. In other words, given a fixed function that one wishes to compute, it is possible to select the parameters of the scheme in a way that allows one to homomorphically compute the specified function, without the use of the costly bootstrapping step. Leveled homomorphic schemes enjoy the flexibility of fully homomorphic schemes, in that they can homomorphically evaluate any function, and are also quite practical (albeit not as practical as somewhat homomorphic schemes). The construction we use in our implementation is a leveled homomorphic encryption scheme.

3.1   The Homomorphic Encryption Scheme

In our implementation we use a modified[7] version of the homomorphic encryption scheme proposed by López-Alt and Naehrig [LN14b], which is based on the schemes [SS11, LTV12, Bra12, BLLN13]. The scheme is a public-key encryption scheme and consists of the following algorithms:

- A key generation algorithm KeyGen(params) that, on input the system parameters params, generates a public/private key pair $(\mathsf{pk}, \mathsf{sk})$.
- An encryption algorithm Encrypt(pk, m) that encrypts a message m using the public key pk.
- A decryption algorithm Decrypt(sk, c) that decrypts a ciphertext c with the private key sk.
- A homomorphic addition function Add(c_1, c_2) that given encryptions c_1 and c_2 of m_1 and m_2, respectively, outputs a ciphertext encrypting the sum $m_1 + m_2$.
- A homomorphic multiplication function Mult(c_1, c_2) that, given encryptions c_1 and c_2 of m_1 and m_2, respectively, outputs a ciphertext encrypting the product $m_1 m_2$.

System Parameters. The scheme operates in the ring $R \stackrel{\text{def}}{=} \mathbb{Z}[X]/(X^n + 1)$, whose elements are polynomials with integer coefficients of degree less than n. All messages, ciphertexts, encryption and decryption keys, etc. are elements in the ring R, and have this form. In more detail, an element $a \in R$ has the form $a = \sum_{i=0}^{n-1} a_i X^i$, with $a_i \in \mathbb{Z}$. Addition in R is done component-wise in the coefficients, and multiplication is simply polynomial multiplication modulo $X^n + 1$.

The scheme also uses an integer modulus q. In what follows, we use the notation $[a]_q$ to denote the operation of reducing the coefficients of $a \in R$ modulo q into the set $\{-\lfloor \frac{q}{2} \rfloor, \ldots, \lfloor \frac{q}{2} \rfloor\}$.

---

[7] The only modification we make to the scheme of López-Alt and Naehrig is removing a step called "relinearization" or "key switching", needed to make decryption independent of the function that was homomorphically evaluated. In our implementation, decryption depends on the number of homomorphic multiplications that were performed. We make this change for efficiency reasons, as relinearization is very costly.

Finally, the scheme uses two probability distributions on R, χ_{key} and χ_{err}, which generate polynomials in R with small coefficients. In our implementation, we let the distribution χ_{key} be the uniform distribution on polynomials with coefficients in $\{-1, 0, 1\}$. Sampling an element according to this distribution means sampling all its coefficients uniformly from $\{-1, 0, 1\}$. For the distribution χ_{err}, we use a discrete Gaussian distribution with mean 0 and appropriately chosen standard deviation (see Sect. 4.4). For clarity of presentation, we refrain from formally describing the specifics of this distribution and instead refer the reader to any of [SS11, LTV12, Bra12, BLLN13, LN14b] for a formal definition.

The system parameters of the scheme are the degree n, modulus q, and distributions $\chi_{\text{key}}, \chi_{\text{err}}$: $\mathsf{params} = (n, q, \chi_{\text{key}}, \chi_{\text{err}})$.

Plaintext Space. The plaintext space of the scheme is the set of integers in the interval $\mathcal{M} = [-2^n, 2^n]$. For the scheme to work correctly, we assume that the initial inputs, the output of the function evaluation, and all intermediate values are all in $\mathcal{M}$.

To encrypt an integer $\mu \in \mathcal{M}$, this integer is first encoded as a polynomial $m \in R$. To do this, we take the bit-decomposition of μ and use these bits as the coefficients in m. Formally, if $\mu = \sum_i \mu_i 2^i$ for $\mu_i \in \{0, 1\}$, then we define

$$m = \sum_i \mu_i X^i.$$

Formal Definition. Below is a formal and detailed definition of the key generation, encryption, decryption, and homomorphic evaluation algorithms.

- KeyGen(params): On input the parameters $\mathsf{params} = (n, q, \chi_{\text{key}}, \chi_{\text{err}})$, the key generation algorithm samples polynomials $f', g \leftarrow \chi_{\text{key}}$ from the key distribution and sets

$$f = [(X - 2)f' + 1]_q.$$

  If f is not invertible modulo q, it chooses a new f'. Otherwise, it computes the inverse f^{-1} of f in R modulo q. Finally, it outputs the key pair:

$$\mathsf{pk} = h \stackrel{\text{def}}{=} [gf^{-1}]_q \in R \quad \text{and} \quad \mathsf{sk} = f \in R.$$

- Encrypt(h, μ): To encrypt an integer $\mu \in \mathcal{M}$, the encryption algorithm first encodes it as a polynomial m, as described above. Then, it samples small error polynomials $s, e \leftarrow \chi_{\text{err}}$, and outputs the ciphertext

$$c \stackrel{\text{def}}{=} [\Delta m + e + hs]_q \in R,$$

  where

$$\Delta \stackrel{\text{def}}{=} \lceil q \Upsilon \rceil \quad \text{and} \quad \Upsilon \stackrel{\text{def}}{=} -\frac{X^{n-1} + 2X^{n-2} + 4X^{n-3} + \ldots + 2^{n-1}}{2^n + 1} \in \mathbb{Q}[X].$$

- Add(c_1, c_2): Given two ciphertexts c_1 and c_2, outputs the ciphertext $c_{\text{add}} \stackrel{\text{def}}{=} [c_1 + c_2]_q$.

- Mult(c_1, c_2): Given two ciphertexts c_1 and c_2, outputs $c_{mult} \stackrel{\text{def}}{=} [c_1 c_2]_q$.
- Decrypt(f, c): Given the private decryption key f and a ciphertext c that is the output of a degree-D function evaluation[8], the decryption algorithm computes $\widetilde{f} \stackrel{\text{def}}{=} f^D \in R$ and

$$\mu = \left(\left\lfloor \frac{(X-2)}{q} \cdot [\widetilde{f}c]_q \right\rceil \mod (X-2) \right) \mod 2^n + 1.$$

We remark that if the function that will be homomorphically computed is known in advance (or even only its degree), then the polynomial f^D can be precomputed when the secret key is generated, simplifying the decryption step to a single polynomial multiplication and some modular operations.

We also note that the modular reduction modulo $(X-2)$ is mathematically equivalent to the evaluation of the polynomial at the point $X = 2$.

4   Computation on Encrypted Data

In this section, we discuss how to run the statistical algorithms described in Sect. 2 on genetic data encrypted using the homomorphic encryption scheme described in Sect. 3. To this end, in Sect. 4.1 we describe how genetic data can be encoded and encrypted. In Sect. 4.2 we discuss how to obtain the genotype and phenotype frequencies that serve as input to the algorithms described in Sect. 2. Additionally, given the constraints of homomorphic computation on encrypted data, we must make some necessary modifications to the statistical algorithms; we describe these in Sect. 4.3. Finally, in Sect. 4.4, we discuss how to choose the parameters of the encryption scheme. In what follows, for a value a, we use $\widehat{a}$ to denote an encryption of a.

4.1   Encoding Genomic Data

Structure of the Data. Data used in genetic association studies consists of individuals' genotypes and phenotypes. The data can be represented in 2 tables or matrices, one for genotype information and the other for phenotype information. In the genotypes table, each row contains information about a single person, and each column specifies a DNA locus. An entry in this table specifies the person's genotype at the given locus. For a bi-allelic gene with alleles A, a, this can be one of 4 possible values: the reference homozygote AA (value 0), the heterozygote Aa (value 1), the non-reference homozygote aa (value 2) and "missing" if that person's genotype at the specified locus is not known.

Similarly, in the phenotypes table, each row contains information about a single person, and each column specifies a single phenotype. An entry in this table specifies the person's given phenotype. For a disease phenotype, this can be one of 3 possible values: unaffected (value 0), affected (value 1) and "missing" if that

---

[8] Informally, a function has degree D if it can be represented as a (possibly multivariate) polynomial of degree D. See Sect. 4.4 for more details.

person's affection status is not known. For continuous phenotypes (e.g. tumor size), the table entry contains a real number. We focus only on phenotypes containing disease affection status.

Genotype Encoding. For each entry (i, j) in the genotype table, we compute 3 ciphertexts, one for each of the possible values 0,1,2 (ie. AA, Aa, aa); we call these ciphertexts $c_0^{(i,j)}, c_1^{(i,j)}, c_2^{(i,j)}$ respectively. A ciphertext encrypts 1 if the entry value is the same as the value it represents, and 0 otherwise. More specifically, the 4 possible genotypes are encoded as follows:

AA (value 0) : $c_0^{(i,j)} \leftarrow \mathsf{Encrypt}(\mathsf{pk}, 1),$ $c_1^{(i,j)} \leftarrow \mathsf{Encrypt}(\mathsf{pk}, 0),$ $c_2^{(i,j)} \leftarrow \mathsf{Encrypt}(\mathsf{pk}, 0),$

Aa (value 1) : $c_0^{(i,j)} \leftarrow \mathsf{Encrypt}(\mathsf{pk}, 0),$ $c_1^{(i,j)} \leftarrow \mathsf{Encrypt}(\mathsf{pk}, 1),$ $c_2^{(i,j)} \leftarrow \mathsf{Encrypt}(\mathsf{pk}, 0),$

aa (value 2) : $c_0^{(i,j)} \leftarrow \mathsf{Encrypt}(\mathsf{pk}, 0),$ $c_1^{(i,j)} \leftarrow \mathsf{Encrypt}(\mathsf{pk}, 0),$ $c_2^{(i,j)} \leftarrow \mathsf{Encrypt}(\mathsf{pk}, 1),$

missing : $c_0^{(i,j)} \leftarrow \mathsf{Encrypt}(\mathsf{pk}, 0),$ $c_1^{(i,j)} \leftarrow \mathsf{Encrypt}(\mathsf{pk}, 0),$ $c_2^{(i,j)} \leftarrow \mathsf{Encrypt}(\mathsf{pk}, 0).$

Phenotype Encoding. For each entry (i, j) in the phenotype table, we compute 2 ciphertexts, one for the "unaffected" phenotype (value 0) and one for the "affected" phenotype (value 1); we call these ciphertexts $c_0^{(i,j)}, c_1^{(i,j)}$ respectively. A ciphertext encrypts 1 if the entry value is the same as the value it represents, and 0 otherwise. More specifically, the 3 possible genotypes are encoded as follows:

unaffected (value 0) : $z_0^{(i,j)} \leftarrow \mathsf{Encrypt}(\mathsf{pk}, 1),$ $z_1^{(i,j)} \leftarrow \mathsf{Encrypt}(\mathsf{pk}, 0),$

affected (value 1) : $z_0^{(i,j)} \leftarrow \mathsf{Encrypt}(\mathsf{pk}, 0),$ $z_1^{(i,j)} \leftarrow \mathsf{Encrypt}(\mathsf{pk}, 1),$

missing : $z_0^{(i,j)} \leftarrow \mathsf{Encrypt}(\mathsf{pk}, 0),$ $z_1^{(i,j)} \leftarrow \mathsf{Encrypt}(\mathsf{pk}, 0).$

4.2   Computing Genotype and Phenotype Counts

Recall that the statistical algorithms described in Sect. 2 take as input genotype and phenotype frequencies or counts. While we are not able to obtain the genotype and phenotype *frequencies*[9], we can obtain the *counts* using a few simple homomorphic additions. Indeed, if the data is encrypted as described in Sect. 4.1, computing the (encrypted) counts $\widehat{N}_k^{(j)}$ of value-k genotypes at locus j can be done by summing all the ciphertexts $c_k^{(i,j)}$ in column j of the genotype table:

$$\widehat{N}_0^{(j)} = \sum_i c_0^{(i,j)}, \quad \widehat{N}_1^{(j)} = \sum_i c_1^{(i,j)}, \quad \widehat{N}_2^{(j)} = \sum_i c_2^{(i,j)}.$$

Finally, we can compute the (encrypted) total number $\widehat{N}^{(j)}$ of available (non-missing genotypes) in column j by summing

$$\widehat{N}^{(j)} = \widehat{N}_0^{(j)} + \widehat{N}_1^{(j)} + \widehat{N}_2^{(j)}.$$

---

[9] Recall from Sect. 3 that we cannot perform homomorphic divisions.

4.3   Modified Algorithms

Unfortunately, since we are not able to compute the genotype and phenotype *frequencies*, we must modify the statistical algorithms to use genotype and phenotype *counts* instead.

Pearson Goodness-of-Fit or Chi-Squared Test. Recall that for a single locus, the Pearson test computes the test statistic

$$X^2 = \sum_{i=0}^{2} \frac{(N_i - E_i)^2}{E_i},$$

where N_i is the observed genotype count, and E_i is the expected genotype count. The expected counts can be computed as

$$E_0 = N \left(\frac{2N_0 + N_1}{2N} \right)^2, \quad E_1 = 2N \left(\frac{2N_0 + N_1}{2N} \right) \left(\frac{2N_2 + N_1}{2N} \right), \quad E_2 = N \left(\frac{2N_2 + N_1}{2N} \right)^2,$$

which can be simplified to

$$E_0 = \frac{(2N_0 + N_1)^2}{4N}, \quad E_1 = \frac{(2N_0 + N_1)(2N_2 + N_1)}{2N}, \quad E_2 = \frac{(2N_2 + N_1)^2}{4N}.$$

The test statistic X^2 can then be computed as

$$
\begin{aligned}
X^2 &= \frac{(N_0 - E_0)^2}{E_0} + \frac{(N_1 - E_1)^2}{E_1} + \frac{(N_2 - E_2)^2}{E_2} \\
&= \frac{(4N_0 N_2 - N_1^2)^2}{2N} \left(\frac{1}{2(2N_0 + N_1)^2} + \frac{1}{(2N_0 + N_1)(2N_2 + N_1)} + \frac{1}{2(2N_2 + N_1)^2} \right).
\end{aligned}
$$

Since we are unable to perform homomorphic divisions, we return encryptions of $\alpha, N, \beta_1, \beta_2, \beta_3$, where

$$\alpha \stackrel{\text{def}}{=} (4N_0 N_2 - N_1^2)^2, \quad \beta_1 \stackrel{\text{def}}{=} 2(2N_0 + N_1)^2, \quad \beta_2 \stackrel{\text{def}}{=} (2N_0 + N_1)(2N_2 + N_1), \quad \beta_3 \stackrel{\text{def}}{=} 2(2N_2 + N_1)^2.$$

From these, the test statistic can be computed as:

$$X^2 = \frac{\alpha}{2N} \left(\frac{1}{\beta_1} + \frac{1}{\beta_2} + \frac{1}{\beta_3} \right).$$

EM Algorithm. To run the EM algorithm on genotypes at loci j and ℓ, we need the 9 genotype counts $N_{xy}^{(j,\ell)}$ for $x, y \in \{0, 1, 2\}$. In other words, we need to know the number of individuals $N_{x,y}^{(j,\ell)}$ in the data set that have genotype x at locus j and genotype y at locus ℓ, for all combinations of x and y. The (encrypted) counts can be computed as

$$\widehat{N}_{xy}^{(j,\ell)} = \sum_i c_x^{(i,j)} \cdot c_y^{(i,\ell)}.$$

Recall that the EM algorithm estimates the haplotype *frequencies*. As before, we are unable to estimate frequencies since we cannot perform homomorphic division, but we are able to estimate haplotype *counts*. Notice that since $N_{xy} = 2N \cdot p_{xy}$, this does not change the fraction in $\mu_{AB/ab}^{(m)}$ and $\mu_{Ab/aB}^{(m)}$ (essentially, this change multiplies both the numerator and the denominator by $4N^2$). This modifies the estimation step as follows.

mTH ESTIMATION STEP

$$E_{AB/ab}^{(m)} = N_{11} \cdot \frac{N_{AB}^{(m-1)} N_{ab}^{(m-1)}}{N_{AB}^{(m-1)} N_{ab}^{(m-1)} + N_{Ab}^{(m-1)} N_{aB}^{(m-1)}} \overset{\text{def}}{=} \frac{\alpha^{(m)}}{\beta^{(i)}},$$

$$E_{Ab/aB}^{(m)} = N_{11} \cdot \frac{N_{Ab}^{(m-1)} N_{aB}^{(m-1)}}{N_{AB}^{(m-1)} N_{ab}^{(m-1)} + N_{Ab}^{(m-1)} N_{aB}^{(m-1)}} \overset{\text{def}}{=} \frac{\gamma^{(m)}}{\beta^{(i)}}.$$

We can also simplify the iteration so that at any given point, we need only remember one numerator and one denominator. Define

$$\zeta_{AB} \overset{\text{def}}{=} 2N_{22} + N_{21} + N_{12}, \qquad \zeta_{ab} \overset{\text{def}}{=} 2N_{00} + N_{01} + N_{10},$$

$$\zeta_{Ab} \overset{\text{def}}{=} 2N_{20} + N_{21} + N_{10}, \qquad \zeta_{aB} \overset{\text{def}}{=} 2N_{02} + N_{12} + N_{01}.$$

Then

$$N_{AB}^{(m)} = \zeta_{AB} + E_{AB/ab}^{(m)} = \zeta_{AB} + \frac{\alpha^{(m)}}{\beta^{(m)}} = \frac{\zeta_{AB} \cdot \beta^{(m)} + \alpha^{(m)}}{\beta^{(m)}},$$

$$N_{ab}^{(m)} = \zeta_{ab} + E_{AB/ab}^{(m)} = \zeta_{ab} + \frac{\alpha^{(m)}}{\beta^{(m)}} = \frac{\zeta_{ab} \cdot \beta^{(m)} + \alpha^{(m)}}{\beta^{(m)}},$$

$$N_{Ab}^{(m)} = \zeta_{Ab} + E_{Ab/aB}^{(m)} = \zeta_{Ab} + \frac{\gamma^{(m)}}{\beta^{(m)}} = \frac{\zeta_{Ab} \cdot \beta^{(m)} + \gamma^{(m)}}{\beta^{(m)}},$$

$$N_{aB}^{(m)} = \zeta_{aB} + E_{Ab/aB}^{(m)} = \zeta_{aB} + \frac{\gamma^{(m)}}{\beta^{(m)}} = \frac{\zeta_{aB} \cdot \beta^{(m)} + \gamma^{(m)}}{\beta^{(m)}}.$$

Following the iteration, at the next estimation step we need to compute:

$$E_{AB/ab}^{(m+1)} = N_{11} \cdot \frac{\left(\frac{\zeta_{AB} \cdot \beta^{(m)} + \alpha^{(m)}}{\beta^{(m)}}\right)\left(\frac{\zeta_{ab} \cdot \beta^{(m)} + \alpha^{(m)}}{\beta^{(m)}}\right)}{\left(\frac{\zeta_{AB} \cdot \beta^{(m)} + \alpha^{(m)}}{\beta^{(m)}}\right)\left(\frac{\zeta_{ab} \cdot \beta^{(m)} + \alpha^{(m)}}{\beta^{(m)}}\right) + \left(\frac{\zeta_{Ab} \cdot \beta^{(m)} + \gamma^{(m)}}{\beta^{(m)}}\right)\left(\frac{\zeta_{aB} \cdot \beta^{(m)} + \gamma^{(m)}}{\beta^{(m)}}\right)}$$

$$= N_{11} \cdot \frac{\left(\zeta_{AB} \cdot \beta^{(m)} + \alpha^{(m)}\right)\left(\zeta_{ab} \cdot \beta^{(m)} + \alpha^{(m)}\right)}{\left(\zeta_{AB} \cdot \beta^{(m)} + \alpha^{(m)}\right)\left(\zeta_{ab} \cdot \beta^{(m)} + \alpha^{(m)}\right) + \left(\zeta_{Ab} \cdot \beta^{(m)} + \gamma^{(m)}\right)\left(\zeta_{aB} \cdot \beta^{(m)} + \gamma^{(m)}\right)}$$

$$\overset{\text{def}}{=} \frac{\alpha^{(m+1)}}{\beta^{(m+1)}}.$$

and similarly,

$$E^{(m+1)}_{Ab/aB} = N_{11} \cdot \frac{\left(\varsigma_{Ab} \cdot \beta^{(m)} + \gamma^{(m)}\right)\left(\varsigma_{aB} \cdot \beta^{(m)} + \gamma^{(m)}\right)}{\left(\varsigma_{AB} \cdot \beta^{(m)} + \alpha^{(m)}\right)\left(\varsigma_{ab} \cdot \beta^{(m)} + \alpha^{(m)}\right) + \left(\varsigma_{Ab} \cdot \beta^{(m)} + \gamma^{(m)}\right)\left(\varsigma_{aB} \cdot \beta^{(m)} + \gamma^{(m)}\right)}$$

$$\stackrel{\text{def}}{=} \frac{\gamma^{(m+1)}}{\beta^{(m+1)}}.$$

In other words, since the denominator $\beta^{(m)}$ always cancels out, we need only remember the numerators. The numerators depend on $\beta^{(m)}$, so we still compute it as part of the numerator computation, but do not need to store it after this computation. Of course, at the last step we must divide by $\beta^{(m)}$ to maintain correctness.

The modified estimation and maximization steps are described below.

mTH ESTIMATION STEP

$$\alpha^{(m)} = N_{11} \cdot N^{(m-1)}_{AB} N^{(m-1)}_{ab}, \quad \gamma^{(m)} = N_{11} \cdot N^{(m-1)}_{Ab} N^{(m-1)}_{aB},$$
$$\beta^{(m)} = N^{(m-1)}_{AB} N^{(m-1)}_{ab} + N^{(m-1)}_{Ab} N^{(m-1)}_{aB}.$$

mTH MAXIMIZATION STEP

$$N^{(m)}_{AB} = \varsigma_{AB} \cdot \beta^{(m)} + \alpha^{(m)}, \quad N^{(m)}_{ab} = \varsigma_{ab} \cdot \beta^{(m)} + \alpha^{(m)},$$
$$N^{(m)}_{Ab} = \varsigma_{Ab} \cdot \beta^{(m)} + \gamma^{(m)}, \quad N^{(m)}_{aB} = \varsigma_{aB} \cdot \beta^{(m)} + \gamma^{(m)}.$$

Measures of LD. The degree of linkage disequilibrium of two bi-allelic genes can be determined by computing the scalar value $D \stackrel{\text{def}}{=} p_{AB} - p_A p_B$ where A and B are the reference alleles of the genes, p_A and p_B are their corresponding population frequencies, and p_{AB} is the population frequency of the haplotype AB. Once more, we need to compute D as a function of counts rather than frequencies, as we cannot compute the latter homomorphically. We have

$$D = \frac{N_{AB}}{2N} - \frac{N_A}{2N} \cdot \frac{N_B}{2N} = \frac{2N \cdot N_{AB} - N_A N_B}{(2N)^2}.$$

The haplotype count N_{AB} is estimated using the EM algorithm, which outputs values α, β such that $N_{AB} = \alpha/\beta$. Thus,

$$D = \frac{2N \cdot \alpha - \beta N_A N_B}{\beta(2N)^2} = \frac{2N \cdot \alpha - \beta(2N_{AA} + N_{Aa})(2N_{BB} + N_{Bb})}{\beta(2N)^2}.$$

Again, since we cannot perform homomorphic division, we return encryptions of δ and β, where

$$\delta \stackrel{\text{def}}{=} 2N \cdot \alpha - \beta(2N_{AA} + N_{Aa})(2N_{BB} + N_{Bb}).$$

The scalar D can be computed as $D = \delta/(\beta(2N)^2)$. To be able to calculate the D' and r^2 statistics, we also return encryptions of N_A, N_a, N_B, N_b, from which they can be computed:

$$D' = \frac{\delta}{\beta D_{\max}}, \quad \text{where} \quad D_{\max} = \begin{cases} \min\{N_A N_b, N_a N_B\} & \text{if } D > 0, \\ \min\{N_A N_B, N_a N_b\} & \text{if } D < 0 \end{cases}$$

and

$$r^2 = \frac{\delta^2}{\beta^2 N_A N_a N_B N_b}.$$

Cochran-Armitage Test for Trend. To run the CATT algorithm on genotype at locus j and phenotype ℓ, we need the 6 genotype-phenotype counts $N_{x,y}^{(j,\ell)}$ for $x \in \{0,1,2\}$ and $y \in \{0,1\}$. In other words, we need to know the number of individuals $N_{x,y}^{(j,\ell)}$ in the data set that have genotype x at locus j and phenotype ℓ with value y, for all combinations of x and y. The (encrypted) counts can be computed as

$$\widehat{N}_{xy}^{(j,\ell)} = \sum_i c_x^{(i,j)} \cdot z_y^{(i,\ell)}.$$

The test statistic can be computed as $X^2 = \alpha/\beta$ where

$$\alpha \overset{\text{def}}{=} N \cdot \left(\sum_{i=0}^{2} w_i(N_{0i}R_1 - N_{1i}R_0) \right)^2,$$

$$\beta \overset{\text{def}}{=} R_0 R_1 \cdot \left(\sum_{i=0}^{2} w_i^2 C_i(N - C_i) - 2 \sum_{i=1}^{k-1} \sum_{j=i+1}^{k} w_i w_j C_i C_j \right).$$

4.4   How to Set Parameters

In order to implement any cryptographic scheme efficiently and securely, one needs to select suitable parameters. For the homomorphic encryption scheme in this work, one needs to find a dimension n, an integer modulus q and the standard deviation σ of the error distribution χ_{err}. These parameters have to satisfy two conditions. The first one guarantees security, more precisely the parameters need to guarantee a desired level λ of security against known attacks on the homomorphic encryption scheme. This means that an attacker is forced to run for at least 2^λ steps in order to break the scheme. The second condition guarantees correctness. Given a desired computation, the encryption scheme must be able to correctly evaluate the computation without the error terms in the ciphertexts growing too large such that the result is decrypted correctly. Subject to these two conditions, we aim to choose the smallest dimension and modulus in order to make computations as efficient as possible. We follow the approach described in [LN14a] for selecting parameters and refer the reader to this paper for more details.

Security. One first picks a desired security level, common values are $\lambda = 80$ or higher. Next, one chooses a dimension n and standard deviation σ. The analysis in [LN14a] shows that, for fixed λ, n and σ, there is an upper bound on the allowed modulus q to achieve the desired security level. In order to increase security, one can either increase the dimension n, or increase the standard deviation σ or a combination of both and then re-evaluate to obtain the new maximal value for q.

Correctness. The correctness condition is in contrast to the security condition in that given a dimension and standard deviation, it demands a lower bound for the modulus q. The complexity of the planned computation and the size of σ influence the error growth throughout the computation. To make correct decryption possible, the relative size of the error compared to the modulus q needs to be small enough, or in other words, q needs to be large enough to accommodate the error that arises during the homomorphic operations.

Efficiency. To maximize the efficiency of our implementation, we select the "smallest" parameter set amongst those that satisfy the security and correctness criteria. Clearly, increasing n or q leads to a decrease in efficiency, so we are interested to keep the dimension and modulus as small as possible. In general, smaller security level and less complex computations allow for smaller parameters, increasing the security and complexity leads to larger, less efficient parameters. In this work, we are contented with a security level of $\lambda = 80$.

5   Performance

In this section, we describe our experiments. We implemented the homomorphic encryption scheme from Sect. 3 and the algorithms described in Sect. 4 in the computer algebra system Magma [BCP97]. Note that specialized implementations in a language such as C/C++ may perform significantly better than our proof-of-concept implementation in Magma. The exact speed-up depends on the optimizations in such an implementation. For example, for the parameters used in [BLLN13], we observe that our Magma implementation of the homomorphic addition, multiplication, and decryption operations is roughly twice as slow as the C/C++ implementation reported in [LN14a], which uses a general purpose C/C++ library for the underlying arithmetic. The decryption operation in the implementation in [LN14a] in turn is roughly twice as slow as the C implementation in [BLLN13]. A completely specialized and optimized implementation will achieve even better efficiency.

Timings for the Scheme. Timings for the basic algorithms of the homomorphic encryption scheme (key generation, encryption, addition, multiplication, and decryption for several degree values D) are shown in Table 1. Timings for both key generation and encryption include the sampling of small elements according to the distributions described in Sect. 3. Depending on the specific implementation scenario, these steps could be precomputed and their cost amortized.

As mentioned in Sect. 3, our implementation does not perform relinearization (a.k.a. key-switching) in homomorphic operations (see any of [BV11a, BGV12, LTV12, Bra12, BLLN13, LN14b]). We choose not to perform this step as an optimization (indeed, the timing for a single multiplication increased more than 50-fold when relinearization was included). The downside to our approach is that decryption depends on the degree of the function that was homomorphically computed (recall from Sect. 3 that the decryption algorithm first computes f^D where f is the secret key and D is the degree of the computed function). Thus, decryption timings depend on the degree of the evaluated function, albeit only logarithmically. We remark that if the function is known in advance (or even only an upper bound on its degree), the element f^D can be precomputed. In this case, the decryption time in *all* cases is the same and equivalent to the decryption time for degree-1 ciphertexts.

Parameters. Table 1 provides timings for two different parameter sets. The first set (I) uses smaller parameters and therefore produces faster timings. All algorithms in this paper can be run correctly with the first parameter set except for the EM algorithm for more than two iterations. The second set (II) uses larger parameters that are suitable to run the EM algorithm for 3 iterations, but the performance is worse due to the larger parameters. Both parameter sets provide 80 bits of security. We refer the reader to Sect. 4.4 for a detailed explanation of how these parameter sets were selected.

In order to increase the security to 128 bits, we must adjust the parameter sizes. For example, this can be done as follows. According to the analysis in Sect. 4.4, when all other parameters are fixed, one can achieve the 128-bit security level by decreasing the modulus q to 149 bits. Such a parameter set can still be used to run the same algorithms as parameter set (I), except for the LD algorithm. In order to run the LD algorithm, one needs to increase the dimension n. If n is restricted to be a power of two, then $n = 8192$ as in parameter set (II). However, q needs to be smaller than in set (II). Arithmetic for such parameters is the same as for the set (II) but with slightly faster arithmetic modulo q. Therefore, the timings in Table 1 give a rough estimate for the upper bound on the performance penalty when moving to 128-bit security.

Testing Correctness. To test the correctness of our homomorphic evaluations, we implemented the statistical algorithms in their original form (as described in Sect. 2) and unencrypted, as well as the modified algorithms described in Sect. 4, also unencrypted. A third implementation ran the modified algorithms (as in Sect. 4) on encrypted data and used the homomorphic operations of the encryption scheme. In each test, we ran all versions of the algorithms and confirmed that their return values were equal.

Data Pre-processing. All algorithms being considered take as input genotype and/or phenotype count tables. Because of this, once the encrypted tables have been computed and appropriate parameters have been chosen, the running times

Table 1. Timings for the operations of the homomorphic encryption scheme. Measurements were done in the computer algebra system Magma [BCP97] V2.17-8 on an Intel(R) Core(TM) i7-3770S CPU @ 3.10 GHz, 8 sGB RAM, running 64-bit Windows 8.1. Values are the mean of 1000 measurements of the respective operation. Decryption depends on the degree of the evaluated function, the timing differences are due to the computation of the respective power of the secret key. Parameter sets are (I) $n = 4096$, $\lceil \log(q) \rceil = 192$ and (II) $n = 8192$, $\lceil \log(q) \rceil = 384$, both use $\sigma = 8$ and provide 80-bit security. A single ciphertext with parameter set (I) is of size slightly less than 100 KB, for parameter set (II), it is less than 400 KB.

Operation	KeyGen	Encrypt	Add	Mult	Decrypt				
					deg 1	deg 2	deg 5	deg 10	deg 20
Parameters I	3.599 s	0.296 s	0.001 s	0.051 s	0.035 s	0.064 s	0.114 s	0.140 s	0.164 s
Parameters II	18.141 s	0.783 s	0.003 s	0.242 s	0.257 s	0.308 s	0.598 s	0.735 s	0.888 s

of the statistical algorithms are independent of the size of the population sample and depend only on the parameter set needed for the computation.[10] Thus, we separate our analysis into two phases: In the first phase, we construct the encrypted genotype and phenotype tables. This includes encoding and encrypting genotype and phenotype data, as well as summing these encryptions (see Sects. 4.1 and 4.2). In the second phase, we run the statistical algorithms on the encrypted tables. Indeed, we view the first phase as a pre-processing of the data, whose cost can be amortized over any number of computations. Moreover, this data can be easily updated by subtracting encryptions if a person's data is no longer needed or desired, by adding new encryptions if new data is collected, or by replacing specific encryptions as a person's record needs to be updated or modified. We emphasize the fact that there is no need to re-encode and re-encrypt the entire data set when modifications are required. Necessary changes will be proportional to the number of entries that need to be modified (inserted, deleted, or updated).

The main cost in pre-processing the data is the computation of the 3 encryptions for each genotype sample and the 2 encryptions for each phenotype sample (see Sect. 4.1). This cost is linear in the size of the data set and can be easily computed from the timings for encryption given in Table 1. For example, encoding and encrypting 1000 genotype data points sequentially using parameter set (I)

---

[10] Admittedly, the size of the parameters needed does depend on the magnitude of the genotype and phenotype counts, which can be as large as the size of the population sample. This is because the size of the message encrypted at any given time (i.e. the size of the counts and all the intermediate values in the computation) cannot grow too large relative to the modulus q. Therefore, larger population sizes (and therefore larger counts) require a larger modulus q, which in turn requires a larger dimension n for security. However, for a fixed parameter set, it is possible to compute an upper bound on the size of the population sample and the homomorphic computations detailed in this work do work correctly for any population sample with size smaller than the given bound.

Table 2. Timings for statistical algorithms. Measurements were done in the computer algebra system Magma [BCP97] V2.17-8 on an Intel(R) Core(TM) i7-3770S CPU @ 3.10 GHz, 8 GB RAM, running 64-bit Windows 8.1. Values are the mean of 100 measurements of the respective algorithm.

Algorithm	Pearson	EM			LD	CATT
		1 iteration	2 iterations	3 iterations		
Parameters I	0.34 s	0.57 s	1.10 s	–	0.19 s	0.94 s
Parameters II	1.36 s	2.29 s	4.54 s	6.85 s	0.74 s	3.63 s

takes roughly 15 min, and encoding and encrypting 1000 phenotype data entries takes roughly 10 min.

Once all genotypes and phenotypes have been encoded and encrypted, we need to construct 3 contingency tables (see Sects. 4.2 and 4.3). The first table contains the genotype counts for a single locus and can be computed by sequential addition of the genotype encryptions. Sequentially adding 1000 ciphertexts takes roughly 1 s; thus, computing all genotype counts for a single locus takes roughly 3 s. Computing the 3×3 contingency table for the counts of individuals having a certain genotype at one locus and another at a second locus requires one multiplication and one addition per individual. Thus, for parameter set (I), each entry in the table can be computed in roughly 1 min and the entire table can be computed in roughly 9 min. Similarly, computing the 2×3 contingency table for the counts of individuals with a certain genotype and a given phenotype requires one multiplication and one addition per individual. Thus, for parameter set (I), the entire table can be computed in roughly 6 min.

Timings for the Statistical Algorithms. As mentioned above, once the data has been processed and the genotype and phenotype tables have been computed, the runtime of the statistical algorithms is independent of the size of the population sample and only depends on the parameter set needed for the computation. Table 2 contains performance numbers for the algorithms after the data has been encoded and encrypted, and population counts have been computed. It includes timings for both parameter sets described above.

Further Specialization. For the case that only one of the statistical algorithms needs to be run, further optimizations are possible, decreasing storage space and runtime significantly. For example, if we focus on running only the Pearson test, we can change the encoding of genotypes from Sect. 4.1 to use only a single ciphertext as follows: value 0 is encrypted as $c^{i,j} = \mathsf{Encrypt}(\mathsf{pk}, 1)$, value 1 as $c^{i,j} = \mathsf{Encrypt}(\mathsf{pk}, x^{100})$, value 2 as $c^{i,j} = \mathsf{Encrypt}(\mathsf{pk}, x^{301})$ and missing values as $c^{i,j} = \mathsf{Encrypt}(\mathsf{pk}, 0)$. By adding up all such ciphertexts, the genotype counts are then contained in a single ciphertexts that encrypts $N_0 + N_1 x^{100} + N_2 x^{301}$. The Pearson test has degree 4 in these counts and can be computed with only two multiplication operations on this ciphertext. Note that needed values encoded in the polynomials $(N_0 + N_1 x^{100} + N_2 x^{301})^i$ for $i \in \{2, 4\}$ can be shifted to the constant coefficient by multiplying with suitable powers of x.

Using this optimization, the storage space for encrypted genotype data is reduced by a factor 3, as is the encryption time. With parameter set (I), the runtime of one Pearson test becomes less than 0.13s.

6   Conclusion and Future Work

In this paper we presented algorithms and proof-of-concept implementations for computing on encrypted genomic data. We showed how to encode genotype and phenotype data for encryption and how to apply the Pearson Goodness-of-Fit test, the D' and r^2-measures of linkage disequilibrium, the Estimation Maximization (EM) algorithm for haplotyping, and the Cochran-Armitage Test for Trend, to the encrypted data to produce encrypted results. These are standard algorithms used in genome wide association studies and our proof-of-concept implementation timings are reasonable. We showed that the timings for evaluating the statistical algorithms do not depend on the population size once the correct parameter sizes are fixed and the encrypted genotype or phenotype counts are input. Timings at the smaller parameter size for the various algorithms vary up to roughly 1 s on a standard PC, indicating that these computations are well within reach of being practical for relevant applications and scenarios.

Homomorphic encryption may well be ripe for deployment, to achieve private outsourcing of computation for simple algorithms such as those presented in this paper when applied to modest-size data sets. That will require increased effort and focus on high-performance implementations for a range of architectures. In addition, many other interesting avenues for research remain. There is still much work to be done to make homomorphic encryption more efficient at scale and to expand the functionality. In addition, to solve a wide-range of practical privacy problems which arise with cloud services, it will be important to consider various cryptographic building blocks such as secure multiparty computation and other more interactive solutions and the trade-offs between storage and interaction costs. One should also consider how homomorphic encryption can be combined with building blocks such as verifiable computation. Currently homomorphic encryption does not provide a practical solution for operating on data encrypted under multiple keys, for example in the setting of a public database where multiple patients upload data under different keys. Finally, the practical homomorphic encryption schemes presented here rely on hardness assumptions for a class of new problems such as RLWE. It is crucial to continue to study the hardness of these new assumptions and to attack the systems to accurately assess parameter bounds to assure security.

Acknowledgments. We thank Tancrède Lepoint for suggesting the encoding in Sect. 4.1.

References

[ADCHT13] Ayday, E., De Cristofaro, E., Hubaux, J.-P., Tsudik, G.: The Chills and Thrills of Whole Genome Sequencing. Technical report (2013). http:// infoscience.epfl.ch/record/186866/files/survey.pdf

[ARH13] Ayday, E., Raisaro, J.L., Hubaux, J.-P.: Personal use of the genomic data: Privacy vs. storage cost. In: Proceedings of IEEE Global Communications Conference, Exhibition and Industry Forum (Globecom) (2013)

[BAFM12] Blanton, M., Atallah, M.J., Frikken, K.B., Malluhi, Q.: Secure and efficient outsourcing of sequence comparisons. In: Foresti, S., Yung, M., Martinelli, F. (eds.) ESORICS 2012. LNCS, vol. 7459, pp. 505–522. Springer, Heidelberg (2012)

[BCP97] Bosma, W., Cannon, J., Playoust, C.: The magma algebra system. I. The user language. J. Symbolic Comput. **24**(3–4), 235–265 (1997). Computational algebra and number theory (London, 1993)

[BGV12] Brakerski, Z., Gentry, C., Vaikuntanathan, V.: Fully homomorphic encryption without bootstrapping. In: ITCS (2012)

[BLLN13] Bos, J.W., Lauter, K., Loftus, J., Naehrig, M.: Improved security for a ring-based fully homomorphic encryption scheme. In: Stam, M. (ed.) IMACC 2013. LNCS, vol. 8308, pp. 45–64. Springer, Heidelberg (2013)

[BLN14] Bos, J.W., Lauter, K., Naehrig, M.: Private predictive analysis on encrypted medical data. J. Biomed. Inform. **50**, 234–243 (2014). MSR-TR-2013-81

[Bra12] Brakerski, Z.: Fully homomorphic encryption without modulus switching from classical GapSVP. In: Safavi-Naini, R., Canetti, R. (eds.) CRYPTO 2012. LNCS, vol. 7417, pp. 868–886. Springer, Heidelberg (2012)

[BV11a] Brakerski, Z., Vaikuntanathan, V.: Efficient fully homomorphic encryption from (standard) LWE. In: Ostrovsky, R. (ed.) FOCS, pp. 97–106. IEEE (2011)

[BV11b] Brakerski, Z., Vaikuntanathan, V.: Fully homomorphic Encryption from Ring-LWE and security for key dependent messages. In: Rogaway, P. (ed.) CRYPTO 2011. LNCS, vol. 6841, pp. 505–524. Springer, Heidelberg (2011)

[BV14] Brakerski, Z., Vaikuntanathan, V.: Lattice-based FHE as secure as PKE. In: Naor, M. (ed.) ITCS, pp. 1–12. ACM (2014)

[dbG] Database of Genotypes and Phenotypes (dbGaP). http://www.ncbi.nlm.nih.gov/gap/

[DCFT13] De Cristofaro, E., Faber, S., Tsudik, G.: Secure genomic testing with size-and position-hiding private substring matching. In: Proceedings of the 2013 ACM Workshop on Privacy in the Electronic Society (WPES 2013). ACM (2013)

[EBI] European Bioinformatics Institute. http://www.ebi.ac.uk/ (Accessed 30 October 2013)

[FSU11] Fienberg, S.E., Slavkovic, A., Uhler, C.: Privacy preserving GWAS data sharing. In: 2011 IEEE 11th International Conference on Data Mining Workshops (ICDMW), pp. 628–635. IEEE (2011)

[FV12] Fan, J., Vercauteren, F.: Somewhat practical fully homomorphic encryption. IACR Cryptology ePrint Archive **2012**, 144 (2012)

[Gen09] Gentry, C.: Fully homomorphic encryption using ideal lattices. In: Mitzenmacher, M. (ed.) STOC, pp. 169–178. ACM (2009)

[GLN13] Graepel, T., Lauter, K., Naehrig, M.: ML confidential: machine learning on encrypted data. In: Kwon, T., Lee, M.-K., Kwon, D. (eds.) ICISC 2012. LNCS, vol. 7839, pp. 1–21. Springer, Heidelberg (2013)

[Glo13] Creating a global alliance to enable responsible sharing of genomic and clinical data, White Paper (2013). http://www.broadinstitute.org/files/news/pdfs/GAWhitePaperJune3.pdf

[GMG+13] Gymrek, M., McGuire, A.L., Golan, D., Halperin, E., Erlich, Y.: Identifying personal genomes by surname inference. Science **339**(6117), 321–324 (2013)

[GSW13] Gentry, C., Sahai, A., Waters, B.: Homomorphic encryption from learning with errors: conceptually-simpler, asymptotically-faster, attribute-based. In: Canetti, R., Garay, J.A. (eds.) CRYPTO 2013, Part I. LNCS, vol. 8042, pp. 75–92. Springer, Heidelberg (2013)

[HAHT13] Humbert, M., Ayday, E., Hubaux, J.-P., Telenti, A.: Addressing the concerns of the lacks family: quantification of kin genomic privacy. In: Proceedings of the 2013 ACM SIGSAC Conference on Computer & Communications Security, pp. 1141–1152. ACM (2013)

[ICG] International cancer genome consortium (ICGC). http://www.icgc.org

[IRD] International rare diseases research consortium (IRDiRC). http://www.irdirc.org

[Jap] DNA Data Bank Of Japan. http://www.ddbj.nig.ac.jp/

[JS13] Johnson, A., Shmatikov, V.: Privacy-preserving data exploration in genome-wide association studies. In: Proceedings of the 19th ACM SIGKDD International Conference on Knowledge Discovery and Data Mining, pp. 1079–1087. ACM (2013)

[LN14a] Lepoint, T., Naehrig, M.: A comparison of the homomorphic encryption schemes FV and YASHE. In: Pointcheval, D., Vergnaud, D. (eds.) AFRICACRYPT. LNCS, vol. 8469, pp. 318–335. Springer, Heidelberg (2014)

[LN14b] López-Alt, A., Naehrig, M.: Large integer plaintexts in ring-based fully homomorphic encryption. In preparation (2014)

[LNV11] Lauter, K., Naehrig, M., Vaikuntanathan, V.: Can homomorphic encryption be practical? In: Proceedings of the 3rd ACM Cloud Computing Security Workshop, pp. 113–124. ACM (2011)

[LTV12] López-Alt, A., Tromer, E., Vaikuntanathan, V.: On-the-fly multiparty computation on the cloud via multikey fully homomorphic encryption. In: Karloff, H.J., Pitassi, T. (eds.) STOC, pp. 1219–1234. ACM (2012)

[MCC+11] McCarty, C.A., Chisholm, R.L., Chute, C.G., Kullo, I.J., Jarvik, G.P., Larson, E.B., Li, R., Masys, D.R., Ritchie, M.D., Roden, D.M., et al.: The emerge network a consortium of biorepositories linked toelectronic medical records data for conducting genomic studies. BMC Med. Genomics **4**(1), 13 (2011)

[PH08] Park, M.Y., Hastie, T.: Penalized logistic regression for detecting gene interactions. Biostatistics **9**(1), 30–50 (2008)

[SS11] Stehlé, D., Steinfeld, R.: Making NTRU as secure as worst-case problems over ideal lattices. In: Paterson, K.G. (ed.) EUROCRYPT 2011. LNCS, vol. 6632, pp. 27–47. Springer, Heidelberg (2011)

[TGP] A map of human genome variation from population-scale sequencing. Nature, 467:1061–1073. http://www.1000genomes.org

[vDGHV10] van Dijk, M., Gentry, C., Halevi, S., Vaikuntanathan, V.: Fully homomorphic encryption over the integers. In: Gilbert, H. (ed.) EUROCRYPT 2010. LNCS, vol. 6110, pp. 24–43. Springer, Heidelberg (2010)

[WLW+09] Wang, R., Li, Y.F., Wang, X.F., Tang, H., Zhou, X.: Learning your identity and disease from research papers: Information leaks in genome wide association study. In: Proceedings of the 16th ACM Conference on Computer and Communications Security, CCS 2009, pp. 534–544. ACM, New York (2009)

[YSK+13] Yasuda, M., Shimoyama, T., Kogure, J., Yokoyama, K., Koshiba, T.: Secure pattern matching using somewhat homomorphic encryption. In: Proceedings of the 2013 ACM Cloud Computing Security Workshop, pp. 65–76. ACM (2013)

Cryptographic Engineering

Full-Size High-Security ECC Implementation on MSP430 Microcontrollers

Gesine Hinterwälder[1,2]([✉]), Amir Moradi[1], Michael Hutter[3], Peter Schwabe[4], and Christof Paar[1,2]

[1] Horst Görtz Institute for IT Security, Ruhr-University Bochum, Bochum, Germany
{gesine.hinterwaelder,amir.moradi,christof.paar}@rub.de
[2] Department of Electrical and Computer Engineering, University of Massachusetts Amherst, Amherst, USA
[3] Institute for Applied Information Processing and Communications (IAIK), Graz University of Technology, Graz, Austria
Michael.Hutter@iaik.tugraz.at
[4] Digital Security Group, Radboud University Nijmegen, Nijmegen, The Netherlands
peter@cryptojedi.org

Abstract. In the era of the Internet of Things, smart electronic devices facilitate processes in our everyday lives. Texas Instrument's MSP430 microcontrollers target low-power applications, among which are wireless sensor, metering and medical applications. Those domains have in common that sensitive data is processed, which calls for strong security primitives to be implemented on those devices. Curve25519, which builds on a 255-bit prime field, has been proposed as an efficient, highly-secure elliptic-curve. While its high performance on powerful processors has been shown, the question remains, whether it is suitable for use in embedded devices. In this paper we present an implementation of Curve25519 for MSP430 microcontrollers. To combat timing attacks, we completely avoid conditional jumps and loads, thus making our software constant time. We give a comprehensive evaluation of different implementations of the modular multiplication and show which ones are favorable for different conditions. We further present implementation results of Curve25519, where our best implementation requires 9.1 million or 6.5 million cycles on MSP430Xs having a 16×16-bit or a 32×32-bit hardware multiplier respectively.

Keywords: MSP430 · Carry-save representation · Karatsuba · Operand-caching multiplication · Curve25519

* This work was supported in part by the German Federal Ministry for Economic Affairs and Energy (Grant 01ME12025 SecMobil), by the Netherlands Organisation for Scientific Research (NWO) through Veni 2013 project 13114, and by the Austrian Science Fund (FWF) under the grant number TRP251-N23. Permanent ID of this document: 0b3f1ea83d48e400ad1def71578c4c66. Date: 2014-10-01.

D.F. Aranha and A. Menezes (Eds.): LATINCRYPT 2014, LNCS 8895, pp. 31–47, 2015.
DOI: 10.1007/978-3-319-16295-9_2

1   Introduction

Implantable medical devices execute services essential for a patient's well-being. Their power consumption must be very low, as they operate either entirely based on harvested power, or contain a battery, which can only be replaced by surgery. Many of them communicate wirelessly over an RF channel, which allows for configuration of those devices without surgical intervention. However, the wireless channel also poses potential attack possibilities, as shown by Halperin et al. in [12]. This calls for strong security mechanisms to be implemented on those very constrained devices.

Texas Instruments designed MSP430 microcontrollers to target low-power applications, and advertises the application of MSP430s in the domain of medical devices [16]. MSP430s can be operated at low voltages (1.8 to 3.3 V). Newer devices of the MSP430 family have AES hardware accelerators that support 256-bit AES. Yet, many security services that are desirable for wireless communication, especially in the domain of medical devices, rely on public-key cryptography. This naturally raises the question about the performance of public-key cryptography on MSP430 microcontrollers.

Bernstein introduced the Curve25519 elliptic-curve Diffie-Hellman key exchange protocol in 2006 [2]. It uses a Montgomery curve defined over a 255-bit prime field and achieves a security level of 128 bits. Montgomery curves are known to allow for very efficient variable-base-point single-scalar multiplication, which makes this curve attractive for elliptic-curve key-agreement schemes.

Our Contribution. In this paper, we present a full implementation of the Curve25519 Diffie-Hellman key-agreement scheme on MSP430X microcontrollers[1]. We differentiate those MSP430Xs with a 16×16-bit and those with a 32×32-bit hardware multiplier and developed our code for both platforms. As all previous implementations of Curve25519, we use projective coordinates for the elliptic-curve point representation. The main performance bottleneck of the variable-base-point single-scalar multiplication are thus modular multiplications in the underlying prime field. We hence put our focus on optimizing the modular multiplication on the MSP430 architecture, and give a comprehensive evaluation of different implementation techniques for MSP430 microcontrollers.

We use the Montgomery powering ladder [24] to implement the scalar multiplication on the elliptic curve, since this is a highly regular algorithm, making the executed computation independent of the scalar. Our software completely avoids input-dependent loads and branches, thus executing in constant time and thus inherently protecting against timing attacks such as [1] or [31].

We evaluate our implementation by executing it on Texas Instrument's MSP-EXP430FR5969 LaunchPad Evaluation Kit. This board integrates an MSP430-FR5969 microcontroller [28] with a 32×32-bit hardware multiplier, which is built into the WISP 5.0 UHF computational RFID tag[2], a device that operates

---

[1] The software is available at http://emsec.rub.de/research/publications/Curve25519 MSPLatin2014/.

[2] http://wisp.wikispaces.com/WISP%205.0.

based on harvested power from the RF field. With a price of a few dollars, this microcontroller is a suitable target for wireless sensor and medical applications.

Related Work. Curve25519 has been implemented on several platforms. In the paper introducing Curve25519 [2], Bernstein presented implementation results for several Intel Pentium and an AMD Athlon processor. In 2009, Costigan and Schwabe presented Curve25519 software for the Cell Broadband Engine [7]. In 2012, Bernstein and Schwabe presented an implementation for ARM processors with NEON vector instructions [5]. Recently, Sasdrich and Güneysu presented an implementation on reconfigurable hardware in [26]. Another recent publication shows an implementation of Curve25519, that fits into 18 tweets [6,20]. So far, only one implementation shows performance results of Curve25519 on constrained devices, namely the implementation for 8-bit AVR microcontrollers by Hutter and Schwabe presented in [13]. No previous work has yet shown implementation results of Curve25519 for 16-bit microcontrollers.

There exist many publications on Elliptic Curve Cryptography (ECC) implementations on the MSP430 microcontroller architecture. One of the first publications of asymmetric cryptography on the MSP430 is by Guajardo, Blümel, Krieger, and Paar in 2001 [11]. They presented an implementation of an elliptic curve with a security level of 64 bits and show that a scalar multiplication can be performed within 3.4 million clock cycles. In 2007, Scott and Szczechowiak presented optimizations for underlying ECC finite-field multiplications [27]. Their 160×160-bit (hybrid) multiplication method requires 1746 cycles. In 2009, Szczechowiak, Kargl, Scott, and Collier presented pairing-based cryptography on the MSP430 [29]. Similar results have been reported by Gouvêa and López in the same year [9]. They reported new speed records for 160-bit and 256-bit finite-field multiplications on the MSP430 needing 1586 and 3597 cycles, respectively. They further presented an implementation of a 256-bit elliptic curve random scalar multiplication needing 20.4 million clock cycles. In 2011, Wenger and Werner compared ECC scalar multiplications on various 16-bit microcontrollers [33]. Their Montgomery-ladder based scalar multiplication needs 23.9 million cycles using a NIST P-256 elliptic curve. Also in 2011, Pendl, Pelnar, and Hutter presented the first ECC implementation running on the WISP UHF RFID tag [25]. Their 192-bit NIST curve implementation achieves an execution time of around 10 million clock cycles. They also reported first multi-precision multiplication results for 192 bits needing 2581 cycles. In 2012, Gouvêa, Oliveira, and López reported new speed records for different MSP430 architectures. They improved their results from [9], namely, for the MSP architecture (with a 16×16 multiplier) their 160-bit and 256-bit finite-field multiplication implementations need 1565 and 3563 cycles, respectively.

Also note that there exist recent works to extend the MSP430 with instruction-set extensions. In 2013, Wenger, Unterluggauer, and Werner [32] presented an MSP430 clone in hardware that implements a special instruction-set extension. For a NIST P-256 elliptic curve, their Montgomery ladder implementation requires 9 million clock cycles – without instruction-set extensions (and to put these numbers in relation), their implementation needs 22.2 million cycles.

There also exist several software libraries for the MSP430 that support ECC. These libraries mainly target sensor nodes such as the Tmote Sky which are equipped with an MSP430 microcontroller. Examples are the NanoECC [30], TinyECC [22], and MIRACL [23] libraries, and the RELIC toolkit [8].

Under the common assumption that the execution time of ECC grows approximately as a cubic function of the field size, our software significantly outperforms all presented ECC implementations on MSP430 microcontrollers in speed, while executing in constant time, thus providing security against timing attacks.

Organization. Section 2 describes specifics about the MSP430 architecture important for our implementation. Section 3 describes general basics about the implementation of Curve25519, Sect. 4 presents a detailed description of the various implementation techniques for modular multiplications that we investigated. Implementation and measurement results are presented in Sect. 5, and we conclude our work with Sect. 6.

2   The MSP430X Microcontroller Architecture

We implemented the modular multiplication operation for MSP430X devices that feature a 16 × 16-bit hardware multiplier as well as for those that feature a 32 × 32-bit multiplier, and show which implementation technique is preferable on either platform. We give cycle count estimations for the MSP430F2618 [19], which has a 16×16-bit hardware multiplier, and cycle count estimations as well as execution results for the MSP430FR5969 [28], which has a 32 × 32-bit hardware multiplier. But, our results can be generalized to other microcontrollers from the MSP430 family. This section describes specifics about the MSP430X architecture that are important for the discussion of the implementation techniques. For more details about the MSP430X architecture, we refer the reader to the MSP430x2xx user's guide [18].

Processing Unit. Both MSP430 microcontrollers that we consider have a 16-bit RISC CPU, with 27 core instructions and 24 emulated instructions. The CPU has 16-bit registers, of which R0 to R3 are special-purpose registers and R4 to R15 are freely usable working registers. The execution time of all register operations is one cycle, but the overall execution time for an instruction depends on the instruction format and the addressing mode.

Addressing Mode. The CPU features 7 addressing modes. Our implementation uses the register mode, indexed mode, absolute mode, indirect auto-increment mode, and immediate mode. It is important to note that while indirect auto-increment mode saves one clock cycle on all operations compared to indexed mode, only indexed mode can be used to store results back to RAM.

Hardware Multiplier. Both devices that we consider feature memory-mapped hardware multipliers, which work in parallel to the CPU. Four types of multiplications, namely signed and unsigned multiply as well as signed and unsigned multiply-and-accumulate are supported. The multiplier registers are peripheral registers, which have to be loaded with CPU instructions. The result is stored in two (in case of 16×16-bit multipliers) or four (in case of 32×32-bit multipliers) 16-bit registers. A register SUMEXT is available, which is similar to the status register in the main CPU. This register shows for the multiply-and-accumulate instructions, whether a multiplication has produced a carry bit. It is not possible to accumulate carries in SUMEXT. The time that is required for the multiplication is determined by the time it takes to load the multiplier registers.

3   Implementation of Curve25519

Curve25519 is an elliptic curve in Montgomery form. This curve has been carefully chosen to provide very high performance for Diffie-Hellman key agreement at the 128-bit security level. It is defined by the equation $y^2 = x^3 + 486662x^2 + x$ over the prime field $\mathbb{F}_{2^{255}-19}$. For details about the choice of curve and security see [2].

The key-agreement scheme computes a 32-byte shared secret Q_x from a 32-byte secret key n and a 32-byte public key P_x. Here Q_x and P_x are x-coordinates of points on the elliptic curve. At its core, the Curve25519 Diffie-Hellman key-agreement scheme executes a variable-base-point single-scalar multiplication on the elliptic curve, multiplying the public key P_x with the secret key n, to obtain the shared secret Q_x. Special conditions are given for the secret scalar n, namely that the 3 least significant bits and the most significant bit are set to zero, and the second-most significant bit is set to 1 [4].

We follow the suggestions of [2] for implementing the variable-base-point single-scalar multiplication on the elliptic curve. We used the Montgomery powering ladder [24] of 255 "ladder steps". Each ladder step computes a differential point addition and a point doubling. Starting with the points R1 and R2, in each ladder step either R2 is added to R1 (R1 ← R1 + R2) and then R2 is doubled (R2 ← 2 · R2), or R1 is added to R2 (R2 ← R2 + R1) and then R1 is doubled (R1 ← 2 · R1). To avoid conditional load addresses that can lead to cache-timing attacks, we execute the same operations (R1 ← R1 + R2 and R2 ← 2 · R2) in each iteration, and swap the contents of R1 and R2 depending on the scalar bit b.

Note that for the conditional swap we do not use branch instructions. Instead, this operation is implemented as follows: An unsigned variable $\hat{b}$ is cleared. Then b is subtracted from $\hat{b}$ leading to $\hat{b}$ being 0 or 0xffff, depending on whether b is 0 or 1. To swap the contents of x and y, an auxiliary variable is used to store $t_{swp} = x \oplus y$. t_{swp} is anded with the value stored in $\hat{b}$, resulting in $t_{swp} = x \oplus y$ for $b = 1$ and $t_{swp} = 0$ otherwise. Then t_{swp} is xored with x and y leading to either the original values being stored in x and y for $b = 0$, or the swapped values for the case of $b = 1$. Together with the constant-time field arithmetic we thus obtain a fully timing-attack protected constant-time implementation.

In [24] Montgomery presented x-coordinate-only doubling and differential-addition formulas for points on a curve defined by an equation of the form $By^2 = x^3 + Ax^2 + x$. He showed the correctness of those formulas, which rely on standard-projective-coordinate representation of the points, for the case of inputs not being equal to the point at infinity. In [2] Bernstein extended the proof of correctness

---

Algorithm 1. x-coordinate-only variable base-point single-scalar point multiplication on Curve25519 based on the Montgomery powering ladder [2, 7].

Input : $n \in \mathbb{Z}$, P_x, x-coordinate of point P.
Output: Q_x, x-coordinate of point $Q \leftarrow n \cdot P$.

1   $X_1 \leftarrow P_x; X_2 \leftarrow 1; Z_2 \leftarrow 0; X_3 \leftarrow P_x; Z_3 \leftarrow 1$

2   **for** $i = 254$ *downto 0* **do**

3   **if** $n_i \neq n_{i-1}$ **then**

4         $\text{swap}(X_2, X_3)$ /* This conditional swapping is implemented */

5         $\text{swap}(Z_2, Z_3)$ /* in constant time (see Sect. 3). */

6   **end**

7      $t_1 \leftarrow X_2 + Z_2$

8      $t_2 \leftarrow X_2 - Z_2$

9      $t_3 \leftarrow X_3 + Z_3$

10     $t_4 \leftarrow X_3 - Z_3$

11     $t_6 \leftarrow t_1^2$

12     $t_7 \leftarrow t_2^2$

13     $t_5 \leftarrow t_6 - t_7$

14     $t_8 \leftarrow t_4 \cdot t_1$

15     $t_9 \leftarrow t_3 \cdot t_2$

16     $X_3 \leftarrow (t_8 + t_9)^2$

17     $Z_3 \leftarrow X_1(t_8 - t_9)^2$

18     $X_2 \leftarrow t_6 \cdot t_7$

19     $Z_2 \leftarrow t_5(t_7 + 121666 t_5)^2$

20 **end**

21 **if** $n_0 == 1$ **then**

22     $\text{swap}(X_2, X_3)$ /* This conditional swapping is implemented */

23     $\text{swap}(Z_2, Z_3)$ /* in constant time (see Sect. 3). */

24 **end**

25 $Z_2 \leftarrow 1/Z_2$

26 **return** $(X_2 \cdot Z_2)$

---

to the case of an input being equal to the point at infinity. Using these formulas, a differential addition of two points requires 4 multiplications and 2 squarings. Point doubling requires 2 multiplications, 2 squarings, and one multiplication by the constant $(486662 + 2)/4 = 121666$. The differential-addition formula requires as input the difference of the input points. If the Z-coordinate of this difference point is one, the addition formula can be reduced to require only 3 multiplications and 2 squarings. Algorithm 1 summarizes the x-coordinate-only variable-base-point single-scalar point multiplication on Curve25519 requiring 255 differential additions and doublings (ladder steps), 255 conditional swaps, and one inversion at the end to transform the result back to affine coordinates [2, 7].

4   Implementation of Modular Multiplication in $\mathbb{F}_{2^{255}-19}$

Many techniques have been proposed to improve the performance of multi-precision multiplication implementations, especially for constrained devices. In the following we describe which techniques we implemented for the MSP430X architecture. To have a fair comparison, all methods were implemented in assembly and were fully unrolled.

Representation of Big Integers. We use an unsigned radix-2^{16} representation for the operand-caching [15] and the Karatsuba multiplication [14,21], and a signed radix-$2^{\lceil 255/26 \rceil}$ representation for the carry-save implementation. In unsigned radix-2^{16} representation, an n-bit integer A is represented as an array of $m = \lceil n/16 \rceil$ words in little-endian order as $(a_0, a_1, \ldots a_{m-1})$, such that $A = \sum_{i=0}^{m-1} a_i 2^{16i}$ where $a_i \in \{0, \ldots, 2^{16} - 1\}$. In the radix-$2^{\lceil 255/26 \rceil}$ representation an n-bit integer B is represented as an array of $\ell = \lceil 26n/255 \rceil$ 16-bit words in little-endian order as $(b_0, b_1, \ldots b_{\ell-1})$, such that $B = \sum_{j=0}^{\ell-1} b_j 2^{\lceil 255j/26 \rceil}$, where $b_j \in \{-2^{15}, \ldots, 2^{15} - 1\}$. Hence, in the radix-$2^{\lceil 255/26 \rceil}$ representation an element in $\mathbb{F}_{2^{255}-19}$ is represented using 26 16-bit words. Since inputs and outputs to the scalar multiplication on Curve25519 are 32-byte arrays, conversions to and from the used representations are executed at the beginning and the end of the complete scalar multiplication.

4.1   Multiplication Using Carry-Save Representation

This implementation follows the fast arithmetic implementation presented in [2]. An integer is represented using the signed radix-$2^{\lceil 255/26 \rceil}$ representation. Beneficial of this representation is that an addition or subtraction can be executed without having to consider carry bits. It only requires pairwise addition or subtraction of the respective coefficients, as long as the result of coefficient additions or subtractions does not exceed the word-length. An element in this representation looks as follows:

$$B = b_0 + b_1 2^{10} + b_2 2^{20} + b_3 2^{30} + b_4 2^{40} + b_5 2^{50} + b_6 2^{59} + b_7 2^{69} + b_8 2^{79} + \cdots + b_{25} 2^{246}.$$

Figure 1 presents the steps executed to compute the first 8 coefficients r_i of the multiplication $r \leftarrow f \times g$. After transforming an integer to radix-$2^{\lceil 255/26 \rceil}$

...	r_7	r_6	r_5	r_4	r_3	r_2	r_1	r_0
...	$f_7 g_0$	$f_6 g_0$	$f_5 g_0$	$f_4 g_0$	$f_3 g_0$	$f_2 g_0$	$f_1 g_0$	$f_0 g_0$
...	$f_6 g_1$	$2 f_5 g_1$	$f_4 g_1$	$f_3 g_1$	$f_2 g_1$	$f_1 g_1$	$f_0 g_1$	$38 f_{24} g_2$
...	$2 f_5 g_2$	$2 f_4 g_2$	$f_3 g_2$	$f_2 g_2$	$f_1 g_2$	$f_0 g_2$	$38 f_{25} g_2$	$38 f_{23} g_3$
...	$2 f_4 g_3$	$2 f_3 g_3$	$f_2 g_3$	$f_1 g_3$	$f_0 g_3$	$38 f_{25} g_3$	$38 f_{24} g_3$	$38 f_{22} g_4$
...	$2 f_3 g_4$	$2 f_2 g_4$	$f_1 g_4$	$f_0 g_4$	$38 f_{25} g_4$	$38 f_{24} g_4$	$38 f_{23} g_4$	$38 f_{21} g_5$
...	$2 f_2 g_5$	$2 f_1 g_5$	$f_0 g_5$	$38 f_{25} g_5$	$38 f_{24} g_5$	$38 f_{23} g_5$	$38 f_{22} g_5$	$38 f_{20} g_6$
...	$f_1 g_6$	$f_0 g_6$	$19 f_{25} g_6$	$19 f_{24} g_6$	$19 f_{23} g_6$	$19 f_{22} g_6$	$19 f_{21} g_6$	$38 f_{19} g_7$
...	$f_0 g_7$	$19 f_{25} g_7$	$19 f_{24} g_7$	$19 f_{23} g_7$	$19 f_{22} g_7$	$19 f_{21} g_7$	$38 f_{20} g_7$	$38 f_{18} g_8$
...								

Fig. 1. Visualisation computation of coefficients for carry-save multiplication.

representation, each coefficient b_i of B is within $(-2^9, 2^9)$ or $(-2^{10}, 2^{10})$. We precompute $2f$ and $19g$ to easily realize constant multiplication with factors 2, 19, and 38. We use the product-scanning technique to compute the coefficients r_i, interleaving the multiplication with the reduction, i.e., we compute a coefficient and reduce it right away. For the computation of each r_i, 26 products of coefficients have to be added.

This type of implementation has two disadvantages on the MSP430X architecture. First of all the MSP430 has very few general-purpose registers, while the inputs have to be loaded from four different arrays $f, g, 2f$ and $19g$. This makes storing inputs in registers difficult, as different operands are loaded for computation of the various coefficients. Further, while we use indirect auto-increment mode to access g and $19g$, there is no indirect auto-decrement mode on the MSP430 and we need to access the other inputs using the costly indexed mode. The other disadvantage is the highly complex reduction of a coefficient, requiring several shift operations, which are expensive on MSP430 devices.

Since we could not achieve good performance results with this type of implementation, we tried to speed things up relying on the refined Karatsuba formulas presented in [3]. A problem occurs when trying to add the low and the high part of B in signed radix-$2^{\lceil 255/26 \rceil}$ representation. For example computing the coefficient of 2^{40} cannot be done by adding b_4 and b_{16} as b_{16} would be input to exponent 2^{39}. Our solution to this was to represent elements using signed radix-$2^{\lceil 256/26 \rceil}$ representation and rely on computations modulo $2^{256} - 38$. Yet still, the disadvantages of this type of implementation on the MSP430 architecture dominate the advantages.

4.2   Operand-Caching Multiplication

Operand-caching was proposed by Hutter and Wenger in 2011 [15]. The idea of this method is to reduce the number of load instructions by organizing the operations in a way that allows the same input operands to be used for multiple computations.

Figure 2 shows a toy-size example of the operand-caching multiplication. Here the execution of computations is divided into the light gray and the dark gray area. First the light gray block is computed followed by the dark gray area.

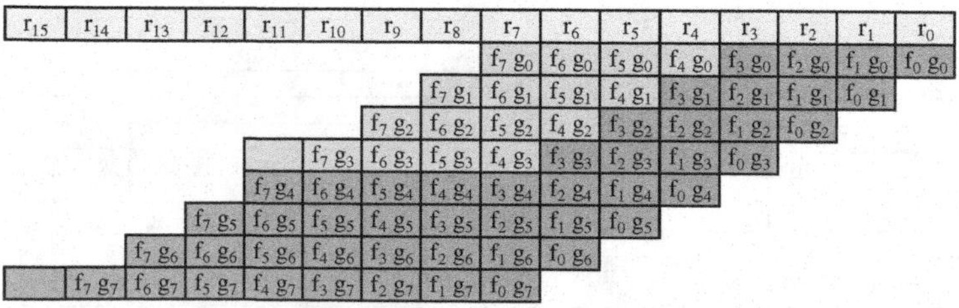

Fig. 2. Visualisation of the operand-caching method for 2 elements consisting of 8 words.

The empty dark gray and light gray boxes represent space that is required for carry-bits.

As we have 8 general-purpose registers available for storing operands during the execution of the multiplication, we chose the row size to be 4. Since each input array has 16 elements, $16/4 = 4$ rows have to be computed. Many loads to the hardware multiplier can be saved when loading operands in a special order. For each operation of the hardware multiplier OP2 has to be loaded to start execution. Yet, MAC does not have to be loaded each time. If it is not loaded, it uses the value that had been loaded to MAC in the previous use of the hardware multiplier. For example, if for the computation of r_1, as the final step f_0 was loaded to MAC and g_1 to OP2, then we start the computation of r_2 by loading g_2 to OP2.

In this multiplication we first multiply both inputs f and g, resulting in a double-sized array and then reduce this result. Since reducing mod $2^{255} - 19$ requires bit shifts, we chose to reduce intermediate results mod $2^{256} - 38$ and only reduce the final result mod $2^{255} - 19$. We implemented two versions of operand-caching multiplication, one making use of the 32×32-bit hardware multiplier (in the following called 32-bit operand-caching) and the other only loading 16-bit inputs to the multiplier (in the following called 16-bit operand-caching). Naturally the implementation that makes use of the 32×32-bit hardware multiplier is faster and also requires less code space, since fewer loads to the multiplier have to be performed.

4.3 Karatsuba Multiplication

This section is based on a very recent paper on the implementation of multi-precision multiplication on AVR microcontrollers [14]. Karatsuba presented a sub-quadratic multiplication method that reduces the number of required word multiplications for multi-precision multiplications [21]. The implementation by Hutter and Schwabe [14] is based on this idea and first demonstrates that this method is more advisable on AVRs even for very small input sizes starting from 48 bits. They implemented what they call *subtractive Karatsuba*. This method avoids having to take extra carry bits into account by computing $|F_l - F_h|$ and $|G_l - G_h|$ instead of $F_l + F_h$ and $G_l + G_h$, which makes it easier to obtain a constant-time implementation. In the following we report the method, as it was presented in [14], adapting it to the case of a 16-bit architecture. The steps for multiplying two n-byte numbers, where in our case $n = 32$, are described in detail. Using a 16-bit architecture, we have to process arrays of $n/2 = 16$ elements. We split those arrays at $k = 16/2 = 8$.

- Write $F = F_\ell + 2^{16k}F_h$ and $G = G_\ell + 2^{16k}G_h$
- compute $L = F_\ell \cdot G_\ell$
- compute $H = F_h \cdot G_h$
- compute $M = |F_\ell - F_h| \cdot |G_\ell - G_h|$ and
- set $t = 0$, if $M = (F_\ell - F_h) \cdot (G_\ell - G_h)$; $t = 1$ otherwise;
- compute $\hat{M} = (-1)^t M$; and
- obtain the result as $FG = L + 2^{16k}(L + H - \hat{M}) + 2^{16n/2}H$.

We use operand-caching multiplication for all multi-precision multiplications within the Karatsuba multiplication, i.e., the computations of L, H, and M. $|F_\ell - F_h|$ is computed as follows: first we subtract with borrow all elements in F_h from those in F_ℓ and subtract with borrow from a register b_F that was cleared before. This results in $b_F = 0$ for $F_\ell > F_h$ and $b_F = $ 0xffff otherwise. We XOR b_F with $F_\ell - F_h$ resulting in the ones-complement of $F_\ell - F_h$. We then compute $t_F = b_F$ AND 1 add this to the ones-complement of $F_\ell - F_h$ and ripple the carry through, resulting in the two's complement of $F_\ell - F_h$, which is equal to $|F_\ell - F_h|$. $|G_\ell - G_h|$ is computed similarly. The value t required for the computation of M is obtained as $t = t_F \oplus t_G$. The same technique that was used to compute the absolute difference above is used for the computation of $\hat{M}$ from M, leaving out the initial subtraction part.

Again we computed the product of the inputs resulting in a double-sized array and reduced the result mod $2^{256} - 38$. Only at the end of the Curve25519 computation we reduced results mod $2^{255} - 19$. In the following we will refer to the implementation making use of the 32×32-bit multiplier as 32-bit Karatsuba and the one for 16×16-bit multiplier as 16-bit Karatsuba. We further implemented this method for 2-level Karatsuba, i.e. using subtractive Karatsuba for the computation of L, H, and M. We will refer to those implementations as 2-Level 32-bit Karatsuba and 2-Level 16-bit Karatsuba, for using 32×32-bit multiplier and 16×16-bit multiplier respectively.

5   Performance and Power Consumption Results

We used IAR Embedded Workbench for MSP430 IDE version 5.60.3 to develop our code and compiled all source code by setting the compiler options to "low". This causes dead code, redundant labels and redundant branches to be eliminated and achieves that variables live only as long as they are needed. It further avoids common subexpression elimination, loop unrolling, function inlining, code motion and type-based alias analysis [17]. Note that all functions implementing arithmetic in $\mathbb{F}_{2^{255}-19}$ were implemented in assembly, while the higher level functions are implemented in C. This section describes our implementation and measurement results.

We first present cycle-count estimates for the modular multiplication implementations given by IAR Embedded Workbench IDE. We compare these results for two devices, namely MSP430FR5969 and MSP430F2618 having a 32×32-bit and a 16×16-bit hardware multiplier, respectively. We further present numbers for the required code space for the multiplication implementations.

For a device that has a 32×32-bit hardware multiplier (MSP430FR5969) we executed the code and measured the execution time using the debugging functionality of IAR Embedded Workbench IDE. We present the cycle count for an execution of the Curve25519 variable-base-point single-scalar multiplication on the MSP430FR5969 for the cases of having a 32×32-bit or a 16×16-bit hardware multiplier on this target. Finally, we present our power measurement results of the execution of different multiplication implementations and the scalar multiplication on the MSP-EXP430FR5969 Launchpad Evaluation Kit.

Table 1. Simulated cycle count for modular multiplication (including reduction) on MSP430F2618 and MSP430FR5969, given by IAR Embedded Workbench IDE version 5.60.3

		MSP430FR5969	MSP430F2618
1	16-bit Operand-caching	3968	3949
2	32-bit Operand-caching	2505	-
3	16-bit Carry-save	7231	7228
4	16-bit Karatsuba	3666	3623
5	32-bit Karatsuba	2501	-
6	16-bit 2-level Karatsuba	3595	3554
7	32-bit 2-level Karatsuba	2705	-

Table 2. Code space (in bytes) required for modular multiplication implementations (including reduction) on MSP430s.

		Code Space (in bytes)
1	16-bit Operand-caching	4762
2	32-bit Operand-caching	2878
3	16-bit Carry-save	8448
4	16-bit Karatsuba	4316
5	32-bit Karatsuba	2826
6	16-bit 2-level Karatsuba	4270
7	32-bit 2-level Karatsuba	3144

5.1 Performance

First we simulated the cycle count and measured the required code space of the different variants of implementation of the modular multiplication that we implemented in IAR Embedded Workbench IDE. Table 1 presents the simulated execution times for the two aforementioned microcontrollers, while Table 2 shows the required code space for each implementation. It seems quite natural that the version making use of the 32×32-bit hardware multiplier is faster and requires less code space since fewer load (and store) operations to (and from) the dedicated registers of the multiplier have to be executed.

We then measured the execution time of all multiplication implementations on the MSP430FR5969 using the debugging functionality of IAR Embedded Workbench IDE (Table 3). During this step we realized that wait cycles must be included when the MSP430FR5969 runs at the frequency of 16 MHz. It is due to the limited access frequency of FRAM, i.e., 8 MHz. So, the speed of the implementation is not doubled by increasing the operation frequency from 8 MHz to 16 MHz. Table 3 displays these results. While in simulation the 32-bit operand-caching multiplication seems to perform similar to the 32-bit Karatsuba implementation, it turns out that, when executing the implementations on the board the 32-bit Karatsuba implementation performs a bit better compared to 32-bit operand-caching (cf. Table 3). This is due to the fact that IAR Embedded Workbench IDE does not correctly simulate the execution time of the hardware multiplier, i.e. the time it takes until the CPU can read out results from the hardware multiplier. Interestingly, the improvement of using 2-level Karatsuba is only given when making use of the 16×16-bit hardware multiplier (MSP430F2618). When making use of the 32×32-bit multiplier, the overhead required for the implementation of 2-level Karatsuba seems to dominate over the improvements in timings. The lowest code space is achieved with 32-bit Karatsuba, but not far from 32-bit operand-caching (Table 2).

Table 3. Execution time (i.e., cycle count) on MSP-EXP430FR5969 Launchpad Evaluation Kit, optimizations set to "low" when running the microcontroller at different frequencies.

		8 MHz	16 MHz
1	16-bit operand-caching	4045	4599
2	32-bit operand-caching	2529	2864
3	16-bit Carry-save	7230	8289
4	16-bit Karatsuba	3696	4203
5	32-bit Karatsuba	2488	2824
6	16-bit 2-level Karatsuba	3606	4119
7	32-bit 2-level Karatsuba	2684	3069

Further we implemented the variable-basepoint single-scalar multiplication for the cases of having a 32×32-bit and having a 16×16-bit hardware multiplier. For the implementation that makes use of the 32×32-bit hardware multiplier we used 32-bit Karatsuba and for the implementation that only requires a 16×16-bit hardware multiplier we used 2-level 16-bit Karatsuba, as those are the fastest implementations for those cases according to Table 3. On the MSP430FR5969 the x-coordinate-only variable-basepoint single-scalar multiplication, which makes use of the 32×32-bit hardware multiplier, executes in 6,513,011 clock cycles and requires 9.1 kB of code space, whereas the 16×16-bit hardware multiplier version, executes in 9,139,739 clock cycles and requires 11.6 kB of code space.

Since there are no implementation results of the plain ECC point multiplication on an MSP430X with a 32×32-bit hardware multiplier given in the literature, we compare the results given in the literature to our result for the 16×16-bit hardware multiplier (Table 4). Note that Gouvêa et al. obtain better performance results for a 128-bit-secure elliptic-curve scalar multiplication on an MSP430X microcontroller with a 32×32-bit hardware multiplier, albeit on a different curve [10], but do not report performance results for the plain scalar multiplication, but instead for the execution of several ECC-based protocols.

Table 4. Execution time (i.e., cycle count) of variable base-point single-scalar multiplications on an elliptic curve providing a security level comparable to 128-bit symmetric security on MSP430 microcontrollers.

	Architecture	Cycle count
Wenger et al. [33]	MSP	23,973,000
Wenger et al. [32]	MSP Clone w/o ISE	22,170,000
Gouvêa et al. [9]	MSP	20,476,234
Our implementation	MSPX	**9,139,739**

5.2   Power Consumption

We further examined our code in terms of power consumption on the MSP-EXP430FR5969 Launchpad Evaluation Kit. We have implemented all multiplications (e.g., listed in Table 1) in such a way that first two random operands are selected then multiplied together by all multiplication algorithms one after another. We also used an I/O pin of the MSP-EXP430FR5969 Launchpad Evaluation Kit to indicate the start and the end of each algorithm thereby being able to identify at which period of time each algorithm is executed.

For the power measurements we made use of a LeCroy WaveRunner HRO 66Zi digital sampling oscilloscope. As the MSP-EXP430FR5969 Launchpad Evaluation Kit has been developed to facilitate power measurements, we could easily place a 2.2 Ω shunt resistor at the Vdd path of the MSP430FR5969 microcontroller while no stabilizing capacitor was placed between the measurement point and the microcontroller. We powered the Evaluation Kit by an external stable power supply and monitored the current passing through the shunt resistor by means of a LeCroy AP 033 differential probe at a sampling rate of 1 GS/s.

Figure 3(a) shows a sample power trace where the parts dedicated to each multiplication are marked. In Fig. 3(b) we also provide a zoomed view of this trace to highlight several—non-periodic—high peaks which we have observed. We have observed the same peaks (but periodic) for a couple of NOP operations as well. The pattern of these high peaks actually differs for different sequence of operations. The source of this high power consumption peaks are not certainly clear to us, but it seems that they are relevant to FRAM accesses. That is because fetching the instructions from the code memory also needs to access the FRAM.

For 1 000 random operand pairs we collected 1000 traces, each of which covers the execution of all 7 multiplications with the same operands. Corresponding to each multiplication, each trace is divided into 7 parts and the voltage observed by the differential probe at each sample point is turned into instantaneous power as $P = V^2/R$, where $R = 2.2\,\Omega$. Average of instantaneous power values over the period of time corresponding to each multiplication gives us the power consumption of the device for that operation. We also can turn this value to amount of energy the device consumed by $P \cdot t$, where t stands for the duration of the multiplication. Figure 4 depicts the average of power and energy consumption of the microcontroller for each multiplication. Note that since the

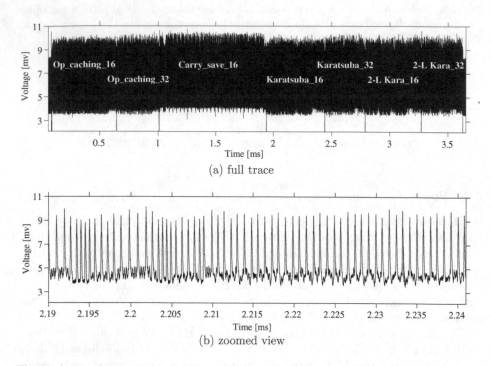

(a) full trace

(b) zoomed view

Fig. 3. A sample power trace measured from MSP-EXP430FR5969 Launchpad Evaluation Kit when running 7 different multiplications

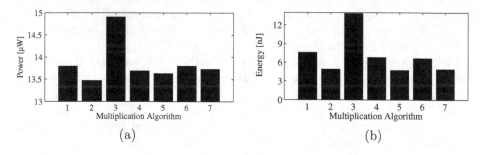

(a)   (b)

Fig. 4. Average of (a) power and (b) energy consumption of different multiplications (the indices for the algorithms fit to the same order shown in Table 1.)

MSP430FR5969 microcontroller on the Evaluation Kit operates by the internal oscillator (8 MHz), the duration of each multiplication was not completely the same for all 1000 measurements due to the small jitter of the oscillator.

As shown by the graphics, 32-bit operand-caching has the lowest power consumption. However, 32-bit Karatsuba consumes less energy as it is the fastest one (see Table 1). As stated above, using 32-bit Karatsuba the debugging functionality of IAR Embedded Workbench IDE reports 6,513,011 clock cycles for the execution of a scalar multiplication on Curve25519 on the board having a

MSP430FR5969. We verified this result measuring the length of the power trace. Based on our practical measurements one full execution of the algorithm takes around 821 ms with operation frequency of 8 MHz. This confirms the cycle count measured with IAR debugging functionality. To measure its power consumption we had to decrease the sampling rate to 200 MS/s due to the length of the trace (825 ms). Based on 100 measurements for random operands, in average the corresponding power consumption and energy consumption is 14.046 μW and 11.623 μJ respectively.

6   Conclusion

This paper is the first that presents a full constant-time implementation of Curve25519 on different MSP430 microcontrollers. In order to evaluate and improve the efficiency, we implemented and analyzed different finite-field multiplication techniques and compared them in terms of speed, code size, and power consumption. Amongst all considered multiplication techniques, the subtractive Karatsuba implementation proposed in [14] performs the best. It turned out that 2-level Karatsuba performs better than 1-level Karatsuba in case a 16×16-bit hardware multiplier is available. This is however not the case if the MSP430 has a 32×32-bit hardware multiplier. We further analyzed our implementation with the MSP-EXP430FR5969 Launchpad Evaluation Kit. We presented numbers for the average power and the energy consumption of Curve25519 on this platform. We showed that with an energy consumption of 11.623 μJ the execution of high-security ECC is feasible on devices operated with battery or harvested power, such as medical implants.

References

1. Acıiçmez, O., Brumley, B.B., Grabher, P.: New results on instruction cache attacks. In: Mangard, S., Standaert, F.-X. (eds.) CHES 2010. LNCS, vol. 6225, pp. 110–124. Springer, Heidelberg (2010). http://www.iacr.org/archive/ches2010/62250105/62250105.pdf. 32
2. Bernstein, D.J.: Curve25519: new Diffie-Hellman speed records. In: Yung, M., Dodis, Y., Kiayias, A., Malkin, T. (eds.) PKC 2006. LNCS, vol. 3958, pp. 207–228. Springer, Heidelberg (2006). http://cr.yp.to/papers.html#curve25519. 32, 33, 35, 36, 37
3. Bernstein, D.J.: Batch binary edwards. In: Halevi, S. (ed.) CRYPTO 2009. LNCS, vol. 5677, pp. 317–336. Springer, Heidelberg (2009). http://cr.yp.to/papers.html#bbe. 38
4. Bernstein, D.J.: Cryptography in NaCl (2009). http://cr.yp.to/highspeed/naclcrypto-20090310.pdf. 35
5. Bernstein, D.J., Schwabe, P.: NEON crypto. In: Prouff, E., Schaumont, P. (eds.) CHES 2012. LNCS, vol. 7428, pp. 320–339. Springer, Heidelberg (2012). http://cryptosith.org/papers/neoncrypto-20120320.pdf. 33
6. Bernstein, D.J., van Gastel, B., Janssen, W., Lange, T., Schwabe, P., Smetsers, S.: TweetNaCl: A crypto library in 100 tweets (to appear). Document ID: c74b5bbf605ba02ad8d9e49f04aca9a2. http://cryptojedi.org/papers/#tweetnacl. 33

7. Costigan, N., Schwabe, P.: Fast elliptic-curve cryptography on the cell broadband engine. In: Preneel, B. (ed.) AFRICACRYPT 2009. LNCS, vol. 5580, pp. 368–385. Springer, Heidelberg (2009). 33, 36
8. Aranha, D.F., Gouvêa, C.P.L.: RELIC is an Efficient LIbrary for Cryptography (2014). http://code.google.com/p/relic-toolkit/. Accessed 06 September 2014. 34
9. Gouvêa, C.P.L., López, J.: Software implementation of pairing-based cryptography on sensor networks using the MSP430 microcontroller. In: Roy, B., Sendrier, N. (eds.) INDOCRYPT 2009. LNCS, vol. 5922, pp. 248–262. Springer, Heidelberg (2009). http://conradoplg.cryptoland.net/files/2010/12/indocrypt09.pdf. 33, 43
10. Gouvêa, C.P.L., Oliveira, L.B., López, J.: Efficient software implementation of public-key cryptography on sensor networks using the MSP430X microcontroller. J. Crypt. Eng. 2(1), 19–29 (2012). http://conradoplg.cryptoland.net/files/2010/12/jcen12.pdf. 42
11. Guajardo, J., Blümel, R., Krieger, U., Paar, C.: Efficient implementation of elliptic curve cryptosystems on the TI MSP430x33x family of microcontrollers. In: Kim, K. (ed.) PKC 2001. LNCS, vol. 1992, pp. 365–382. Springer, Heidelberg (2001). 33
12. Halperin, D., Heydt-Benjamin, T.S., Ransford, B., Clark, S.S., Defend, B., Morgan, W., Fu, K., Kohno, T., Maisel, W.H.: Pacemakers and implantable cardiac defibrillators: Software radio attacks and zero-power defenses. In: IEEE Symposium on Security and Privacy - IEEE S&P 2008d, pp. 129–142. IEEE Computer Society (2008). http://www.secure-medicine.org/public/publications/icd-study.pdf. 32
13. Hutter, M., Schwabe, P.: NaCl on 8-Bit AVR microcontrollers. In: Youssef, A., Nitaj, A., Hassanien, A.E. (eds.) AFRICACRYPT 2013. LNCS, vol. 7918, pp. 156–172. Springer, Heidelberg (2013). http://cryptojedi.org/papers/avrnacl-20130220.pdf. 33
14. Hutter, M., Schwabe, P.: Multiprecision multiplication on AVR revisited (2014). http://cryptojedi.org/papers/#avrmul. 37, 39, 45
15. Hutter, M., Wenger, E.: Fast multi-precision multiplication for public-key cryptography on embedded microprocessors. In: Preneel, B., Takagi, T. (eds.) CHES 2011. LNCS, vol. 6917, pp. 459–474. Springer, Heidelberg (2011). https://online.tugraz.at/tug_online/voe_main2.getvolltext?pCurrPk=58138. 37, 38
16. T.I. Incorporated: Enabling secure portable medical devices with TI's MSP430 MCU and wireless technologies (2012). http://www.ti.com/lit/wp/slay027/slay027.pdf. 32
17. T.I. Incorporated: MSP430FR58xx, MSP430FR59xx, MSP430FR68xx, and MSP430FR69xx family user's guide (2012). 40
18. T.I. Incorporated: MSP430x2xx family - user's guide, July 2013. http://www.ti.com/lit/ug/slau144j/slau144j.pdf. 34
19. T.I. Incorporated: MSP430F261x datasheet (rev. K) (2014). http://www.ti.com/lit/ds/symlink/msp430f2618.pdf. 34
20. Janssen, W.: Curve25519 in 18 tweets. Bachelor's thesis, Radboud University Nijmegen (2014). http://www.cs.ru.nl/bachelorscripties/2014/Wesley_Janssen___4037332___Curve25519_in_18_tweets.pdf. 33
21. Karatsuba, A., Ofman, Y.: Multiplication of multidigit numbers on automata. Soviet Physics Doklady, 7, 595–596 (1963). Translated from Doklady Akademii Nauk SSSR, Vol. 145, No. 2, pp. 293–294, July 1962. 37, 39
22. Liu, A., Ning, P.: TinyECC: a configurable library for elliptic curve cryptography in wireless sensor networks. In: International Conference on Information Processing in Sensor Networks - IPSN 2008, pp. 245–256. IEEE (2008). discovery.csc.ncsu.edu/pubs/ipsn08-TinyECC-IEEE.pdf. 34

23. C.U. Ltd.: MIRACL cryptographic SDK (2011). http://www.certivox.com/miracl/ (Accessed 06 September 2014). 34
24. Montgomery, P.L.: Speeding the pollard and Elliptic Curve methods of factorization. Math. Comput. **48**(177), 243–264 (1987). 32, 35
25. Pendl, C., Pelnar, M., Hutter, M.: Elliptic curve cryptography on the WISP UHF RFID tag. In: Juels, A., Paar, C. (eds.) RFIDSec 2011. LNCS, vol. 7055, pp. 32–47. Springer, Heidelberg (2012). 33
26. Sasdrich, P., Güneysu, T.: Efficient elliptic-curve cryptography using curve25519 on reconfigurable devices. In: Goehringer, D., Santambrogio, M.D., Cardoso, J.M.P., Bertels, K. (eds.) ARC 2014. LNCS, vol. 8405, pp. 25–36. Springer, Heidelberg (2014). https://www.hgi.rub.de/media/sh/veroeffentlichungen/2014/03/25/paper_arc14_curve25519.pdf. 33
27. Scott, M., Szczechowiak, P.: Optimizing multiprecision multiplication for public key cryptography. Cryptology ePrint Archive, Report 2007/299 (2007). http://eprint.iacr.org/2007/299/. 33
28. I. Systems: IAR C/C++ Compiler reference guide for texas instruments' msp430 microcontroller family (2011). 32, 34
29. Szczechowiak, P., Kargl, A., Scott, M., Collier, M.: On the application of pairing based cryptography to wireless sensor networks. In: Basin, D.A., Capkun, S., Lee, W. (eds.) Proceedings of the Second ACM Conference on Wireless Network Security - WiSec 2009, pp. 1–12. ACM (2009). 33
30. Szczechowiak, P., Oliveira, L.B., Scott, M., Collier, M., Dahab, R.: NanoECC: testing the limits of elliptic curve cryptography in sensor networks. In: Verdone, R. (ed.) EWSN 2008. LNCS, vol. 4913, pp. 305–320. Springer, Heidelberg (2008). http://www.ic.unicamp.br/ leob/publications/ewsn/NanoECC.pdf. 34
31. Tromer, E., Osvik, D.A., Shamir, A.: Efficient cache attacks on AES, and countermeasures. J. Cryptol. **23**(1), 37–71 (2010). http://www.tau.ac.il/tromer/papers/cache-joc-20090619.pdf. 32
32. Wenger, E., Unterluggauer, T., Werner, M.: 8/16/32 shades of elliptic curve cryptography on embedded processors. In: Paul, G., Vaudenay, S. (eds.) INDOCRYPT 2013. LNCS, vol. 8250, pp. 244–261. Springer, Heidelberg (2013). 33, 43
33. Wenger, E., Werner, M.: Evaluating 16-bit processors for elliptic curve cryptography. In: Prouff, E. (ed.) CARDIS 2011. LNCS, vol. 7079, pp. 166–181. Springer, Heidelberg (2011). 33, 43

Efficient Integer Encoding for Homomorphic Encryption via Ring Isomorphisms

Matthias Geihs[1]($\boxtimes$) and Daniel Cabarcas[2]

[1] Technische Universität Darmstadt, Darmstadt, Germany
mgeihs@cdc.informatik.tu-darmstadt.de
[2] Universidad Nacional de Colombia Sede Medellín, Medellín, Colombia

Abstract. Homomorphic encryption allows computation on encrypted data at the cost of a significant loss in efficiency. In this paper we propose a powerful integer encoding for homomorphic encryption. The proposed encoding offers more efficient and convenient homomorphic computations on integers compared to previously used methods. This is possible by making the message space of the encryption scheme isomorphic to an integer quotient ring. The encoding can be used across various lattice-based homomorphic encryption schemes such as NTRU and various ring-LWE based schemes. We analyse the efficiency of our proposed encoding, which shows a significant gain compared to a naive integer encoding for a ring-LWE based scheme.

Keywords: Integer encoding · Fully homomorphic encryption · Lattice based cryptography · Privacy

1 Introduction

In 2009, Craig Gentry proposed the first fully homomorphic encryption scheme [7]. Generally speaking, homomorphic encryption allows performing operations on encrypted data. The potential use of homomorphic encryption in privacy applications is huge [1,8,10,11,14,16]. But, despite numerous improvements over Gentry's original framework [2–6,9,19,19–21], computing on encrypted data is still much less efficient than directly computing on the data. For this technology to become really practical, we need to squeeze as much computation as possible on every single homomorphic computation cycle. In this paper we propose a way to encode integers as plaintext of a homomorphic encryption scheme so that we can perform integer operations efficiently in the encrypted domain.

Homomorphic encryption schemes do not operate on bits but rather on a message space that is often a complex algebraic structure. A common message space is the quotient ring $R_2 = \mathbb{Z}_2[x]/\langle x^n + 1\rangle$. Meaningful data needs to be encoded onto such a structure in order to compute homomorphically on it. It is possible to encode a single bit in R_2, but it would be wasteful. It is rather desirable to encode several bits of information in a single message. However, care must be taken because operations obey the algebraic rules of the message space.

© Springer International Publishing Switzerland 2015
D.F. Aranha and A. Menezes (Eds.): LATINCRYPT 2014, LNCS 8895, pp. 48–63, 2015.
DOI: 10.1007/978-3-319-16295-9_3

It is thus an important question how to encode as much information as possible on a message space, in a way that the operations are meaningful.

We focus on how to encode a fundamental data type, the integer. Previously proposed integer encodings are problematic because they operate on polynomials of restricted coefficients and degree. Once an encoding polynomial's coefficient or degree grows too large, sensible decoding is not possible anymore.

We observe that it is possible to construct homomorphic encryption schemes for which the message space is isomorphic to the ring $\mathbb{Z}_t$, for a large integer t. This leads to an integer encoding that is efficient and convenient to use. Our encoding assures that operations reside in $\mathbb{Z}_t$ and hence computing on encoded integers is similar to computing on $\log_2(t)$-bit integers. Furthermore, the proposed encoding is better in computational complexity with $O(l^2 \log l)$ compared to $O(l^3)$ for the previously used encoding, where l is the number of layered multiplications.

The proposed encoding can also be enhanced to encode multiple integers into a single message which allows for *single instruction multiple data* (SIMD) style homomorphic computation [19]. We can also obtain an integer finite field encoding by constructing a scheme whose message space is isomorphic to $\mathbb{Z}_t$, where t is a prime. The encoding can be used across different homomorphic encryption schemes whose message space is isomorphic to certain quotient rings. For example, NTRU based schemes are supported [12,16], as well as a variety of ring-LWE based schemes.

We show how to modify the ring-LWE scheme by Brakerski and Vaikunthanatan [6] (BV-RLWE) to allow our integer encoding. The message space of BV-RLWE is a quotient ring of the form $R_p = \mathbb{Z}[x]/\langle x^n + 1, p\rangle$, where p is an integer. The key observation to achieve our encoding is that R_p is isomorphic to an integer quotient ring $\mathbb{Z}_t$ if we allow p to be a polynomial. Then, we can encode integers using a straightforward mapping from the integer quotient ring to the message space.

Choosing p to be a polynomial does not affect the security of the scheme, and has only minor implications on its efficiency. In particular, we show that modulus switching, the noise reduction technique proposed by Brakerski and Vaikuntanathan [5], caries over to this setting. This makes the scheme much more efficient, to the point that the size of the public key grows only linearly in the depth of the circuits it can evaluate. Moreover, depending on the choice of p, the scheme can be turned into a leveled fully homomorphic encryption scheme, i.e., a scheme that can evaluate its decryption function homomorphically.

We develop a methodology to evaluate the efficiency of integer encodings. Our approach consists on calculating the scheme parameters necessary to correctly evaluate a benchmark function. Using this methodology we compare our encoding to the previously proposed integer encoding by Lauter et al. [14]. The analysis yields that our encoding is significantly more efficient than the one by Lauter et al.

1.1 Related Work

Another integer encoding was proposed by Lauter et al. [14], in which they straightforwardly encode integer bits as polynomial coefficients. The message space in their case is $\mathbb{Z}_t[x]/\langle x^n + 1 \rangle$, for a small integer t. They propose to encode an integer $a = \sum_{i=0}^k a_i 2^i$ as a polynomial $\sum_{i=0}^k a_i x^i$. A major drawback of this encoding is that the operations allowed are restricted by t and n. Additions increase the size of the coefficients which must not reach t, while multiplications also increase the degree k, which must not reach n. In Sect. 3 we give a more detailed description of this encoding.

In another related work, Gentry et al. [10] propose an encoding for elements in the finite field $\mathbb{F}_{2^8}$. In their case, the message space is $\mathbb{Z}_2[x]/\langle \Phi_m(x) \rangle$, where $\Phi_m(x)$ is the m-th cyclotomic polynomial. Their goal is to efficiently encode the AES-128 state, which consists of 16 $\mathbb{F}_{2^8}$ elements. For that, they choose m so that $\Phi_m(x)$ factors modulo 2 into k degree d polynomials, such that 8 divides d. Thus $\mathbb{Z}_2[x]/\langle \Phi_m(x) \rangle$ is isomorphic to k copies of $\mathbb{F}_{2^d}$, which contains $\mathbb{F}_{2^8}$ as a subfield. By doing this they can encode k elements of $\mathbb{F}_{2^8}$ in a single message and operate on them in parallel. Our encoding of integers is similar in the sense that we can pack k integers in a single message and operate on them in parallel. However, we use different structures and mappings to be able to encode integers. The structure of the ring of integers cannot be efficiently emulated on a field of the form $\mathbb{F}_{2^d}$.

1.2 Organization

In Sect. 2 we describe the homomorphic encryption scheme BV-RLWE and analyze the conditions for its correctness. In Sect. 3 we present two integer encodings, the one by Lauter et al. [14] and our isomorphism-based encoding. In Sect. 4 we propose a methodology to evaluate the efficiency of integer encodings and compare our proposed encoding to the one by Lauter et al.

2 BV-RLWE

We describe a modified version of the homomorphic encryption scheme originally proposed by Brakerski and Vaikuntanathan [6]. We will later use this scheme for evaluating efficiency of integer encodings.

2.1 Notation

The scheme, which is described in Sect. 2.2, relies on switching between certain quotient ring structures. In order to describe the scheme properly it is important to describe how elements from one quotient ring are carried over to another. We consider three types of quotient rings, which are $R = \mathbb{Z}[x]/\langle x^n + 1 \rangle$ and $R_p = \mathbb{Z}[x]/\langle x^n + 1, p \rangle$, $p \in \mathbb{N}$ and $R_{x-a} = \mathbb{Z}[x]/\langle x^n + 1, x - a \rangle$, $a \in \mathbb{Z}$. When carrying over an element from one ring to the other we will use a hybrid unique representation of the element over the underlying polynomial ring $\mathbb{Z}[X]$. For $f(x) + \langle x^n + 1 \rangle \in R$, we denote by $[f(x)]_R \in \mathbb{Z}[x]$ the unique polynomial with $[f(x)]_R \equiv f(x) \pmod{x^n + 1}$, and $\deg [f(x)]_R < n$. For $f(x) + \langle x^n + 1, p \rangle \in R_p$, we denote by $[f(x)]_{R_p} \in \mathbb{Z}[x]$ the unique polynomial with

$[f(x)]_{R_p} = \sum_i f_i * x^i$, $[f(x)]_{R_p} \equiv f(x) \pmod{x^n + 1, p}$, $f_i \in (-p/2, \ldots, p/2]$, and $\deg \lfloor f(x) \rfloor_{R_p} < n$. For $f(x) + \langle x^n + 1, x - a \rangle \in R_{x-a}$, we denote by $[f(x)]_{R_{x-a}} \in \mathbb{Z}$ the unique integer with $[f(x)]_{R_{x-a}} \equiv f(x) \pmod{x^n + 1, x - a}$, and $[f(x)]_{R_{x-a}} \in \{0, \ldots, |a^n + 1| - 1\}$.

Furthermore, sampling an element a from distribution χ over R is denoted by $a \leftarrow_\chi R$ and $a \leftarrow_\$ R$ denotes sampling an element uniformly at random from R. The infinity norm of a polynomial $a = \sum_i a_i * x^i \in \mathbb{Z}[x]$ is denoted by $\|a\|_\infty = \max_i |a_i|$. We say a distribution χ over R is bounded by B if $\|[a]_R\|_\infty \leq B$ for all a in the support of χ.

2.2 Scheme

The BV-RLWE homomorphic encryption scheme is parametrized by $n, q \in \mathbb{N}$, and $p \in \mathbb{Z}[x]$. Note that in contrast to the original scheme, we allow p to be a polynomial to support our proposed integer encoding later on. Plaintext space and ciphertext space are derived from the quotient ring $R := \mathbb{Z}[x]/\langle x^n + 1 \rangle$. The plaintext space is the quotient ring $R_p := \mathbb{Z}[x]/\langle x^n + 1, p \rangle$ and the ciphertext space is the polynomial ring $R_q[X]$ over the quotient ring $R_q := \mathbb{Z}[x]/\langle x^n + 1, q \rangle$. The scheme is also parametrized by an error distribution χ over R bounded by B_χ.

- **Key Generation.** Sample $s, e \leftarrow_\chi R$, $a_1 \leftarrow_\$ R_q$ and compute $a_0 := [-(a_1 * s + p * e)]_{R_q}$. The secret key is $\mathbf{sk} = s$, the public key is $\mathbf{pk} = (a_0, a_1)$.
- **Encryption.** A message $\mathbf{m} \in R_p$ is encrypted using the public key $\mathbf{pk} = (a_0, a_1)$. First, sample $u, f, g \leftarrow_\chi R$. Then, compute $c_0 := [a_0 * u + p * g + \mathbf{m}]_{R_q}$ and $c_1 := [a_1 * u + p * f]_{R_q}$. The ciphertext is $\mathbf{ct} := c_0 + c_1 * X$.
- **Decryption.** A ciphertext $\mathbf{ct}(X) \in R_q[X]$ is decrypted by computing $\mathbf{m} := [[\mathbf{ct}(\mathbf{sk})]_{R_q}]_{R_p}$.

We can easily check that the plaintext space $(R_p, +, *)$ is somewhat homomorphic to the ciphertext space $(R_q[X], +, *)$ by $Enc : R_p \rightarrow R_q[X]$ and $Dec : R_q[X] \rightarrow R_p$. We denote homomorphic evaluation of a function f on ciphertexts $\mathbf{ct}_1, \ldots, \mathbf{ct}_l$ by $Eval(f, \mathbf{ct}_1, \ldots, \mathbf{ct}_l)$.

2.3 Noise Analysis

With the BV-RLWE scheme, decryption works as long as ciphertext noise is small. However, homomorphic operations increase ciphertext noise. In this section we give upper bounds on ciphertext noise for fresh ciphertexts and for ciphertexts which result from homomorphic operations. The following lemma gives an upper bound on polynomial coefficient size for multiplication over $R = \mathbb{Z}[x]/\langle x^n + 1 \rangle$.

Lemma 1. *Let $a, b \in \mathbb{Z}[x]$, then $\|[a * b]_R\|_\infty \leq n \|[a]_R\|_\infty \|[b]_R\|_\infty$.*

Proof. Write $a = \sum_{i=0}^{n-1} a_i x^i$, $b = \sum_{i=0}^{n-1} b_i x^i$. Using $x^n \equiv -1 \pmod{x^n + 1}$ we have

$$[a * b]_R = \left[\sum_{i,j=0}^{n-1} \gamma_{i,j} * a_{(i+j \mod n)} * b_{(-j \mod n)} * x^i \right]_R,$$

with $\gamma_{i,j} = 1$ if $i + j < n$ or else $\gamma_{i,j} = -1$. We realize that every term occurs exactly n times. $\square$

For our noise analysis it is important to realize that a BV-RLWE ciphertext can be written in the form

$$\mathbf{ct} = \left[\mathbf{m} + p * e + \sum_{i=1}^{k} c_i * (X^i - s^i) \right]_{R_q} .$$

It is easy to check that this form is preserved upon ciphertext addition and multiplication. Note that the decryption algorithm given the secret key $\mathbf{sk} = s$ is based on the equality $[\mathbf{ct}(\mathbf{sk})]_{R_q} = [\mathbf{m} + p * e]_{R_q}$. We next introduce the notion of decryption noise which will be useful to describe noise properties of the scheme. Decryption noise of a ciphertext $\mathbf{ct}$ is defined as

$$\mathrm{DN}(\mathbf{ct}) := \|[\mathbf{m} + p * e]_R\|_\infty .$$

Decryption works as long as $\mathrm{DN}(\mathbf{ct}) < q/2$ because then it holds that

$$\mathrm{Dec}_{\mathbf{sk}}(\mathbf{ct}) = \left[[\mathbf{m} + p * e]_{R_q} \right]_{R_p} = [\mathbf{m} + p * e]_{R_p} = [\mathbf{m}]_{R_p} .$$

Next, we evaluate how homomorphic operations affect decryption noise. We state decryption noise of a ciphertext resulting from a homomorphic operation as a function of the decryption noise of each of the input ciphertexts. For homomorphic addition, $\mathbf{ct}_{\mathrm{sum}} = \mathbf{ct}_1 + \mathbf{ct}_2$, we have

$$\begin{aligned}
\mathrm{DN}(\mathbf{ct}_{\mathrm{sum}}) &= \|[(\mathbf{m}_1 + p * e_1) + (\mathbf{m}_2 + p * e_2)]_R\|_\infty \\
&= \|[\mathbf{m}_1 + p * e_1]_R\|_\infty + \|[\mathbf{m}_2 + p * e_2]_R\|_\infty \\
&= \mathrm{DN}(\mathbf{ct}_1) + \mathrm{DN}(\mathbf{ct}_2) .
\end{aligned}$$

For homomorphic multiplication, $\mathbf{ct}_{\mathrm{prod}} = \mathbf{ct}_1 * \mathbf{ct}_2$, using Lemma 1 we obtain an upper bound on decryption noise,

$$\begin{aligned}
\mathrm{DN}(\mathbf{ct}_{\mathrm{prod}}) &= \|[(\mathbf{m}_1 + p * e_1) * (\mathbf{m}_2 + p * e_2)]_R\|_\infty \\
&\leq n \, \|[\mathbf{m}_1 + p * e_1]_R\|_\infty \, \|[\mathbf{m}_2 + p * e_2]_R\|_\infty \\
&= n \mathrm{DN}(\mathbf{ct}_1) \mathrm{DN}(\mathbf{ct}_2) .
\end{aligned}$$

2.4   Security

Security of the scheme is based on the *Ring Learning With Errors* problem [17]. More precisely, following the approach by Brakerski and Vaikuntanathan, it can be shown that the scheme is secure assuming the PLWE assumption holds [6]. Choosing p to be a polynomial instead of an integer does not affect the security argument as long as p and q are coprime in R. The argument boils down to the PLWE assumption by noting that with $s, e \leftarrow_\chi R$, and $a, u \leftarrow_\$ R_q$, we have that if $(a, a * s + e)$ is indistinguishable from (a, u) then $(a, a * s + p * e)$ is indistinguishable from (a, u).

Choosing secure parameters is not trivial and depends on the best known attack. We will not go into any details here, but simply give some recommendations mainly based on the works of Micciancio and Regev [18] and Lindner and Peikert [15]. For more detailed information see also [6].

- The ring dimension parameter **n** must be a power of two.
- The ciphertext space modulus **q** must be a prime over $\mathbb{Z}$.
- The message space modulus **p** must be coprime with q over R.
- The error distribution χ is the discrete Gaussian distribution $D_{\mathbb{Z}^n, \sigma}$.

According to Gentry et al. [10], a security of at least κ bits is achieved if

$$n > \log\left(\frac{q}{\sigma}\right) \cdot \left(\frac{\kappa + 110}{7.2}\right) .$$

3   Integer Encoding

In order to use homomorphic encryption for meaningful computation on encrypted data, we must encode data and operations onto the native message space of the cryptosystem. The native message space to many of the recently proposed homomorphic encryption schemes is the quotient ring $R_p := \mathbb{Z}[x]/\langle x^n + 1, p\rangle$. If we want to do efficient computation on encrypted data using such a scheme we thus need to find an efficient encoding of data and operations onto the message space R_p.

Here we focus on the encoding of integers, a fundamental data type in computation. A commonly used integer encoding encodes an integer onto R_p by reinterpreting its binary representation as a polynomial with binary coefficients [14]. However, working with this encoding is problematic. When coefficients of an encoding polynomial wrap around modulo p, the encoding polynomial cannot be decoded to the correct integer anymore.

We present a powerful integer encoding utilizing the fact that we can choose $p \in \mathbb{Z}[x]$ such that the quotient ring $\mathbb{Z}[x]/\langle x^n + 1, p\rangle$ is isomorphic to an integer quotient ring $\mathbb{Z}_t$, with $t \in \mathbb{N}$. Previously, Hoffstein and Silverman used a similar isomorphism for encoding binary data [13].

3.1   Bit Coefficient Encoding

For completeness we first present the bit coefficient encoding (BCE) by Lauter et al. [14]. We describe the encoding and decoding steps and how integer operations on encoding elements are carried out. Note that for BCE the message space is $R_p = \mathbb{Z}[x]/\langle x^n + 1, p\rangle$, with $p \in \mathbb{N}$.

$m = $ **BCE.Encode(z)**. We encode an $z \in \mathbb{Z}$, with $|z| < 2^n$, as a polynomial $m \in R_p$ using BCE. We first compute the binary representation $|z| = \sum_{i=0}^{n-1} z_i \cdot 2^i$, with $z_i \in \{0,1\}$. The encoding message polynomial is

$$m := \text{sign}(z) * \sum_{i=0}^{n-1} z_i * x^i .$$

$z = \textbf{BCE.Decode}(m)$. We decode a polynomial $m \in R_p$ to an integer $z \in \mathbb{Z}$. Write $[m]_{R_p} = \sum_{i=0}^{n-1} m_i * x^i$, then compute

$$z := \sum_{i=0}^{n-1} m_i * 2^i \ .$$

Integer Operations. Addition and multiplication of encoded integers are evaluated by polynomial addition and multiplication of the encoding polynomials over the message space R_p. It is easy to check that as long as $\deg(m) < n$ and $\|m\|_\infty \leq p/2$, the encoding polynomial m decodes to the correct integer.

3.2   Ring Isomorphism Encoding

In the following we describe our proposed ring isomorphism integer encoding (RIE). The encoding is established through an isomorphism between the message space $\mathbb{Z}[x]/\langle x^n + 1, p \rangle$, with $p \in \mathbb{Z}[x]$, and an integer quotient ring $\mathbb{Z}_t$, $t \in \mathbb{N}$. To encode an integer onto the message space we use a straightforward mapping from the integer quotient ring $\mathbb{Z}_t$ to the native message space R_p. Since the message space is isomorphic to $\mathbb{Z}_t$, when using this encoding we can think of computing on encrypted integers almost as if we were computing on $\log_2(t)$ bit integers. In contrast to BCE, we do not have to worry about coefficients of encoding polynomials exceeding the modulus range p. When computing on encoded integers using RIE, we are not just emulating an integer message space, but we are in fact operating on a structure that truly behaves like one.

Theorem 1 establishes the isomorphism we use for the encoding. A slightly different version of this theorem has already been proven by Hoffstein and Silverman [13]. Here we present a slightly more general result.

Theorem 1. *For $a \in \mathbb{Z}$ the map $\phi : \mathbb{Z}[x]/\langle x^n + 1, x - a \rangle \to \mathbb{Z}/\langle a^n + 1 \rangle$ given by*

$$f(x) + \langle x^n + 1, x - a \rangle \mapsto f(a) + \langle a^n + 1 \rangle$$

is an isomorphism.

Proof. Consider the evaluation map $\phi' : \mathbb{Z}[x] \to \mathbb{Z}$ given by $f(x) \mapsto f(a)$. The map ϕ' is clearly a surjective homomorphism. Moreover, since for all $f(x) \in \mathbb{Z}[x]$ we have $f(x) \equiv f(a) \pmod{x - a}$, the kernel of ϕ' is $\langle x - a \rangle$. Thus, by the first isomorphism theorem $\mathbb{Z}[x]/\langle x - a \rangle$ is isomorphic to $\mathbb{Z}$ via the map $f(x) + \langle x - a \rangle \mapsto f(a)$. Moreover, we have $x^n + 1 \equiv a^n + 1 \pmod{x - a}$ and hence $\mathbb{Z}[x]/\langle x^n + 1, x - a \rangle$ is isomorphic to $\mathbb{Z}/\langle a^n + 1 \rangle$.   □

In the following we describe the encoding and decoding procedures in detail. The choice of the modulus $p = x - a$ is fundamental to the functioning of the encoding because it determines the structure of the integer quotient ring we will be working with. To demonstrate the ring isomorphism integer encoding we make the convenient choice $p = x - 2$. However, different values for p may result in slightly more efficient encodings or might produce other interesting message space properties, such as making the message space isomorphic to a finite field or to a useful product ring (see Sect. 3.3 for an application of the latter).

According to Theorem 1, for $p = x - 2$ the map $\phi : R_p \to \mathbb{Z}/\langle 2^n + 1 \rangle$, $f(x) + \langle x^n + 1, p \rangle \mapsto f(2) + \langle 2^n + 1 \rangle$ is an isomorphism. To efficiently encode integers modulo $2^n + 1$ as elements of R_p we choose the set of ternary polynomials with degree less than n denoted by

$$T_n := \left\{ a = \sum_{i=0}^{n-1} a_i * x^i \in \mathbb{Z}[x] : a_i \in \{-1, 0, 1\} \right\} .$$

Clearly, T_n suffices to represent the $2^n + 1$ elements of $\mathbb{Z}/\langle 2^n + 1 \rangle$. However, the mapping between $\mathbb{Z}/\langle 2^n + 1 \rangle$ and T_n is not one-to-one. Thus, encoding an integer does not yield an unique ternary polynomial. On the other hand, it is easy to see that decoding a ternary polynomial yields indeed a unique integer modulo $2^n + 1$. In the following we describe the encoding and decoding algorithms.

$m = \mathbf{RIE.Encode}(z)$. We encode an integer $z \in \mathbb{Z}$, with $|z| \leq 2^{n-1}$, as a polynomial $m \in R_p$. Determine $z_0, \ldots, z_{n-1} \in \{-1, 0, 1\}$ with $z \equiv \sum_{i=0}^{n-1} z_i \cdot 2^i$ (mod $2^n + 1$), and $\sum_{i=0}^{n-1} |z_i|$ minimal. The encoding polynomial is

$$m := \sum_{i=0}^{n-1} z_i * x^i .$$

$z = \mathbf{RIE.Decode}(m)$. We decode a polynomial $m \in R_p$ to an integer $z \in \mathbb{Z}$. Write $m = \sum_{i=0}^{n-1} m_i * x^i$ and compute

$$z' := \left(\sum_{i=0}^{n-1} m_i * 2^i \right) \bmod (2^n + 1) .$$

The decoded integer is

$$z := \begin{cases} z' - (2^n + 1) & \text{if } z' > 2^{n-1} , \\ z' & \text{otherwise.} \end{cases}$$

Integer Operations. Since $R_p \cong \mathbb{Z}_{2^n+1}$, addition and multiplication over R_p are isomorphically represented as addition and multiplication over $\mathbb{Z}_{2^n+1}$.

3.3 SIMD

Using the Chinese Remainder Theorem, our proposed ring isomorphism integer encoding can be enhanced in order to do single instruction multiple data (SIMD) computation. Suppose $a^n + 1$ factors into pair-wise coprimes $n_1, \ldots, n_k$. Then, by the Chinese Remainder Theorem $\mathbb{Z}_{a^n+1}$ is isomorphic to $\mathbb{Z}_{n_1} \times \cdots \times \mathbb{Z}_{n_k}$ via the map $m + \langle a^n + 1 \rangle \mapsto (m + \langle n_1 \rangle, \ldots, m + \langle n_k \rangle)$. The inverse map is also known and easy to compute. Hence, we can encode k integers into a single element of our encryption scheme message space. A single homomorphic ciphertext space operation would then correspond to k integer operations.

3.4 Modulus Switching

So far we described a somewhat homomorphic encryption (SWHE) scheme that uses the plaintext modulus $p = x - 2$ to ease the encoding of integers. Unfortunately, noise grows exponentially with the depth of the circuit, thus the scheme can only perform a small number of prescribed operations. It would be desirable to control the noise growth, to improve efficiency. Among different ways to control noise growth, the modulus switching technique, proposed by Brakerski and Vaikuntanathan [5], is often preferred because there is no need for additional assumptions [4]. Here, we adapt modulus switching to work in the case that $p = x - 2$.

Modulus switching, as described by Brakerski et al. [4], transforms a ciphertext c that decrypts under a modulus q into a ciphertext c' that decrypts under a smaller modulus q', while the noise level essentially remains constant. To achieve this, they propose a $\text{scale}(c, q, q', p)$ function (Definition 6 [4]) that returns $c' \equiv c \mod p$ close enough to $(q'/q)c$. We define a scaling function with equivalent properties suitable for the case $p = x - 2$.

Definition 1. *Let $c \in \mathbb{Z}[x]$ be of degree less than n, $p = x - 2$, and $q', q \in \mathbb{Z}$ such that $q' < q$ and $q \equiv q' \equiv 1 \mod p$. Let $y = (q'/q)c \in \mathbb{Q}[x]$. Define $\text{scale}(c, q, q', p)$ to be an algorithm that chooses $c' \in \mathbb{Z}[x]$ such that $c' \equiv c \mod \langle x^n + 1, x - 2 \rangle$, and as close as possible to y in the infinity norm.*

Lemma 2. *Let $c \in \mathbb{Z}[x]$ be of degree less than n, $p = x - 2$, and $q', q \in \mathbb{Z}$ such that $q' < q$ and $q \equiv q' \equiv 1 \mod p$. Then $c' = \text{scale}(c, q, q', p)$ satisfies that $c' \equiv c \mod \langle x^n + 1, x - 2 \rangle$ and $\|c' - (q'/q)c\|_\infty \leq 1.5$.*

Proof. Let T_n be the set of all ternary polynomials of degree less than n. It is easy to see that the projection map $\pi : \mathbb{Z}[x] \to R_p$ maps T_n onto R_p. Moreover, for $y \in \mathbb{Q}[x]$ of degree less than n, if we define the set

$$S_y = \left\{ \sum_{i=0}^{n-1} h_i x^i \mid h_i \in \{\lfloor y_i \rceil - 1, \lfloor y_i \rceil, \lfloor y_i \rceil + 1\} \right\},$$

then, also π maps S_y onto R_p. Therefore, it is possible to choose $c' \in S_y$ such that $c' \equiv c \mod \langle x^n + 1, x - 2 \rangle$. Finally, notice that for $c' \in S_y$, $\|c' - (q'/q)c\|_\infty \leq 1.5$.

The previous result can easily be adapted to $\mathbb{Z}[x]$-vectors, and thus be used to establish a modulus switching procedure like the one described by Brakerski et al. [4, Lemma 4]. It is also worth noting that in the simplified BV-RLWE scheme presented above, the size of the ciphertext grows exponentially with the depth of the circuit. This is easily fixed by relinearization [6] which is not affected at all by the choice of p.

Therefore, through modulus switching, the BV-RLWE scheme equipped with our efficient integer encoding, can be transformed into a leveled FHE scheme. Modulus switching improves the efficiency so that the size of the public key is linear in the depth of the circuits that the scheme can evaluate. Moreover, our proposed scheme evaluates depth L circuits of multiplications and additions in the ring $\mathbb{Z}/\langle a^n + 1 \rangle$, thus it compactly evaluate all depth L circuits of a complete set of gates.

4   Integer Encoding Efficiency

We evaluate and compare efficiency of the integer encodings BCE and RIE described in Sect. 3.2 when used with the BV-RLWE scheme described in Sect. 2.2. In order to establish an efficiency measurement we consider two observations.

- There exist lower bounds on encryption scheme parameters n and q that guarantee the correct homomorphic evaluation of a chosen function with respect to the employed scheme encoding configuration.
- Evaluation performance of BV-RLWE scales with the size of the parameters n and q which define the ciphertext space $R_q = \mathbb{Z}[x]/\langle x^n + 1, q \rangle$. More precisely, the ciphertext size is linear in n and $\log_2(q)$ and the time required to perform ciphertext operations like addition and multiplication is at least linear in the ciphertext size. We thus consider $n \log_2(q)$ to be a reasonable indicator for measuring homomorphic computation performance of different instances of BV-RLWE.

Our approach to measure integer encoding efficiency uses a prototype benchmark function with respect to which we find scheme parameters such that the employed scheme encoding is guaranteed to correctly evaluate the benchmark function on encrypted inputs. We thereby obtain lower bounds on the parameters n, q and hence a lower bound on the evaluation performance.

4.1   Benchmark Function

We choose a prototype benchmark function f that shall represent arbitrary integer functions involving a given number of layered additions and multiplications. The benchmark function is parametrized by $n_1, l_{\text{add}}, l_{\text{mul}} \in \mathbb{N}$, where n_1 roughly corresponds to the precision of input integers in bits and $l_{\text{add}}, l_{\text{mul}}$ correspond to the number of involved integer additions and multiplications. The input integer space is chosen as $\mathbb{Z}/\langle 2^{n_1} - 1 \rangle$ and input integers are assumed to lie in $\{-(2^{n_1-1} - 1), \ldots, 2^{n_1-1} - 1\}$. The benchmark function f is defined as

$$f : \mathbb{Z}_{2^{n_1}-1}^{l_{\text{mul}} \times l_{\text{add}}} \to \mathbb{Z} \;\; ; \;\; (z_{i,j}) \mapsto \prod_{i=1}^{l_{\text{mul}}} \sum_{j=1}^{l_{\text{add}}} z_{i,j} \;\; .$$

4.2   Evaluating Lower Bounds

In the following, we find lower bounds on BV-RLWE parameters n, q, and p to guarantee the successful evaluation of the benchmark function f on encrypted inputs for BCE and RIE. More precisely, the lower bound on q guarantees successful decryption and the lower bounds on n and p guarantee successful decoding. More formally, we try to find encryption scheme parameters n, q, and p for which

$$\text{Decode}(\text{Decrypt}(\text{Eval}(f, [\mathbf{ct}_{i,j}]))) = f([z_{i,j}])$$

with $\mathbf{ct}_{i,j} = \text{Encrypt}(\text{Encode}(z_{i,j}))$ holds and $n \log_2 q$ is minimal.[1] We first determine lower bounds on n and p separately for each of the encodings and then evaluate a lower bound on q as a function of n and p.

BCE. Finding lower bounds on n and p for BCE is problematic because p depends on the coefficient size of the resulting encoding polynomial $\mathbf{m}_f = \prod_{i=1}^{l_{\text{mul}}} \sum_{j=1}^{l_{\text{add}}} \mathbf{m}_{i,j}$ evaluated over $\mathbb{Z}[x]$, where $\mathbf{m}_{i,j} = \text{BCE.Encode}(z_{i,j})$. We thus first need to determine the maximum coefficient size of $\mathbf{m}_f$. According to Sect. 3.1 we have $\|\mathbf{m}_{i,j}\|_\infty \leq 1$ and $\deg(\mathbf{m}_{i,j}) < n_1$. In order to guarantee $\text{BCE.Decode}([\mathbf{m}_f]_{R_p}) = f([z_{i,j}])$ we need that $n > \deg(\mathbf{m}_f)$ and $p/2 > \|\mathbf{m}_f\|_\infty$. Let for $i = 1, \ldots, l_{\text{mul}}$, $\mathbf{m}_i = \sum_{j=1}^{l_{\text{add}}} \mathbf{m}_{i,j}$. It clearly holds $\deg(\mathbf{m}_i) < n_1$ and $\|\mathbf{m}_i\|_\infty \leq l_{\text{add}}$. Furthermore, since $\mathbf{m}_f = \prod_{i=1}^{l_{\text{mul}}} \mathbf{m}_i$, it holds $\deg(\mathbf{m}_f) \leq l_{\text{mul}} \cdot (n_1 - 1)$ and using Lemma 1 we obtain $\|\mathbf{m}_f\|_\infty \leq (n_1)^{l_{\text{mul}}-1} \cdot (l_{\text{add}})^{l_{\text{mul}}}$. This gives us the parameter lower bounds

$$n > l_{\text{mul}} \cdot (n_1 - 1) \quad, \quad p > 2 \cdot n_1^{l_{\text{mul}}-1} \cdot l_{\text{add}}^{l_{\text{mul}}} \quad.$$

RIE. For RIE lower bounds on parameters n, q can be obtained more easily and precisely. Using RIE we work on integers $\{-2^{n-1}, \ldots, 2^{n-1}\}$ and thus we just need to make sure that our computation result does not fall out of this range, that is $2^{n-1} \geq \left| \prod_{i=1}^{l_{\text{mul}}} \left(\sum_{j=1}^{l_{\text{add}}} z_{i,j} \right) \right|$. Rewriting the expression we obtain a lower bound on n given by

$$n > l_{\text{mul}} \cdot ((n_1 - 1) + \log_2 l_{\text{add}}) \quad.$$

Lower Bound on q. After estimating n and p, we are ready to compute a lower bound on q. As described in Sect. 2.3, in order to guarantee the correct decryption of the resulting ciphertext $\mathbf{ct}_f = \text{Eval}(f, [\mathbf{ct}_{i,j}])$, we need $q > 2 \cdot \text{DN}(\mathbf{ct}_f)$. According to the decryption noise growth analysis in Sect. 2.3 we have $\text{DN}(\mathbf{ct}_f) \leq n^{l_{\text{mul}}-1} \cdot (\text{B}_{\text{fresh}} \cdot l_{\text{add}})^{l_{\text{mul}}}$. As a result, in order to guarantee correct decryption we need to choose q such that

$$q > 2 \cdot n^{l_{\text{mul}}-1} \cdot (\text{B}_{\text{fresh}} \cdot l_{\text{add}})^{l_{\text{mul}}} \quad,$$

where $\text{B}_{\text{fresh}} \geq \text{DN}(\text{Enc}(\mathbf{m}))$ denotes a bound on the decryption noise of a fresh ciphertext. It remains to estimate B_{fresh}. We have that $\text{DN}(\text{Enc}(\mathbf{m})) = \|\mathbf{m} + p * (-e * u + g + f * s)\|_\infty$. We assume that the distributions from which e, f, g, s, and u are sampled are all bounded by B_χ, hence

$$\|e\|_\infty, \|f\|_\infty, \|g\|_\infty, \|s\|_\infty, \|u\|_\infty \leq B_\chi \quad.$$

---

[1] We use the notation $[a_{i,j}]$ to represent a matrix of input values.

Additionally, we know that a message polynomial $\mathbf{m}$ encoded with BCE or RIE has ternary coefficients, hence $\|\mathbf{m}\|_\infty \leq 1$. We obtain

$$B_{\text{fresh}} = 1 + p' \cdot \left(2 \cdot n \cdot B_\chi^2 + B_\chi\right) \quad,$$

where[2]

$$p' = \begin{cases} p & \text{if } p \in \mathbb{N} \;, \\ 3 & \text{if } p = x - 2 \;. \end{cases}$$

4.3   Results

Based on our reasoning at the beginning of Sect. 4, as a performance indiciator to measure encoding efficiency we use the lower bound on $n \log_2 q$. In the following we will refer to this bound by $\lfloor n \log_2 q \rfloor$. We start with calculating the complexity classes of $\lfloor n \log_2 q \rfloor$ for each of the encodings using the results from Sect. 4.2. For BCE we have $\lfloor n \rfloor \sim \mathcal{O}(l_{\text{mul}} \cdot n_1)$ and $\lfloor p \rfloor \sim \mathcal{O}((l_{\text{add}} \cdot n_1)^{l_{\text{mul}}})$. The complexity class of the lower bound on $\log q$ as a function of $l_{\text{mul}}, l_{\text{add}}, n_1$ hence evaluates to

$$\begin{aligned}
\lfloor \log q \rfloor_{\text{BCE}} &\sim \mathcal{O}\left(l_{\text{mul}} \cdot \log(n \cdot B_{\text{fresh}} \cdot l_{\text{add}})\right) \\
&\sim \mathcal{O}\left(l_{\text{mul}} \cdot \log(l_{\text{mul}}{}^2 \cdot l_{\text{add}}{}^{1+l_{\text{mul}}} \cdot n_1{}^{2+l_{\text{mul}}})\right) \\
&\sim \mathcal{O}\left(l_{\text{mul}}{}^2 \cdot \log(l_{\text{add}} \cdot n_1)\right) \;.
\end{aligned}$$

For RIE we have $\lfloor n \rfloor \sim \mathcal{O}(l_{\text{mul}} \cdot (n_1 + \log l_{\text{add}}))$ and $\lfloor p \rfloor \sim O(1)$. Then for $\lfloor \log q \rfloor$ we get

$$\begin{aligned}
\lfloor \log q \rfloor_{\text{RIE}} &\sim \mathcal{O}\left(l_{\text{mul}} \cdot \log(n \cdot B_{\text{fresh}} \cdot l_{\text{add}})\right) \\
&\sim \mathcal{O}\left(l_{\text{mul}} \cdot \log((l_{\text{mul}} \cdot (n_1 + \log l_{\text{add}}))^2 \cdot l_{\text{add}})\right) \\
&\sim \mathcal{O}\left(l_{\text{mul}} \cdot \log(l_{\text{mul}} \cdot l_{\text{add}} \cdot n_1)\right) \;.
\end{aligned}$$

In summary, we obtain the resulting complexity classes for our performance indicator $\lfloor n \cdot \log q \rfloor$ as a function of l_{mul}, l_{add}, and n_1,

$$\lfloor n \cdot \log q \rfloor_{\text{BCE}} \sim \mathcal{O}\left(l_{\text{mul}}{}^3 \cdot n_1 \cdot \log(l_{\text{add}} \cdot n_1)\right) \;,$$

$$\lfloor n \cdot \log q \rfloor_{\text{RIE}} \sim \mathcal{O}\left(l_{\text{mul}}{}^2 \cdot (n_1 + \log l_{\text{add}}) \cdot \log(l_{\text{mul}} \cdot l_{\text{add}} \cdot n_1)\right) \;.$$

The plots in Fig. 1 visualize how these differences in complexity reflect in computation performance. In the three plots, the horizontal axis holds one of the benchmark function parameters, that is l_{add}, l_{mul}, or n_1. The plots show for both encodings how computation performance is affected by modifying the supported number of additions, multiplications or the precision of input integers. The vertical axis holds the resulting lower bound on $n \log_2(q)$ for the given parameters. Recall that this lower bound on $n \log_2(q)$ is an indicator of the

---

[2] We set $p' = 3$ for $p = x - 2$ because for $a \in R$ it holds that $\|[(x-2) * a]_R\|_\infty \leq 3 \cdot \|a\|_\infty$.

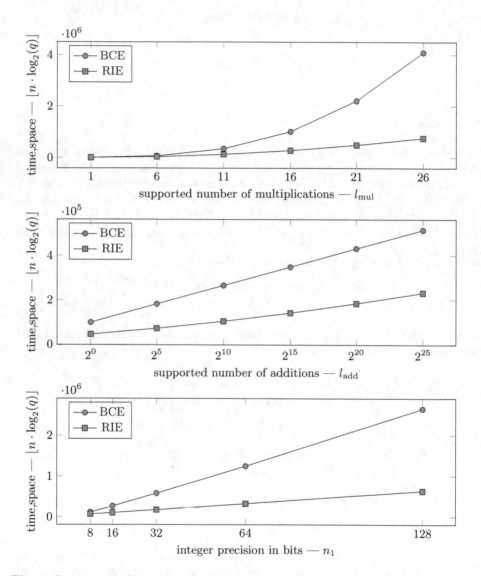

Fig. 1. Integer encoding efficiency of BCE versus RIE. Efficiency is measured in terms of $\lfloor n \cdot \log_2(q) \rfloor$ as described in Sect. 4. The benchmark function is $f([x_{i,j}]) = \prod_{i=1}^{l_{\mathrm{mul}}} \sum_{j=1}^{l_{\mathrm{add}}} x_{i,j}$ with inputs $x_{i,j}$ being integers of bitlength n_1. If not stated differently, parameters are set to $l_{\mathrm{add}} = 2^{10}$, $l_{\mathrm{mul}} = 10$, $n_1 = 16$, and $B_\chi = 2^7$.

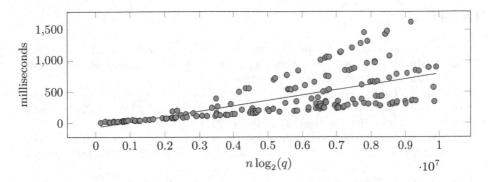

Fig. 2. Computation time for the multiplication of two random elements over R_q.

runtime and memory required to do homomorphic computation as explained at the beginning of Sect. 4.

The first plot shows how performance is affected when increasing the number of supported multiplications. Here, RIE clearly outperforms BCE because of its smaller lower bound complexity of $\mathcal{O}(l_{\mathrm{mul}}^2 \log(l_{\mathrm{mul}}))$ compared to $\mathcal{O}(l_{\mathrm{mul}}^3)$. In the second plot we look at supported number of additions over performance. For addition, BCE has complexity $\log(l_{\mathrm{add}})$ and BCE has complexity $\log^2(l_{\mathrm{add}})$. However, because we chose $l_{\mathrm{mul}} = 10$ as a default parameter value, this advantage in efficiency of BCE over RIE is outruled by the effect of supporting a small number of multiplications. The third plot then shows how performance behaves for modifying the input integer precision n_1. In this case complexity for both encodings is the same, but we observe a difference in the slope of the two curves. For the chosen parameter settings RIE performs better than BCE.

To give a better feeling what specific values for $n \log_2(q)$ mean in terms of real computation performance, in Fig. 2 we present timings for multiplying two random elements over R_q as a function of $n \log_2(q)$. In this plot each mark represents a specific configuration of n and q. The black line represents a fit of a linear function to the data using linear regression.[3]

It is worth mentioning that the efficiency of our encoding can further be improved by modifying the RIE parameter a, which defines the plaintext space quotient ring. For the sake of simplicity, we chose $a = 2$ and use encoding polynomials with binary coefficients. If we choose a different value for a we would adjust the coefficient range to $(-|a|/2, |a|/2]$. Increasing the coefficient range obviously increases ciphertext noise, but we have also seen that the lower bound on the ring dimension parameter n is given by $n > l_{\mathrm{mul}} \cdot \log_a(l_{\mathrm{add}} \cdot 2^{n_1 - 1})$ and thus decreases with increasing a. Hence, RIE efficiency can be fine tuned by finding the optimal balance between the chosen plaintext quotient ring and the evaluated functions.

---

[3] Benchmarks were written in C++ using NTL, a library for doing number theory, and run on a machine with an Intel Core i5-2557M CPU and 4GB of RAM.

References

1. Asharov, G., Jain, A., Wichs, D.: Multiparty computation with low communication, computation and interaction via threshold FHE. Cryptology ePrint Archive, Report 2011/613 (2011). http://eprint.iacr.org/
2. Bos, J.W., Lauter, K., Loftus, J., Naehrig, M.: Improved security for a ring-based fully homomorphic encryption scheme. In: Stam, M. (ed.) IMACC 2013. LNCS, vol. 8308, pp. 45–64. Springer, Heidelberg (2013)
3. Brakerski, Z.: Fully homomorphic encryption without modulus switching from classical gapSVP. Cryptology ePrint Archive, Report 2012/078 (2012). http://eprint.iacr.org/
4. Brakerski, Z., Gentry, C., Vaikuntanathan, V.: (leveled) fully homomorphic encryption without bootstrapping. In: Proceedings of the 3rd Innovations in Theoretical Computer Science Conference, ITCS 2012, pp. 309–325. ACM, New York (2012)
5. Brakerski, Z., Vaikuntanathan, V.: Efficient fully homomorphic encryption from (standard) LWE. In: 2011 IEEE 52nd Annual Symposium on Foundations of Computer Science (FOCS), pp. 97–106 (2011)
6. Brakerski, Z., Vaikuntanathan, V.: Fully homomorphic encryption from ring-LWE and security for key dependent messages. In: Rogaway, P. (ed.) CRYPTO 2011. LNCS, vol. 6841, pp. 505–524. Springer, Heidelberg (2011)
7. Gentry, C.: A fully homomorphic encryption scheme. Ph.D. thesis, Stanford University (2009). crypto.stanford.edu/craig
8. Gentry, C., Goldman, K.A., Halevi, S., Julta, C., Raykova, M., Wichs, D.: Optimizing ORAM and using it efficiently for secure computation. In: De Cristofaro, E., Wright, M. (eds.) PETS 2013. LNCS, vol. 7981, pp. 1–18. Springer, Heidelberg (2013)
9. Gentry, C., Halevi, S., Smart, N.P.: Better bootstrapping in fully homomorphic encryption. In: Fischlin, M., Buchmann, J., Manulis, M. (eds.) PKC 2012. LNCS, vol. 7293, pp. 1–16. Springer, Heidelberg (2012)
10. Gentry, C., Halevi, S., Smart, N.P.: Homomorphic evaluation of the AES circuit. In: Safavi-Naini, R., Canetti, R. (eds.) CRYPTO 2012. LNCS, vol. 7417, pp. 850–867. Springer, Heidelberg (2012)
11. Graepel, T., Lauter, K., Naehrig, M.: ML confidential: Machine learning on encrypted data. Cryptology ePrint Archive, Report 2012/323 (2012). http://eprint.iacr.org/
12. Hoffstein, J., Pipher, J., Silverman, J.H.: NTRU: a ring-based public key cryptosystem. In: Buhler, J.P. (ed.) ANTS 1998. LNCS, vol. 1423, pp. 267–288. Springer, Heidelberg (1998)
13. Hoffstein, J., Silverman, J.: Optimizations for NTRU. In: Public-Key Cryptography and Computational Number Theory: Proceedings of the International Conference organized by the Stefan Banach International Mathematical Center Warsaw, Poland, September 11–15, 2000, p. 77. De Gruyter (2001)
14. Lauter, K., Naehrig, M., Vaikuntanathan, V.: Can homomorphic encryption be practical? In: Proceedings of the 3rd ACM Workshop on Cloud Computing Security Workshop, CCSW 2011, pp. 113–124. ACM, New York (2011). http://doi.acm.org/10.1145/2046660.2046682
15. Lindner, R., Peikert, C.: Better key sizes (and Attacks) for LWE-based encryption. In: Kiayias, A. (ed.) CT-RSA 2011. LNCS, vol. 6558, pp. 319–339. Springer, Heidelberg (2011)

16. López-Alt, A., Tromer, E., Vaikuntanathan, V.: On-the-fly multiparty computation on the cloud via multikey fully homomorphic encryption. In: Proceedings of the 44th Symposium on Theory of Computing, STOC 2012, pp. 1219–1234. ACM, New York (2012)

17. Lyubashevsky, V., Peikert, C., Regev, O.: On ideal lattices and learning with errors over rings. In: Gilbert, H. (ed.) EUROCRYPT 2010. LNCS, vol. 6110, pp. 1–23. Springer, Heidelberg (2010)

18. Micciancio, D., Regev, O.: Lattice-based cryptography. In: Bernstein, D.J., Buchmann, J., Dahmen, E. (eds.) Post-Quantum Cryptography, pp. 147–191. Springer, Heidelberg (2009)

19. Smart, N.P., Vercauteren, F.: Fully homomorphic encryption with relatively small key and ciphertext sizes. In: Nguyen, P.Q., Pointcheval, D. (eds.) PKC 2010. LNCS, vol. 6056, pp. 420–443. Springer, Heidelberg (2010)

20. Smart, N., Vercauteren, F.: Fully homomorphic SIMD operations. Cryptology ePrint Archive, Report 2011/133 (2011). http://eprint.iacr.org/

21. Stehlé, D., Steinfeld, R.: Faster fully homomorphic encryption. In: Abe, M. (ed.) ASIACRYPT 2010. LNCS, vol. 6477, pp. 377–394. Springer, Heidelberg (2010)

TweetNaCl: A Crypto Library in 100 Tweets

Daniel J. Bernstein[1,2]([✉]), Bernard van Gastel[3], Wesley Janssen[3],
Tanja Lange[2], Peter Schwabe[3], and Sjaak Smetsers[3]

[1] Department of Computer Science, University of Illinois at Chicago,
Chicago, IL 60607–7053, USA
`djb@cr.yp.to`
[2] Department of Mathematics and Computer Science,
Technische Universiteit Eindhoven,
P.O. Box 513, 5600 MB Eindhoven, The Netherlands
`tanja@hyperelliptic.org`
[3] Digital Security Group, Radboud University Nijmegen,
P.O. Box 9010, 6500 GL Nijmegen, The Netherlands
`B.vanGastel@cs.ru.nl, w.janssen@student.ru.nl,`
`peter@cryptojedi.org, s.smetsers@science.ru.nl`

Abstract. This paper introduces TweetNaCl, a compact reimplementation of the NaCl library, including all 25 of the NaCl functions used by applications. TweetNaCl is published on Twitter and fits into just 100 tweets; the tweets are available from anywhere, any time, in an unsuspicious way. Distribution via other social media, or even printed on a sheet of A4 paper, is also easily possible.

TweetNaCl is human-readable C code; it is the smallest readable implementation of a high-security cryptographic library. TweetNaCl is the first cryptographic library that allows correct functionality to be verified by auditors with reasonable effort, making it suitable for inclusion into the trusted code base of a secure computer system. This paper uses two examples of formally verified correctness properties to illustrate the impact of TweetNaCl's conciseness upon auditability.

TweetNaCl consists of a single C source file, accompanied by a single header file generated by a short Python script (1811 bytes). The library can be trivially integrated into a wide range of software build processes.

Portability and small code size come at a loss in efficiency, but TweetNaCl is sufficiently fast for most applications. TweetNaCl's cryptographic implementations meet the same security and reliability standards as NaCl: for example, complete protection against cache-timing attacks.

Keywords: Trusted code base · Source-code size · Auditability · Software implementation · Timing-attack protection · NaCl · Twitter

This work was supported by the National Science Foundation under grant 1018836 and by the Netherlands Organisation for Scientific Research (NWO) under grant 639.073.005 and through Veni 2013 project 13114. Permanent ID of this document: `c74b5bbf605ba02ad8d9e49f04aca9a2`. Date: 2014.10.04.

D.F. Aranha and A. Menezes (Eds.): LATINCRYPT 2014, LNCS 8895, pp. 64–83, 2015.
DOI: 10.1007/978-3-319-16295-9_4

1   Introduction

> *OpenSSL is the space shuttle of crypto libraries. It will get you to space, provided you have a team of people to push the ten thousand buttons required to do so. NaCl is more like an elevator—you just press a button and it takes you there. No frills or options.*
>
> *I like elevators.*   —Matthew D. Green, 2012 [15]

Cryptographic libraries form the backbone of security applications. The Networking and Cryptography library (NaCl) [10], see www.nacl.cr.yp.to, is rapidly becoming the crypto library of choice for a new generation of applications. NaCl is used, for example, in BitTorrent Live [12]; in DNSCrypt [22] from OpenDNS; in the secure mobile messaging app Threema [24]; and in the "new (faster and safer) NTor" protocol [14], the new default for Tor [25].

There are several reasons that NaCl has attracted attention. NaCl presents the developer with a high-level API: for example, all of the work necessary for signing a message is integrated into NaCl's `crypto_sign` function, and all of the work necessary for public-key authenticated encryption is integrated into NaCl's `crypto_box` function. For each of these functionalities NaCl provides exactly one default combination of cryptographic primitives selected for high security and easy protection against timing attacks. For comparison, OpenSSL [23] provides the implementor with a minefield of options, including many combinations that are broken by timing attacks and many combinations that provide no security at all.

NaCl is also much faster than OpenSSL. For example, on one core of a 2.5 GHz Intel Core i5-3210M Ivy Bridge CPU, OpenSSL's RSA-2048 encryption takes 0.13 million cycles but RSA-2048 decryption takes 4.2 million cycles and elliptic-curve encryption/decryption (DH) takes 0.7 million cycles. NaCl's elliptic-curve encryption/decryption takes just 0.18 million cycles. Both NaCl and OpenSSL include optimized assembly-language implementations, but NaCl uses state-of-the-art primitives that inherently allow higher speed than the primitives included in OpenSSL: in this case, the Curve25519 elliptic curve rather than the NIST P-256 elliptic curve or lower-security RSA-2048. This performance gap is not limited to high-end Intel CPUs: see [11] for a performance analysis of the same primitives on the ARM Cortex-A8 CPU core used in the iPad 1 and iPhone 4 three years ago and in the low-cost BeagleBone Black today.

However, NaCl's performance comes at a price. A single NaCl function usually consists of several different implementations, often including multiple implementations in assembly optimized for different CPUs. NaCl's compilation system is correspondingly complicated. Auditing the NaCl source is a time-consuming job. For example, four implementations of the `ed25519` signature system have been publicly available and waiting for integration into NaCl since 2011, but in total they consist of 5521 lines of C code and 16184 lines of `qhasm` code. Partial audits have revealed a bug in this software (`r1 += 0 + carry` should be `r2 += 0 + carry` in `amd64-64-24k`) that would not be caught by random tests; this illustrates the importance of audits. There has been some progress towards computer verification of formal proofs of correctness of software, but this progress is still far from complete verification of a usable high-security cryptographic library.

TweetNaCl: a small reimplementation of NaCl. This paper introduces TweetNaCl (pronounced "tweet salt"), a reimplementation of all 25 C NaCl functions used by applications. Each TweetNaCl function has exactly the same interface and semantics as the C NaCl function by the same name. (NaCl also includes an alpha-test networking component and support for languages other than C; TweetNaCl does not attempt to imitate these features.)

What distinguishes TweetNaCl from NaCl, and from other cryptographic libraries, is TweetNaCl's conciseness. We have posted TweetNaCl at https:// twitter.com/TweetNaCl as a sequence of just 100 tweets. The tweets are also shown in Appendix A of this paper. The tweets, plus 1 byte at the end of each line, occupy a total of 13456 bytes.

What we actually wrote was a slightly less compact 809-line 16637-byte `tweetnacl.c`. We then wrote a simple Python script, shown in Appendix B, to remove unnecessary spaces and produce the tweet form of TweetNaCl shown in Appendix A. Developers using TweetNaCl are expected to feed the tweet form of TweetNaCl through any standard indentation program, such as the UNIX `indent` program, to produce something similar to the original `tweetnacl.c`.

An accompanying 1811-byte Python script, shown in Appendix C, prints a `tweetnacl.h` that declares all the functions in `tweetnacl.c`, together with the same set of macros provided by NaCl. NaCl actually splits these declarations and macros into a moderately large collection of `.h` files such as `crypto_box.h`, `crypto_box_curve25519xsalsa20poly1305.h`, etc.; we have a similar Python script that creates the same collection of `.h` files, but switching to `tweetnacl.h` is minimal effort for developers.

TweetNaCl is not "obfuscated C": in indented form it is easily human-readable. It does use two macros and five `typedef`s, for example to abbreviate `for (i = 0;i < n;++i)` as `FOR(i,n)` and to abbreviate `unsigned char` as `u8`, but we believe that these abbreviations improve readability, and any readers who disagree can easily remove the abbreviations.

TweetNaCl is auditable. TweetNaCl is not merely readable; we claim that it is *auditable*. TweetNaCl is short enough and simple enough to be audited against a mathematical description of the functionality in NaCl such as [2]. TweetNaCl makes it possible to comprehensively audit the complete cryptographic portion of the trusted code base of a computer system. Of course, compilers also need to be audited (or to produce proofs of correct translations), as do other critical system components.

Section 6 explains how we have efficiently verified two memory-safety properties of TweetNaCl. Of course, this is far from a complete audit, but it already illustrates the impact of TweetNaCl's conciseness upon auditability: verifying the same properties for NaCl would be beyond current technology, and verifying the same properties for OpenSSL would be inconceivable.

TweetNaCl is secure and reliable. TweetNaCl is a C library containing the same protections as NaCl against simple timing attacks, cache-timing attacks, etc. It has no branches depending on secret data, and it has no array indices depending on secret data. We do not want developers to be faced with a choice between TweetNaCl's conciseness and NaCl's security.

TweetNaCl is also thread-safe, and has no dynamic memory allocation. Tweet-NaCl, like C NaCl, stores all temporary variables in limited areas of the stack. There are no hidden failure cases: TweetNaCl reports forgeries in the same way as C NaCl, and is successful in all other cases.

TweetNaCl's functions compute the same outputs as C NaCl: the libraries are compatible. We have checked all TweetNaCl functions against the NaCl test suite.

TweetNaCl is portable and easy to integrate. Another advantage of Tweet-NaCl's conciseness is that developers can simply add the files `tweetnacl.c` and `tweetnacl.h` into their applications, without worrying about complicated configuration systems or dependencies upon external libraries. TweetNaCl works straightforwardly with a broad range of compilation systems, including cross-compilation systems, and runs on any device that can compile C. We comment that TweetNaCl also provides another form of portability, namely literal portability, while maintaining literal readability: TweetNaCl fits onto a single sheet of paper in a legible font size, see Appendix A.

For comparison, the Sodium library from Denis [13] is a "portable, cross-compilable, installable, packageable fork of NaCl, with a compatible API"; current `libsodium-0.6.0.tar.gz` has 1546246 bytes and unpacks into 447 files totaling 5525939 bytes. Many NaCl applications (e.g., DNSCrypt), and 26 NaCl bindings for various languages, are actually using Sodium. TweetNaCl is similar to Sodium in being portable, cross-compilable, installable, and packageable; but TweetNaCl has the added advantage of being so small that it can be trivially incorporated into applications by inclusion rather than by reference. We have placed TweetNaCl into the public domain, and we encourage applications to make use of it.

The first version of Sodium was obtained by reducing NaCl to its reference implementations, removing all of the optimized implementations, and simplifying the build system accordingly. We emphasize that this does not produce anything as concise as TweetNaCl. Sections 2–5 of this paper describe the techniques we used to reduce the complexity of the TweetNaCl code, compared to the NaCl reference implementations.

TweetNaCl is fast enough for typical applications. TweetNaCl's focus on code size means that TweetNaCl cannot provide optimal run-time performance; NaCl's optimized assembly is often an order of magnitude faster. However, Tweet-NaCl is sufficiently fast for most cryptographic applications. Most applications can tolerate the 4.2 million cycles that OpenSSL uses on an Ivy Bridge CPU for RSA-2048 decryption, for example, so they can certainly tolerate the 2.5 million cycles that TweetNaCl uses for higher-security decryption (Curve25519). Note that, at a typical 2.5 GHz CPU speed, this is 1000 decryptions per second per CPU core. One can of course find examples of busy applications that need the higher performance of NaCl, but those examples do not affect the usability of TweetNaCl in typical lower-volume cryptographic applications.

Of course, it would be better for compilers to turn concise source code into optimal object code, so that there is no need for optimized assembly in the first place. We leave this as a challenge for language designers and compiler writers.

TweetNaCl is also small after compilation. TweetNaCl remains reasonably small when compiled, even though this was not its primary goal. For example, when TweetNaCl is compiled with `gcc -Os` on an Intel CPU, it takes only 11512 bytes. Small compiled code has several benefits: perhaps most importantly, it avoids instruction-cache misses, both for its own startup and for other code that would otherwise have been kicked out of cache. Note that typical cryptographic benchmarks ignore these costs.

For some C compilers, putting all of TweetNaCl into a single .c file prevents separate linking: the final binary will include all TweetNaCl functions even if not all of those functions are used. Any developers who care about the penalty here could comment out the unused code, but TweetNaCl is so small that this penalty is negligible in the first place.

On some platforms, code is limited in total size, not just in the amount that can be cached. This was the motivation for Hutter and Schwabe to reimplement NaCl to fit into the limited flash storage and RAM available on AVR microcontrollers [18]. Their low-area implementation consists of several thousand lines written in assembly and compiles to 17366 bytes; they also have faster implementations using somewhat more area. TweetNaCl compiles to somewhat more code, 29682 bytes on the same platform, but is much easier to read and to verify, especially since the verification work for TweetNaCl is shared across platforms.

TweetNaCl is a full library, not just isolated functions. In June 2013, Green [16] announced a new contest to "identify useful cryptographic algorithms that can be formally described in one Tweet." TweetNaCl is inspired by, but not a submission to, this contest. Unlike the submissions in that Twitter thread, later submissions using #C1T on Twitter, or TweetCipher [1] (authenticated encryption in 6 tweets, but with an experimental cryptosystem cobbled together for the sole purpose of being short), TweetNaCl provides exactly NaCl's high-level high-security cryptographic operations. TweetNaCl includes all necessary conversions to and from wire format, modular arithmetic from scratch, etc., using nothing but the C language.

TweetNaCl provides extremely high source-code availability. In 1995, at the height of the crypto wars, the United States government regarded cryptographic software as arms and subjected it to severe export control. In response, Zimmermann published the PGP software as a printed book [28]. The export-control laws did not cover printed material, so the book could be shipped abroad. Producing usable PGP software from the printed copies (see [27]) required hours of volunteer work to OCR and proofread over 6000 pages of code.

TweetNaCl fits onto just 1 page. This conciseness opens up many new possibilities for software distribution, ensuring the permanent availability of Tweet-NaCl to users worldwide, even users living under regimes that have decided to censor our 100 tweets. Of course, PGP is a full-fledged cryptographic application rather than just a cryptographic library, but we expect TweetNaCl to enable a broad spectrum of *small* high-security cryptographic applications.

Functions supported by TweetNaCl. Simple NaCl applications need only six high-level NaCl functions: `crypt_box` for public-key authenticated encryption; `crypto_box_open` for verification and decryption; `crypto_box_keypair` to create a public key in the first place; and similarly for signatures (Fig. 1) `crypto_sign`, `crypto_sign_open`, and `crypto_sign_keypair`.

A minimalist implementation of the NaCl API would provide just these six functions. TweetNaCl is more ambitious, supporting all 25 of the NaCl functions listed in Table 1, which as mentioned earlier are all of the C NaCl functions used by applications. This list includes all of NaCl's "default" primitives except for `crypto_auth_hmacsha512256`, which was included in NaCl only for compatibility with standards and is superseded by `crypto_onetimeauth`.

```
crypto_box = crypto_box_curve25519xsalsa20poly1305
crypto_box_open
crypto_box_keypair
crypto_box_beforenm
crypto_box_afternm
crypto_box_open_afternm
crypto_core_salsa20
crypto_core_hsalsa20
crypto_hashblocks = crypto_hashblocks_sha512
crypto_hash = crypto_hash_sha512
crypto_onetimeauth = crypto_onetimeauth_poly1305
crypto_onetimeauth_verify
crypto_scalarmult = crypto_scalarmult_curve25519
crypto_scalarmult_base
crypto_secretbox = crypto_secretbox_xsalsa20poly1305
crypto_secretbox_open
crypto_sign = crypto_sign_ed25519
crypto_sign_open
crypto_sign_keypair
crypto_stream = crypto_stream_xsalsa20
crypto_stream_xor
crypto_stream_salsa20
crypto_stream_salsa20_xor
crypto_verify_16
crypto_verify_32
```

Fig. 1. Functions supported by TweetNaCl.

As mentioned earlier, the Ed25519 signature system has not yet been integrated into NaCl, since the Ed25519 software has not yet been fully audited; NaCl currently provides an older signature system. However, NaCl has announced that it will transition to Ed25519, so TweetNaCl provides Ed25519.

In surveying NaCl applications we have found two main reasons that applications go beyond the minimal list of six functions. First, many NaCl applications split (e.g.) `crypto_box` into `crypto_box_beforenm` and `crypto_box_afternm` to improve speed. Second, some NaCl applications are experimenting with variations of NaCl's high-level operations but continue to use lower-level NaCl functions such as `crypto_secretbox` and `crypto_hash`.

It is important for all of these applications to continue to work with Tweet-NaCl. The challenge here is the code size required to provide many functions. Even a single very simple function such as

```
int crypto_box_beforenm(u8 *k,const u8 *y,const u8 *x)
{
  u8 s[32];
  crypto_scalarmult(s,x,y);
  return crypto_core_hsalsa20(k,z,s,sigma);
}
```

costs us approximately 1 tweet. We could use shorter function names internally, but we would then need further wrappers to provide all the external function names listed in Table 1. We have many such functions, and a limited tweet budget, limiting the space available for actual cryptographic computations.

2   Salsa20, HSalsa20, and XSalsa20

NaCl encrypts messages by xor'ing them with the output of Bernstein's Salsa20 [5] stream cipher. The Salsa20 stream cipher generates 64-byte output blocks using the Salsa20 "core function" in counter mode. The main loop in NaCl's reference implementation of this core function, `crypto_core/salsa20/ref/core.c`, transforms 16 32-bit words x0, x1, ..., x15 as follows, where ROUNDS is 20:

```
for (i = ROUNDS;i > 0;i -= 2) {
   x4 ^= rotate( x0+x12, 7);   x8 ^= rotate( x4+ x0, 9);
  x12 ^= rotate( x8+ x4,13);   x0 ^= rotate(x12+ x8,18);
   x9 ^= rotate( x5+ x1, 7);  x13 ^= rotate( x9+ x5, 9);
   x1 ^= rotate(x13+ x9,13);   x5 ^= rotate( x1+x13,18);
  x14 ^= rotate(x10+ x6, 7);   x2 ^= rotate(x14+x10, 9);
   x6 ^= rotate( x2+x14,13);  x10 ^= rotate( x6+ x2,18);
   x3 ^= rotate(x15+x11, 7);   x7 ^= rotate( x3+x15, 9);
  x11 ^= rotate( x7+ x3,13);  x15 ^= rotate(x11+ x7,18);
   x1 ^= rotate( x0+ x3, 7);   x2 ^= rotate( x1+ x0, 9);
   x3 ^= rotate( x2+ x1,13);   x0 ^= rotate( x3+ x2,18);
   x6 ^= rotate( x5+ x4, 7);   x7 ^= rotate( x6+ x5, 9);
   x4 ^= rotate( x7+ x6,13);   x5 ^= rotate( x4+ x7,18);
  x11 ^= rotate(x10+ x9, 7);   x8 ^= rotate(x11+x10, 9);
   x9 ^= rotate( x8+x11,13);  x10 ^= rotate( x9+ x8,18);
  x12 ^= rotate(x15+x14, 7);  x13 ^= rotate(x12+x15, 9);
  x14 ^= rotate(x13+x12,13);  x15 ^= rotate(x14+x13,18);
}
```

Notice that this loop involves 96 x indices: x4, x0, x12, x8, x4, etc. TweetNaCl handles the same loop much more concisely:

```
FOR(i,20) {
  FOR(j,4) {
    FOR(m,4) t[m] = x[(5*j+4*m)
    t[1] ^= rotate(t[0]+t[3], 7); t[2] ^= rotate(t[1]+t[0], 9);
    t[3] ^= rotate(t[2]+t[1],13); t[0] ^= rotate(t[3]+t[2],18);
    FOR(m,4) w[4*j+(j+m)
  }
  FOR(m,16) x[m] = w[m];
}
```

We emphasize two levels of Salsa20 symmetry that appear in the Salsa20 specification and that are expressed explicitly in this TweetNaCl loop. First, the 20 rounds in Salsa20 alternate between "column rounds" and "row rounds", with column rounds operating on columns of the 4×4 matrix

$$\begin{pmatrix} \texttt{x[0]} & \texttt{x[1]]} & \texttt{x[2]} & \texttt{x[3]} \\ \texttt{x[4]} & \texttt{x[5]} & \texttt{x[6]} & \texttt{x[7]} \\ \texttt{x[8]} & \texttt{x[9]} & \texttt{x[10]} & \texttt{x[11]} \\ \texttt{x[12]} & \texttt{x[13]} & \texttt{x[14]} & \texttt{x[15]} \end{pmatrix}$$

and row rounds operating in exactly the same way on rows of the matrix. Tweet-NaCl computes a row round as a transposition of the matrix followed by a column round followed by another transposition; i.e., the 20 rounds consist of 20 iterations of "compute a column round and transpose the output". The transposed result of each round is built in a separate array w to avoid overwriting the round input; it is then copied from w back to x. One can easily see that the indices 4*j+(j+m)%4 for w are the transposes of the indices (5*j+4*m)%16 for x.

Second, the column round operates on the column down from x[0], operates in the same way on the column down from x[5] (wrapping around to x[1]), operates in the same way on the column down from x[10], and operates in the same way on the column down from x[15]. TweetNaCl has j loop over the 4 columns; the x index (5*j+4*m)%16 is m columns down from the starting point in column j.

For comparison, the indices in the second half of the NaCl loop shown above are the transposes of the indices in the first half, and the indices in the first half have these symmetries across columns. Verifying these 96 indices is of course feasible but takes considerably more time than verifying the corresponding segment of TweetNaCl code—and this is just the first of many ways in which NaCl's reference implementations consume more code than TweetNaCl.

Stream generation and stream encryption. NaCl actually has two ways to use Salsa20: crypto_stream_salsa20 produces any desired number of bytes of the Salsa20 output stream; crypto_stream_salsa20_xor produces a ciphertext from a plaintext. Both functions are wrappers around crypto_core_salsa20; both functions handle initialization and updates of the block counter, and output lengths that are not necessarily multiples of 64. The difference is that the second function xors each block with a plaintext block, moving along the plaintext accordingly.

TweetNaCl's implementation of `crypto_stream_salsa20` calls the function `crypto_stream_salsa20_xor` with a null pointer for the plaintext. This eliminates essentially all duplication of code between these functions, at the expense of three small tweaks to `crypto_stream_salsa20_xor`, such as replacing

```
FOR(i,64) c[i] = m[i] ^ x[i];
```

with

```
FOR(i,64) c[i] = (m?m[i]:0) ^ x[i];
```

to treat a null pointer `m` as if it were a pointer to an all-0 block.

XSalsa20 and HSalsa20. NaCl's `crypto_stream` actually uses Bernstein's XSalsa20 stream cipher (see [6]) rather than the Salsa20 stream cipher. The difference is that XSalsa20 supports 32 bytes of nonce/counter input while Salsa20 supports only 16 bytes of nonce/counter input. XSalsa20 uses the original 32-byte key and the first 16 bytes of the nonce to generate an intermediate 32-byte key, and then uses Salsa20 with the intermediate key and the remaining 16 bytes of nonce/counter to generate each output block.

The intermediate key generation, called "HSalsa20", is similar to Salsa20 but slightly more efficient, and has a separate implementation in NaCl. For our purposes this is a problem: it means almost doubling the code size.

TweetNaCl does better by viewing HSalsa20 as (1) generating a 64-byte Salsa20 output block, (2) extracting 32 bytes from particular output positions, and then (3) transforming those 32 bytes in a public invertible way. The transformation is much more concise than a separate HSalsa20 implementation, allowing us to implement both `crypto_core_salsa20` and `crypto_core_hsalsa20` as wrappers around a unified `core` function in TweetNaCl.

We do not claim novelty for this view of HSalsa20: the same structure is exactly what allowed the proof in [6] that the security of Salsa20 implies the security of HSalsa20 and XSalsa20. What is new is the use of this structure to simplify a unified Salsa20/HSalsa20 implementation.

3   Poly1305

Secret-key authentication in NaCl uses Bernstein's Poly1305 [3] authenticator. The Poly1305 code in the NaCl reference implementation is already quite concise. For elements of $\mathbb{F}_{2^{130}-5}$ it uses a radix-2^8 representation; we use the same representation for TweetNaCl.

The NaCl reference implementation uses a `mulmod` function for multiplication in $\mathbb{F}_{2^{130}-5}$, a `squeeze` function to perform two carry chains after multiplication and a `freeze` function to produce a unique representation of an element of $\mathbb{F}_{2^{130}-5}$. Each of these functions is called only once in the Poly1305 main loop; we inline those functions to remove code for the function header and the call.

The reference implementation also uses an `add` function which is called once in the main loop, once during finalization and once inside the `freeze` function. We keep the function, but rename it to `add1305` to avoid confusion with the `add` function used (as described in Sect. 5) for elliptic-curve addition.

We furthermore shorten the code of modular multiplication. NaCl's reference implementation performs multiplication of `h` by `r` with the result in `hr` as follows:

```
for (i = 0;i < 17;++i) {
  u = 0;
  for (j = 0;j <= i;++j)
    u += h[j] * r[i - j];
  for (j = i + 1;j < 17;++j)
    u += 320 * h[j] * r[i + 17 - j];
  hr[i] = u;
}
```

This piece of code exploits the fact that $2^{136} \equiv 320 \pmod{2^{130} - 5}$ for modular reduction on the fly. TweetNaCl merges the two inner loops:

```
FOR (i, 17) {
  x[i] = 0;
  FOR (j, 17)
    x[i] += h[j] * ((j <= i) ? r[i - j] : 320 * r[i + 17 - j]);
}
```

4   SHA-512

The default hash function in NaCl and the hash function used within the Ed25519 signature scheme (see Sect. 5) is SHA-512 [26]. The SHA-512 code in the NaCl reference implementation consists of two main portions of code:

- The function `crypto_hash`, which performs initialization of the hash value with the IV and computation of message padding; and
- the `crypto_hashblocks` function which performs hashing of full blocks.

Message padding. Outside of `crypto_hashblocks`, message padding is the most complex part of `crypto_hash`. The reference padding code, with Tweet-NaCl's choices of variable names substituted for the original choices, is as follows:

```
for (i = 0;i < n;++i) x[i] = m[i];
x[n] = 0x80;
if (n < 112) {
  for (i = n + 1;i < 119;++i) x[i] = 0;
  x[119] = b >> 61;
  x[120] = b >> 53; x[121] = b >> 45;
  x[122] = b >> 37; x[123] = b >> 29;
  x[124] = b >> 21; x[125] = b >> 13;
  x[126] = b >> 5;  x[127] = b << 3;
  crypto_hashblocks(h,x,128);
} else {
  for (i = n + 1;i < 247;++i) x[i] = 0;
  x[247] = b >> 61;
  x[248] = b >> 53; x[249] = b >> 45;
```

```
x[250] = b >> 37; x[251] = b >> 29;
x[252] = b >> 21; x[253] = b >> 13;
x[254] = b >> 5;  x[255] = b << 3;
crypto_hashblocks(h,x,256);
}
```

This segment handles two possibilities for processing the final partial block of SHA-512 input: if the block has fewer than 112 bytes then it is padded to 128 bytes; otherwise it is padded to 256 bytes. The padding ends with a 9-byte big-endian encoding of the number of message bits.

TweetNaCl simplifies this code in three ways. First, it eliminates the two separate lines of zero-padding x in favor of initializing the whole array to 0. Second, elsewhere in TweetNaCl there is a ts64 function (used at the end of the SHA-512 compression function) that stores 64 bits in big-endian form; TweetNaCl reuses this function inside the padding. Third, TweetNaCl merges the two branches, reusing n (which has no later use) for the number of bytes in the padded block. The final padding code is much more concise than the original:

```
FOR(i,256) x[i] = 0;
FOR(i,n) x[i] = m[i];
x[n] = 128;
n = 256-128*(n<112);
x[n-9] = b >> 61;
ts64(x+n-8,b << 3);
crypto_hashblocks(h,x,n);
```

Hashing blocks. SHA-512 performs 80 rounds of computation per block. The NaCl reference implementation has 80 lines for these 80 rounds. Each round is just one invocation of an F macro (interruped by invocations of an EXPAND macro after every 16 rounds), but this still results in a significant amount of code. TweetNaCl instead uses a loop over the 80 rounds. With such a "rolled" loop there is only one invocation of each of the macros, so TweetNaCl inlines those.

In NaCl the 16 64-bit message words are loaded into variables w0, w1, ..., w15; the internal temporary state is kept in variables a, b, ..., h. TweetNaCl uses arrays u64 w[16] and u64 a[8] instead. This allows us to also roll all initialization and copy loops. The final code for processing one 128-byte block is the following:

```
FOR(i,16) w[i] = dl64(m + 8 * i);

FOR(i,80) {
  FOR(j,8) b[j] = a[j];
  t = a[7] + Sigma1(a[4]) + Ch(a[4],a[5],a[6]) + K[i] + w[i
  b[7] = t + Sigma0(a[0]) + Maj(a[0],a[1],a[2]);
  b[3] += t;
  FOR(j,8) a[(j+1)
  if (i
    FOR(j,16)
      w[j] += w[(j+9)
}

FOR(i,8) { a[i] += z[i]; z[i] = a[i]; }
```

Obviously there is still some complexity in this code, but this directly reflects the inherent complexity of the SHA-512 function; the SHA-512 specification [26] is easily verified to match TweetNaCl's implementation. The functions `Sigma1`, `Ch`, `Sigma0`, `Maj`, `sigma0`, and `sigma1` are one-line implementations of the functions Σ_1, Ch, Σ_0, Maj, σ_0, and σ_1 from the SHA-512 specification.

5   Curve25519 and Ed25519

Asymmetric cryptography in NaCl uses Bernstein's Curve25519 elliptic-curve Diffie-Hellman key exchange [4] and will use the Ed25519 elliptic-curve signature scheme from Bernstein, Duif, Lange, Schwabe, and Yang [7,8]. This section explains the techniques we use for our compact implementation of these two schemes.

Arithmetic in $\mathbb{F}_{2^{255}-19}$. Both Curve25519 and Ed25519 require arithmetic in the field $\mathbb{F}_{2^{255}-19}$. We represent an element of this finite field as an array of 16 signed 64-bit integers (datatype `signed long long`) in radix 2^{16}:

```
typedef i64 gf[16];
```

Additions and subtractions do not have to worry about carries or modular reduction; they simply turn into a loop that performs 16 coefficient additions or subtractions.

Multiplication performs simple "operand scanning" schoolbook multiplication in two nested loops. We then reduce modulo $2^{256} - 38$:

```
i64 i,j,t[31];
FOR(i,31) t[i]=0;
FOR(i,16) FOR(j,16) t[i+j]+=a[i]*b[j];
FOR(i,15) t[i]+=38*t[i+16];
FOR(i,16) o[i]=t[i];
```

The 16 result coefficients in o are too large to be used as input to another multiplication. We use two calls to a `car25519` carry function to solve this problem. This carry function modifies the result o in place as follows:

```
FOR(i,16) {
  o[i] += (1LL<<16);
  c = o[i]>>16;
  o[(i+1)*(i<15)] += c-1+37*(c-1)*(i==15);
  o[i] -= c<<16;
}
```

Aside from carrying from one limb to the next, the function also adds 2^{16} to each limb and subtracts 1 from the next highest limb before performing the carry. This ensures that repeated application of the function brings all limbs into the interval $[0, 2^{16} - 1]$. Without this addition, repeated application of the carry chain would bring all limbs into the interval $[-2^{16} - 1, 2^{16} - 1]$. We use this additional functionality to "freeze" field elements to a unique representation at the very end of the Curve25519 or Ed25519 computations.

We reuse the multiplication for squarings, but make squarings explicit by spending a few bytes of source code for a separate function that simply calls multiplication. This makes it easy during code audit to compare the code to elliptic-curve addition formulas, for example from the Explicit Formulas Database [9]. To match the notation of [9] we use the names M and S for functions that multiply and square in the field $\mathbb{F}_{2^{255}-19}$; we also use A for addition and Z for subtraction.

Inversion uses Fermat's little theorem and is implemented through exponentation with $2^{255} - 21$. We use a simple square-and-multiply algorithm and avoid storing the exponent by making use of its special shape: it has all bits set but the bits at position 4 and position 2. We perform the square-and-multiply loop for inversion as follows:

```
for(a=253;a>=0;a--) {
  S(c,c);
  if(a!=2&&a!=4) M(c,c,i);
}
```

The square-root computation for point decompression in Ed25519 uses exponentiation by $2^{252} - 3$. See [7, Sect. 5]. Observe that this exponent has all bits set but the bit at position 1; we use the same approach as for inversion.

Curve arithmetic. The typical Curve25519 implementation computes a Montgomery ladder [20] on the Montgomery curve $M : y^2 = x^3 + 486662x^2 + x$. The Ed25519 signature scheme performs arithmetic on the birationally equivalent twisted Edwards curve $E : -x^2 + y^2 = 1 - \frac{121665}{121666}x^2y^2$. More specifically, Ed25519 key generation and signing perform a fixed-basepoint scalar multiplication; verification performs a double-scalar multiplication.

In principle, TweetNaCl could use the same scalar-multiplication code for both Curve25519 and Ed25519. This would require conversion of points on M to points on E and back. If we used the x-coordinate-based differential addition ladder of Curve25519 also for Ed25519, we would additionally need code to recover the y-coordinate as described by Okeya and Sakurai in [21]. Conversion code is not substantially shorter than the code required for the Curve25519 Montgomery ladder, so we decided to *not* use the same code for scalar multiplication in Curve25519 and Ed25519.

Curve25519 uses the same Montgomery ladder as the reference implementation, except that we do not use a dedicated function for multiplication by the constant 121666. For Ed25519 we decided to use only one scalar-multiplication routine that can be used in key generation, signing, and verification. We represent points on E in extended coordinates as described in [17] and implement the complete addition law in an **add** function. We use this function for both addition and doubling of points. The scalar multiplication then performs a ladder of 256 steps; each step performs an addition and a doubling:

```
set25519(p[0],gf0);
set25519(p[1],gf1);
set25519(p[2],gf1);
```

```
set25519(p[3],gf0);
for (i = 255;i >= 0;--i) {
  u8 b = (s[i/8]>>(i&7))&1;
  cswap(p,q,b);
  add(q,p);
  add(p,p);
  cswap(p,q,b);
}
```

The first four lines set the point p to the neutral element. The cswap function performs a constant-time conditional swap of p and q depending on the scalar bit that has been extracted into b before. The constant-time swap calls sel25519 for each of the 4 coordinates of p and q. The function sel25519 is reused in conditional swaps for the Montgomery ladder in Curve25519 and performs a constant-time conditional swap of field elements as follows:

```
sv sel25519(gf p,gf q,int b) {
  i64 t,i,c=~(b-1);
  FOR(i,16) {
    t = c & (p[i]^q[i]);
    p[i] ^= t;
    q[i] ^= t;
  }
}
```

Arithmetic modulo the group order. Signing requires reduction of a 512-bit integer modulo the order of the Curve25519 group, a prime $p = 2^{252} + \delta$ where $\delta \approx 2^{124.38}$. We store this integer as a sequence of limbs in radix 2^8. We eliminate the top limb of the integer, say $2^{504}b$, by subtracting $2^{504}b$ and also subtracting $2^{252}\delta b$; we then perform a partial carry so that 20 consecutive limbs are each between -2^7 and 2^7. We repeat this procedure to eliminate subsequent limbs from the top. This is considerably more concise than typical reduction methods:

```
for (i = 63;i >= 32;--i) {
  carry = 0;
  for (j = i - 32;j < i - 12;++j) {
    x[j] += carry - 16 * x[i] * L[j - (i - 32)];
    carry = (x[j] + 128) >> 8;
    x[j] -= carry << 8;
  }
  x[j] += carry;
  x[i] = 0;
}
```

We similarly eliminate any remaining multiple of 2^{252}, leaving an integer between $-1.1 \cdot 2^{251}$ and $1.1 \cdot 2^{251}$. We then multiply the final carry bit by p and add, obtaining an integer between 0 and $p - 1$, and carry in the traditional way so that each limb is between 0 and 255.

6   Auditability: Two Case Studies

This section explains how we verified two memory-safety properties of Tweet-NaCl: first, all of TweetNaCl's array accesses are within bounds for all inputs whose *lengths* meet certain requirements; second, TweetNaCl makes no use of uninitialized data. Most of this verification was formal (i.e., comprehensively checked by the computer), except for small parts that were carried out by hand.

Our basic bounds-checking strategy is as follows. Recall that TweetNaCl, like NaCl, protects against cache-timing attacks by avoiding all data flow from input *contents* to pointers. Langley's `ctgrind` tool [19] is adequate to verify this property. Consequently all pointers are determined by input *lengths*. Systematically monitoring all pointers for a single input of each length, for example with `valgrind`, is thus equivalent to monitoring all pointers for arbitrary inputs of those lengths. (Of course, this is not the same as arbitrary inputs of arbitrary lengths. However, our main target is applications that cryptographically protect every packet. These applications normally impose packet-length limits, such as the 16384-byte limit imposed by TLS.)

We decided to use C++ overloading instead of `valgrind`, so that we would have a framework for formally verifying further TweetNaCl properties beyond `valgrind`'s capabilities. There are some C language features for which C++ broke compatibility with C, but TweetNaCl does not use any of those features. We had to change some variable definitions and parameter definitions, but because TweetNaCl is so concise this was easy to do by hand. TweetNaCl's declaration structure is highly regular, so scripting this translation would also be straightforward without any compiler patches.

Specifically, we changed array definitions and pointer-parameter definitions such as

```
int crypto_stream(u8 *c,u64 d,const u8 *n,const u8 *k)
{
  u8 s[32];
  ...
}
```

to

```
int crypto_stream(a<u8> c,u64 d,const a<u8> n,const a<u8> k)
{
  da<u8,32> s;
  ...
}
```

where a and da are defined as follows:

```
template <typename Primitive>
struct a {
  Primitive  *content;
  int        size;
  mutable int index;

  a(Primitive *content, int size, int index)
  : content(content), size(size), index(index) {
    assert(index <= size);
  }
```

```
Primitive& operator[](int i) {
  assert((index+i) >= 0 && (index+i) < size);
  return content[index+i];
}
}

template <typename Primitive, int Size>
struct da {
  mutable Primitive content[Size];

  operator const a<Primitive>() const {
    return a<Primitive>(&content[0], Size, 0);
  }
}
```

This is only an illustrative excerpt from the complete definitions of a and da. The complete definitions use two methods (one for mutables, one for constants) for each pointer operation used in TweetNaCl. We note that

```
a operator+(int i) {
  assert((index+i) < size);
  return {content, size, index+i};
}
```

would have been too restrictive: there is no problem in C with using a pointer just past the end of an array as long as the pointer is not dereferenced.

We used overloading in a similar way to check for uninitialized array elements. We created an auxiliary structure with a flag for each array element stating whether the element was initialized; initialized each flag to false; set each flag to true upon write; and checked upon read whether the flag was true.

We close by emphasizing that TweetNaCl's simplicity was essential for this verification. We do not claim to have completed an audit of TweetNaCl, but we do claim that a complete audit will be feasible, and that TweetNaCl is the first cryptographic library for which this is true.

References

1. Aumasson, J.-P.: Tweetcipher! (crypto challenge) (2013). http://cybermashup. com/2013/06/12/tweetcipher-crypto-challenge/. Accessed 06 Sept. 2014, 71
2. Bernstein, D.J.: Cryptography in NaCl. http://cr.yp.to/highspeed/naclcrypto-20090310.pdf. Accessed 06 Sept. 2014, 66
3. Bernstein, D.J.: The Poly1305-AES message-authentication code. In: Gilbert, H., Handschuh, H. (eds.) FSE 2005. LNCS, vol. 3557, pp. 32–49. Springer, Heidelberg (2005). http://cr.yp.to/papers.htmlpoly#1305, 72
4. Bernstein, D.J.: Curve25519: New Diffie-Hellman speed records. In: Yung, M., Dodis, Y., Kiayias, A., Malkin, T. (eds.) PKC 2006. LNCS, vol. 3958, pp. 207–228. Springer, Heidelberg (2006). http://cr.yp.to/papers.htmlcurve#25519, 75
5. Bernstein, D.J.: The Salsa20 family of stream ciphers. In: Robshaw, M., Billet, O. (eds.) New Stream Cipher Designs. LNCS, vol. 4986, pp. 84–97. Springer, Heidelberg (2008). http://cr.yp.to/papers.htmlsalsafamily, 70
6. Bernstein, D.J.: Extending the Salsa20 nonce. In: Workshop Record of Symmetric Key Encryption Workshop 2011 (2011). http://cr.yp.to/papers.html#xsalsa, 72

7. Bernstein, D.J., Duif, N., Lange, T., Schwabe, P., Yang, B.-Y.: High-speed high-security signatures. In: Preneel, B., Takagi, T. (eds.) CHES 2011. LNCS, vol. 6917, pp. 124–142. Springer, Heidelberg (2011). See also full version 75, 76, 80
8. Bernstein, D.J., Duif, N., Lange, T., Schwabe, P., Yang, B.-Y.: High-speed high-security signatures. J. Cryptographic Eng. **2**(2), 77–89 (2012). http://cryptojedi.org/papers/#ed25519. See also short version 75, 80
9. Bernstein, D.J., Lange, T.: Explicit-formulas database. http://www.hyperelliptic.org/EFD/ Accessed 06 Sept. 2014, 76
10. Bernstein, D.J., Lange, T., Schwabe, P.: The security impact of a new cryptographic library. In: Hevia, A., Neven, G. (eds.) Latin-Crypt 2012. LNCS, vol. 7533, pp. 159–176. Springer, Heidelberg (2012). http://cryptojedi.org/papers/#coolnacl, 65
11. Bernstein, D.J., Schwabe, P.: NEON crypto. In: Prouff, E., Schaumont, P. (eds.) CHES 2012. LNCS, vol. 7428, pp. 320–339. Springer, Heidelberg (2012). http://cryptojedi.org/papers/#neoncrypto, 65
12. BitTorrent Live. http://live.bittorrent.com/. Accessed 06 Sept. 2014, 65
13. Denis, F.: Introducing Sodium, a new cryptographic library (2013). http://labs.opendns.com/2013/03/06/announcing-sodium-a-new-cryptographic-library/. Accessed 06 Sept. 2014, 67
14. Dingledine, R.: Tor 0.2.4.17-rc is out. Posting in [tor-talk] (2013). https://lists.torproject.org/pipermail/tor-talk/2013-September/029857.html, 65
15. Green, M.: The anatomy of a bad idea (2012). http://blog.cryptographyengineering.com/2012/12/the-anatomy-of-bad-idea.html. Accessed 06 Sept. 2014, 65
16. Green, M.: Announcing a contest: identify useful cryptographic algorithms that can be formally described in one Tweet (2013). https://twitter.com/matthew_d_green/status/342755869110464512. Accessed 06 Sept. 2014, 68
17. Hisil, H., Wong, K.K.-H., Carter, G., Dawson, E.: Twisted edwards curves revisited. In: Pieprzyk, J. (ed.) ASIACRYPT 2008. LNCS, vol. 5350, pp. 326–343. Springer, Heidelberg (2008). http://eprint.iacr.org/2008/522/, 76
18. Hutter, M., Schwabe, P.: NaCl on 8-Bit AVR microcontrollers. In: Youssef, A., Nitaj, A., Hassanien, A.E. (eds.) AFRICACRYPT 2013. LNCS, vol. 7918, pp. 156–172. Springer, Heidelberg (2013). http://cryptojedi.org/papers/#avrnacl, 68
19. Langley, A.: ctgrind–checking that functions are constant time with Valgrind (2010). https://github.com/agl/ctgrind, 78
20. Montgomery, P.L.: Speeding the Pollard and elliptic curve methods of factorization. Math. Comput. **48**(177), 243–264 (1987). http://www.ams.org/journals/mcom/1987-48-177/S0025-5718-1987-0866113-7/S0025--5718-1987-0866113-7.pdf, 76
21. Okeya, K., Sakurai, K.: Efficient elliptic curve cryptosystems from a scalar multiplication algorithm with recovery of the y-coordinate on a montgomery-form elliptic curve. In: Koç, Ç.K., Naccache, D., Paar, C. (eds.) CHES 2001. LNCS, vol. 2162, p. 126. Springer, Heidelberg (2001). 76
22. Introducing DNSCrypt (preview release). http://www.opendns.com/technology/dnscrypt/. Accessed 06 Sept. 2014, 65
23. OpenSSL: OpenSSL: The open source toolkit for SSL/TLS. http://www.openssl.org/. Accessed 06 Sept. 2014, 65
24. Threema: seriously secure mobile messaging. https://threema.ch/en/. Accessed 06 Sept. 2014, 65
25. Tor project: Anonymity online. https://www.torproject.org/. Accessed 06 Sept. 2014, 65

26. U.S. Department OF COMMERCE/National Institute of Standards and Technology. Secure Hash Standard (SHS) (2012). Federal Information Processing Standards Publication 180–4. http://csrc.nist.gov/publications/fips/fips180-4/fips-180-4.pdf, 73, 75
27. Ytteborg, S.S.: The PGPi scanning project. http://www.pgpi.org/pgpi/project/scanning/. Accessed 06 Sept. 2014, 68
28. Zimmermann, P.: PGP Source Code and Internals. MIT Press, Cambridge (1995). 68

A   The 100 Tweets

```
#include "tweetnacl.h"
#define FOR(i,n) for (i = 0;i < n;++i)
#define sv static void
typedef unsigned char u8;typedef unsigned long u32;typedef unsigned long long u64;typedef long long i64;typedef i64 gf[16];extern void
randombytes(u8*,u64);static const u8 _0[16],_9[32]={9};static const gf gf0,gf1={1},_121665={0xDB41,1},D={0x78a3,0x1359,0x4dca,0x75eb,0xd8ab,
0x4141,0x0a4d,0x0070,0xe898,0x7779,0x4079,0x8cc7,0xfe73,0x2b6f,0x6cee,0x5203},D2={0xf159,0x26b2,0x9b94,0xebd6,0xb156,0x8283,0x149a,0x00e0,
0xd130,0xeef3,0x80f2,0x198e,0xfce7,0x56df,0xd9dc,0x2406},X={0xd51a,0x8f25,0x2d60,0xc956,0xa7b2,0x9525,0xc760,0x692c,0xdc5c,0xfdd6,0xe231,
0xc0a4,0x53fe,0xcd6e,0x36d3,0x2169},Y={0x6658,0x6666,0x6666,0x6666,0x6666,0x6666,0x6666,0x6666,0x6666,0x6666,0x6666,0x6666,0x6666,0x6666,
0x6666,0x6666},I={0xa0b0,0x4a0e,0x1b27,0xc4ee,0xe478,0xad2f,0x1806,0x2f43,0xd7a7,0x3dfb,0x0099,0x2b4d,0xdf0b,0x4fc1,0x2480,0x2b83};static
u32 L32(u32 x,int c){return(x<<c)|((x&0xffffffff)>>(32-c));}static u32 ld32(const u8*x){u32 u=x[3];u=(u<<8)|x[2];u=(u<<8)|x[1];return(u<<8)|
x[0];}static u64 dl64(const u8*x){u64 i,u=0;FOR(i,8)u=(u<<8)|x[i];return u;}sv st32(u8*x,u32 u){int i;FOR(i,4){x[i]=u;u>>=8;}}sv ts64(u8*x,
u64 u){int i;for(i=7;i>=0;--i){x[i]=u;u>>=8;}}static int vn(const u8*x,const u8*y,int n){u32 i,d=0;FOR(i,n)d|=x[i]^y[i];return(1&((d-1)>>8))
-1;}int crypto_verify_16(const u8*x,const u8*y){return vn(x,y,16);}int crypto_verify_32(const u8*x,const u8*y){return vn(x,y,32);}sv core(u8
*out,const u8*in,const u8*k,const u8*c,int h){u32 w[16],x[16],y[16],t[4];int i,j,m;FOR(i,4){x[5*i]=ld32(c+4*i);x[1+i]=ld32(k+4*i);x[6+i]=
ld32(in+4*i);x[11+i]=ld32(k+16+4*i);}FOR(i,16)y[i]=x[i];FOR(i,20){FOR(j,4){FOR(m,4){t[m]=x[(5*i+4*m)%16];}t[1]^=L32(t[0]+t[3],7);t[2]^=L32(t[1
]+t[0],9);t[3]^=L32(t[2]+t[1],13);t[0]^=L32(t[3]+t[2],18);FOR(m,4)w[4*j+(j+m)%4]=t[m];}FOR(m,16)x[m]=w[m];}if(h){FOR(i,16)x[i]+=y[i];FOR(i,4
){x[5*i]-=ld32(c+4*i);x[6+i]-=ld32(in+4*i);}FOR(i,4){st32(out+4*i,x[5*i]);st32(out+16+4*i,x[6+i]);}}else FOR(i,16)st32(out+4*i,x[i]+y[i]);}
int crypto_core_salsa20(u8*out,const u8*in,const u8*k,const u8*c){core(out,in,k,c,0);return 0;}int crypto_core_hsalsa20(u8*out,const u8*in,
const u8*k,const u8*c){core(out,in,k,c,1);return 0;}static const u8 sigma[16]="expand 32-byte k";int crypto_stream_salsa20_xor(u8*c,const u8
*m,u64 b,const u8*n,const u8*k){u8 z[16],x[64];u32 u,i;if(!b)return 0;FOR(i,16)z[i]=0;FOR(i,8)z[i]=n[i];while(b>=64){crypto_core_salsa20(x,z
,k,sigma);FOR(i,64)c[i]=(m?m[i]:0)^x[i];u=1;for(i=8;i<16;++i){u+=(u32)z[i];z[i]=u;u>>=8;}b-=64;c+=64;if(m)m+=64;}if(b){crypto_core_salsa20(x
,z,k,sigma);FOR(i,b)c[i]=(m?m[i]:0)^x[i];}return 0;}int crypto_stream_salsa20(u8*c,u64 d,const u8*n,const u8*k){return
crypto_stream_salsa20_xor(c,0,d,n,k);}int crypto_stream(u8*c,u64 d,const u8*n,const u8*k){u8 s[32];crypto_core_hsalsa20(s,n,k,sigma);return
crypto_stream_salsa20(c,d,n+16,s);}int crypto_stream_xor(u8*c,const u8*m,u64 d,const u8*n,const u8*k){u8 s[32];crypto_core_hsalsa20(s,n,k,
sigma);return crypto_stream_salsa20_xor(c,m,d,n+16,s);}sv add1305(u32*h,const u32*c){u32 j,u=0;FOR(j,17){u+=h[j]+c[j];h[j]=u&255;u>>=8;}}
static const u32 minusp[17]={5,0,0,0,0,0,0,0,0,0,0,0,0,0,0,0,252};int crypto_onetimeauth(u8*out,const u8*m,u64 n,const u8*k){u32 s,i,j,u,x[
17],r[17],h[17],c[17],g[17];FOR(j,17)r[j]=h[j]=0;FOR(j,16)r[j]=k[j];r[3]&=15;r[4]&=252;r[7]&=15;r[8]&=252;r[11]&=15;r[12]&=252;r[15]&=15;
while(n>0){FOR(j,17)c[j]=0;for(j=0;(j<16)&&(j<n);++j)c[j]=m[j];c[j]=1;m+=j;n-=j;add1305(h,c);FOR(i,17){x[i]=0;FOR(j,17)x[i]+=h[j]*((j<=i)?r[
i-j]:320*r[i+17-j]);}FOR(i,17)h[i]=x[i];u=0;FOR(j,16){u+=h[j];h[j]=u&255;u>>=8;}u+=h[16];h[16]=u&3;u=5*(u>>2);FOR(j,16){u+=h[j];h[j]=u&255;u
>>=8;}u+=h[16];h[16]=u;}FOR(j,17)g[j]=h[j];add1305(h,minusp);s=-(h[16]>>7);FOR(j,17)h[j]^=s&(g[j]^h[j]);FOR(j,16)c[j]=k[j+16];c[16]=0;
add1305(h,c);FOR(j,16)out[j]=h[j];return 0;}int crypto_onetimeauth_verify(const u8*h,const u8*m,u64 n,const u8*k){u8 x[16];
crypto_onetimeauth(x,m,n,k);return crypto_verify_16(h,x);}int crypto_secretbox(u8*c,const u8*m,u64 d,const u8*n,const u8*k){int i;if(d<32)
return-1;crypto_stream_xor(c,m,d,n,k);crypto_onetimeauth(c+16,c+32,d-32,c);FOR(i,16)c[i]=0;return 0;}int crypto_secretbox_open(u8*m,const u8
*c,u64 d,const u8*n,const u8*k){int i;u8 x[32];if(d<32)return-1;crypto_stream(x,32,n,k);if(crypto_onetimeauth_verify(c+16,c+32,d-32,x)!=0)
return-1;crypto_stream_xor(m,c,d,n,k);FOR(i,32)m[i]=0;return 0;}sv set25519(gf r,const gf a){int i;FOR(i,16)r[i]=a[i];}sv car25519(gf o){int
i;i64 c;FOR(i,16){o[i]+=(1LL<<16);c=o[i]>>16;o[(i+1)*(i<15)]+=c-1+37*(c-1)*(i==15);o[i]-=c<<16;}}sv sel25519(gf p,gf q,int b){i64 t,i,c=~(b-
1);FOR(i,16)t=c&(p[i]^q[i]);p[i]^=t;q[i]^=t;}}sv pack25519(u8*o,const gf n){int i,j,b;gf m,t;FOR(i,16)t[i]=n[i];car25519(t);car25519(t);
car25519(t);FOR(j,2){m[0]=t[0]-0xffed;for(i=1;i<15;i++){m[i]=t[i]-0xffff-((m[i-1]>>16)&1);m[i-1]&=0xffff;}m[15]=t[15]-0x7fff-((m[14]>>16)&1)
;b=(m[15]>>16)&1;m[14]&=0xffff;sel25519(t,m,1-b);}FOR(i,16){o[2*i]=t[i]&0xff;o[2*i+1]=t[i]>>8;}}static int neq25519(const gf a,const gf b){
u8 c[32],d[32];pack25519(c,a);pack25519(d,b);return crypto_verify_32(c,d);}static u8 par25519(const gf a){u8 d[32];pack25519(d,a);return d[0
]&1;}sv unpack25519(gf o,const u8*n){int i;FOR(i,16)o[i]=n[2*i]+((i64)n[2*i+1]<<8);o[15]&=0x7fff;}sv A(gf o,const gf a,const gf b){int i;FOR
(i,16)o[i]=a[i]+b[i];}sv Z(gf o,const gf a,const gf b){int i;FOR(i,16)o[i]=a[i]-b[i];}sv M(gf o,const gf a,const gf b){i64 i,j,t[31];FOR(i,
31)t[i]=0;FOR(i,16)FOR(j,16)t[i+j]+=a[i]*b[j];FOR(i,15)t[i]+=38*t[i+16];FOR(i,16)o[i]=t[i];car25519(o);car25519(o);}sv S(gf o,const gf a){M(
o,a,a);}sv inv25519(gf o,const gf i){gf c;int a;FOR(a,16)c[a]=i[a];for(a=253;a>=0;a--){S(c,c);if(a!=2&&a!=4)M(c,c,i);}FOR(a,16)o[a]=c[a];}sv
pow2523(gf o,const gf i){gf c;int a;FOR(a,16)c[a]=i[a];for(a=250;a>=0;a--){S(c,c);if(a!=1)M(c,c,i);}FOR(a,16)o[a]=c[a];}int
crypto_scalarmult(u8*q,const u8*n,const u8*p){u8 z[32];i64 x[96],r,i;gf a,b,c,d,e,f;FOR(i,31)z[i]=n[i];z[31]=(n[31]&127)|64;z[0]&=248;
unpack25519(x,p);FOR(i,16){b[i]=x[i];d[i]=a[i]=0;}a[0]=d[0]=1;for(i=254;i>=0;--i){r=(z[i>>3]>>(i&7))&1;sel25519(a,b,r);sel25519(c,d,r);
A(e,a,c);Z(a,a,c);A(c,b,d);Z(b,b,d);S(d,e);S(f,a);M(a,c,a);M(c,b,e);A(e,a,c);Z(a,a,c);S(b,a);Z(c,d,f);M(a,c,_121665);A(a,a,d);M(c,c,a);M(a,d
,f);M(d,b,x);S(b,e);sel25519(a,b,r);sel25519(c,d,r);}FOR(i,16){x[i+32]=a[i];x[i+48]=c[i];x[i+64]=b[i];x[i+80]=d[i];}inv25519(x+48,x+48);M(x+
32,x+32,x+48);pack25519(q,x+32);return 0;}int crypto_scalarmult_base(u8*q,const u8*n){return crypto_scalarmult(q,n,_9);}int
crypto_box_keypair(u8*y,u8*x){randombytes(x,32);return crypto_scalarmult_base(y,x);}int crypto_box_beforenm(u8*k,const u8*y,const u8*x){u8 s
[32];crypto_scalarmult(s,x,y);return crypto_core_hsalsa20(k,_0,s,sigma);}int crypto_box_afternm(u8*c,const u8*m,u64 d,const u8*n,const u8*k)
{return crypto_secretbox(c,m,d,n,k);}int crypto_box_open_afternm(u8*m,const u8*c,u64 d,const u8*n,const u8*k){return crypto_secretbox_open(m
```

```
,c,d,n,k);}int crypto_box(u8*c,const u8*m,u64 d,const u8*n,const u8*y,const u8*x){u8 k[32];crypto_box_beforenm(k,y,x);return
crypto_box_afternm(c,m,d,n,k);}int crypto_box_open(u8*m,const u8*c,u64 d,const u8*n,const u8*y,const u8*x){u8 k[32];crypto_box_beforenm(k,y,
x);return crypto_box_open_afternm(m,c,d,n,k);}static u64 R(u64 x,int c){return(x>>c)|(x<<(64-c));}static u64 Ch(u64 x,u64 y,u64 z){return(x&
y)^(~x&z);}static u64 Maj(u64 x,u64 y,u64 z){return(x&y)^(x&z)^(y&z);}static u64 Sigma0(u64 x){return R(x,28)^R(x,34)^R(x,39);}static u64
Sigma1(u64 x){return R(x,14)^R(x,18)^R(x,41);}static u64 sigma0(u64 x){return R(x,1)^R(x,8)^(x>>7);}static u64 sigma1(u64 x){return R(x,19)^
R(x,61)^(x>>6);}static const u64 K[80]={0x428a2f98d728ae22ULL,0x7137449123ef65cdULL,0xb5c0fbcfec4d3b2fULL,0xe9b5dba58189dbbcULL,
0x3956c25bf348b538ULL,0x59f111f1b605d019ULL,0x923f82a4af194f9bULL,0xab1c5ed5da6d8118ULL,0xd807aa98a3030242ULL,0x12835b0145706fbeULL,
0x243185be4ee4b28cULL,0x550c7dc3d5ffb4e2ULL,0x72be5d74f27b896fULL,0x80deb1fe3b1696b1ULL,0x9bdc06a725c71235ULL,0xc19bf174cf692694ULL,
0xe49b69c19ef14ad2ULL,0xefbe4786384f25e3ULL,0x0fc19dc68b8cd5b5ULL,0x240ca1cc77ac9c65ULL,0x2de92c6f592b0275ULL,0x4a7484aa6ea6e483ULL,
0x5cb0a9dcbd41fbd4ULL,0x76f988da831153b5ULL,0x983e5152ee66dfabULL,0xa831c66d2db43210ULL,0xb00327c898fb213fULL,0xbef597fc7beef0ee4ULL,
0xc6e00bf33da88fc2ULL,0xd5a79147930aa725ULL,0x06ca6351e003826fULL,0x142929670a0e6e70ULL,0x27b70a8546d22ffcULL,0x2e1b21385c26c926ULL,
0x4d2c6dfc5ac42aedULL,0x53380d139d95b3df5ULL,0x650a73548baf63deULL,0x766a0abb3c77b2a8ULL,0x81c2c92e47edaee6ULL,0x92722c851482353bULL,
0xa2bfe8a14cf10364ULL,0xa81a664bbc423001ULL,0xc24b8b70d0f89791ULL,0xc76c51a30654 be30ULL,0xd192e819d6ef5218ULL,0xd69906245565a910ULL,
0xf40e35855771202aULL,0x106aa07032bbd1b8ULL,0x19a4c116b8d2d0c8ULL,0x1e376c085141ab53ULL,0x2748774cdf8eeb99ULL,0x34b0bcb5e19b48a8ULL,
0x391c0cb3c5c95b63ULL,0x4ed8aa4ae3418acbULL,0x5b9cca4f7763e373ULL,0x682e6ff3d6b2b8a3ULL,0x748f82ee5defb2fcULL,0x78a5636f43172f60ULL,
0x84c87814a1f0ab72ULL,0x8cc702081a6439ecULL,0x90befffa23631e28ULL,0xa4506cebde82bde9ULL,0xbef9a3f7b2c67915ULL,0xc67178f2e372532bULL,
0xca273eceea26619cULL,0xd186b8c721c0c207ULL,0xeada7dd6cde0eb1eULL,0xf57d4f7fee6ed178ULL,0x06f067aa72176fbaULL,0x0a637dc5a2c898a6ULL,
0x113f9804bef90daeULL,0x1b710b35131c471bULL,0x28db77f523047d84ULL,0x32caab7b40c72493ULL,0x3c9ebe0a15c9bebcULL,0x431d67c49c100d4cULL,
0x4cc5d4becb3e42b6ULL,0x597f299cfc657e2aULL,0x5fcb6fab3ad6faecULL,0x6c44198c4a475817ULL};int crypto_hashblocks(u8*x,const u8*m,u64 n){u64 z[
8],b[8],a[8],w[16],t;int i,j;FOR(i,8)z[i]=a[i]=d164(x+8*i);while(n>=128){FOR(i,16)w[i]=d164(m+8*i);FOR(i,80)FOR(j,8)b[j]=a[j];t=a[7]+Sigma1
(a[4])+Ch(a[4],a[5],a[6])+K[i]+w[i%16];b[7]=t+Sigma0(a[0])+Maj(a[0],a[1],a[2]);b[3]+=t;FOR(j,8)a[(j+1)%8]=b[j];if(i%16==15)FOR(j,16)w[j]+=w[
(j+9)%16]+sigma0(w[(j+1)%16])+sigma1(w[(j+14)%16]);}FOR(i,8){a[i]+=z[i];z[i]=a[i];}m+=128;n-=128;}FOR(i,8)ts64(x+8*i,z[i]);return n;}static
const u8 iv[64]={0x6a,0x09,0xe6,0x67,0xf3,0xbc,0xc9,0x08,0xbb,0x67,0xae,0x85,0x84,0xca,0xa7,0x3b,0x3c,0x6e,0xf3,0x72,0xfe,0x94,0xf8,0x2b,
0xa5,0x4f,0xf5,0x3a,0x5f,0x1d,0x36,0xf1,0x51,0x0e,0x52,0x7f,0xad,0xe6,0x82,0xd1,0x9b,0x05,0x68,0x8c,0x2b,0x3e,0x6c,0x1f,0x1f,0x83,0xd9,0xab,
0xfb,0x41,0xbd,0x6b,0x5b,0xe0,0xcd,0x19,0x13,0x7e,0x21,0x79};int crypto_hash(u8*out,const u8*m,u64 n){u8 h[64],x[256];u64 i,b=n;FOR(i,64)h[i
]=iv[i];crypto_hashblocks(h,m,n);m+=n;n&=127;m-=n;FOR(i,256)x[i]=0;FOR(i,n)x[i]=m[i];x[n]=128;n=256-128*(n<112);x[n-9]=b>>61;ts64(x+n-8,b<<3
);crypto_hashblocks(h,x,n);FOR(i,64)out[i]=h[i];return 0;}sv add(gf p[4],gf q[4]){gf a,b,c,d,t,e,f,g,h;Z(a,p[1],p[0]);Z(t,q[1],q[0]);M(a,a,t
);A(b,p[0],p[1]);A(t,q[0],q[1]);M(b,b,t);M(c,p[3],q[3]);M(c,c,D2);M(d,p[2],q[2]);A(d,d,d);Z(e,b,a);Z(f,d,c);A(g,d,c);A(h,b,a);M(p[0],e,f);M(
p[1],h,g);M(p[2],g,f);M(p[3],e,h);}sv cswap(gf p[4],gf q[4],u8 b){int i;FOR(i,4)sel25519(p[i],q[i],b);}sv pack(u8*r,gf p[4]){gf tx,ty,zi;
inv25519(zi,p[2]);M(tx,p[0],zi);M(ty,p[1],zi);pack25519(r,ty);r[31]^=par25519(tx)<<7;}sv scalarmult(gf p[4],gf q[4],const u8*s){int i;
set25519(p[0],gf0);set25519(p[1],gf1);set25519(p[2],gf1);set25519(p[3],gf0);for(i=255;i>=0;--i){u8 b=(s[i/8]>>(i&7))&1;cswap(p,q,b);add(q,p)
;add(p,p);cswap(p,q,b);}}sv scalarbase(gf p[4],const u8*s){gf q[4];set25519(q[0],X);set25519(q[1],Y);set25519(q[2],gf1);M(q[3],X,Y);
scalarmult(p,q,s);}int crypto_sign_keypair(u8*pk,u8*sk){u8 d[64];gf p[4];int i;randombytes(sk,32);crypto_hash(d,sk,32);d[0]&=248;d[31]&=127;
d[31]|=64;scalarbase(p,d);pack(pk,p);FOR(i,32)sk[32+i]=pk[i];return 0;}static const u64 L[32]={0xed,0xd3,0xf5,0x5c,0x1a,0x63,0x12,0x58,0xd6,
0x9c,0xf7,0x0a2,0xde,0xf9,0xde,0x14,0,0,0,0,0,0,0,0,0,0,0,0,0,0,0,0x10};sv modL(u8*r,i64 x[64]){i64 carry,i,j;for(i=63;i>=32;--i){carry=0;for
(j=i-32;j<i-12;++j){x[j]+=carry-16*x[i]*L[j-(i-32)];carry=(x[j]+128)>>8;x[j]-=carry<<8;}x[j]+=carry;x[i]=0;}carry=0;FOR(j,32){x[j]+=carry-(x
[31]>>4)*L[j];carry=x[j]>>8;x[j]&=255;}FOR(j,32)x[j]-=carry*L[j];FOR(i,32){x[i+1]+=x[i]>>8;r[i]=x[i]&255;}}sv reduce(u8*r){i64 x[64],i;FOR(i
,64)x[i]=(u64)r[i];FOR(i,64)r[i]=0;modL(r,x);}int crypto_sign(u8*sm,u64*smlen,const u8*m,u64 n,const u8*sk){u8 d[64],h[64],r[64];i64 i,j,x[
64];gf p[4];crypto_hash(d,sk,32);d[0]&=248;d[31]&=127;d[31]|=64;*smlen=n+64;FOR(i,n)sm[64+i]=m[i];FOR(i,32)sm[32+i]=d[32+i];crypto_hash(r,sm
+32,n+32);reduce(r);scalarbase(p,r);pack(sm,p);FOR(i,32)sm[i+32]=sk[i+32];crypto_hash(h,sm,n+64);reduce(h);FOR(i,64)x[i]=0;FOR(i,32)x[i]=(
u64)r[i];FOR(i,32)FOR(j,32)x[i+j]+=h[i]*(u64)d[j];modL(sm+32,x);return 0;}static int unpackneg(gf r[4],const u8 p[32]){gf t,chk,num,den,den2
,den4,den6;set25519(r[2],gf1);unpack25519(r[1],p);S(num,r[1]);M(den,num,D);Z(num,num,r[2]);A(den,r[2],den);S(den2,den);S(den4,den2);M(den6,
den4,den2);M(t,den6,num);M(t,t,den);pow2523(t,t);M(t,t,num);M(t,t,den);M(t,t,den);M(r[0],t,den);S(chk,r[0]);M(chk,chk,den);if(neq25519(chk,
num))M(r[0],r[0],I);S(chk,r[0]);M(chk,chk,den);if(neq25519(chk,num))return-1;if(par25519(r[0])==(p[31]>>7))Z(r[0],gf0,r[0]);M(r[3],r[0],r[1]
);return 0;}int crypto_sign_open(u8*m,u64*mlen,const u8*sm,u64 n,const u8*pk){int i;u8 t[32],h[64];gf p[4],q[4];*mlen= -1;if(n<64)return-1;
if(unpackneg(q,pk))return-1;FOR(i,n)m[i]=sm[i];FOR(i,32)m[i+32]=pk[i];crypto_hash(h,m,n);reduce(h);scalarmult(p,q,h);scalarbase(q,sm+32);add
(p,q);pack(t,p);n-=64;if(crypto_verify_32(sm,t)){FOR(i,n)m[i]=0;return-1;}FOR(i,n)m[i]=sm[i+64];*mlen=n;return 0;}
```

B A Python script to convert `tweetnacl.c` into the 100 tweets

```python
import re
import sys

output = ''

while True:
  line = sys.stdin.readline()
  if not line: break
  if line[0] == '#':
    if output:
      print output
      output = ''
    print line.strip()
  else:
    x = re.findall('\w+|\W',line)
    for u in x:
      if not u.isspace():
        if len(output) + len(u) > 140:
          print output
          output = ''
        if (re.match('\w',output[-1:]) and re.match('\w',u[:1])) or (output[-1:] == '=' and u[:1] == '-'):
          if len(output) + 1 + len(u) > 140:
            print output
            output = ''
          else:
            output += ' '
        output += u

print output
```

C   A Python script to print `tweetnacl.h`

```
print '#ifndef TWEETNACL_H'
print '#define TWEETNACL_H'

for z in [
'auth:hmacsha512256/32/32:BYTES,KEYBYTES:,_verify:qpup,ppup',
'box:curve25519xsalsa20poly1305/32/32/32/24/32/16:PUBLICKEYBYTES,SECRETKEYBYTES,BEFORENMBYTES,NONCEBYTES,ZEROBYTES,BOXZEROBYTES:'
+ ',_open,_keypair,_beforenm,_afternm,_open_afternm:qpuppp,qpuppp,qq,qpp,qpupp,qpupp',
'core:salsa20/64/16/32/16,hsalsa20/32/16/32/16:OUTPUTBYTES,INPUTBYTES,KEYBYTES,CONSTBYTES::qppp',
'hashblocks:sha512/64/128,sha256/32/64:STATEBYTES,BLOCKBYTES::qpu',
'hash:sha512/64,sha256/32:BYTES::qpu',
'onetimeauth:poly1305/16/32:BYTES,KEYBYTES:,_verify:qpup,ppup',
'scalarmult:curve25519/32/32:BYTES,SCALARBYTES:,_base:qpp,qp',
'secretbox:xsalsa20poly1305/32/24/32/16:KEYBYTES,NONCEBYTES,ZEROBYTES,BOXZEROBYTES:,_open:qpupp,qpupp',
'sign:ed25519/64/32/64:BYTES,PUBLICKEYBYTES,SECRETKEYBYTES:,_open,_keypair:qvpup,qvpup,qq',
'stream:xsalsa20/32/24,salsa20/32/8:KEYBYTES,NONCEBYTES:,_xor:qupp,qpupp',
'verify:16/16,32/32:BYTES::pp'
]:
  x,q,s,f,g = [i.split(',') for i in z.split(':')]
  o = 'crypto_'+x[0]
  sel = 1
  for p in q:
    p = p.split('/')
    op = o+'_'+p[0]
    opi = op+'_'+'tweet'
    if sel:
      print '#define '+o+'_PRIMITIVE "'+p[0]+'"'
      for m in f+['_'+m for m in s+['IMPLEMENTATION','VERSION']]: print '#define '+o+m+' '+op+m
      sel = 0
    for j in range(len(s)): print '#define '+opi+'_'+s[j]+' '+str(p[j+1])
    for j in range(len(f)):
      a = g[j].replace('v','u *').replace('u','unsigned long long').replace('q','unsigned char *').replace('p',',const unsigned char *')
      print 'extern int '+opi+f[j]+'('+a[1:]+');'
    print '#define '+opi+'_VERSION "-"'
    for m in f+['_'+m for m in s+['VERSION']]: print '#define '+opi+m+' '+opi+m
    print '#define '+op+'_IMPLEMENTATION "'+o+'/'+p[0]+'/tweet'+'"'

print '#endif'
```

High-Speed Signatures from Standard Lattices

Özgür Dagdelen[1], Rachid El Bansarkhani[1], Florian Göpfert[1],
Tim Güneysu[2], Tobias Oder[2], Thomas Pöppelmann[2(✉)],
Ana Helena Sánchez[3], and Peter Schwabe[3]

[1] Technische Universität Darmstadt, Darmstadt, Germany
oezguer.dagdelen@cased.de,
{elbansarkhani,fgoepfert}@cdc.informatik.tu-darmstadt.de
[2] Horst Görtz Institute for IT-Security, Ruhr-University Bochum,
Bochum, Germany
thomas.poeppelmann@rub.de
[3] Digital Security Group, Radboud University Nijmegen,
Nijmegen, The Netherlands
ahsanchez@cs.ru.nl, peter@cryptojedi.org

Abstract. At CT-RSA 2014 Bai and Galbraith proposed a lattice-based
signature scheme optimized for short signatures and with a security
reduction to hard standard lattice problems. In this work we first refine
the security analysis of the original work and propose a new 128-bit
secure parameter set chosen for software efficiency. Moreover, we increase
the acceptance probability of the signing algorithm through an improved
rejection condition on the secret keys. Our software implementation targeting Intel CPUs with AVX/AVX2 and ARM CPUs with NEON vector
instructions shows that even though we do not rely on ideal lattices, we
are able to achieve high performance. For this we optimize the matrix-
vector operations and several other aspects of the scheme and finally
compare our work with the state of the art.

Keywords: Signature scheme · Standard lattices · Vectorization · Ivy
bridge

1 Introduction

Most practical lattice-based signatures [7,16,21], proposed as post-quantum [9]
alternatives to RSA and ECDSA, are currently instantiated and implemented
using structured ideal lattices [30] corresponding to ideals in rings of the form

P. Schwabe—This work was supported by the German Research Foundation
(DFG) through the DFG Research Training Group GRK 1817/1, by the German Federal Ministry of Economics and Technology through Grant 01ME12025
SecMobil), by the Netherlands Organisation for Scientific Research (NWO) through
Veni 2013 project 13114, and by the German Federal Ministry of Education
and Research (BMBF) through EC-SPRIDE. Permanent ID of this document:
c5e2da3f0d05a056a5490a5c9b88baa9. Date: 2014-09-04.

D.F. Aranha and A. Menezes (Eds.): LATINCRYPT 2014, LNCS 8895, pp. 84–103, 2015.
DOI: 10.1007/978-3-319-16295-9_5

$\mathbb{Z}[x]/\langle \mathbf{f} \rangle$, where $\mathbf{f}$ is a degree-n irreducible polynomial (usually $\mathbf{f} = x^n + 1$). With those schemes one is able to achieve high speeds on several architectures as well as reasonably small signatures and key sizes. However, while no attacks are known that perform significantly better against schemes based on ideal lattices, it is still possible that further cryptanalysis will be able to exploit the additional structure[1]. Especially, if long-term security is an issue, it seems that standard lattices and the associated problems—e.g., the Learning With Errors (LWE) [34] or the Small Integer Solution (SIS) problem—offer more confidence than their ring counterparts.

The situation for code-based cryptography [9] is somewhat similar. The use of more structured codes, such as quasi-dyadic Goppa codes [31], has been the target of an algebraic attack [15] which is effective against certain (but not all) proposed parameters. This is an indication that the additional structure used to improve the efficiency of such cryptosystems might be also used by adversaries to improve their attack strategies. Moreover, basing a scheme on the plain LWE or SIS problem seems much more secure than using stronger assumptions on top of ideal lattices like the discrete-compact-knapsack (DCK) [21] or NTRU-related assumptions [16] that have not been studied extensively so far.

While results for ideal-lattice-based signatures have been published recently [11,22,32,33], currently no research is available dealing with implementation and performance issues of standard-lattice-based signatures. While the large keys of such schemes might prevent their adoption on constrained devices or reconfigurable hardware, the size of the keys is much less an issue on current multi-core CPUs which have access to large amounts of memory. In this context, the scheme by Bai and Galbraith [6] (from now on referred to as BG signature) is an interesting proposal as it achieves small signatures and is based on the standard LWE and SIS problems.

An interesting question arising is also the performance of schemes based on standard lattices and how to choose parameters for high performance. While FFT-techniques have been used successfully for ideal lattices on various architectures [22,35] there are no fast algorithms to speed up the necessary matrix-vector arithmetic. However, matrix-vector operations can be parallelized very efficiently and there are no direct restrictions on the parameters (for efficiency of ideal lattices n is usually chosen as power of two) so that there is still hope for high speed. The only results currently available dealing with the implementation of standard lattice-based instantiations rely on arithmetic libraries [7,20] and can thus not fully utilize the power of their target architectures.

An additional feature of the BG signature is that sampling of Gaussian noise is only needed during the much less performance-critical key-generation phase but not for signing[2]. While there was some progress on techniques for efficient

---

[1] There exists sieving algorithms which can exploit the ideal structure, but the speed-up is of no significance [24,36]. Some first ideas towards attacks with lower complexity were sketched by Bernstein in his blog [8].

[2] Omitting costly Gaussian sampling was also the motivation for the design of the GLP signature [21].

discrete Gaussian sampling [16, 17, 33] it is still not known how to implement the sampling efficiently[3] without leaking information on the sampled values through the runtime of the signing process (contrary to uniform sampling [22]).

While we cannot present a direct attack, careful observation of the runtime of software implementations (even remotely over a network) has led to various attacks in the past and thus it is desirable to achieve constant runtime or at least a timing independent from secret data [13, 25].

Our Contribution. The contribution of this paper is twofold. First, we study the parameter selection of the BG signature scheme in more detail than in the original paper and assess its security level[4]. Based on our analysis of the currently most efficient attack we provide a new 128-bit security parameter set chosen for efficient software implementation and long-term security. We compare the runtimes of several attacks on LWE with and without a limit on the number of samples available. Since the behavior of the attacks in a suboptimal attack dimension is not well understood at this point, our analysis may be of independent interest for the hardness assessment of other LWE instances. Additionally, we introduce an optimized rejection sampling procedure and rearrange operations in the signature scheme.

The second part of the paper deals with the implementation of this parameter set on the ARM NEON and Intel AVX architectures optimized for high speed. By using parallelization, interleaving, and vectorization we achieve on average 1203924 cycles for signing and 335072 cycles for verification on the Haswell architecture. This corresponds to roughly 2824 signing and 10147 verification operations per second on one core of a CPU clocked with 3.4 GHz. While we do not set a speed record for general lattices, we are able to present the currently fastest implementation of a lattice-bases signature scheme that relies solely on standard assumptions and is competitive in terms of performance compared to classical and post-quantum signature schemes.

Availability of Software. We will place all software described in this paper into the public domain to maximize reusability of our results. We will submit the software to the eBACS benchmarking project [10] for public benchmarking.

Road Map. The paper is organized as follows: In Sect. 3 we introduce the original BG signature scheme and our modifications for efficiency. The security analysis is revisited and appropriate parameters are selected in Sect. 4. In Sect. 5 we discuss our NEON and AVX software implementation and finish with results and a comparison in Sect. 6.

---

[3] A software implementation of a constant time discrete Gaussian sampler using the Cumulative Distribution Table (CDT) approach was recently proposed by Bos et al. [12]. However, even for the small standard deviation required for lattice-based encryption schemes, the constant time requirement leads to a significant overhead.

[4] We note here that there was some vagueness in the parameter selection in the original work [6], also noticed later by the authors of the paper [5].

2   Preliminaries

Notation. We mainly follow the notation of [6] and denote column vectors by bold lower case letters (e.g., $\mathbf{v} = (v_1, \ldots, v_n)^T$ where $\mathbf{v}^T$ is the transpose) and matrices by bold upper case letters (e.g., $\mathbf{M}$). The centered discrete Gaussian distribution $\mathcal{D}_\sigma$ for $\sigma > 0$ associates the probability $\rho_\sigma(x)/\rho_\sigma(\mathbb{Z})$ to $x \in \mathbb{Z}$ for $\rho_\sigma(x) = \exp(\frac{-x^2}{2\sigma^2})$ and $\rho_\sigma(\mathbb{Z}) = 1 + 2 \sum_{x=1}^{\infty} \rho_\sigma(x)$. We denote by $d \xleftarrow{\$} \mathcal{D}_\sigma$ the process of sampling a value d randomly according to $\mathcal{D}_\sigma$. In case S is a finite set, then $s \xleftarrow{\$} S$ means that the value s is sampled according to a uniform distribution over the set S. For an integer $c \in \mathbb{Z}$, we define $[c]_{2^d}$ to be the integer in the set $(-2^{d-1}, 2^{d-1}]$ such that $c \equiv [c]_{2^d} \mod 2^d$ which is basically extraction of the least significant bits. For $c \in \mathbb{Z}$ we define $\lfloor c \rfloor_d = (c - [c]_{2^d})/2^d$ to drop the d least significant bits. Both operators can also be applied to vectors.

Lattices. A k-dimensional lattice Λ is a discrete additive subgroup of $\mathbb{R}^m$ containing all integer linear combinations of k linearly independent vectors $\mathbf{b}_1, \ldots, \mathbf{b}_k$ with $k \leq m$ and $m \geq 0$. More formally, we have $\Lambda = \{ \mathbf{B} \cdot \mathbf{x} \mid \mathbf{x} \in \mathbb{Z}^k \}$. Throughout this paper we are mostly concerned with q-ary lattices $\Lambda_q^\perp(\mathbf{A})$ and $\Lambda_q(\mathbf{A})$, where $q = poly(n)$ denotes a polynomially bounded modulus and $\mathbf{A} \in \mathbb{Z}_q^{n \times m}$ is an arbitrary matrix. $\Lambda_q^\perp(\mathbf{A})$ resp. $\Lambda_q(\mathbf{A})$ are defined by

$$\Lambda_q^\perp(\mathbf{A}) = \{\mathbf{x} \in \mathbb{Z}^m \mid \mathbf{A}\mathbf{x} \equiv \mathbf{0} \mod q\}$$
$$\Lambda_q(\mathbf{A}) = \{\mathbf{x} \in \mathbb{Z}^m \mid \exists \mathbf{s} \in \mathbb{Z}^m \text{ s.t. } \mathbf{x} = \mathbf{A}^\top \mathbf{s} \mod q\}.$$

By $\lambda_i(\Lambda)$ we denote the *i-th successive minimum*, which is the smallest radius r such there exist i linearly independent vectors of norm r (typically l_2 norm) in Λ. For instance, $\lambda_1(\Lambda) = \min_{\mathbf{x} \neq 0} ||\mathbf{x}||_2$ denotes the minimum distance of a lattice determined by the length of its shortest nonzero vector.

The SIS and LWE Problem. In the following we recall the main problems used in order to construct secure lattice-based cryptographic schemes.

Definition 1 (SIS-Problem). *Given a matrix* $\mathbf{A} \in \mathbb{Z}_q^{n \times m}$, *a modulus* $q > 0$, *and a real* β, *the small-integer-solution problem (l-norm typically* $l = 2$*)* $SIS_{n,m,\beta}$ *asks to find a vector* $\mathbf{x}$ *such that* $\mathbf{A}\mathbf{x} \equiv \mathbf{0} \mod q$ *and* $||\mathbf{x}||_l \leq \beta$.

Let χ be a distribution over $\mathbb{Z}$. We define by $A_{\mathbf{s},\chi}$ the distribution of $(\mathbf{a}, \mathbf{a}^\top \cdot \mathbf{s} + e) \in \mathbb{Z}_q^n \times \mathbb{Z}_q$ for $n, q > 0$, where $\mathbf{a} \xleftarrow{\$} \mathbb{Z}_q^n$ is chosen uniformly at random and $e \leftarrow \chi$.

Definition 2 (LWE-Problem). *For a modulus* $q = poly(n)$ *and given vectors* $(\mathbf{a}_i, b_i) \in \mathbb{Z}_q^n \times \mathbb{Z}_q$ *sampled according to* $A_{\mathbf{s},\chi}$ *the learning-with-errors problem* $LWE_{\chi,q}$ *asks to distinguish* $A_{\mathbf{s},\chi}$, *where* $\mathbf{s}$ *is chosen uniformly at random, from the uniform distribution on* $\mathbb{Z}_q^n \times \mathbb{Z}_q$.

It is also possible to sample $\mathbf{s}$ according to the error distribution χ^n [3].

Departing from the original definition of LWE, that gives access to arbitrary many samples, an attacker has often only access to a maximum number of samples. Typically, this number of samples is denoted by m. In this case, one typically "collects" all samples $\mathbf{a}_i, b_i \in \mathbb{Z}_q^n \times \mathbb{Z}_q$ to $\mathbf{A}, \mathbf{b} \in \mathbb{Z}_q^{m \times n} \times \mathbb{Z}_q^m$, and the LWE problem is to decide whether the entries of $\mathbf{b}$ were sampled uniformly at random and independently from $\mathbf{A}$ or according to the LWE distribution.

3   The Bai-Galbraith Signature Scheme

The Bai-Galbraith digital signature scheme [6] (BG signature) is based on the Fiat-Shamir paradigm which transforms an identification scheme into a signature scheme [18] and closely follows previous proposals by Lyubashevsky et al. [16,21,28,29]. The hardness of breaking the BG signature scheme, in the random oracle model, is reduced to the hardness of solving standard worst-case computational assumptions on lattices. The explicit design goal of Bai and Galbraith is having short signatures.

3.1   Description of the BG Signature Scheme

For easy reference, the key generation, signing, and the verification algorithm of the BG signature scheme are given in Fig. 1. Our proposed parameter set is summarized in Table 1. An analysis of the original parameter sets can be found in the full online version of this paper. However, the algorithms have been simplified and redundant definitions have been removed (e.g., we just use σ as standard deviation and do not differentiate between $\sigma_{\mathbf{E}}, \sigma_{\mathbf{S}}$ and set $n = k$).

During key generation two secret matrices $\mathbf{S} \in \mathbb{Z}^{n \times n}, \mathbf{E} \in \mathbb{Z}^{m \times n}$ are sampled from a discrete Gaussian distribution $D_\sigma^{n \times n}$ and $D_\sigma^{m \times n}$, respectively. A rejection condition CHECK_E enforces certain constraints on $\mathbf{E}$, which are necessary for correctness and short signatures (see Sect. 3.2). Finally, the public key $\mathbf{T} = \mathbf{AS} + \mathbf{E}$ and the secret key matrices $\mathbf{S}, \mathbf{E}$ are returned where $\mathbf{AS}$ is the only matrix-matrix multiplication necessary in the scheme. As we choose $\mathbf{A} \in \mathbb{Z}^{m \times n}$ as a global constant, it does not have to be sampled during key generation and is also not included in the public key and secret key.

For signing, the global constant $\mathbf{A}$ as well as secret keys $\mathbf{S}, \mathbf{E}$ are required (no usage of $\mathbf{T}$ in this variant). The vector $\mathbf{y}$ is sampled uniformly random from $[-B, B]^n$. For the instantiation of the random oracle H (using a hash function) only the higher order bits of $\mathbf{Ay}$ are taken into account and hashed together with the message μ. The algorithm $F(c)$ takes the binary output of the hash c and produces a vector $\mathbf{c}$ of weight ω (see [16] for a definition of $F(c)$). In a different way than [6] $\mathbf{w}$ is computed following an idea that has also been applied in [21]. Instead of computing $\mathbf{w} = \mathbf{Az} - \mathbf{Tc} \ (\bmod \ q)$ we calculate $\mathbf{w} = \mathbf{v} - \mathbf{Ec} \ (\bmod \ q)$, where $\mathbf{v} = \mathbf{Ay} \ (\bmod \ q)$. This is also the reason why $\mathbf{E}$ has to be included into the secret key $sk = (\mathbf{S}, \mathbf{E}) \in \mathbb{Z}^{n \times n} \times \mathbb{Z}^{m \times n}$. Thus, the large public key $\mathbf{T} \in \mathbb{Z}^{m \times n}$ is not needed anymore for signing and the operations become simpler (see further discussion in Sect. 5). The test whether

Algorithm KeyGen	Algorithm Sign	Algorithm Verify
INPUT:	INPUT:	INPUT:
$\mathbf{A}, n, m, q, \sigma$	$\mu, \mathbf{A}, \mathbf{S}, \mathbf{E}, B, U, d, w, \sigma$	$\mu, \mathbf{z}, c, \mathbf{A}, \mathbf{T}, B, U, d$
OUTPUT: $sk = (\mathbf{S}, \mathbf{E}), pk = (\mathbf{T})$	OUTPUT: $(\mathbf{z}, c)$	OUTPUT: Accept/Reject
1. $\mathbf{S} \xleftarrow{\$} D_\sigma^{n \times n}$	1. $\mathbf{y} \xleftarrow{\$} [-B, B]^n$	1. $\mathbf{y} \xleftarrow{\$} [-B, B]^n$
2. $\mathbf{E} \xleftarrow{\$} D_\sigma^{m \times n}$	2. $\mathbf{v} = \mathbf{Ay} \,(\bmod\, q)$	2. $\mathbf{v} = \mathbf{Ay} \,(\bmod\, q)$
3. if CHECK_E($\mathbf{E}$) = 0	3. $c = H(\lfloor \mathbf{v} \rfloor_d, \mu)$	3. $c = H(\lfloor \mathbf{v} \rfloor_d, \mu)$
then Restart	4. $c = F(c)$	4. $c = F(c)$
4. $\mathbf{T} = \mathbf{AS} + \mathbf{E} \,(\bmod\, q)$	5. $\mathbf{z} = \mathbf{y} + \mathbf{Sc}$	5. $\mathbf{z} = \mathbf{y} + \mathbf{Sc}$
5. return $sk = (\mathbf{S}, \mathbf{E}), pk = (\mathbf{T})$	6. $\mathbf{w} = \mathbf{v} - \mathbf{Ec} \,(\bmod\, q)$	6. $\mathbf{w} = \mathbf{v} - \mathbf{Ec} \,(\bmod\, q)$
	7. if $|[\mathbf{w}_i]_{2^d}| > 2^{d-1} - L$	7. if $|[\mathbf{w}_i]_{2^d}| > 2^{d-1} - L$
	then Restart	then Restart
	8. return $(\mathbf{z}, c)$	8. return $(\mathbf{z}, c)$
	if $\|\mathbf{z}\|_\infty \le B - U$	if $\|\mathbf{z}\|_\infty \le B - U$

Fig. 1. The BG signature scheme [6]; see Sect. 3.2 for implementations of CHECK_E.

$|[\mathbf{w}_i]_{2^d}| > 2^{d-1} - L_{BG}$ ($L_{BG} = 7w\sigma$ in [6]) ensures that the signature verification will not fail on a generated signature ($\mathbf{w}$ is never released) and the last line ensures that the signature is uniformly distributed within the allowed range $[-B + U, B - U]^n$ for $U = 14 \cdot \sigma\sqrt{w}$.

For verification the higher order bits of $\mathbf{w} = \mathbf{Az} - \mathbf{Tc} = \mathbf{Ay} - \mathbf{Ec}$ are hashed and a valid signature $(\mathbf{z}, c)$ is accepted if and only if $\mathbf{z}$ is small, i.e., $\|\mathbf{z}\|_\infty \le B - U$, and $c = c'$ for $c' := H(\lfloor \mathbf{w} \rfloor_d, \mu)$. For the security proof and standard attacks we refer to the original work [6].

3.2 Optimizing Rejection Sampling

In the original signature scheme [6] CHECK_E_{BG} restarts the key generation if $|\mathbf{E}_{i,j}| > 7\sigma$ for any (i, j) and the rejection condition in Line 7 of Sign is $|[\mathbf{w}_i]_{2^d}| > 2^{d-1} - L_{BG}$ for $L_{BG} = 7w\sigma$. This ensures that it always holds that $\lfloor \mathbf{Ay} \rfloor_d = \lfloor \mathbf{Ay} - \mathbf{Ec} \rfloor_d$ and thus verification works even for the short signature. However, in practice the acceptance probability of $(1 - 14w\sigma/2^d)^m$ has a serious impact on performance and leaves much room for improvement. On first sight it would seem most efficient to test during signing whether $\lfloor \mathbf{Ay} \rfloor_d = \lfloor \mathbf{Ay} - \mathbf{Ec} \rfloor_d$ and just reject signatures that would not be verifiable. However, in this case the proof structure given in the full version of [6] does not work anymore. In Game 1, sign queries are replaced by a simulation (in the random oracle model) which is not allowed to use the secret key and later on has to produce valid signatures even for an invalidly chosen public key (Game 2).

Our optimization (similar to [16]) is to reject $\mathbf{E}$ during key generation only if the error generated by $\mathbf{Ec}$ in $\lfloor \mathbf{Ay} \rfloor_d = \lfloor \mathbf{Ay} - \mathbf{Ec} \rfloor_d$ for the worst-case c is larger than a threshold L. Thus, our CHECK_E_{new} algorithm works the following: Using $\max_k(\cdot)$ which returns the k-th largest value of a vector we compute thresholds $t_h = \sum_{k=1}^{w} \max_k(|\mathbf{E}_h|), \forall h \in [0, m]$ where $\mathbf{E}_h$ is the h-th row of $\mathbf{E}$ and reject if one or more t_h are larger than L. Thus the rejection probability for the close-to-uniform $\mathbf{w}$ is independent of $\mathbf{c}$ and $\mathbf{E}$ and does not leak any information. When L is chosen such that only a small percentage of secret keys are rejected the LWE instances generated by the public key are still hard due to the same argument on the bounded number of samples as in [6,16]. The acceptance probability of $\mathbf{w}$ in Line 7 of Sign is $(1 - 2L/2^d)^m$. Table 1 shows concrete values for our choice of L_{new} and the original L_{BG}.

4 Security Analysis and Parameter Selection

In the original work [6], Bai and Galbraith proposed five different parameter sets to instantiate their signature scheme. In this section we revisit their security analysis and propose a new instantiation that is optimized for software implementations on modern server and desktop computers (Intel/AMD) and also mobile processors (ARM). The security analysis has been refined due to the following reasons: First, a small negligence in the assessment of the underlying LWE instances leads to a slightly wrong hardness estimation, which was acknowledged by the authors after publication [5]. Second, an important attack, namely the decoding attack, was not considered in [6]. We justify that indeed the decoding attack is less efficient than the one considered if one takes into account the limited number of samples m given to the attack algorithms.

In Table 1 we propose a parameter set for an instantiation of the signature scheme from Sect. 3 with 128 bits of security, for which we provide evidence in the next section.

4.1 Hardness of LWE

The decoding attack dates back to the nearest-plane algorithm by Babai [4] and was further improved by Lindner and Peikert in [26] and Liu and Nguyen in [27]. While it is often the fastest known approach, it turns out that it is not very suitable for our instances, because an attacker has only access to a few samples. Thus we concentrate on the embedding approach here and an analysis of the behavior of the decoding attack can be found in Appendix A.

The embedding approach solves LWE via a reduction to the unique-shortest-vector problem (uSVP). We will analyze two variants, the standard embedding approach [26] and the variant that is very suitable for LWE instances with small m that was already considered in [6]. Unfortunately, it is necessary to re-do the analysis, because the hardness evaluation in the original work [6] set some constant – namely τ – in the attack wrong yielding up to 17 bits more security for their parameters than actually offered. We will focus on the security of our parameter set in this section. Updated values for some of the parameter sets proposed in the original paper can be found in the full version of this paper.

Embedding Approach. Given an LWE instance $(\mathbf{A}, \mathbf{b})$ such that $\mathbf{As} = \mathbf{b} \mod q$, the idea of the embedding approach proposed in [19] is to use the

Table 1. The parameter set we use for 128 bits of security. Note that signature and key sizes refer to fully compressed signature and keys. Our software uses slightly a larger (padded) signature and keys to support faster loads and stores aligned to byte boundaries.

Parameter Selection		
Parameter	Bound	Value
n		532
m		840
σ		43
ω	$2^{\omega}\binom{n}{\omega} \geq 2^{128}$	18
d	d is s.t. $(1 - 14\sigma\omega/2^d)^m \geq 1/3$	23
B	power of two $\geq 14\sqrt{\omega}\sigma(n-1)$	$2^{21} - 1$
q	$\geq \left(2^{(d+1)m+\kappa}/(2B)^n\right)^{1/(m-n)}$	$2^{29} - 3$
U	$14 \cdot \sigma\sqrt{\omega}$ (Prob. of acceptance Line 8 of Sign: 0.51)	2554.1
L_{BG}	$7w\sigma$ (Prob. of acceptance Line 3 of KeyGen: ≈ 1) (Prob. of acceptance Line 7 of Sign: 0.337)	5418
L_{new}	$3w\sigma$ (Prob. of acceptance Line 3 of KeyGen: 0.99) (Prob. of acceptance Line 7 of Sign: 0.628)	2322
public-key size	$m \cdot n \cdot \lceil\log_2(q)\rceil$	1.54 Mb
secret-key size	$(n^2 + n \cdot m)\lceil\log_2(14 \cdot \sigma)\rceil$	0.87 Mb
signature size	$n \cdot \lceil\log_2(2B)\rceil + 256$	11960 bits

embedding lattice $\Lambda_q(\mathbf{A}_e)$ defined as

$$\Lambda_q(\mathbf{A}_e) = \{\mathbf{v} \in \mathbb{Z}^m \mid \exists \mathbf{x} \in \mathbb{Z}^n : \mathbf{A}_e \cdot \mathbf{x} = \mathbf{v} \mod q\},$$

where $\mathbf{A}_e = \begin{pmatrix} \mathbf{A} & \mathbf{b} \\ \mathbf{0} & 1 \end{pmatrix}$. Throughout the paper the subscript stands for the technique used in an attack such as e denoting the standard embedding approach. Since

$$\mathbf{A}_e \begin{pmatrix} -\mathbf{s} \\ 1 \end{pmatrix} = \begin{pmatrix} \mathbf{A} & \mathbf{b} \\ \mathbf{0} & 1 \end{pmatrix}\begin{pmatrix} -\mathbf{s} \\ 1 \end{pmatrix} = \begin{pmatrix} -\mathbf{As} + \mathbf{b} \\ \mathbf{0} \cdot \mathbf{s} + 1 \cdot 1 \end{pmatrix} = \begin{pmatrix} \mathbf{e} \\ 1 \end{pmatrix} =: \mathbf{v}$$

is a very short lattice vector, one can apply a solver for uSVP to recover $\mathbf{e}$. We estimate the norm of $\mathbf{v}$ via $||\mathbf{v}|| \approx ||\mathbf{e}|| \approx \sqrt{m}\sigma_E$, and for the determinant of the lattice we have $\det(\Lambda_q(\mathbf{A}_e)) = q^{m+1-n}$ with very high probability [9].

It is known that the hardness of uSVP depends on the gap between the first and the second successive minimum $\lambda_1(\Lambda)$ and $\lambda_2(\Lambda)$, respectively. Gama and Nguyen [19] claim that an attack with a lattice-reduction algorithm that achieves Hermite factor δ succeeds with high probability if $\lambda_2(\Lambda)/\lambda_1(\Lambda) \geq \tau \cdot \delta^{\dim(\Lambda)}$,

Table 2. Security of our parameter set

Security level		
Problem	Attack	Bit security
LWE	Decoding [26]	271
	Embedding [2]	192
	Embedding [6]	130
SIS	Lattice reduction [6]	159

where $\tau \approx 0.4$ is a constant that depends on the reduction algorithm used. In fact, this factor is missing in the analysis by Bai and Galbraith, which causes too optimistic (i.e., too large) runtime predictions.

The successive minima of a random lattice Λ can be predicted by the Gaussian heuristic via

$$\lambda_i(\Lambda) \approx \frac{\Gamma(1 + \dim(\Lambda)/2)^{1/\dim(\Lambda)}}{\sqrt{\pi}} \det(\Lambda)^{1/\dim(\Lambda)}.$$

Consequently, a particular short vector $\mathbf{v}$ of length $||\mathbf{v}|| = l$ can be found if

$$\delta^{\dim(\Lambda)} \leq \frac{\lambda_2(\Lambda)}{\lambda_1(\Lambda) \cdot \tau} \approx \frac{\Gamma(1 + \dim(\Lambda)/2)^{1/\dim(\Lambda)}}{l \cdot \sqrt{\pi} \cdot \tau} \det(\Lambda)^{1/\dim(\Lambda)}. \tag{1}$$

We can therefore estimate the necessary Hermite delta to break LWE with the embedding approach to be

$$\delta \approx \left(\frac{\Gamma(1 + \frac{m+1}{2})^{1/(m+1)}}{\sqrt{\pi} \cdot m \cdot \tau \cdot \sigma_E} q^{\frac{m+1-n}{m+1}} \right)^{1/(m+1)},$$

where the dimension is set to $\dim(\Lambda_q(\mathbf{A}_e)) = m + 1$. Note that it is possible to apply this attack in a smaller subdimension. In fact, there exists an optimal dimension that minimizes δ in Eq. (1). Our parameters, however, do not provide enough LWE samples to allow an attack in the optimal dimension, and in this case choosing the highest possible dimension seems to be optimal.

To achieve a small Hermite delta, it is necessary to run a basis-reduction algorithm like BKZ [37] or its successor BKZ 2.0 [14]. Lindner and Peikert [26] proposed the function

$$\log_2(T(\delta)) = 1.8/\log_2(\delta) - 110$$

to predict the time necessary to achieve a given Hermite delta by BKZ. More recently, Albrecht et al. [2] proposed the prediction

$$\log_2(T(\delta)) = 0.009/\log_2(\delta)^2 - 27$$

based on data taken from experiments with BKZ 2.0 [27]. We will stick to this estimation in the following, since it takes more recent improvements into consideration. Combining it with the fact that they run their experiments on a machine that performs about $2.3 \cdot 10^9$ operations per second, we estimate the number of operations necessary to achieve a given Hermite factor with

$$T(\delta) = \frac{2.3 \cdot 10^9}{2^{27}} \cdot 2^{0.009/\log(\delta)^2}. \tag{2}$$

We can therefore conclude that our LWE instance provides about 192 bits of security against the embedding attack, which corresponds to a Hermite delta of approximately 1.0048.

The efficacy of the standard embedding approach decreases significantly if the instance does not provide enough samples for the attack to run in the optimal dimension. Another attack, which is very suitable for LWE instances with few samples, reduces LWE to an uSVP instance defined by the lattice $\Lambda_q^\perp(\mathbf{A}_o) = \{\mathbf{v} \in \mathbb{Z}^{m+n+1} \mid \mathbf{A}_o \cdot \mathbf{v} = \mathbf{0} \mod q\}$ for $\mathbf{A}_o = [\mathbf{A} \mid \mathbf{I} \mid \mathbf{b}]$ (we use the index o because this attack runs in the lattice of the vectors that are orthogonal to $\mathbf{A}_o$). The main advantage of this attack is that it runs in dimension $n + m + 1$ (recall that the standard embedding approach runs in dimension $m + 1$). For $\mathbf{v} = (\mathbf{s}, \mathbf{e}, -1)^T$, we have $\mathbf{A}_o \cdot \mathbf{v} = \mathbf{A} \cdot \mathbf{s} + \mathbf{e} - \mathbf{b} = \mathbf{0}$ and therefore $\mathbf{v} \in \Lambda_q^\perp(\mathbf{A}_o)$ is a small vector in the lattice. We estimate its length via $||\mathbf{v}|| \approx \sqrt{||\mathbf{s}||^2 + ||\mathbf{e}||^2} \approx \sqrt{m+n} \cdot \sigma$. Since $\det(\Lambda_q(\mathbf{A}_o)) = q^m$ with high probability [9], Eq. (1) predicts the necessary Hermite delta to be approximately

$$\delta \approx \left(\frac{\Gamma(1 + \frac{n+m+1}{2})^{1/(n+m+1)}}{\sqrt{n + m}\sigma \cdot \sqrt{\pi} \cdot \tau} q^{\frac{m}{n+m+1}} \right)^{1/(n+m+1)}.$$

Using Eq. (2), we can estimate the hardness of our instance against this attack to be about 130 bits (the Hermite delta is close to 1.0059).

4.2 Hardness of SIS

Instead of recovering the secret key, which corresponds to solving an instance of LWE, an attacker could also try to forge a signature directly and thus solve an SIS instance. We predict the hardness of SIS for the well-known lattice-reduction attack (see for example [9]) like it was done in [6]. This attack views SIS as a variant of the (approximate) shortest-vector problem and finds the short vector by applying a basis reduction. Forging a signature through this attack requires to find a reduced basis with Hermite factor

$$\delta = (D/q^{m/(m+n)})^{1/(n+m+1)}, \tag{3}$$

with $D = (\max(2B, 2^{d-1}) + 2E'\omega)$ for E' satisfying $(2E')^{m+n} \geq q^m 2^{132}$. Applying Eq. (2), we estimate that a successful forger requires to perform about 2^{159} operations (see Table 2).

4.3   An Instantiation for Software Efficiency

Choosing optimal parameters for the scheme is a non-trivial multi-dimensional optimization problem and our final parameter set is given in Table 1. Since the probability that the encoding function F maps two random elements to the same value must be negligible (i.e. smaller than 2^{-128}), we choose ω such that $2^{\omega}\binom{n}{\omega} \geq 2^{128}$. Since $\mathbf{Sc}$ is distributed according to a Gaussian distribution with parameter $\sqrt{\omega}\sigma$, we can bound its entries by $14\sqrt{\omega}\sigma$. Consequently, $B-U$ is lower bounded by $14\sqrt{\omega}\sigma(n-1)$ such that the acceptance probability of a signature P_{acc} (Line 8 in Fig. 1) is at least

$$P_{acc} = \left(\frac{2(B-U)+1}{2B}\right)^{m} = \left(\frac{2\cdot 14\sqrt{\omega}\sigma(n-1)+1}{2\cdot 14\sqrt{\omega}\sigma n+1}\right)^{m} \approx \left(1-\frac{1}{n}\right)^{m} \approx 1/e.$$

The next important choice to be made is the value for the parameter d. It has a determining influence on the trade-off between runtime and key sizes: The success probability in the signing algorithm (Line 7 in Fig. 1) is given by $(1-2L/2^d)^m$, which means that large values for d lead to a high success probability, and thereby to fewer rejections implying better running times. On the other hand, the security proof requires $(2B)^n q^{m-n} \geq 2^{(d+1)m+\kappa}$ to be satisfied, which means that increasing d implies larger values for q, hence, worsening runtime and key sizes.

Our goal is to come up with a parameter set that ensures at least 128 bits of security. We will focus on n,m and σ in this paragraph, since the other parameters depend on them. For easy modular reduction we choose a modulus slightly smaller than a power of two (like $2^{29}-3$). Furthermore, dimensions n resp. m are multiples of 4 to support four parallel operations in vector registers. In a way, n determines the overall security level, and the choice of σ and n can be used to balance the security of the scheme and the size of the second-order parameters q and B. Using our parameters we have set $L = L_{new} = 3\omega\sigma$ and thus reject a secret key with probability 0.025 and accept with probability $(1-2L/2^d)^m$ where we get ≈ 0.63 instead of ≈ 0.34 for $L_{BG} = 7\sigma\omega$.

For instance, Fig. 2 shows for $n = 532$ how the lower bound on q depends on σ for various values of m. Since too small values of σ lead to LWE-instances that are significantly easier than 128 bits, the best possible choice that allows $q = 2^{29}-3$ is $m = 840$ and $\sigma = 43$. We further choose $n = 532$ which leads to $\omega = 18$. This results in the lower bound $\log_2(B) \geq 20.4$, which allows our choice $B = 2^{21}-1$.

5   Implementation Details

In this section we discuss our techniques used for high performance on modern desktop and mobile CPUs with fast vector units. More specifically, we optimized the signature scheme for Intel Ivy Bridge CPUs with AVX, for Intel Haswell CPUs with AVX2 and for ARMv7 CPUs with NEON vector instructions. We first describe various high-level (platform-independent) optimizations for signing and

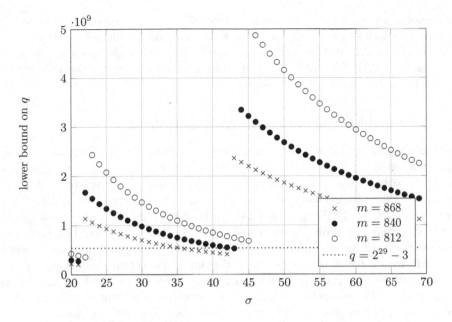

Fig. 2. Lower bound on q for $n = 532$ and various values of m

verification and then detail the low-level implementation techniques for the three target platforms. Our implementation only optimizes signing and verification speeds; our implementation includes a (slow) routine for key generation but we will not discuss key generation here.

5.1   High-Level Optimizations

Regarding platform independent high-level optimizations we follow the approach from [22] and would like to emphasize the changes to the algorithm (adding $\mathbf{E}$ to the private key and choosing $\mathbf{A}$ as global constant) and improved rejection sampling (usage of L_{new}) as discussed in Sect. 3. For uniform sampling of $\mathbf{y} \xleftarrow{\$} [-B, B]^n$ during signing we rely on the hybrid approach of seeding the Salsa20 stream cipher using true randomness from the Linux random number [22]. As $B = 2^{21} - 1$ we sample $3n + 68$ uniform bytes at once using Salsa20 and construct a sample r' from 3 bytes each. By computing $r = r' \bmod 2^{22}$ we bring r into the range $[0, 2^{22} - 1]$, reject if $r = 2^{22} - 1$ and return $r - (2^{22} - 1)$. The probability to discard an element is 2^{-22} and by oversampling 68 bytes it is highly unlikely that we have to sample additional randomness. We also exploit that $\mathbf{c}$ is sparse with weight ω. Thus, we store $\mathbf{c}$ not as a vector but as list with ω tuples containing the position and sign bits of entries which are non zero. Additionally, when multiplying $\mathbf{c}$ with $\mathbf{S}$ and $\mathbf{E}$, only a small subset of coefficients from $\mathbf{S}, \mathbf{E}$ is actually needed. As a consequence, we do not unpack the whole matrices $\mathbf{S}, \mathbf{E}$ from the binary representation of the secret key (which is the usual approach) but just the coefficients that are required in this case. Additionally, during signing we perform rejection sampling on $\mathbf{w}$ before we actually compute $\mathbf{v}$ in order to be able to abort as early as possible (without leaking timing information). For hashing $H(\lfloor\mathbf{v}\rceil_d, \mu)$ and $H(\lfloor\mathbf{w}\rceil_d, \mu)$, respectively, we pack the input to the hash function after extraction of higher-order bits in order to keep the input buffer to the hash function as small as possible.

5.2   Low-Level Optimizations in AVX and AVX2

With the Sandy Bridge microarchitecture, Intel introduced the AVX instruction set. AVX extends the 16 128-bit XMM vector registers of the SSE instruction set to 256-bit YMM registers. Arithmetic instructions treat these registers either as vectors of 4 double-precision or 8-single precision floating-point numbers. Each cycle, the Intel Sandy Bridge and later Ivy Bridge CPUs can issue one addition instruction and one multiplication instruction on those vectors. The power of these vector-arithmetic units was exploited by [22] to achieve very high speeds for GLP signatures. We also use these floating-point vector operations for our software. With the Haswell microarchitecture, Intel introduced AVX2, which extends the AVX instruction set. There are two notable additions. One is that vector registers can now also be treated as vectors of integers (of various sizes); the other is that Intel added floating-point multiply-accumulate instructions. Haswell CPUs can issue two floating-point multiply-accumulate vector instructions per cycle.

The basic approach for our implementation is that all elements of $\mathbb{Z}_q$ are represented as double-precision floating-point numbers. The mantissa of a double-precision float has 53 bits and a 29-bit integer can thus obviously be represented exactly. One might think that 53 bits are still not enough, because *products* of elements of $\mathbb{Z}_q$ do not fit into the mantissa. However, the signature scheme never computes the product of two full-size field elements. The largest products appear in the matrix-vector multiplications $\mathbf{Ay}$ and $\mathbf{Az}$. The coefficients of $\mathbf{A}$ are full-size $\mathbb{Z}_q$ elements in the interval $[-(q-1)/2, (q-1)/2]$, but the coefficients of $\mathbf{y}$ are in $[-B, B]$ and the coefficients of $\mathbf{z}$ are in $[-(B-U), B-U]$. With $B = 2^{21} - 1$ each coefficient multiplication in $\mathbf{Ay}$ produces a result of at most 49 bits.

Matrix-vector multiplication. The matrix-vector multiplications $\mathbf{Ay}$ and $\mathbf{Az}$ are not only the operations which produce the largest intermediate results, they are also the operations which dominate the cost for signing and verification, respectively. The AVX and AVX2 implementations store the matrix $\mathbf{A}$ in transposed form which allows more efficient access to the elements of $\mathbf{A}$ in vector registers. One can think of the whole computation as a sequence of multiply-accumulate instructions, where one factor is a vector register containing 4 coefficients of $\mathbf{A}$, the other factor is a vector register containing 4 copies of the same coefficient of $\mathbf{y}$ (or $\mathbf{z}$) and the accumulator is a vector register containing 4 result coefficients. Loading the same coefficient of $\mathbf{y}$ into all 4 elements of a vector register can be done efficiently through the vbroadcastsd instruction. Latencies can be hidden by interleaving instructions from the computation of independent vectors of result coefficients.

One might think that $n \cdot m = 532 \cdot 840 = 446880$ multiplications and accumulations translate into 111720 AVX and 55860 AVX2 cycles (because AVX handles 4 vectorized multiplications and 4 vectorized additions per cycle and AVX2 handles 2×4 vectorized multiply-accumulates per cycle), but this is not the case. It turns out that arithmetic is not the bottleneck but access to matrix coefficients. Note that if we store $\mathbf{A}$ as 446880 double-precision floats, the matrix occupies about 3.5 MB of storage – way too much for the 32 KB L1 cache. Also note that each matrix coefficient is used exactly once, which is the most cache-unfriendly access pattern possible. We overcome this bottleneck to some extent by storing the coefficients of $\mathbf{A}$ as 32-bit integers. We then load 4 coefficients (and convert to double-precision floats on the fly) using the vcvtdq2pd instruction of the AVX instruction set. An additional cost stems from reductions modulo q of coefficients. We can use lazy-reduction, i.e., we do not have to reduce after every multiply-accumulate. For example in the computation of $\mathbf{Ay}$ we have to reduce after 16 multiply-accumulate operations. Our software is currently overly conservative and reduces after 7 multiply-accumulates in both cases. We perform modular reduction of floating-point coefficients in the same way as [22]: We produce a "carry" by multiplying by a floating-point approximation of q^{-1}, then use the vroundpd instruction to round that carry to the nearest integer, multiply by q and then subtract the carry from the original value.

In total, the matrix-vector multiplication takes 278912 cycles on a Haswell CPU and 488474 cycles on an Ivy Bridge CPU.

5.3 Low-Level Optimization in NEON

Fast vector units are not only present in large desktop and server CPUs but also in mobile CPUs. Most ARM Cortex-A processors include the NEON vector extensions. These extensions add 16 128-bit vector registers. The most powerful arithmetic instructions are addition and subtraction of vectors of 4 32-bit integers or 2 64-bit integers (one per cycle) and multiplication of vectors of 2 32-bit integers producing as a result a vector of 2 64-bit integers (one every two cycles). The NEON instruction set also includes multiply-accumulate at the same cost of a multiplication.

For our optimized NEON implementation we represent elements of $\mathbb{Z}_q$ as 32-bit signed integers. Products of coefficients in the matrix-vector multiplications $\mathbf{Ay}$ and $\mathbf{Az}$ are represented as 64-bit signed integers. Lazy reduction can go much further than in AVX and AVX2; we only have to perform one reduction modulo q at the very end of the computation.

In most aspects, the NEON implementation follows the ideas of the AVX and AVX2 implementations, but two aspects are different. One aspect is that simply storing the transpose of $\mathbf{A}$ is not sufficient for efficient vectorized access to the elements of $\mathbf{A}$. The reason is that the ARM-NEON addressing modes are by far not as flexible as the Intel addressing modes. Therefore, we store the matrix $\mathbf{A}$ such that each vector load instruction can simply pick up the next 4 coefficients of $\mathbf{A}$ and then increment the pointer to $\mathbf{A}$ as part of the load instruction.

The other aspect is modular reduction. In NEON we are operating on integers so the modular reduction technique we use for floats in AVX and AVX2 does not work. This is where the special shape of $q = 2^{29} - 3$ comes into play. Reduction modulo q on integers can be achieved with various different approaches, we currently use one shift, a logical and, and three additions to reduce modulo q. Obviously we always reduce two coefficients in parallel using vector instructions.

The penalty for access to coefficients of **A** is even higher than on the Intel platforms. Instead of 446880 cycles which one might expect from an arithmetic lower bound, matrix-vector multiplication takes 2448008 cycles.

6   Results and Comparison

Our software follows the eBACS API [10] and we will submit the software to eBACS for public benchmarking. In this section we do *not* report cycle counts obtained by running the eBACS benchmarking framework SUPERCOP. The reason is the same as in [22]: eBACS reports median cycle counts which is much too optimistic for the signing procedure which includes rejection sampling. Instead, we benchmark 10,000 signature generations and report the average of those measurements. Verification does not include any rejection sampling and we thus report the more stable median of 10,000 measurements.

We benchmarked our software on three machines, namely

- A machine with an Intel Core i7-4770K (Haswell) CPU running Debian GNU/Linux with gcc 4.6.3. Compilation used compiler flags `-msse2avx -march=corei7-avx -O3 -std=gnu99`.
- A machine with an Intel Core i5-3210M (Ivy Bridge) CPU running Ubuntu GNU/Linux with gcc 4.6.3. Compilation used compiler flags `-msse2avx -march=corei7-avx -O3 -std=gnu99`.
- A Beaglebone Black development board with a TI Sitara AM335x (ARM Cortex-A8) CPU running Debian GNU/Linux with gcc 4.6.3. Compilation used compiler flags `-O3 -flto -march=armv7-a -Ofast -funroll-all-loops -marm -mfpu=neon -fprefetch -loop-arrays-mvectorize-with-neon-quad -mthumb-interwork -mtune=cortex-a15`.

All benchmarks were carried out on just one core of the CPU and we followed the standard practice of turning off TurboBoost and hyperthreading.

Table 3 reports performance results of our software and compares it to previous implementations of lattice-based signatures. As an additional contribution of this paper we improved the performance of the software presented in [22]. We report both the original and the improved cycle counts in Table 3. For details on the improvement we refer to the full version of this paper. Compared with our work it becomes clear that usage of standard lattices only incurs a small performance penalty. This is remarkable, as no efficient and quasi-logarithmic-runtime arithmetic like the number-theoretic transform (NTT) is available for standard lattices. Moreover, for a security level matching the security level of GLP we expect our implementation to be much faster (m, n, q could be decreased). For BLISS performance we rely on the implementation given in [16]. However, an implementation of BLISS which uses techniques similar to those described in [22], should be much faster due to smaller parameters and lower rejection rates than in GLP. The main problem of BLISS is that it requires efficient (and secure) sampling of Gaussian noise not only for key generation but also for signing. All efficient techniques for Gaussian sampling rely heavily on secret branch conditions or lookup tables, which are both known to create timing leaks (see [12]).

Table 3. Comparison of lattice-based-signature software performance

Software	CPU	Security	Cycles	
Software using standard lattices				
This work	Intel Core i7-4770K (Haswell)	128 bits	**sign:** **verify:**	1203924 335072
This work	Intel Core i5-3210M (Ivy Bridge)	128 bits	**sign:** **verify:**	1973610 608870
This work	TI Sitara AM335x (ARM Cortex-A8)	128 bits	**sign:** **verify:**	10264721 2796433
`GPV-matrix` [7] ($n = 512, k = 27$)	AMD Opteron 8356 (Barcelona)	100 bits	**sign:** **verify:**	287500000 48300000
Software using ideal lattices				
GLP [22]	Intel Core i5-3210M (Ivy Bridge)	75–80 bits	**sign:** **verify:**	634988 45036
GLP [22] (see full version)	Intel Core i5-3210M (Ivy Bridge)	75–80 bits	**sign:** **verify:**	452223 34004
`GPV-poly` [7] ($n = 512, k = 27$)	AMD Opteron 8356 (Barcelona)	100 bits	**sign:** **verify:**	71300000 9200000
`BLISS` [16] (BLISS-I)	Intel Core i7	128 bits	**sign:** **verify:**	≈ 421600 ≈ 102000
`PASS` [23] ($N = 1153$)	Intel Core i7-2640M (Sandy Bridge)	130 bits	**sign:** **verify:**	584230 172641

Conclusion and future work. With this work we have shown that the performance impact of using standard lattices over ideal lattices for short digital signatures is only small for signing and manageable for verification. Possible future work might consist in evaluating the performance of an independent time implementation of vectorized BLISS or PASS. Moreover, NTRUsign might become interesting again if it is possible to fix the security issues efficiently, as proposed in [1].

Acknowledgment. We would like to thank Patrick Weiden, Rafael Misoczki, Shi Bai, and Steven Galbraith for useful discussions. We would further like to thank the anonymous reviewers for their suggestions and comments.

References

1. Melchor, C.A., Boyen, X., Deneuville, J.-C., Gaborit, P.: Sealing the leak on classical NTRU signatures. In: Mosca, M. (ed.) PQCrypto 2014. LNCS, vol. 8772, pp. 1–21. Springer, Heidelberg (2014). 99
2. Albrecht, M.R., Fitzpatrick, R., Göpfert, F.: On the efficacy of solving LWE by reduction to unique-SVP. Cryptology ePrint Archive, Report 2013/602 (2013). http://eprint.iacr.org/2013/602/. 92
3. Applebaum, B., Cash, D., Peikert, C., Sahai, A.: Fast cryptographic primitives and circular-secure encryption based on hard learning problems. In: Halevi, S. (ed.) CRYPTO 2009. LNCS, vol. 5677, pp. 595–618. Springer, Heidelberg (2009). 87
4. Babai, L.: On Lovász' lattice reduction and the nearest lattice point problem. Combinatorica 6(1), 1–13 (1986). http://www.csie.nuk.edu.tw/~cychen/Lattices/Onlovaszlatticereductionandthenearestlatticepointproblem.pdf. 90, 102
5. Bai, S., Galbraith, S.: Personal communication and e-mail exchanges (2014). 86, 90
6. Bai, S., Galbraith, S.D.: An improved compression technique for signatures based on learning with errors. In: Benaloh, J. (ed.) CT-RSA 2014. LNCS, vol. 8366, pp. 28–47. Springer, Heidelberg (2014). 85, 86, 87, 88, 89, 90, 92, 93, 102
7. El Bansarkhani, R., Buchmann, J.: Improvement and efficient implementation of a lattice-based signature scheme. In: Lange, T., Lauter, K., Lisoněk, P. (eds.) SAC 2013. LNCS, vol. 8282, pp. 48–67. Springer, Heidelberg (2014). 84, 85, 99
8. Bernstein, D.J.: A subfield-logarithm attack against ideal lattices, Feb 2014. http://blog.cr.yp.to/20140213-ideal.html. 85
9. Bernstein, D.J., Buchmann, J., Dahmen, E. (eds.): Post-Quantum Cryptography. Mathematics and Statistics. Springer, Heidelberg (2009). 84, 85, 91, 93
10. Bernstein, D.J., Lange, T.: eBACS: ECRYPT benchmarking of cryptographic systems. http://bench.cr.yp.to. Accessed 25 Jan 2013. 86, 98
11. Boorghany, A., Jalili, R.: Implementation and comparison of lattice-based identification protocols on smart cards and microcontrollers. IACR Cryptology ePrint Archive, 2014. http://eprint.iacr.org/2014/078/. 85
12. Bos, J.W., Costello, C., Naehrig, M., Stebila, D.: Post-quantum key exchange for the TLS protocol from the ring learning with errors problem. IACR Cryptology ePrint Archive (2014). http://eprint.iacr.org/2014/599. 86, 99
13. Brumley, D., Boneh, D.: Remote timing attacks are practical. In: SSYM 2003 Proceedings of the 12th Conference on USENIX Security Symposium. USENIX Association (2003). http://crypto.stanford.edu/dabo/pubs/papers/ssl-timing.pdf. 86
14. Chen, Y., Nguyen, P.Q.: BKZ 2.0: Better lattice security estimates. In: Lee, D.H., Wang, X. (eds.) ASIACRYPT 2011. LNCS, vol. 7073, pp. 1–20. Springer, Heidelberg (2011). 92
15. Couvreur, A., Otmani, A., Tillich, J.P.: Polynomial time attack on wild McEliece over quadratic extensions. In: Nguyen, P.Q., Oswald, E. (eds.) EUROCRYPT 2014. LNCS, vol. 8441, pp. 17–39. Springer, Heidelberg (2014). 85
16. Ducas, L., Durmus, A., Lepoint, T., Lyubashevsky, V.: Lattice signatures and bimodal Gaussians. In: Canetti, R., Garay, J.A. (eds.) CRYPTO 2013, Part I. LNCS, vol. 8042, pp. 40–56. Springer, Heidelberg (2013). 85, 86, 88, 89, 99
17. Dwarakanath, N.C., Galbraith, S.D.: Sampling from discrete Gaussians for lattice-based cryptography on a constrained device. Appl. Algebra Eng. Commun. Comput. 25(3), 159–180 (2014). 86

18. Fiat, A., Shamir, A.: How to prove yourself: practical solutions to identification and signature problems. In: Odlyzko, A.M. (ed.) CRYPTO 1986. LNCS, vol. 263, pp. 186–194. Springer, Heidelberg (1987). 88

19. Gama, N., Nguyen, P.Q.: Predicting lattice reduction. In: Smart, N.P. (ed.) EURO-CRYPT 2008. LNCS, vol. 4965, pp. 31–51. Springer, Heidelberg (2008). 90, 91

20. Göttert, N., Feller, T., Schneider, M., Buchmann, J., Huss, S.: On the design of hardware building blocks for modern lattice-based encryption schemes. In: Prouff, E., Schaumont, P. (eds.) CHES 2012. LNCS, vol. 7428, pp. 512–529. Springer, Heidelberg (2012). 85

21. Güneysu, T., Lyubashevsky, V., Pöppelmann, T.: Practical lattice-based cryptography: a signature scheme for embedded systems. In: Prouff, E., Schaumont, P. (eds.) CHES 2012. LNCS, vol. 7428, pp. 530–547. Springer, Heidelberg (2012). 84, 85, 88

22. Güneysu, T., Oder, T., Pöppelmann, T., Schwabe, P.: Software speed records for lattice-based signatures. In: Gaborit, P. (ed.) PQCrypto 2013. LNCS, vol. 7932, pp. 67–82. Springer, Heidelberg (2013). 85, 86, 95, 96, 97, 98, 99

23. Hoffstein, J., Pipher, J., Schanck, J.M., Silverman, J.H., Whyte, W.: Practical signatures from the Partial Fourier recovery problem. In: Boureanu, I., Owesarski, P., Vaudenay, S. (eds.) ACNS 2014. LNCS, vol. 8479, pp. 476–493. Springer, Heidelberg (2014). 99

24. Ishiguro, T., Kiyomoto, S., Miyake, Y., Takagi, T.: Parallel Gauss Sieve algorithm: solving the SVP challenge over a 128-Dimensional ideal lattice. In: Krawczyk, H. (ed.) PKC 2014. LNCS, vol. 8383, pp. 411–428. Springer, Heidelberg (2014). 85

25. Kocher, P.C.: Timing attacks on implementations of Diffie-Hellman, RSA, DSS, and other systems. In: Koblitz, N. (ed.) CRYPTO 1996. LNCS, vol. 1109, pp. 104–113. Springer, Heidelberg (1996). 86

26. Lindner, R., Peikert, C.: Better key sizes (and attacks) for LWE-based encryption. In: Kiayias, A. (ed.) CT-RSA 2011. LNCS, vol. 6558, pp. 319–339. Springer, Heidelberg (2011). 90, 92, 102, 103

27. Liu, M., Nguyen, P.Q.: Solving BDD by enumeration: an update. In: Dawson, E. (ed.) CT-RSA 2013. LNCS, vol. 7779, pp. 293–309. Springer, Heidelberg (2013). 90, 93, 102, 103

28. Lyubashevsky, V.: Fiat-Shamir with aborts: applications to lattice and factoring-based signatures. In: Matsui, M. (ed.) ASIACRYPT 2009. LNCS, vol. 5912, pp. 598–616. Springer, Heidelberg (2009). 88

29. Lyubashevsky, V.: Lattice signatures without trapdoors. In: Pointcheval, D., Johansson, T. (eds.) EUROCRYPT 2012. LNCS, vol. 7237, pp. 738–755. Springer, Heidelberg (2012). 88

30. Lyubashevsky, V., Peikert, C., Regev, O.: On Ideal Lattices and Learning with Errors over Rings. In: Gilbert, H. (ed.) EUROCRYPT 2010. LNCS, vol. 6110, pp. 1–23. Springer, Heidelberg (2010). 84

31. Misoczki, R., Barreto, P.S.L.M.: Compact McEliece keys from Goppa codes. In: Jacobson Jr, M.J., Rijmen, V., Safavi-Naini, R. (eds.) SAC 2009. LNCS, vol. 5867, pp. 376–392. Springer, Heidelberg (2009). 85

32. Oder, T., Pöppelmann, T., Güneysu, T.: Beyond ECDSA and RSA: Lattice-based digital signatures on constrained devices. In: DAC 2014 Proceedings of the The 51st Annual Design Automation Conference on Design Automation Conference, pp. 1–6. ACM (2014). https://www.sha.rub.de/media/attachments/files/2014/06/bliss_arm.pdf. 85

33. Pöppelmann, T., Ducas, L., Güneysu, T.: Enhanced lattice-based signatures on reconfigurable hardware. In: Batina, L., Robshaw, M. (eds.) CHES 2014. LNCS, vol. 8731, pp. 353–370. Springer, Heidelberg (2014). 85, 86
34. Regev, O.: On lattices, learning with errors, random linear codes, and cryptography. In: Gabow, H.N., Fagin, R. (eds.) STOC 2005 Proceedings of the Thirty-Seventh Annual ACM Symposium on Theory of computing, pp. 84–93. ACM (2005). http://www.cims.nyu.edu/~regev/papers/qcrypto.pdf. 85
35. Roy, S.S., Vercauteren, F., Mentens, N., Chen, D.D., Verbauwhede, I.: Compact ring-LWE cryptoprocessor. In: Batina, L., Robshaw, M. (eds.) CHES 2014. LNCS, vol. 8731, pp. 371–391. Springer, Heidelberg (2014). 85
36. Schneider, M.: Sieving for shortest vectors in ideal lattices. In: Youssef, A., Nitaj, A., Hassanien, A.E. (eds.) AFRICACRYPT 2013. LNCS, vol. 7918, pp. 375–391. Springer, Heidelberg (2013). 85
37. Schnorr, C.-P., Euchner, M.: Lattice basis reduction: Improved practical algorithms and solving subset sum problems. Math. Program. **66**, 181–199 (1994). http://www.csie.nuk.edu.tw/~cychen/Lattices/LatticeBasisReductionImproved PracticalAlgorithmsandSolvingSubsetSumProblems.pdf. 92

A Decoding Attack

An approach for solving LWE that has not been considered in the original work [6] is the decoding attack. It is inspired by the nearest plane algorithm proposed by Babai [4]. For a given lattice basis and a given target vector, it returns a lattice vector that is relatively close to the target vector. Hence, improving the quality of the lattice basis yields a vector that is closer to the target vector. Lindner and Peikert [26] proposed the nearest planes algorithm, a generalization of the former that returns more than one vector and thereby enhances the previous algorithm with a trade-off between its runtime and the probability of returning the actual closest vector within the set of obtained vectors.

There is a continuous correspondence between the success probability of this attack and the Hermite delta. We follow the approach proposed by Lindner and Peikert [26] to predict this success probability. In short, they show how one can use the Geometric Series Assumption (GSA) in order to predict the length of the Gram-Schmidt vectors of a reduced basis, and this estimation in turn serves to predict the success probability of the attack. Together with an estimation of the running time of nearest plane – the authors propose 2^{-16} s – and the runtime estimation for basis reduction (see Eq. (2)), it is possible to predict the runtime and success probability of nearest planes.

Optimizing the trade-offs between the time spent on the attack and its success probability is not trivial, but simulations of the attack show that it is in most cases preferable to run multiple attacks with small success probabilities. This technique is called randomization and was investigated by Liu and Nguyen (see [27]), together with a further improvement called pruning. In comparison to the big improvement achieved with randomization, pruning leads only to a moderate speedup. The maximal speedup achieved in [27] is about 2^6, while randomization can reduce the cost by a factor of 2^{32}. Since it turned out that the decoding-attack is outperformed by other attacks by far (and pruning is furthermore very hard to analyze), we focused on the randomized version.

Briefly speaking, [26] provides the tools necessary to estimate the expected runtime of the attack for a given set of attack parameters, and [27] proposed to minimize the expected runtime (i.e. the time for one attack divided by the success probability of the attack). We applied this technique to our instance (cf. Table 2).

Block Cipher Speed and Energy Efficiency Records on the MSP430: System Design Trade-Offs for 16-Bit Embedded Applications

Benjamin Buhrow[✉], Paul Riemer, Mike Shea, Barry Gilbert,
and Erik Daniel

Mayo Clinic, Rochester, MN, USA
{buhrow.benjamin,riemer.paul,shea.michael,
gilbert.barry,daniel.erik}@mayo.edu

Abstract. Embedded microcontroller applications often experience multiple limiting constraints: memory, speed, and for a wide range of portable devices, power. Applications requiring encrypted data must simultaneously optimize the block cipher algorithm and implementation choice against these limitations. To this end we investigate block cipher implementations that are optimized for speed and energy efficiency, the primary metrics of devices such as the MSP430 where constrained memory resources nevertheless allow a range of implementation choices. The results set speed and energy efficiency records for the MSP430 device at 132 cycles/byte and 2.18 μJ/block for AES-128 and 103 cycles/byte and 1.44 μJ/block for equivalent block and key sizes using the lightweight block cipher SPECK. We provide a comprehensive analysis of size, speed, and energy consumption for 24 different variations of AES and 20 different variations of SPECK, to aid system designers of microcontroller platforms optimize the memory and energy usage of secure applications.

Keywords: AES · SPECK · Lightweight · Encryption · MSP430 · Speed · Energy · Efficient · Measurements · Trade-offs

1 Introduction

Many lightweight block ciphers have been established in recent years in response to the growing use of resource-constrained electronic devices in a wide variety of embedded applications. Examples include TWINE [1], Piccolo [2], Lblock [3], LED [4], PRESENT [5], SIMON, and SPECK [6], in addition to the mainstay AES [7]. Lightweight block ciphers are largely targeted or optimized for small hardware implementations although some are specifically architected to admit software-friendly designs for microcontrollers (e.g., TWINE and SPECK).

Microcontroller based software applications occupy an interesting middle ground: resources are very constrained relative to general purpose 32- or 64-bit processors but they are abundant relative to devices like RFID tags or smart cards. For example, sensor nodes like the MicaZ [8] or TelosB [9] utilize

© Springer International Publishing Switzerland 2015
D.F. Aranha and A. Menezes (Eds.): LATINCRYPT 2014, LNCS 8895, pp. 104–123, 2015.
DOI: 10.1007/978-3-319-16295-9_6

microcontroller devices that offer enough ROM and RAM to implement large lookup tables or to unroll program loops. Further, the programmable nature of microcontrollers provides support for diverse applications, each with a different set of resource requirements, of which the block cipher is typically only a small part. Outside of choosing different block cipher algorithms, the ability to tailor a particular algorithm to the device resources at hand is desirable; for example when an algorithm provides exceptional security or energy efficiency.

Overall, many parameters of block ciphers are important to an embedded system designer such as size, security, speed, and energy efficiency, depending on the application. Varying a block cipher's implementation strategy within embedded devices is a relatively unexplored topic, and one that can provide much in the way of trade-off data to designers. Survey authors do a thorough evaluation over many different ciphers (e.g., [10–12]), but in many cases one implementation strategy (e.g., small size versus high speed or C language versus assembly language [hereafter, abbreviated to assembler]) is chosen per cipher without much discussion. In this paper we quantitatively discuss this issue by measuring multiple implementations of two algorithms: AES and the lightweight block cipher SPECK. These were chosen for two primary reasons: both are expected to have good performance on the MSP430 [13] and each can be implemented in a variety of ways on 16-bit platforms that trade off size for speed. We focus on SPECK over other lightweight block ciphers because (1) it is a very recently proposed cipher and its implementation has not been fully explored on the MSP430 platform and (2) the wide range of block and key sizes in SPECK are interesting from the standpoint of configurablity. The intent is not to promote one algorithm as "better" or "worse" than the other - their designs are sufficiently different to preclude such a comparison. Nor is the intent to provide security analysis of either algorithm, other than by increasing key sizes. The goals are to provide system designers data and analysis over a wider range of size, speed, and energy efficiency than could be obtained from either block cipher alone, and to discuss efficient implementations of each algorithm.

The primary contributions of this paper are the presentation of a matrix of size, speed, and energy consumption data for 8 different implementation strategies of AES coupled with its 3 different key sizes and a first look at similarly thorough results for the entire family of 10 different SPECK parameterizations, for both C and assembler implementations. The thorough analysis across AES implementation strategies on the MSP430 presented here is unavailable elsewhere in the literature, to our knowledge. SPECK is a relatively new cipher and the implementations here represent the most thorough to date. Our fastest 128-bit implementations operate at 132 cycles/byte for AES-128 and 103 cycles/byte for SPECK-128, both setting speed records for 128-bit block ciphers on the MSP430. Measured energy of 2.18 μJ/block for AES and 1.44 μJ/block for SPECK fit in at numbers 5 and 6 in Guneysu's list of top energy efficient AES implementations for any platform [14], but notably the results are considerably more efficient than all other microcontroller platforms tested in that report.

In addition, we provide C and assembler implementation tactics for our AES designs as well as a detailed review of SPECK implementations in C that improve performance relative to conventional approaches. None of the implementation strategies used here for AES are new; however the 16-bit optimization of Gouvea [15] is fairly recent and led to the record speed and energy efficiency results. For all implementations we concentrate solely on encryption and omit decryption.

The remainder of the paper is organized as follows. In Sect. 2 we discuss related work, in Sect. 3 the algorithms of study and their implementation variations, in Sect. 4 efficient implementation details, in Sect. 5 the experimental setup, metrics, and results, and in Sect. 6 our conclusions.

2   Related Work

To a system designer choosing a block cipher for adoption in a microcontroller-based application, several relevant works exist. As previously mentioned, many lightweight block ciphers have been recently proposed and their authors typically offer performance results and/or implementation tactics although the scope of these efforts varies. Several surveys help distill the relative performance of these lightweight and other block ciphers. For example Eisenbarth et al. in [10] provides results of several ciphers on an 8-bit ATtiny45 device. The authors concentrate on small size as a design goal and provide energy consumption data but the results are of limited relevance to this study given the differences in target platform. Law et al. in [11] compares block ciphers on an MSP430F149 device. They adopt source code from public sources such as OpenSSL [16]. This approach ensures quality code, but fixes the implementation strategy to that of the public source that is not necessarily optimized for embedded devices. Cazorla et al. in [12] compare 12 lightweight and 5 conventional block ciphers on an MSP430F1611 device. The authors compare many ciphers, but understandably chose a single implementation for each and do not state any particular optimization goals. Didla in [17] investigates implementation tactics of AES in a MSP430F1611 device; however, all are variations of AES for 8-bit platforms. Finally, in [18] the authors compare AES with other block ciphers on both MSP430- and ATmega-based platforms. They address the variable key size of AES but otherwise choose a single implementation (unstated, but from their provided ROM size it appears to be a table-based one).

Concerning speed records for AES on microcontroller devices, Hyncica in [19] presents optimized AES results of 172 cycles/byte for the MSP430 platform that is based on 32-bit table-based code ported from LibTomCrypt [20]. Gouvea [15] first presented the 16-bit lookup table strategy for AES in which they reported 180 cycles/byte on a MSP430 platform. On an AVR device the current speed record is described by Bos in [21], previously held by Poettering in [22].

Implementation and analysis results have begun to appear for SPECK. The designers present implementation results for SPECK on Atmels ATmega128 8-bit processor and the BLOC project [13] provides preliminary performance data on the MSP430. Cryptanalysis of SPECK can be found in [23,24], and [25].

3   Algorithms and Implementation Variations

3.1   AES

The AES algorithm uses a substitution-permutation approach and operates on a block size of 128-bits organized as a 4x4 array of bytes [7]. Four basic transformations are iteratively applied over a variable number of rounds (depending on key size) to complete each block encryption. These operations are SubBytes, ShiftRows, MixColumns, and AddRoundKey. Of these, MixColumns is the most complex operation requiring multiplication of state bytes by constants over the Galois Field $GF(2)^8$. Most of the AES implementation variations in common use concern themselves with optimizing this transformation.

Daemen and Rijmen in [26] discuss implementation aspects for both 8-bit and 32-bit processors. In the 8-bit approach, SubBytes, ShiftRows, and AddRoundKey can be easily combined and executed byte-by-byte for each of the 16 input bytes and the MixColumns step can also be implemented efficiently. In this paper the 8-bit approach is implemented in 4 different ways. The first is optimized for speed by unrolling all transformations within each round. It was adapted from the implementation provided by Texas Instruments (TI) in [27]. The second also follows [27], but condenses the transformations into nested loops to reduce ROM size. The third and fourth variations further optimize for speed by introducing extra 256-byte lookup tables to speed up the field multiplications within MixColumns. In the sections below, these four 8-bit variations are referred to as 8-BIT-UNROLL, 8-BIT-LOOPED, 8-BIT-2T, and 8-BIT-2SBOX, respectively.

The 32-bit approach discussed by Daemen and Rijmen is also practical using the 16-bit instruction set of the MSP430. The matrix formulation of the round transformation can be used to define a set of four 256-entry 32-bit lookup tables, known as T-tables, for a total of 4096 precomputed and stored bytes. One iteration of the round function amounts to 16 table lookups and 16 XORs. Three of the T-tables are byte rotations of the first T-table, thus as a space/speed tradeoff, 1024 bytes of storage can be used together with cyclic 8-bit shifts of the single table. Both of these 32-bit variations are implemented; in the sections below they are referred to as 32-BIT-4T and 32-BIT-1T, respectively.

A new optimization was proposed by Gouvea [15] targeting 16-bit processors. This new approach is a variation of the 32-bit table lookup approach where 4 tables of 16-bit entries are defined such that each of the original 32-bit tables can be constructed by concatenating two of the 16-bit tables. This formulation reduces the memory requirement by a factor of 2. After initial tests showed that this variation was the best performing in terms of speed and energy consumption, it was also implemented in assembler. In the sections below these variations are referred to as 16-BIT-4T and 16-BIT-4T-ASM, respectively.

3.2   SPECK

SPECK is a family of lightweight block ciphers with a wide range of block and key size choices and hence is potentially interesting to system designers

Table 1. SPECK parameters

Block size	Key size	Word size	Key words	Rotation	Rotation	Rounds	Version
$2n$	mn	n	m	α	β	T	Name
32	64	16	4	7	2	22	32-BIT
48	72	24	3	8	3	22	48-BIT
	96		4			23	
64	96	32	3	8	3	26	64-BIT
	128		4			27	
96	96	48	2	8	3	28	96-BIT
	144		3			29	
128	128	64	2	8	3	32	128-BIT
	192		3			33	
	256		4			34	

desiring trade-offs between size, speed, and security. SPECK is a Feistel-like algorithm that uses the map $R_k : \mathrm{GF}(2)^N \times \mathrm{GF}(2)^N \to \mathrm{GF}(2)^N \times \mathrm{GF}(2)^N$, where $k \in \mathrm{GF}(2)^N$, defined by

$$R_k(x,y) = ((S^{-\alpha}x + y) \oplus k, S^{\beta}y \oplus (S^{-\alpha}x + y) \oplus k) \tag{1}$$

where $\oplus$ denotes bitwise XOR, $+$ denotes addition modulo 2^N, S^j, S^{-j} denote left and right circular shifts by j bits, and α, β are constants defined according to the block size chosen.

The family of SPECK algorithms is defined according to Table 4.1 in [6], reproduced here for convenience in Table 1. We implemented each of the 10 parameterizations of SPECK shown in Table 1 in both C and assembler. In the sections below we refer to these implementations by the version name with an -ASM or -C suffix for assembler or C, respectively.

4   Implementation Details

In all cases the interface to the block ciphers consists of two byte-pointer arguments? to an array of bytes to be encrypted and to the expanded key, respectively. We use IAR Embedded Workbench version 5.51 as a development platform. The target device is the MSP430F5528.

4.1   AES 8-BIT

The first 8-bit version of AES, 8-BIT-UNROLL, is based on the implementation by TI for the MSP430 [27] that makes use of the efficient 8-bit implementation hints given in [26].

The 8-BIT-LOOPED version replaces the unrolled MixColumns step with a loop over the 4 columns of the state, and replaces the unrolled AddRoundKey step with another loop over the 16 bytes of the state.

The 8-BIT-2T version replaces each AES $GF(2)^8$ multiply-by-2, requiring test, branch, shift, and XOR instructions with a single table lookup. The goal with this version is to increase speed at the expense of program size, and increase side-channel timing attack resistance (see Sect. 4.5).

The 8-BIT-2SBOX version precomputes the 256-byte table 2Sbox $= 2 \otimes$ Sbox$[a]$, where $\otimes$ denotes multiplication in the Galois Field, for each input byte a. To see how this is effective, recall that the MixColumns step computes a vector-matrix multiplication, for example,

$$
\begin{pmatrix} b_0 \\ b_1 \\ b_2 \\ b_3 \end{pmatrix} = \begin{pmatrix} 02\ 03\ 01\ 01 \\ 01\ 02\ 03\ 01 \\ 01\ 01\ 02\ 03 \\ 03\ 01\ 01\ 02 \end{pmatrix} \begin{pmatrix} \mathrm{Sbox}[a_0] \\ \mathrm{Sbox}[a_1] \\ \mathrm{Sbox}[a_2] \\ \mathrm{Sbox}[a_3] \end{pmatrix} \tag{2}
$$

Also recall that in multiplication over $GF(2)^8$ we have that $3 \otimes \mathrm{SBox}[a_x] ==$ $(2 \otimes \mathrm{SBox}[a_x]) \oplus \mathrm{SBox}[a_x]$. Therefore multiplication of Sbox$[a_x]$ by 1, 2, and 3 can be done in a straightforward way by application of Sbox and 2Sbox. Once the row shifts are folded into each column's matrix-vector multiplication, clever ordering of the resulting systems of equations yields a very efficient implementation, as realized by Pottering in [22]. We adapted his 8-bit AVR assembler code for the MSP430.

4.2   AES 32-BIT

The two 32-bit versions of AES use the T-table approach as described by Daemen and Rijmen in Sect. 4.2 of [26]. On 8-bit architectures this approach may not be practical because each of the 32-bit table lookups would require 4 byte-lookups. In total, 64 byte-lookups and 64 XORs per round would be required, which is comparable to the instruction count of a non-table-lookup approach. However on 16-bit architectures 32 word-lookups and 32 XORs per round are required, thus the overall instruction count is reduced quite a bit compared to 8-bit approaches.

In our implementation, the input array of bytes to be encrypted is used directly as the state matrix. The state bytes are used to index the table lookups and the resulting new columns are stored in four temporary 32-bit variables. These are XORed with a 32-bit pointer aliased to the expanded key byte-array. The results are stored back into the state matrix using a 32-bit pointer aliased to the state byte-array. Aliasing the pointer allows access to the same data bytes at different granularity without use of temporary storage. Reducing temporary storage is an important strategy in fast designs, (discussed further in Sect. 4.5). Pointer aliasing is accomplished using casting, e.g.:

```
state32 = (uint32_t *)state;
```

32-BIT-1T is very similar to the above, except T1, T2, and T3 are replaced with T0 and macros to perform left circular byte shifts by 1, 2, and 3 bytes respectively.

4.3   AES 16-BIT

As described by Gouvea in [15], the 32-bit lookup tables can be reduced to 16-bit lookup tables such that concatenations of two of the 16-bit tables can produce any of the original 32-bit tables. The pointer aliasing approach used in the 32-bit case applies similarly to the 16-BIT-4T version. Of note, by directly using 16-bit operations on 16-bit data types the compiler is no longer relied upon to synthesize 16-bit instructions from source code using 32-bit operations on 32-bit data types. Compilers are not perfect in this regard, so in addition to reducing the memory footprint by a factor of two, this approach was found to be faster as well. There is no equivalent byte-rotation-based memory/speed tradeoff in the 16-bit approach as there is in the 32-bit approach.

After initial testing, the 16-BIT-4T version of AES was found to be the fastest and lowest energy of all of the AES implementation variations tested. A complete listing of 16-BIT-4T can be found in Listing A in the Appendix. To further enhance the performance, 16-BIT-4T was also implemented in assembler, using the IAR generated assembler as a starting point. In the generated assembler, the compiler was unable to utilize the 12 general purpose registers (R4-R15 [28]) of the MSP430 efficiently enough and thus required temporary data to be loaded from and stored to the stack (in RAM). As discussed in Sect. 4.5, accessing temporary data in RAM is detrimental to speed.

The following register scheme in our assembler version avoids storing temporary data to RAM, thus increasing the speed and further reducing the energy consumption of the 16-BIT-4T-ASM code version. Registers R4 through R11 are used to hold the 8 temporary 16-bit column results, R12 and R13 hold pointers to the state array and expanded key respectively, R14 is the loop counter, and R15 is used to compute offsets into the tables. An excerpt of the resulting assembler round function is shown in the Appendix, Listing B.

It is likely that other AES versions implemented in assembly language would also see performance improvements. For instance, an assembler implementation of the 32-BIT-4T version could likely be made identical to the 16-BIT-4T-ASM version, speed-wise, since in assembler the only difference would be different offsets into the larger 32-bit tables. Based on the preceeding reasoning, we limit our AES assembly language analysis to a single implementation variation; we do not anticipate that 32-bit table-based variations re-implemented in assembly language would exceed the performance of 16-BIT-4T-ASM.

4.4   SPECK

The round function of SPECK is very succinct; therefore all implementations fully unroll the round function, leaving a single loop over the required number of rounds. Beyond decisions like unrolling or not, or inlining or not, Beaulieu in [6] provides no guidance on implementation of the round function. In this section implementations in C and assembly language are discussed.

SPECK performs all operations modulo n, the word size. Whenever the C data type (e.g., uint32_t or uint64_t) is larger than the 16-bit processor word size, then the compiler must translate XOR, addition $(+)$, and shift operations within C code into multi-precision operations over the native 16-bit instructions of the MSP430 [28]. For example, let $X = \{X_3, X_2, X_1, X_0\}$ and $Y = \{Y_3, Y_2, Y_1, Y_0\}$ be 64-bit integers composed of 16-bit words X_i and Y_i $(0 <= i < 4)$. Adding $X + Y$ in C would ideally result in the following sequence of operations in a 16-bit instruction set:

```
add X0, YO         ; add X0 to Y0
adc X1, Y1         ; add X1 and previous carry to Y1
adc X2, Y2         ; add X2 and previous carry to Y2
adc X3, Y3         ; add X3 and previous carry to Y3
```

C implementations may be preferred by developers (e.g., to simplify the coding effort), and our particular compiler generated efficient multi-precision code for the XOR and addition operations within SPECK. However, the generated code for the circular shifts was not very efficient. Since at least one compiler appears to have difficulty generating efficient code for the SPECK round function, we also implement the round function in assembler to quantify the trade-off.

The SPECK round function in most cases requires both a left circular shift (LCS) by 3 bits and a right circular shift (RCS) by 8 bits. The LCS can be implemented in an efficient way in assembler using three one bit circular shifts, as follows, where 64 bits of data are stored in the four 16-bit registers R4 through R7:

```
; assembly language 1-bit LCS
; each instruction takes one clock cycle
; (in register addressing mode)
rla r4             ; shift first 16-bit word
rlc r5             ; shift with carry second 16-bit word
rlc r6             ; shift with carry third 16-bit word
rlc r7             ; shift with carry second 16-bit word
adc r4             ; rotate final carry back to first word
```

The 8-bit RCS can be performed in assembler in an efficient way using swap-byte and XOR operations, as shown in Listing C on a 64-bit word held in registers R4-R7. Equivalent LCS and RCS operations in C were not as efficient. For example the RCS implemented as $x = (((x) << 56)|((x) >> 8))$, did not use an extra temporary register and final swap-byte/XOR, as in Listing C, instead using two AND operations (in immediate mode), a swap-byte, and an OR operation. (The immediate addressing mode is slower than the register addressing mode on the MSP430, two clock cycles versus one [28].)

For developers who do not want to proceed to assembler there unfortunately may be limited options to optimize the multi-precision LCS/RCS operations. Typically there is not enough direct access to machine status words, for example to access/modify carry flags, or access to specialized instructions like "rotate left through carry" (rlc), from within high level languages such as C. Listing D in the Appendix shows the full implementation of SPECK-128 in C. The listing illustrates different ways to implement LCS and RCS that resulted in an 8 % speedup over versions that used the C language methods shown above.

4.5   MSP430 Features and Capabilities

Several features of the MSP430 family of microcontrollers have direct bearing on the implementation results presented in Sect. 5. Chiefly, these are (1) the instruction set, (2) the addressing modes, and (3) the register set. The MSP430 Family User Guide [28] provides detailed information on all of these features. In this section, we offer comments on specific use of several of the features as they pertain to SPECK and AES implementations.

Instructions on the MSP430 allow operations on either bytes or 16-bit words (via .b or .w suffixes). The swap bytes (swpb) instruction is very useful to SPECK implementations (for the RCS operation). Byte operations to registers clear the most significant byte of the word; this effect is also used during the RCS operation (Appendix, Listing C).

The addressing modes of the MSP430 include register modes (operations on data held in processor registers) and several memory modes (operations on data held in processor memory, RAM or ROM). Operations on data held in memory are generally much slower than data held in registers. For example, to XOR two words held in processor registers takes one clock cycle, but to XOR two words held in memory takes five or six clock cycles, depending on the specific addressing mode employed. As such, whenever possible AES and SPECK code is structured to attempt to minimize loading from and storing to memory. Unfortunately there are only 12 general purpose registers (designated R4 through R15) in which to hold data. A consequence of the limited register set is that temporary variables must be used very sparingly in C code. As the compiler encounters "larger" numbers of temporary variables (e.g., function locals) it will utilize stack memory (physically stored in RAM) to hold them. Accessing these temporary values will therefor incur a speed penalty due to slower memory addressing modes on the MSP430. We do not attempt to quantify "larger" in this study, since detailed examination of the compiler is not our goal (and will be different, for other compilers). However, as the number of temporary variables grows it becomes more difficult for the compiler to avoid temporary use of RAM-based stack.

The MSP430's addressing modes have the advantage that memory accesses are constant time. There are no cache hierarchy effects or interactions with other concurrently running processes to worry about [29]. AES versions that use table lookups thus do not have key- or input-dependent timing variability and appear to have resistance to timing attacks on the MSP430. In our suite of implementations, the only AES versions that do not use table lookups are 8-BIT-UNROLL and 8-BIT-LOOPED. In these implementations, the computation of multiply-by-2 over $GF(2)^8$ depends on the input (a branch containing an extra instruction may or may not be taken). We have not investigated the feasibility of a timing attack on these AES implementations. The SPECK round function involves no branches and on the MSP430 takes constant time. Based on the constant time property of the round function, we expect SPECK to be resistant to timing-based side-channel attacks on the MSP430, although this has not been investigated.

5 Results and Discussion

5.1 Experimental Setup and Procedure

The 8 variations of AES were evaluated for each of the 3 AES key sizes along with the 20 variations of SPECK (10 in C and 10 in assembler). The metrics for each test were speed of encryption, code size, and energy consumption. Speed was measured using the IAR debugger and function profiler tools in simulation mode. (Speed was also independently verified using timing information obtained from the measured waveforms described below, running released code.) Code size is provided by the IAR linker, broken down into CODE, DATA, and CONST segment sizes. Since all CONST segment data is stored in ROM along with the CODE segment, below we have grouped CODE and CONST together as a total ROM size, reported along with total RAM size (DATA segments). Energy consumption was calculated by first measuring the voltage drop across a 10 ohm resistor in series with the MSP430 digital voltage supply, V_{dvcc}, on a custom evaluation board (nominally $V_{dvcc} = 2.85$ V). Voltage drop was measured using a National Instruments PXI-1024Q chassis, PXI-8108 controller, and PXI-4071 7 digit, 26-bit digitizer. Custom MATLAB scripts then converted the voltage to current and performed integration of the current waveforms over the encryption time-period to get charge, Q. Finally, energy is calculated as $E = QV_{dvcc}$.

In every case key expansion was performed and all round keys were stored in RAM (code to perform key expansion is included in our ROM figures; however, we omit key expansion speed results). (In most cases the key expansion speed is within a factor of 2 of the number of cycles for a block encryption.) Stack utilization also consumes RAM; stack usage was determined by careful examination of compiler generated code.

5.2 Results and Discussion

The results are shown in Figs. 1 through 5 below. Figure 1 shows the speed data for each algorithm, arranged right-to-left from fastest to slowest. Figures 2 through 5 are presented in the same x-axis order as Fig. 1, i.e., all results are sorted according to speed. Figures 2 through 5 show energy consumption per byte, ROM size, RAM size, and a combined metric, the code size × cycle count product normalized by block size [10]. In all figures smaller bars are better. In the SPECK charts, "Small Key" refers to the smaller of the key options for each block size shown in Table 1. Similarly, "Large Key" refers to the larger of the key options. The 256-bit key only applies to the 128-bit block size.

For AES, the fastest and most energy efficient C implementation is the 16-BIT-4T variation at 152 cycles/byte and 2.46 μJ/block. The speedup obtained over Gouvea's implementation [30] is due to the avoidance of storing the state matrix in temporary stack space. This was accomplished via the pointer aliasing technique discussed in Sect. 4.3: aliasing the input state array (uint8_t *), used to index the lookup tables, with a word-array pointer (uint16_t *), used for assignments to the state matrix. The 16 extra stack bytes in Gouvea's round

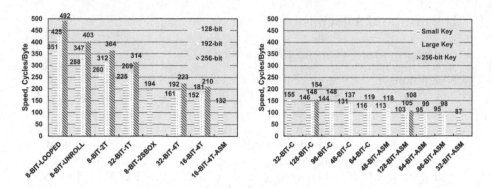

Fig. 1. Speed of AES (left) and SPECK (right) (Block encryption only) (44520)

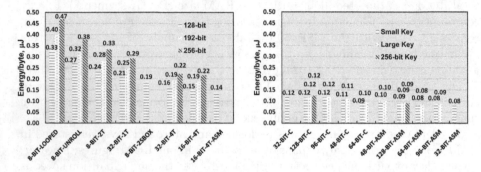

Fig. 2. Energy consumption per byte of AES (left) and SPECK (right) (44522)

function implementation cause more data movement to and from RAM that in addition to adding instructions, incurs the memory addressing mode cycle-count penalty discussed in Sect. 4.5.

The assembler implementation 16-BIT-4T-ASM gives a further 14 % speedup and 12 % decrease in energy usage over the C implementation, to 132 cycles/byte and 2.18 μJ/block. This improvement is again a direct consequence of improving register utilization (and thus reducing the memory addressing mode cycle-count penalty). The register utilization scheme that was employed is discussed in Sect. 4.3.

Of the lighter weight 8-bit AES versions, 8-BIT-2SBOX is the fastest at 194 cycles/byte but has the disadvantage of needing an assembler implementation to realize its performance. The 8-BIT-2T version is 25 % slower but much simpler to implement. The combined metric shows that 8-BIT-LOOPED provides very good overall performance due to its reasonable throughput and small code size, 8-BIT-2SBOX is exceptional for the same reason, and 16-BIT-4T is also good due to its high speed. Figure 4 shows that table-driven C versions of AES consume slightly more RAM (beyond that required to hold the expanded key) because of temporary storage to the stack; however the 16-BIT-4T-ASM version does not require stack as discussed in Sect. 4.3. The ROM size of AES

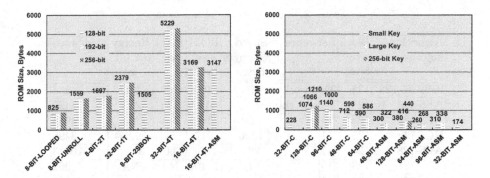

Fig. 3. ROM usage of AES (left) and SPECK (right) (Including key schedule code) (44521)

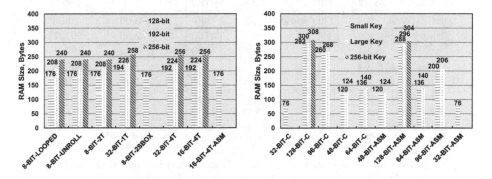

Fig. 4. RAM usage of AES (left) and SPECK (right) (44523)

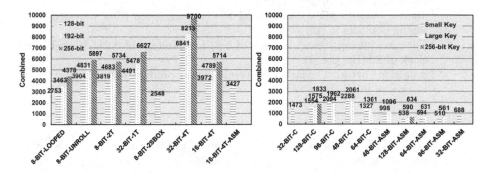

Fig. 5. Combined metric of AES (left) and SPECK (right) (code size cycle count/block size) (44524)

increases slightly for 256-bit key versions because the compiler optimizes away the portion of the AES key schedule that is only valid for 256-bit keys (see, for example, Sect. 5.2 of [7]). Energy consumption generally tracks speed quite well.

Table 2. Related work comparison

Algorithm-key size	Block size, bits	Reference	Speed cycle/byte bytes	ROM size, bytes	RAM size,	Energy per blk, μJ	Combined[5]
SPECK-128	128	This work	103	380	288	1.44	538
AES-128	128	This work	132	3147	176	2.18	3427
AES-128	128	[15]	180	2904	NA	NA	4084
RC6-128	128	[11][1]	1120	917	54	19.654	8496
AES-128	128	[11][1]	204	6555	60	3.584	10543
AES-128	128	[19]	172	12400	NA	NA	16662
AES-128	128	[12][2]	1891	2230	19	NA	33225
Camellia-128	128	[11][1]	393	11769	85	6.894	36395
AES-128	128	[18][3]	765	4500	1800	28.16	37652
CLEFIA-128	128	[12][2]	6134	4780	180	NA	237693
SPECK-96	64	This work	96	260	136	0.66	594
Skipjack-80	64	[18][3]	350	3750	40	2.63	20727
XXTEA-128	64	[18][3]	2340	1900	200	17.48	76781
TWINE-128	64	[12][2]	5125	1108	23	NA	90568
Piccolo-128	64	[12][2]	4562	1255	91	NA	95945
Lblock-80	64	[12][2]	5369	1784	13	NA	150751
LED-128	64	[12][2]	21382	1132	41	NA	391892
PRESENT	64	[12][2]	45573	4814	142	NA	3529059
SPECK-64	32	This work	87	174	76	0.31	680

[1] ROM and RAM figures do not include expanded key bytes; all metrics include overhead in Output Feedback mode of operation;

[2] Includes key expansion in encryption speed (separated key expansion speed data not provided in [12])

[3] TelosB Data (8 MHz Clock) used to compute speed from provided timing numbers in [18]

[4] Computed using average current of 2.93 mA, voltage of 2.994 V, and clock of 8 MHz, per [11]

[5] code size (bytes) × cycle count (cycles/byte) product normalized by block size (bits) [10]

Slightly varying average current levels for different implementations (not shown here) is a second order effect, confirming the observations of other authors [11].

For SPECK, the fastest C implementation is the 64-BIT block size version at 116 cycles/byte ($0.74\,\mu J$/block) and the fastest assembler implementation was the 32-BIT version at 87 cycles/byte ($0.31\,\mu J$/block). Of note, all of the assembler implementations are faster than the fastest C implementation. The speed-up from C to assembler is due to two factors. First, in assembly language the multi-precision rotation operations can be implemented more efficiently (as shown in Sect. 4.4). And second, in the C implementation of the larger block size versions of SPECK (96- and 128-bit) there is inefficient register utilization. The two temporary variables x and y (shown in Listing D) fill 8 of 12 available general purpose registers. This was enough to cause some temporary storage to stack (RAM) with an associated speed penalty. The assembly implementations were able to avoid use of extra RAM and the associated speed penalty.

Larger key sizes for similar versions of SPECK use incrementally more code and energy. However the effect is much less pronounced than in AES. In AES, the number of extra rounds increases from 10 to 14 when going from 128-bit keys to 256-bit keys (a 40 % increase) while in SPECK the number of rounds increases from 32 to 34 (just over 6 %). Block size has a much stronger impact than key size on both ROM and RAM, as seen in Figs. 3 and 4.

Although we chose to limit our in-depth study to the two tailorable block ciphers AES and SPECK, related work provides figures on AES and other block ciphers for the MSP430 platform that can be compared to our results. Related work is summarized in Table 2. In Table 2, implementations are first sorted by block size and then by the combined metric, to facilitate comparison of overall performance between ciphers of similar block size. Note that comparisons such as these can be difficult to interpret due to differing measurement conditions or techniques. Our measurement conditions are stated at the beginning of this section; we have indicated known differences between our approach and the various references as footnotes to Table 2. In cases where the reference indicated that the authors implemented both encrypt and decrypt, but provided only one code size result, the ROM for encrypt only is estimated by dividing the reported ROM by 2.

6   Conclusions

We have implemented and measured 24 different variations of AES and 20 different variations of the new lightweight block cipher SPECK on the low power MSP430 platform, in both C and assembler. Many of these implementations represent records for speed and energy efficiency among lightweight and traditional block ciphers on that device, e.g. 132 cycles/byte and 2.18 µJ/block for AES and 103 cycles/byte and 1.44 µJ/block for SPECK, both with 128-bit block and key sizes. The 32-bit block size of SPECK with a 64-bit key produced even lower numbers at 87 cycles/byte and 0.31 µJ/block. We provide implementation tactics for both AES and SPECK in both C and assembler for the 16-bit MSP430 platform. Finally, we provide a thorough analysis of measured results across algorithm, implementation strategy, and key size to aid system designers needing to incorporate block ciphers into their designs.

References

1. Suzaki, T., Minematsu, K., Morioka, S., Kobayashi, E.: Twine: a lightweight block cipher for multiple platforms. In: Knudsen, L.R., Wu, H. (eds.) SAC 2012. LNCS, vol. 7707, pp. 339–354. Springer, Heidelberg (2013)
2. Shibutani, K., Isobe, T., Hiwatari, H., Mitsuda, A., Akishita, T., Shirai, T.: *Piccolo*: an ultra-lightweight blockcipher. In: Preneel, B., Takagi, T. (eds.) CHES 2011. LNCS, vol. 6917, pp. 342–357. Springer, Heidelberg (2011)

3. Wu, W., Zhang, L.: LBlock: a lightweight block cipher. In: Lopez, J., Tsudik, G. (eds.) ACNS 2011. LNCS, vol. 6715, pp. 327–344. Springer, Heidelberg (2011)
4. Guo, J., Peyrin, T., Poschmann, A., Robshaw, M.: The LED block cipher. In: Preneel, B., Takagi, T. (eds.) CHES 2011. LNCS, vol. 6917, pp. 326–341. Springer, Heidelberg (2011)
5. Bogdanov, A.A., Knudsen, L.R., Leander, G., Paar, C., Poschmann, A., Robshaw, M., Seurin, Y., Vikkelsoe, C.: PRESENT: An ultra-lightweight block cipher. In: Paillier, P., Verbauwhede, I. (eds.) CHES 2007. LNCS, vol. 4727, pp. 450–466. Springer, Heidelberg (2007)
6. Beaulieu, R., Shors, D., Smith, J., Treatman-Clark, S., Weeks, B., Wingers, L.: The SIMON and SPECK families of lightweight block ciphers. Cryptology ePrint Archive, June 2013. http://eprint.iacr.org/2013/404
7. National Institute of Standards and Technology (NIST). FIPS-197: Advanced Encryption Standard (AES) (2001). http://www.csrc.nist.gov/publications/ps/ps197/ps-197.pdf
8. MICAz wireless measurement system. http://www.memsic.com/userfiles/files/Datasheets/WSN/micaz_datasheet-t.pdf
9. TelosB   Platform.   http://www.memsic.com/userfiles/files/Datasheets/WSN/telosb_datasheet.pdf
10. Eisenbarth, T., Gong, Z., Güneysu, T., Heyse, S., Indesteege, S., Kerckhof, S., Koeune, F., Nad, T., Plos, T., Regazzoni, F., Standaert, F.-X., van Oldeneel tot Oldenzeel, L.: Compact implementation and performance evaluation of block ciphers in ATtiny devices. In: Mitrokotsa, A., Vaudenay, S. (eds.) AFRICACRYPT 2012. LNCS, vol. 7374, pp. 172–187. Springer, Heidelberg (2012)
11. Law, Y.W., Doumen, J., Hartel, P.: Survey and benchmark of block ciphers for wireless sensor networks. ACM Trans. Sens. Netw. (TOSN) 2(1), 65–93 (2006). ACM, New York
12. Cazorla, M., Marquet, K., Minier, M.: Survey and benchmark of lightweight block ciphers for wireless sensor networks. In: Proceedings of the 10th International Conference on Security and Cryptography, SECRYPT 2013, pp. 543–548. SciTePress, Reykjavk, Iceland, 29–31 July (2013)
13. BLOC project performance evaluations, June 2014. http://bloc.project.citi-lab.fr/library.html
14. Guneysu, T.: Implementing AES on a bunch of processors. ECRYPT AES day, Bruges, Belgium (2012). https://www.cosic.esat.kuleuven.be/ecrypt/AESday/slides/AES-DAY-Gueneysu.pdf
15. Gouvêa, C.P.L., López, J.: High speed implementation of authenticated encryption for the MSP430X microcontroller. In: Hevia, A., Neven, G. (eds.) LatinCrypt 2012. LNCS, vol. 7533, pp. 288–304. Springer, Heidelberg (2012)
16. OpenSSL Cryptography and SSL/TLS toolkit. http://www.openssl.org/
17. Didla, S., Ault, A., Bagchi, S.: Optimizing AES for embedded devices and wireless sensor networks. In: Proceedings of the 4th International Conference on Testbeds and research infrastructures for the development of networks and communities (TridenCOM), Article No. 4, 2008. ICST (Institute for Computer Sciences, Social-Informatics and Telecommunications Engineering), Brussels, Belgium (2008)
18. Lee, J., Kapitanova, K., Son, S.H.: The price of security in wireless sensor networks. Comput. Netw. 54(17), 2967–2978 (2010). Elsevier, New York
19. Hyncica, O., Kucera, P., Honzik, P., Fiedler, P.: Performance evaluation of symmetric cryptography in embedded systems. In: Proceedings of the 6th International Conference on Intelligent Data Acquistion and Advanced Computing Systems: Technology and Applications, pp. 277–282, Prague (2011)

20. St. Denis, T., LibTomCrypt (source code). http://libtom.org/?page=features& newsitems=5&whatfile=crypt
21. Osvik, D.A., Bos, J.W., Stefan, D., Canright, D.: Fast software AES encryption. In: Hong, S., Iwata, T. (eds.) FSE 2010. LNCS, vol. 6147, pp. 75–93. Springer, Heidelberg (2010)
22. Poettering, B.: AVRAES: The AES block cipher on AVR controllers (2006). http://point-at-infinity.org/avraes/
23. Abed, F., List, E., Wenzel, J., Lucks, S.: Differential cryptanalysis of round-reduced SIMON and SPECK. In: FSE 2014. LNCS (2014, to appear)
24. Biryukov, A., Roy, A., Velichkov, V.: Differential analysis of block ciphers SIMON and SPECK. In: FSE 2014. LNCS (2014, to appear)
25. Dinur, I.: Improved differential cryptanalysis of round-reduced speck. In: Joux, A., Youssef, A. (eds.) SAC 2014. LNCS, vol. 8781, pp. 147–164. Springer, Heidelberg (2014)
26. Daemen, J., Rijmen, V.: The Design of Rijndael. Springer, Berlin (2002)
27. Kretzschmar, U.: AES software support for encryption and decryption. MSP430 Systems. http://www.ti.com/litv/zip/slaa397a
28. MSP430 Family, Instruction Set Summary. http://www.ti.com/sc/docs/products/micro/msp430/userguid/as_5.pdf
29. Bernstein, D.J.: Cache-timing attacks on AES (2005). http://cr.yp.to/antiforgery/cachetiming-20050414.pdf
30. Gouvea, C.: Authenticated Encryption on the MSP430 (source code). http://conradoplg.cryptoland.net/software/authenticated-encryption-for-the-msp430/

7   Appendix A Code Listings

Listing A C code for 16-bit AES using four lookup tables T0, T1, T2, and T3:

```
// ===============================================================
void encrypt_aes(uint8_t *s, uint8_t *expanded_key) {
    int round;
    uint16_t *state16;    // 16-bit alias of input array
    uint16_t *key16;      // 16-bit alias of input array
    uint16_t tmp[8];      // 16-bit temporary columns
    uint8_t buf1;
    uint8_t buf2;

    state16 = (uint16_t *)s;
    key16 = (uint16_t *)(expanded_key);

    state16[0] ^= key16[0]; state16[1] ^= key16[1];
    state16[2] ^= key16[2]; state16[3] ^= key16[3];
    state16[4] ^= key16[4]; state16[5] ^= key16[5];
    state16[6] ^= key16[6]; state16[7] ^= key16[7];

    key16 += 8;

    for (round = 1; round < NR; round++) {
        // use the state array to access byte indices to the tables
        tmp[0] = T0[s[0]] ^ T2[s[5]] ^ T1[s[10]] ^ T3[s[15]];
        tmp[1] = T1[s[0]] ^ T3[s[5]] ^ T0[s[10]] ^ T2[s[15]];

        tmp[2] = T0[s[4]] ^ T2[s[9]] ^ T1[s[14]] ^ T3[s[3]];
        tmp[3] = T1[s[4]] ^ T3[s[9]] ^ T0[s[14]] ^ T2[s[3]];

        tmp[4] = T0[s[8]] ^ T2[s[13]] ^ T1[s[2]] ^ T3[s[7]];
        tmp[5] = T1[s[8]] ^ T3[s[13]] ^ T0[s[2]] ^ T2[s[7]];

        tmp[6] = T0[s[12]] ^ T2[s[1]] ^ T1[s[6]] ^ T3[s[11]];
        tmp[7] = T1[s[12]] ^ T3[s[1]] ^ T0[s[6]] ^ T2[s[11]];

        state16[0] = tmp[0] ^ key16[0];
        state16[1] = tmp[1] ^ key16[1];
        state16[2] = tmp[2] ^ key16[2];
        state16[3] = tmp[3] ^ key16[3];
        state16[4] = tmp[4] ^ key16[4];
        state16[5] = tmp[5] ^ key16[5];
        state16[6] = tmp[6] ^ key16[6];
        state16[7] = tmp[7] ^ key16[7];
```

```
        key16 += 8;
    }

    //substitution and shift using Bytes
    // row 0
    s[ 0]  = sbox[s[ 0]];
    s[ 4]  = sbox[s[ 4]];
    s[ 8]  = sbox[s[ 8]];
    s[12]  = sbox[s[12]];
    // row 1
    buf1 = s[1];
    s[ 1]  = sbox[s[ 5]];
    s[ 5]  = sbox[s[ 9]];
    s[ 9]  = sbox[s[13]];
    s[13]  = sbox[buf1];
    // row 2
    buf1 = s[2];
    buf2 = s[6];
    s[ 2]  = sbox[s[10]];
    s[ 6]  = sbox[s[14]];
    s[10]  = sbox[buf1];
    s[14]  = sbox[buf2];
    // row 3
    buf1 = s[15];
    s[15]  = sbox[s[11]];
    s[11]  = sbox[s[ 7]];
    s[ 7]  = sbox[s[ 3]];
    s[ 3]  = sbox[buf1];

    state16[0] ^= key16[0]; state16[1] ^= key16[1];
    state16[2] ^= key16[2]; state16[3] ^= key16[3];
    state16[4] ^= key16[4]; state16[5] ^= key16[5];
    state16[6] ^= key16[6]; state16[7] ^= key16[7];

    return;
}
```

Listing B Assembler code snippet for 16-bit AES round function:

```
// ================================================================
; column 0
mov.b   @R12,R15        ; state[0]
rla.w   R15             ; address into 16-bit T0
mov.w   T0(R15),r4      ; table lookup into temp column
mov.w   T1(R15),r5      ; table lookup into temp column
mov.b   0x5(R12),R15    ; state[5]
```

```
rla.w    R15           ; address into 16-bit T2
xor.w    T2(R15),r4    ; accumulate table lookup into temp column
xor.w    T3(R15),r5    ; accumulate table lookup into temp column
mov.b    0xA(R12),R15  ; state[10]
rla.w    R15           ; address into 16-bit T1
xor.w    T1(R15),r4    ; accumulate table lookup into temp column
xor.w    T0(R15),r5    ; accumulate table lookup into temp column
mov.b    0xF(R12),R15  ; state[15]
rla.w    R15           ; address into 16-bit T3
xor.w    T3(R15),r4    ; accumulate table lookup into temp column
xor.w    T2(R15),r5    ; accumulate table lookup into temp column

; other columns similar (omitted)

; key add
xor.w    @R13+,r4      ; xor key bytes 0-1 with temp bytes
mov.w    r4,0x0(R12)   ; update state bytes 0-1
xor.w    @R13+,r5      ; xor key bytes 2-3 with temp bytes
mov.w    r5,0x2(R12)   ; update state bytes 2-3
xor.w    @R13+,r6      ; xor key bytes 4-5 with temp bytes
mov.w    r6,0x4(R12)   ; update state bytes 4-5
xor.w    @R13+,r7      ; xor key bytes 6-7 with temp bytes
mov.w    r7,0x6(R12)   ; update state bytes 6-7
xor.w    @R13+,r8      ; xor key bytes 8-9 with temp bytes
mov.w    r8,0x8(R12)   ; update state bytes 8-9
xor.w    @R13+,r9      ; xor key bytes 10-11 with temp bytes
mov.w    r9,0xA(R12)   ; update state bytes 10-11
xor.w    @R13+,r10     ; xor key bytes 12-13 with temp bytes
mov.w    r10,0xC(R12)  ; update state bytes 12-13
xor.w    @R13+,r11     ; xor key bytes 14-15 with temp bytes
mov.w    r11,0xE(R12)  ; update state bytes 14-15
```

Listing C Assembler code for multi-word Right Circular Shift by 8 bits:

```
// ============================================================
                ; register contents (Byte numbers):
                ; R4          R5         R6         R7         R9
mov.b r4, r9    ; B0    B1 B2    B3 B4    B5 B6    B7 B0  00
swpb  r9        ; B0    B1 B2    B3 B4    B5 B6    B7 00  B0
swpb  r4        ; B1    B0 B2    B3 B4    B5 B6    B7 00  B0
swpb  r5        ; B1    B0 B3    B2 B4    B5 B6    B7 00  B0
swpb  r6        ; B1    B0 B3    B2 B5    B4 B6    B7 00  B0
swpb  r7        ; B1    B0 B3    B2 B5    B4 B7    B6 00  B0
xor.b r5, r4    ; B1^B3 00 B3    B2 B5    B4 B7    B6 00  B0
xor   r5, r4    ; B1    B2 B3    B2 B5    B4 B7    B6 00  B0
xor.b r6, r5    ; B1    B2 B3^B5 00 B5    B4 B7    B6 00  B0
```

```
xor    r6, r5    ; B1    B2 B3    B4 B5    B4 B7    B6 00 B0
xor.b  r7, r6    ; B1    B2 B3    B4 B5^B7 00 B7    B6 00 B0
xor    r7, r6    ; B1    B2 B3    B4 B5    B6 B7    B6 00 B0
xor.b  r9, r7    ; B1    B2 B3    B4 B5    B6 B7^00 00 00 B0
xor    r9, r7    ; B1    B2 B3    B4 B5    B6 B7    B0 00 B0
```

Listing D C language code for SPECK-128:

```
// =============================================================
void encrypt_speck(uint8_t * pointer, uint8_t * expanded_key) {
    uint64_t *p64 = (uint64_t*)pointer;
    uint64_t *k64 = (uint64_t*)expanded_key;
    uint64_t  y;
    uint64_t  x;

    // Copy values to be encrypted into x and y
    y = p64[0];
    x = p64[1];

    for (int round = 0; round < SPECK_T; round++) {
        uint64_t t;
        uint8_t y8;

        // RCS, addition with y, and key addition
        t = x >> 8;
        x = x << 56;
        x = x + t + y;
        x = x ^ k64[round];

        // LCS and XOR with x
        t = y << 3;
        y8 = (uint8_t)(y >> 56);
        y8 >>= 5;
        y = (t | y8) ^ x;
    }

    // Copy encrypted values back into ram
    p64[0] = y;
    p64[1] = x;
}
```

Side-Channel Attacks
and Countermeasures

On Efficient Leakage-Resilient Pseudorandom Functions with Hard-to-Invert Leakages

Fabrizio De Santis[1]([⊠]) and Stefan Rass[2]

[1] Technische Universität München, Munich, Germany
desantis@tum.de
[2] Alpen-Adria Universität, Klagenfurt, Austria

Abstract. Side-channel attacks have grown into a central threat to the security of nowadays cryptographic devices. The set of implementation countermeasures constantly competes with the set of known attack strategies, however, systematic ways to protect against information leakage are uncommon. Despite many achievements in the field of secure implementations, side-channel countermeasures only offer ad-hoc remedies which do not conform to the idea of provably secure cryptosystems. On the other side, leakage-resilient constructions often hinge on assumptions which can be hardly translated into practice. This work is an attempt to provide a theoretical, yet practical, modeling of side-channels that aids in identifying spots and making design choices towards a comprehensive side-channel security treatment from theoretical proofs down to hardware implementations. More precisely, we illustrate a simple sufficient condition for building physically secure hardware that follows directly from the decomposition of the side-channel into an algorithmic-related part and a physical-related part, *and* hardness of inversion. We put forward that our simple modeling allows to commit clear security goals to cryptographers and hardware designers and preserve the security of theoretical constructions all the way down to final chip fabrication. As a showcase application, we consider the security of the Goldwasser-Goldreich-Micali (GGM) construction scheme for *efficient* pseudorandom functions with and without leakages. These security proofs have been left open in previous literature and here serve to demonstrate the feasibility of our modeling approach.

1 Introduction

Guarding against side-channel information leakage is a notoriously difficult and challenging engineering task, as conflicting constraints such as performance and security have to be met simultaneously. The standard adversarial model in the physical setting assumes an adversary with physical access to a cryptographic device, from which it seeks to extract secret data based on the observation of side-channel information gathered during (possibly adaptively many) usages of the device. The way this information is gathered (e.g., power consumption or electromagnetic field radiation), in connection with the number of observed variables, the amount of available observations and how the inference towards secret data

© Springer International Publishing Switzerland 2015
D.F. Aranha and A. Menezes (Eds.): LATINCRYPT 2014, LNCS 8895, pp. 127–145, 2015.
DOI: 10.1007/978-3-319-16295-9_7

is done differentiates competing attack strategies [5, 6, 12, 22, 27, 30]. Over the last two decades, several ad-hoc implementation countermeasures have been developed to protect cryptographic hardware against known side-channel attacks. These countermeasures can be roughly classified into three categories depending on their abstraction level: (1) *logic-level countermeasures*, such as masked logic styles [17, 25, 29] protect the logic gates of the physical layer, (2) *algorithmic-level countermeasures*, such as secret sharing schemes [7, 15, 23, 26] break the processing of secrets into multiple shares and (3) *protocol-level countermeasures*, such as fresh re-keying schemes [2, 3, 20] wrap cryptographic algorithms into side-channel resistant protocols by frequent key-updates. Common to any such practice is the attempt to either hide or make the side-channel information independent from the internal secret information, in order to thwart inference attacks [19]. A complementary approach to prevent side-channel attacks has been taken only in recent years by leakage-resilient cryptography [9] to devise cryptographic primitives which are secure even in presence of side-channel information leakage. Unfortunately, the theoretical security proofs of leakage-resilient primitives often become void upon transition to real-world implementations, as this approach typically fails to capture realistic *physical* models of computation [31].

The crux of this work is to show a simple sufficient criterion for side-channel security that allows to preserve the validity of theoretical security proofs all the way down to hardware implementations. Our criterion is built upon the logical decomposition of the side-channel leakages into an algorithmic function f, which is specified by the particular cryptographic algorithm, and a physical function ℓ which depends on the physical properties of the technology realizing the considered cryptographic device. In this decomposed view, we require that the *composition* of the two functions f and ℓ is hard-to-invert, i.e., we can allow information flow through either one of them, but not through both. We put forward that our explicit leakage modeling is advantageous for at least three reasons: *First*, it allows to preserve the security of theoretical proofs by committing the hardness of inversion across lower abstraction levels until the final implementation in a clearly defined way. *Second*, it allows to identify and assign precise roles in the development of secure cryptographic hardware devices: theoreticians are in charge of devising secure constructions and provide security proofs from hardness of inversion, while cryptographers, hardware and physical designers are committed to preserve this property until final physical fabrication, by acting on the design of cryptographic algorithms, hardware architectures and the final physical chip layout, respectively. *Third*, it outlines a valuable asset (degree of freedom) in the development of secure cryptographic devices: in such cases where it is difficult to achieve hard-to-invert physical functions (modeled by the ℓ function), our security condition can be met by proper algorithm and hardware design (the f function), or by a combination thereof. Conversely, if the algorithmic step f cannot be prevented from leaking information, then stronger hardware protection in the physical layer are demanded (e.g., secure logic and layout design [17]). To illustrate our modeling, we look at Standaert et al.'s [33] *efficient* construction of leakage-resilient pseudorandom functions as showcase application. Their proposal is an efficiency-improved version of the well-known GGM-tree construction for pseudorandom functions due to Goldreich et al. [14]. This particular use-case is interesting in a double mean: First, the security proofs for Standaert et al.'s *efficient* GGM-tree were omitted in previous work [33], which we now present here based on our new modeling, together with its *concrete* security analysis [4]. Second, GGM-tree hardware implementations [3, 21] still provide today an exceptional example of implementations with super-exponentially hard-to-invert leakages which support our modeling arguments.

Organization. We start introducing the necessary technical background in Sect. 2. Then, we introduce our leakage model based upon the logical decomposition between algorithmic and leakage functions in Sect. 3. We continue discussing the assumptions and limitations of different approaches to security in presence of leakage, which finally culminate in the hardness of inversion property, defined in Sect. 4. Towards a demonstration of our modeling, we prove security of efficient GGM-tree PRFs with and without leakages in Sect. 5. Conclusions are drawn in Sect. 6.

2 Background

Before going into concrete constructions, we use this section to briefly introduce the concepts necessary to soundly and consistently explain what follows in the next sections.

2.1 Pseudorandom Generators

Informally, a pseudorandom generator (PRG) is an efficient deterministic function that expands a short and *truly* random seed into a polynomially long *pseudorandom* sequence, which cannot be efficiently distinguished from a truly random one. Throughout this work, we use $\kappa \in \mathbb{N}$ to denote the security parameter that is used to set up all the cryptographic engines. By $x \xleftarrow{\$} \mathcal{U}_\kappa$, we denote a uniformly random draw of a bitstring $x \in \{0,1\}^\kappa$. The bitlength of a string y is written as $|y|$.

Definition 1 (Pseudorandom Generator (PRG)). *Let an integer $\kappa \in \mathbb{N}$ be given, and take $\varepsilon, \tau > 0$. A function $\mathsf{G} : \{0,1\}^\kappa \to \{0,1\}^{s(\kappa)}$ is an (s, τ, ε)-secure PRG if:*

1. *G is computable in deterministic polynomial time in the input length κ.* [Efficiency]
2. *There exists a polynomial s, called the* stretch, *such that:* [Regular stretch]
   - *$s(\kappa) > \kappa$*
   - *$|\mathsf{G}(x)| = s(\kappa)$ for all strings $x \in \{0,1\}^\kappa$.*
3. *The distribution $\{x \xleftarrow{\$} \mathcal{U}_\kappa : \mathsf{G}(x)\}$ is computationally indistinguishable from the uniform distribution $\mathcal{U}_{s(\kappa)}$ on $s(\kappa)$-bit strings to any efficient adversary $\mathsf{A}(x)$ running in time at most τ:* [Pseudorandomness]

$$|\Pr[x \xleftarrow{\$} \mathcal{U}_\kappa : \mathsf{A}(\mathsf{G}(x)) = 1] - \Pr[x \xleftarrow{\$} \mathcal{U}_{s(\kappa)} : \mathsf{A}(x) = 1]| \leq \varepsilon \qquad (1)$$

2.2 Pseudorandom Functions

Informally, a pseudo-random function (PRF) is a function that is computationally indistinguishable from a random function in the sense that no efficient adversary can distinguish a PRF from a *truly* random function. Let a PRF family be denoted as $\mathcal{F}$, where $k \xleftarrow{\$} \mathcal{U}_\kappa$ is an integer key that selects a random member from $\mathcal{F}$, being a function $F_k : \{0,1\}^m \rightarrow \{0,1\}^n$, where m, n are fixed input and output sizes. The security of pseudorandom functions can be defined in a gamed-based fashion, where an adversary A is asked to distinguish between the output of a pseudorandom function F_k and a truly random value. More precisely, the game asks an adversary A, which is composed of two algorithms (A_1, A_2), to take a decision by running two formal experiments $\mathbf{Exp}_A^{\text{PRF-IND-0}}$ and $\mathbf{Exp}_A^{\text{PRF-IND-1}}$. The two experiments consist of a profiling phase, where the algorithm A_1 adaptively generates the input values for which PRF evaluations are collected, and a decision phase, where the algorithm A_1 outputs yet another fresh input to F_k and the algorithm A_2 is requested to distinguish between the evaluation of the PRF on this last input value and a truly random value. The two experiments are formalized in Definition 1. Notice that the two experiments are identical up to the point where the final response in the decision phase is generated (the difference is bold-printed for convenience of the reader).

Definition 2 (Pseudorandom function (PRF)). *A function family $\mathcal{F}$ is (q, τ, ε)-secure if an adversary A $= (A_1, A_2)$ that runs in time τ, given an oracle for a function F_k and allowed to make at most q queries to F_k has advantage at most ε in distinguishing the output of F_k from random:*

$$\mathsf{Adv}(\mathbf{Exp}_A^{PRF\text{-}IND}) := \left| \Pr(\mathbf{Exp}_A^{PRF\text{-}IND\text{-}0} = 1) - \Pr(\mathbf{Exp}_A^{PRF\text{-}IND\text{-}1} = 1) \right| < \varepsilon \tag{2}$$

$\mathbf{Exp}_A^{\text{PRF-IND-0}}$:	$\mathbf{Exp}_A^{\text{PRF-IND-1}}$:	
$k \xleftarrow{\$} \{0,1\}^\kappa$	$k \xleftarrow{\$} \{0,1\}^\kappa$	[Initialization]
$S, I \leftarrow \emptyset$	$S, I \leftarrow \emptyset$	
for $i = 1, 2, \ldots, q$	for $i = 1, 2, \ldots, q$	[Profile with at most q queries]
$\quad x_i \leftarrow A_1(S, I)$	$\quad x_i \leftarrow A_1(S, I)$	[Generate inputs]
$\quad S \leftarrow S \cup \{x_i\}$	$\quad S \leftarrow S \cup \{x_i\}$	[Collect inputs in S]
$\quad I \leftarrow I \cup \{(x_i, F_k(x_i))\}$	$\quad I \leftarrow I \cup \{(x_i, F_k(x_i))\}$	[Collect input/output pairs in I]
endfor	endfor	
$x_{q+1} \notin S \leftarrow A_1(S, I)$	$x_{q+1} \notin S \leftarrow A_1(S, I)$	[Generate a new challenge]
$\mathbf{z} \leftarrow \mathbf{F_k(x_{q+1})}$	$\mathbf{z} \xleftarrow{\$} \mathcal{U}_\kappa$	[**Generate response**]
$b \leftarrow A_2(x_{q+1}, z, I)$	$b \leftarrow A_2(x_{q+1}, z, I)$	[Take a decision]
return b	return b	

3 The Leakage Model

In this section, we describe the leakage model considered in this work. Similar to [24], we model the computation of cryptographic primitives by the means of abstract machines (e.g., Turing-machines, circuits, or similar) with an auxiliary function that provides some information about the internal configuration of the computational device whenever it becomes active. A *configuration* here is understood as the entirety of internal data (Turing-machine tape contents and state, current logical states of all circuit wires and gates, etc.) that is processed in each step (or stage) of a computation. A *state* is generally represented by a word $\{0,1\}^{\nu}$ e.g., a string whose length ν is polynomial in the security parameter κ of the algorithm, while the transition from one state to the next is called a *step*. In this view, the computation of a cryptographic primitive is split up into a sequence of n (fixed) steps $f_{n-1} \circ f_{n-2} \circ \cdots \circ f_1 \circ f_0$, each one computing an *intermediate function* f_i and emitting some information on its input and output values through a *physical function* ℓ. Let us now detail more about our assumptions on such functions and the adversary capabilities to obtain a representative modeling of physical reality:

Adjustable Granularity. The specific look of intermediate functions f_i is determined by the given cryptographic instance and the granularity of its security analysis, where by *granularity*, we mean the number of steps into which the primitive can be broken up, meaningfully. In practice, when a complexity-theoretic approach to security is taken, intermediate functions correspond to the invocation of a primitive underlying the analyzed construction, say a PRG for the analysis of PRFs [10]. When concrete hardware implementations are considered, then intermediate functions may correspond to very simple functions, say a b-bit S-box for the analysis of a block-cipher mode of operation, depending on the particular hardware architecture. This different level of abstraction actually represents the main gap between security proofs obtained in the complexity-theoretic approach and the actual security margins achieved by real-world implementations [31].

Only Computation Leaks Information. This axiom has been originally given in [24] and states that a cryptographic device can leak information only on the data which are manipulated during a transition step. This assumption is widely used in literature [1,11,24,33,35] and implies that no leakage can occur on data which are stored in memory or on data which has not been manipulated yet.

Physical functions. In accordance with physical experience, we assume that physical functions are fixed, deterministic and *cannot* be adaptively chosen by the adversary [1,11,33,35]. This last condition is a necessary in all the practical cases where the adversary has (stateless) reinitialization capabilities [35], whereas otherwise the adversary could extract a "bit of secret information" in every iteration by adaptively choosing the leakage function accordingly. Notice that no constraints on the specific form of physical functions are set.

Hard-to-Invert Leakages. Early constructions in leakage-resilient cryptography [9,28], imposed quantitative bounds on the amount of leaked information to show that, e.g., the output of a PRG is indistinguishable from a distribution with high min-entropy even in presence of λ bit of information [9]. Unfortunately, prior works did not spend many words on the practical interpretation of these bounds. Yet understanding how to achieve such conditions by proper algorithm and hardware design is hard to translate in practice. In this work, we achieve leakage-resiliency by requiring that the *composition* of intermediate functions and physical functions is hard-to-invert. So, we allow information leaking through any one of the two functions, but not through both, so that no advantage is obtained by the adversary from observing the composition only. We stress that, no matter which abstraction level is chosen in the security analysis, we require the composition of intermediate and physical functions to remain hard-to-invert. Notice that hard-to-invert leakages were introduced in [8] without the explicit decomposition given by our modeling. We believe that hardness of inversion and the logical decomposition of leakages is one important conceptual contribution and a necessary step towards closing the gap to practice.

(q, s)-*Limited Gaussian Adversary.* We limit the adversary to a bounded number of noisy leakages s obtained from a bounded number of queries q to the cryptographic device operating on the same secret [33,34]. This practically models an attacker who can only (passively) monitor some varying (physical) property of a cryptographic device over a side-channel (e.g., power consumption or electromagnetic (EM) field radiation), which is naturally affected by Gaussian measurement noise (s practically quantifies the estimation effort necessary for properly estimating the leakage distribution on a given input [32], that is, the adversary may require more than one single observation to learn a certain leakage).

Global Adversary. Finally, we assume that the adversary obtains the leakages associated to the entirety of data which are processed and not to any of its subpart, that is, the leakage is a function of *all* the data which are active in a transition step and the adversary is not able to tell them apart (e.g., observe the leakages associated with the computation of just a few bits). This models the vast majority of attackers who can observe the global activity of a cryptographic device (e.g., the global power consumption of a cryptographic device), but it does not include attackers that can observe the activity of small portions of the circuitry, hence focusing only on the processing of small portions of the active state (e.g., by the means of localized electro-magnetic field measurements [16]).

4   Security with Leakages

In this section, we take a step-by-step approach towards reaching a practically useful property for leakages (hardness of inversion) that allows transferring theoretical security proofs down to implementations and aiding secure cryptographic algorithm- and hardware-design. To this end, we give a series of "approaches" that – in different ways – extend classical security notions by providing the adversary with some additional leakage information, and *could* be made into security definitions. We let each approach to be followed by arguments outlining its peculiarities. This line of arguments will eventually culminate in the definition of hard-to-invert leakages (Definition 4).

Following the classical road to security, one would first aim at defining security in a leakage model in its broadest acceptation as "indistinguishable computations despite leakages", in the very same way as security in the standard model can be rested on outputs indistinguishability. Since the leakage is a property of physical cryptographic implementations, we think of an implementation as a pair $P = (F_k, \mathcal{L}_k)$, where F_k is a PRF instance and $\mathcal{L}_k$ is the concatenation of the leakages $\{\ell(f_i)\}$ produced by its execution. Informally, the concept "indistinguishability despite leakages" can be understood as: the output of a leakage-resilient PRF looks like random even if the adversary has access to some leakage information. This security definition is formally defined by Approach 1, which basically updates the experiments $\mathbf{Exp}_{A,P}^{PRF\text{-}IND\text{-}b}$ of Definition 2 as $\mathbf{Exp}_{A,P}^{LRPRF\text{-}IND\text{-}b}$ by including the leakage obtained via oracle calls to $\mathcal{L}_k$ in the profiling phase. It is worth noting that no leakages can be actually generated in the response phase, so to avoid trivial distinguishing in the decision phase based upon e.g., the identification of known plaintext/ciphertext values in the leakages as shown in [33].

Approach 1 (Indistinguishability despite Leakages). *A function family* $\mathcal{F}$ *is* $(q, s, \tau, \varepsilon)$-*secure leakage-resilient PRF if an adversary* $A = (A_1, A_2)$ *that runs in time at most* τ, *given access to a physical implementation* $P = (F_k, \mathcal{L}_k)$ *and allowed to make at most* q *queries to* F_k *and repeat at most* s *queries to* $\mathcal{L}_k$, *has advantage at most* ε *in distinguishing the output of* F_k *from random:*

$$\mathsf{Adv}(\mathbf{Exp}_{A,P}^{LRPRF\text{-}IND}) := \left| \Pr(\mathbf{Exp}_{A,P}^{LRPRF\text{-}IND\text{-}0} = 1) - \Pr(\mathbf{Exp}_{A,P}^{LRPRF\text{-}IND\text{-}1} = 1) \right| < \varepsilon$$

This approach captures practically relevant side-channel attacks such as non-profiled attacks [5,12]. However, this plain formulation does not cover profiled side-channel attacks [6,30], as the adversary can only sample the leakages from $\mathcal{L}_k$ (for a random unknown key k), but it can not perform a profiling phase of the leakage distribution using a key of its choice k^*, that is, sampling from $\mathcal{L}_{k^*}$ in the profiling phase of experiments. This issue could be solved by providing the adversary with oracle access to the whole family of leakage functions $\mathcal{L}$. We observe that Approach 1 requires establishing the same implication that yields semantic security of encryptions from ciphertext indistinguishability via recycling standard hybrid arguments on the leakages (as available in [13]). In other words, this approach, just as its counterpart without leakages, only states that the leakages should not contain any information which enable distinguishing pseudorandomness from true randomness.

A more recent and practical variation of "indistinguishability despite leakages" has been recently proposed in [32]. Their work introduces the concept of *simulatable leakage*, calling for a "simulator" that produces indistinguishable leakages when provided with the same input-output values on different keys. This concept models the practical case of adversaries who own identical copies of the cryptographic device under attack and therefore have the opportunity to get the leakages for different keys. This updated definition is formalized by

Approach 2, where the experiments $\mathbf{Exp}_{A,P,S_{\mathcal{L}}}^{LRPRF\text{-}SIM\text{-}b}$ update the previous experiments $\mathbf{Exp}_{A,P}^{LRPRF\text{-}IND\text{-}b}$ by considering the leakages produced by the simulator during the profiling and decision phase (cf. [32, Sect. 2.1]).

Approach 2 (Simulatable Leakages). *A function family $\mathcal{F}$ has $(s_S, \tau_S, s_A, \tau_A, \varepsilon)$ q-simulatable leakages if there is a simulator $S = (F, \mathcal{L})$ running in time at most τ_s and making at most s_S queries to the leakage function $\mathcal{L}$, yielding a distinguishing advantage at most ε for any adversary A running in time at most τ_A and making no more than s_A queries to $\mathcal{L}$:*

$$\mathsf{Adv}(\mathbf{Exp}_{A,S}^{LRPRF\text{-}SIM}) := \left| \Pr(\mathbf{Exp}_{A,S}^{LRPRF\text{-}SIM\text{-}0} = 1) - \Pr(\mathbf{Exp}_{A,S}^{LRPRF\text{-}SIM\text{-}1} = 1) \right| < \varepsilon$$

Interestingly, this setting is *empirically verifiable* by hardware designers which can reproduce and anticipate the adversary working conditions during the development of cryptographic devices by running "simulations" to verify the side-channel security of their implementations. It is worth noting that this approach demands indistinguishability against *every* efficient adversary, but only for *specific* bounded simulators. Therefore, this approach actually poses the problem of how to meaningfully instantiate good simulators to verify contingent security claims for a cryptographic device.

Yet a more practical approach consists to consider generic key-recovery attacks having additional access to the information obtained by physical observations of a cryptographic device, as follows:

Approach 3 (Key-Recovery Attacks with Leakages). *A function family $\mathcal{F}$ is a $(q, s, \tau, \varepsilon)$-secure PRF implementation if an adversary A running in time at most τ, given access to a physical implementation $P = (F_k, \mathcal{L}_k)$ and allowed to make at most q queries to P and to repeat at most s queries to $\mathcal{L}_k$, has advantage at most ε in recovering the secret k:*

$$\left| \Pr(k \leftarrow A_P(\{x_1, \ldots, x_q\})) - \Pr(k \overset{\$}{\leftarrow} \mathcal{U}_\kappa) \right| \leq \varepsilon.$$

This approach basically extends the black-box security model by considering the success probability of generic key-recovery attacks exploiting the input-output characteristic of F_k (black-box model) as well as the information associated to its physical leakages $\mathcal{L}_k$ (leakage model). Similarly to before, this approach defines security against generic non-profiled distinguishers A by stating that no information about the key can be recovered either from the input-output characteristic of F_k nor from its associated leakages $\mathcal{L}_k$.

This leads us now to introduce the hardness of inversion property that establishes a formal relationship to the auxiliary leakage model of Dodis et al. [8]:

Definition 3 (Hardness of Inversion). *We say that a function f is (τ, ε)-hard to invert, if every algorithm A, of running time $\leq \tau$, on input of a value $y \in Range(f)$ satisfies*

$$\Pr[x' \leftarrow A(y) : y = f(x')] < \varepsilon.$$

For rigor, we provide now the following lemma stating that, given two functions f and g, the hardness to invert either of f or g implies difficulties in the inversion of their composition $f(g(\cdot))$:

Lemma 1. *Let f be (τ_1, ε_1)-hard to invert, and let g be (τ_2, ε_2)-hard to invert. Furthermore, let f, g both be computable in time no more than μ. Then the composition $h(x) := f(g(x))$ is $(\tau - \mu, \varepsilon)$-hard to invert, where $\tau = \min\{\tau_1, \tau_2\}$ and $\varepsilon = \max\{\varepsilon_1, \varepsilon_2\}$.*

Proof. Assume the opposite, i.e., let there be an algorithm A that runs in time $< \tau - \mu$, having success-rate of $p > \varepsilon = \max\{\varepsilon_1, \varepsilon_2\}$ to output a pre-image x to a given image $h = h(x) = f(g(x))$. A preimage to f is obtained by computing a guess $x = g(\mathsf{A}(x))$ in time $< \tau - \mu + \mu = \tau$, which with probability $p > \varepsilon$ corresponds to a pre-image y that satisfies $f(y) = h$ for given h. Hence, we can invert f in time $\tau < \tau_1$ with probability $> \varepsilon$. It remains to show that g must as well be easy to invert: to this end, let y be given, compute $h = f(y)$ in time μ and run $\mathsf{A}(h)$ to obtain a value x in total time $\tau - \mu + \mu = \tau < \tau_2$. By construction, with probability $p > \varepsilon$, the pre-image x satisfies $g(x) = y$, as $f(g(x)) = f(y) = h$, and the inversion of g is again easy.   ∎

Lemma 1 lead us to conclude that, given a cryptographically secure primitive (in a mathematical sense), we can achieve leakage resiliency in our considered leakage model if either one between the physical functions or the intermediate algorithmic functions, or the combination thereof, is hard-to-invert. We capture this security notion by the following definition:

Definition 4 (Hard-to-Invert Leakages). *A function family $\mathcal{F}$ has $(s, q, \tau, \varepsilon)$-hard-to-invert leakages if any adversary A running in time at most τ, given access to a physical implementation $\mathsf{P} = (\mathsf{F}_k, \mathcal{L}_k)$ and allowed to make q queries to F_k, repeat at most s queries to $\mathcal{L}_k$, has advantage at most ε in inverting the leakages:*

$$\Pr[k \leftarrow \mathsf{A}_\mathsf{P}(\{y_1, \ldots, y_q\}) : \{y_i = \mathcal{L}_k(x_i)\}_{i=1}^q] < \varepsilon.$$

Definition 4 provides a simple, clear and sufficient criterion to side-channel security by stating that the composition of intermediate and physical functions must be hart-to-invert to achieve security in presence of leakages. We put forward that, despite many achievements in the field of secure layout and logic style design, designing secure physical functions basically reduces to the design of constant switching logic styles and some more ad-hoc layout protections to secure the chip die. These solutions are typically unsatisfactory as they are error-prone and too expensive to realize. Furthermore, the task of designing hard-to-invert physical functions belongs completely to the engineering domain and therefore out-of-scope in the context of (leakage-resilient) cryptography, whereas intermediate functions should be considered indeed. This fact directly translates into committing cryptographers and hardware designers to the task of designing and implementing cryptographic algorithmic, whose intermediate states remain hard-to-invert in our (realistic) leakage model. In this way, we (possibly) relieve physical designers from the complex task of designing hard to invert circuits and allow for the fabrication of cryptographic device using standard cell libraries and layout rules without affecting the final physical security. In other words, although standard fabrication processes might lead to simple physical functions (such as the Hamming- weight or distance), the resistance to physical attacks is preserved as long as their combination with algorithmic functions yield hard-to-invert leakages by careful design and implementation of intermediate functions and hardware architectures.

5   Efficient Pseudorandom Functions

Having defined the leakage model and our approach to side-channel security (Definition 4), let us illustrate how we can work with it, using the efficient GGM construction as our showcase. For the sake of self-containment, we use the next paragraphs to refresh the reader's memory about GGM-trees for pseudorandom functions.

In practice, GGM-trees PRFs can be instantiated from any length doubling PRG using the binary tree construction [14] or the more efficient b-ary tree construction [33], as formalized by Definition 5. The interest for *efficient* pseudorandom function constructions lies in the reduced number of PRG evaluations which reduces the computational burden down to more practicable levels (m/b evaluations on the b-ary tree as opposed to m evaluations of the original binary tree construction).

Definition 5 (Efficient GGM-tree PRF). *Let κ be a security parameter and $G: \{0,1\}^\kappa \to \{0,1\}^{2^b\kappa}$ be a $(2^b\kappa, \tau, \varepsilon)$-secure PRG. Let $g_j(x)$ denote the $(j+1)^{th}$ κ-bit block of $G(x)$ for $j = 0, \ldots, 2^b - 1$, and fix a positive constant $m \in \mathbb{N}$. Let $x \in \{0,1\}^m$ be an input message, partitioned into b-bit blocks $x_1, \ldots, x_n$ ($m = nb$) and $k \in \{0,1\}^\kappa$ a secret value. A PRF $F: \{0,1\}^\kappa \times \{0,1\}^m \to \{0,1\}^\kappa$ is constructed as follows, using the efficient GGM:*

$$\begin{cases} y_0 & = k & \text{[Initialization]} \\ y_i & = g_{x_i}(y_{i-1}) \quad i = 1, \ldots, n & \text{[Iteration]} \\ F_k(x) := y_n & & \text{[Output]} \end{cases} \tag{3}$$

5.1   Concrete Security Analysis

In this section, we define the concrete security of efficient GGM constructions in terms of time complexity τ and data complexity q, but without considering any leakage information yet. The purpose of this analysis is to formalize the security versus performance trade-off as a function of the efficiency parameter b and input size m in a theoretical setting.

We partition the output of the PRG $G: \{0,1\}^\kappa \to \{0,1\}^{2^b\kappa}$ into 2^b blocks. This defines 2^b functions $g_1, \ldots, g_{2^b}: \{0,1\}^\kappa \to \{0,1\}^\kappa$, so that $G(x) = g_1(x)\| g_2(x)\| \ldots \|g_{2^b}(x)$, where $\|$ denotes the string concatenation (or more generally, any encoding from which the individual parts can be extracted uniquely and efficiently). We claim that if G is a $(2^b\kappa, \varepsilon, \tau)$-secure PRG, then F is a $(q, \tau - q \cdot t_{\mathsf{PRF}}, m/b\varepsilon)$-secure PRF, when constructed according to the efficient GGM construction of Definition 5. Hence, the security of F can be traded for performance by convenient tuning of its input size m and the efficiency parameter b, being the adversary's advantage actually regulated by the ratio m/b.

Definition 6. *Let $\delta_\tau^{\mathsf{A}}(X, Y)$ denote the advantage of an adversary* A *running in time at most τ in distinguishing X from Y as:*

$$\delta_\tau^{\mathsf{A}}(X, Y) := |\Pr[x \xleftarrow{\$} X : \mathsf{A}(x) = 1] - \Pr[y \xleftarrow{\$} Y : \mathsf{A}(y) = 1]|.$$

The computational distance is defined as

$$\delta_\tau(X, Y) := \max_{\mathsf{A}} \delta_\tau^{\mathsf{A}}(X, Y) = \max_{\mathsf{A}} \left| \Pr[x \xleftarrow{\$} X : \mathsf{A}(x) = 1] - \Pr[y \xleftarrow{\$} Y : \mathsf{A}(y) = 1] \right|$$

Lemma 2. *Let* G $: \{0, 1\}^\kappa \to \{0, 1\}^{2^b\kappa}$ *be an $(2^b\kappa, \tau, \varepsilon)$-secure PRG, and let $g_i(x)$ denote the $(i+1)^{th}$ κ-bit block of the output G(x). Then each g_i satisfies*

$$\delta_\tau(g_i(\mathcal{U}_\kappa), \mathcal{U}_\kappa) \leq \varepsilon.$$

Proof. Assume the opposite, i.e., for some $1 \leq i \leq 2^b$ there is an algorithm A for which $\delta_\tau^{\mathsf{A}}(g_i(\mathcal{U}_\kappa), \mathcal{U}_\kappa)) > \varepsilon$. From A, we can construct an algorithm A' that distinguishes G$(\mathcal{U}_\kappa)$ from $\mathcal{U}_{2^b\kappa}$, by extracting the i-th b-bit block from its input (either G$(\mathcal{U}_\kappa)$ or $\mathcal{U}_{2^b\kappa}$), feeding it into A' and outputting whatever A computes. Observe that extracting the same i-th b-bit block from $\mathcal{U}_{2^b\kappa}$ yields another uniformly distributed random variable $\mathcal{U}_\kappa$. So, by construction, $\delta_\tau^{\mathsf{A}'}(\mathsf{G}(\mathcal{U}_\kappa), \mathcal{U}_{2^b\kappa}) = \delta_\tau^{\mathsf{A}}(g_i(\mathcal{U}_\kappa), \mathcal{U}_\kappa) > \varepsilon$, thus contradicting the $(2^b\kappa, \tau, \varepsilon)$-security of G. ∎

Proposition 1 (Concrete Security of Efficient GGM-tree PRF). *If* G $: \{0, 1\}^\kappa \to \{0, 1\}^{2^b\kappa}$ *is a $(2^b\kappa, \tau, \varepsilon)$-secure PRG, then* F$: \{0, 1\}^\kappa \times \{0, 1\}^{nb} \to \{0, 1\}^\kappa$, *constructed according to the efficient GGM-tree construction, is a $(q, \tau - q \cdot t_{\mathsf{PRF}}, m\varepsilon/b)$-secure PRF, where t_{PRF} is the (constant) time to evaluate any member of the PRF family $\mathcal{F}$ induced by the efficient GGM-tree construction.*

Proof By the hypothesis and Lemma 2, we have

$$\delta_\tau(g_{x_i}(\mathcal{U}_\kappa), \mathcal{U}_\kappa) \leq \varepsilon \tag{4}$$

for $i = 1, 2, \ldots, 2^b$. Since F$(k, x) = g_{x_n}(g_{x_{n-1}}(\cdots g_{x_2}(g_{x_1}(k)) \cdots))$, we can repeatedly apply (4) $n = m/b$ times to infer that

$$\delta_\tau(\mathsf{F}(\mathcal{U}_\kappa, x), \mathcal{U}_\kappa) \leq \frac{m}{b}\varepsilon \quad \text{for all } x. \tag{5}$$

Suppose towards a contradiction that F would not be a $(q, \tau - q \cdot t_{\mathsf{PRF}}, m\varepsilon/b)$-secure PRF, i.e., there are (adaptively chosen) values $x_1^*, x_2^*, \ldots, x_q^*, x_{q+1}^*$ and an algorithm A_2 with the property that

$$\left| \Pr(A_2(x_{q+1}^*, \mathsf{F}(k, x_{q+1}^*), I^*) = 1) - \Pr(A_2(x_{q+1}^*, \mathcal{U}_\kappa, I^*) = 1) \right| > \frac{m}{b}\varepsilon,$$

where $I^* := \{(x_i^*, F(k, x_i^*)) | i = 1, 2, \ldots, q\}$. We define an algorithm $A_2'(z) := A_2(x_{q+1}^*, z, I^*)$. The complexity of A_2' is τ, since it first recovers I^* by q evaluations of $F_k = F(k, \cdot)$, each in t_{PRF} steps, and finally invokes A_2 in time $\tau - q \cdot t_{PRF}$. By construction, we have

$$\left| \Pr(A_2'(F(k, x_{q+1}^*)) = 1) - \Pr(A_2'(\mathcal{U}_\kappa) = 1) \right| > \frac{m}{b}\varepsilon,$$

and therefore[1] also

$$\delta_\tau(F(\mathcal{U}_\kappa, x_{q+1}^*), \mathcal{U}_\kappa) = \max_A \delta_\tau^A(F(\mathcal{U}_\kappa, x_{q+1}^*), \mathcal{U}_\kappa)$$
$$= \max_A \left| \Pr(A(F(\mathcal{U}_\kappa, x_{q+1}^*)) = 1) - \Pr(A(\mathcal{U}_\kappa) = 1) \right|$$
$$\geq \left| \Pr(A_2'(F(\mathcal{U}_\kappa, x_{q+1}^*)) = 1) - \Pr(A_2'(\mathcal{U}_\kappa) = 1) \right| > \frac{m}{b}\varepsilon,$$

thus contradicting (5). ∎

5.2   Theoretical Analysis with Hard-to-Invert Leakages

In this section, we extend the security assertion of Proposition 1 to security under hard-to-invert compositions of leakages and algorithmic functions in the sense of Lemma 1 and Definition 3.

Similarly to [10], every algorithmic step in the analysis corresponds to a PRG invocation, so that the intermediate functions f_i correspond to the functions g_i, as defined in Lemma 2. Assuming that the leakages draws information at the point where some secret goes into the transformation (input leakage), and at the point where the information is used in the algorithm (output leakage), the total leakage for a efficient GGM construction upon the input $x = x_1 \| x_2 \| \cdots \| x_n$ and the secret key k specializes to $\{\ell(f_{i-1}(x, k), f_i(x, k)) = \ell(k_{i-1}, g_{x_i}(k_i))\}$ for all $i = 1, 2, \ldots, n$. Therefore, as per Definition 5, the evaluation of a PRF F_k with leakage $\mathcal{L}_k$ amounts to the following chain:

$$[k_0 = k] \to [k_1 = g_{x_1}(k_0)] \to [k_2 = g_{x_2}(k_1)] \to \cdots \to [g_{x_n}(k_{n-1}) = F_k(x)] \tag{6}$$

$$\ell(k_0, g_{x_1}(k_0)) \quad \ell(k_1, g_{x_2}(k_1)) \quad \cdots \quad \ell(k_{n-1}, g_{x_n}(k_{n-1}))$$

Next, we demonstrate that the proof of [33, Theorem 2] remains valid under the even weaker condition of hardness of inversion (which implies Standaert et al.'s *symmetric seed-preserving* condition). This is especially interesting as hardness of inversion, unlike symmetric seed-preservation, is *not* limited to the particular application to 2PRGs and can be applied to general transformations like 2^bPRGs to deduce security. This generalization is enabled by the observation that 2^bPRGs do not actually *compute* the fully stretched $2^b\kappa$-bit output when

---

[1] Note that the computational distance satisfies the triangle inequality, i.e., $\delta_\tau(X, Y) \leq \delta_\tau(X, Z) + \delta_\tau(Z, Y)$.

plugged into the GGM-tree PRF, rather they only compute the κ-bit output necessary for the next iteration. Hence, there is no leakage occurring on the $(2^b - 1)\kappa$-bit portion of the output which is not either computed or given to the adversary.

Proposition 2 (Leakage Resilience of Efficient GGM-tree PRF). *A PRF constructed upon a $(2^b\kappa, \tau, \varepsilon/(p \cdot (n+1)))$-secure 2^bPRG according to the efficient GGM-tree construction $(m = nb)$ and having $(\tau + (q+1) \cdot t_{PRF}, \varepsilon/(2^b p \cdot (n+1)))$-hard-to-invert leakages is a $(q, s, \tau, \varepsilon)$-secure efficient leakage-resilient PRF in the following sense: an adversary $\mathsf{A}^{2^b \mathsf{PRG}} = (A_1, A_2)$ that runs in time at most τ, given access to a physical implementation $\mathsf{P} = (\mathsf{F}_k, \mathcal{L}_k)$ and an (additional) access to 2^bPRG and allowed to make at most q queries to F_k, repeat at most s queries to $\mathcal{L}_k$ and make at most p queries to 2^bPRG, has advantage at most ε in distinguishing the output of F_k from random:*

$$\mathsf{Adv}(\mathbf{Exp}_{A^{2^b \mathsf{PRG}, \mathsf{P}}}^{LRPRF\text{-}IND}) := \left| \Pr(\mathbf{Exp}_{A^{2^b \mathsf{PRG}, \mathsf{P}}}^{LRPRF\text{-}IND\text{-}0} = 1) - \Pr(\mathbf{Exp}_{A^{2^b \mathsf{PRG}, \mathsf{P}}}^{LRPRF\text{-}IND\text{-}1} = 1) \right| < \varepsilon$$

Our proof partially works along analogous lines as the proof of Theorem 2 in [33], by showing that upon a given set of q independent instances of the form $\ell(k_{i-1}, g_{x_i}(k_{i-1}))$, where the challenge is to recover k_i, an attacker with significant advantage in $\mathbf{Exp}_{A^{2^b \mathsf{PRG}, \mathsf{P}}}^{LRPRF\text{-}IND}$ could invert at least one of the given instances with significant probability $> \varepsilon/(2^b p(n+1))$. Please note that differently from [33]: (1) our proof rests on the hardness of inversion[2], instead of the stronger symmetric seed-preserving definition, (2) it is valid for the more generic case of 2^bPRGs, rather than for the specific case of 2PRG only and (3) our argument is concrete rather than asymptotic.

Proof. By construction, the outcome of $\mathbf{Exp}_{A^{2^b \mathsf{PRG}, \mathsf{P}}}^{LRPRF\text{-}IND\text{-}0}$ is the output of A_2 (when $\mathsf{A}^{2^b \mathsf{PRG}, \mathsf{P}} = (A_1, A_2)$) with the first q adaptive queries to $(\mathsf{F}_k, \mathcal{L}_k)$, and the $(q+1)^{th}$ query to F_k alone. Let us denote this fact by $A_2^{(\mathsf{F}_k, \mathcal{L}_k)_{[1:q]}, (\mathsf{F}_k, \emptyset)}$. The random function R is constructed using the same chain structure as (6), except that all values k_i are chosen independently and uniformly at random. The output of R is the last random value $k_n \xleftarrow{\$} \mathcal{U}_\kappa$. By the triangle inequality, we have

$$\mathsf{Adv}(\mathbf{Exp}_{A^{2^b \mathsf{PRG}, \mathsf{P}}}^{LRPRF\text{-}IND}) = \left| \Pr[A_2^{(\mathsf{F}_k, \mathcal{L}_k)_{[1:q]}, (\mathsf{F}_k, \emptyset)} = 1] - \Pr[A_2^{(\mathsf{F}_k, \mathcal{L}_k)_{[1:q]}, (R, \emptyset)} = 1] \right|$$

$$\leq \left| \Pr[A_2^{(\mathsf{F}_k, \mathcal{L}_k)_{[1:q]}, (\mathsf{F}_k, \emptyset)} = 1] - \Pr[A_2^{(R, \mathcal{L}_R)_{[1:q]}, (R, \emptyset)} = 1] \right|$$

$$+ \left| \Pr[A_2^{(R, \mathcal{L}_R)_{[1:q]}, (R, \emptyset)} = 1] - \Pr[A_2^{(\mathsf{F}_k, \mathcal{L}_k)_{[1:q]}, (R, \emptyset)} = 1] \right|$$

$$\leq \left| \Pr[A_2^{(\mathsf{F}_k, \mathcal{L}_k)_{[1:q+1]}} = 1] - \Pr[A_2^{(R, \mathcal{L}_R)_{[1:q+1]}} = 1] \right|$$

$$+ \left| \Pr[A_2^{(R, \mathcal{L}_R)_{[1:q]}} = 1] - \Pr[A_2^{(\mathsf{F}_k, \mathcal{L}_k)_{[1:q]}} = 1] \right|,$$

---

[2] Our proof exclusively uses the compositions of intermediate and physical functions in their arguments. Hence by Lemma 1, we are safe to assume at least one of them is hard to invert.

so, we can reduce the problem to show the indistinguishability between $(F_k, \mathcal{L}_k)$ and $(R, \mathcal{L}_R)$. Towards a contradiction, assume that there were an attacker $A^{2^b PRG, P}$ with oracle (physical) access to the $2^b PRG$, who queries the pseudorandom generator at most p times and his $(F_k, \mathcal{L}_k)$-oracle no more than q times, and over a maximal running time of τ has an advantage of

$$\text{Adv}(\mathbf{Exp}^{\text{LRPRF-IND}}_{A^{2^b PRG, P}}) > \varepsilon \tag{7}$$

in distinguishing the PRF output in $\mathbf{Exp}^{\text{LRPRF-IND}}_{A^{2^b PRG, P}}$ from random, by taking advantage from the leakages. From such an attacker, we will construct an algorithm that extracts the unknown k_{i-1} from at least one out of a system of q equations of the form $y_i = \ell(k_{i-1}, g_{x_i}(k_{i-1}))$ for some x. We consider hybrids $(H_0, L_0), \ldots, (H_n, L_n)$, where each pair (H_j, L_j) is constructed using the chain (6) in $\mathbf{Exp}^{\text{LRPRF-IND-0}}_{A^{2^b PRG, P}}$, with random intermediate values $k_i \xleftarrow{\$} \mathcal{U}_\kappa$ for $i < j$, and the remaining intermediate values k_j are computed by invocations of $2^b PRG$, whenever F_k is evaluated in the experiment. We set H_j as the final output F_k, and L_j is the collection of all leakages (formerly denoted as $\mathcal{L}_k$) along the (so-modified) computation of F_k, including those obtained from the q queries during the learning phase. It is easy to see that (H_0, L_0) equals $(F_k, \mathcal{L}_k)$ and (H_n, L_n) equals $(R, \mathcal{L}_R)$. Notice that (7) implies that there must be at least one pair of consecutive hybrids $(H_j, L_j), (H_{j+1}, L_{j+1})$ that $A^{2^b PRG, P}$ can distinguish with an advantage of at least $> \varepsilon/(n+1)$, for otherwise, we would have a total advantage of $\le \varepsilon$, contradicting (7). Call $\{k'_1, k'_2, \ldots, k'_p\}$ the intermediate values that $A^{2^b PRG, P}$ obtains internally over its computation with at most p invocations of its oracle. Call $\mathcal{L}_k(x) = \{\ell(k_{i-1}, g_{x_i}(k_{i-1}))\}^n_{i=1}$ the entirety of leakages harvested from the computation of $F_k(x)$. Assume that, given $\mathcal{L}_k$, none of $A^{2^b PRG, P}$'s interim results would match any intermediate value the real computation of F_k, i.e.,

$$\{k'_1, k'_2, \ldots, k'_p\} \cap \{k_0, k_1, \ldots, k_n\} = \emptyset. \tag{8}$$

Then, every of the p invocations of $2^b PRG$ on the values $k'_1, k'_2, \ldots, k'_p$ (note that there are no other values to which $A^{2^b PRG, P}$ could query $2^b PRG$ on) is indistinguishable from random (by the security of the $2^b PRG$ and Lemma 2), thus giving an advantage of no more than $\delta = \varepsilon/(p \cdot (n+1))$ over a computation that would not use this helping oracle. Moreover, we stress that (H_j, L_j) and (H_{j+1}, L_{j+1}) are identically distributed for anyone without access to $2^b PRG$. Hence, any such algorithm would have zero advantage. However, we could replace $A^{2^b PRG, P}$ by another algorithm A' that avoids oracle-access to $2^b PRG$ by using random values wherever $A^{2^b PRG, P}$ would invoke $2^b PRG$, and whose advantage differs from that of $A^{2^b PRG, P}$ by a magnitude of no more than $p \cdot \delta$, thus implying that the advantage of $A^{2^b PRG, P}$ is less than $p \cdot \delta = p\varepsilon/(p(n+1)) = \varepsilon/(n+1)$, which is a contradiction to the $\varepsilon/(n+1)$ advantage implied by (7). Hence, (8) cannot hold, and $A^{2^b PRG, P}$ must obtain its $\varepsilon/(n+1)$ advantage by the ability of inverting at least one of the leakage values along the chain. In that case, it simply completes the chain by completing the computation of F_k and verifies whether or not the outcome corresponds to H_j. Now, given q independent instances $y_i = \ell(k_{i-1}, g_{x_i}(k_{i-1}))$ for $i = 1, 2, \ldots, q$, with the challenge to recover k_i, an inversion algorithm Inv that solves one of these instances can be constructed as follows from $(H_j, L_j), (H_{j+1}, L_{j+1})$ and $\mathsf{A}^{2^b\mathsf{PRG},\mathsf{P}}$: notice that (H_j, L_j) and (H_{j+1}, L_{j+1}) differ only at the j-th position, where we have

$$(H_j, L_j) \begin{cases} [r_0 \xleftarrow{\$} \mathcal{U}_\kappa] \to \cdots \to [k_j = g_{x_j}(r_{j-1})] \to \left| [\mathsf{k_{j+1}} = \mathsf{g_{x_{j+1}}}(\mathsf{k_j})] \to \cdots \to [g_{x_n}(k_{n-1}) = F_k(x)] \right. \\ \qquad\qquad\qquad\qquad\quad \updownarrow \qquad\qquad\quad \left| \quad \updownarrow \qquad\qquad\qquad\qquad\qquad \updownarrow \right. \\ \qquad\qquad\qquad\quad \ell(r_{j-1}, g_{x_j}(r_{j-1})) \quad \left| \ell(\mathsf{k_j}, \mathsf{g_{x_{j+1}}}(\mathsf{k_j})) \quad \cdots \quad \ell(k_{n-1}, g_{x_n}(k_{n-1})) \right. \end{cases}$$

$$(H_{j+1}, L_{j+1}) \begin{cases} [r_0 \xleftarrow{\$} \mathcal{U}_\kappa] \to \cdots \to [k_j = g_{x_j}(r_{j-1})] \to [\mathsf{k_{j+1}} = \mathsf{g_{x_{j+1}}}(\mathsf{r_j})] \left| \to \cdots \to [g_{x_n}(k_{n-1}) = F_k(x)] \right. \\ \qquad\qquad\qquad\qquad\quad \updownarrow \qquad\qquad\qquad\quad \updownarrow \qquad\qquad \left| \qquad\qquad\qquad \updownarrow \right. \\ \qquad\qquad\qquad\quad \ell(r_{j-1}, g_{x_j}(r_{j-1})) \quad \ell(\mathsf{r_j}, \mathsf{g_{x_{j+1}}}(\mathsf{r_j})) \left| \quad \cdots \quad \ell(k_{n-1}, g_{x_n}(k_{n-1})) \right. \end{cases}$$

The inversion algorithm Inv simulates the hybrids in time $\leq (q+1) \cdot t_{PRF}$, and in each of the $q + 1$ rounds (taking $\leq n$ executions of $2^b\mathsf{PRG}$) within the experiment $\mathbf{Exp}_{\mathsf{A}^{2^b\mathsf{PRG},\mathsf{P}}}^{\mathsf{LRPRF\text{-}IND}}$, and puts another of the input instances $y_i = \ell(k_{i-1}, g_{x_i}(k_{i-1}))$ (for $i = 1, 2, \ldots, q$) in place of the j-th leakage y_j in both, H_j and H_{j+1}, giving the modified hybrids for a replacement $y_j \leftarrow y_i$ (the i-th input to Inv) among q input instances,

$$(H_j, L_j') \begin{cases} [r_0 \xleftarrow{\$} \mathcal{U}_\kappa] \to \cdots \to [k_j = g_{x_j}(r_{j-1})] \to \left| [\mathsf{k_{j+1}} = \mathsf{g_{x_{j+1}}}(\mathsf{k_j})] \to \cdots \to [g_{x_n}(k_{n-1}) = F_k(x)] \right. \\ \qquad\qquad\qquad\qquad\quad \updownarrow \qquad\qquad\qquad\quad \left| \qquad\qquad\qquad\qquad\qquad\qquad \updownarrow \right. \\ \qquad\qquad\qquad\quad \ell(r_{j-1}, g_{x_j}(r_{j-1})) \quad \left| \qquad \mathsf{y_i} \qquad\qquad \cdots \quad \ell(k_{n-1}, g_{x_n}(k_{n-1})) \right. \end{cases}$$

$$(H_{j+1}, L_{j+1}') \begin{cases} [r_0 \xleftarrow{\$} \mathcal{U}_\kappa] \to \cdots \to [k_j = g_{x_j}(r_{j-1})] \to [\mathsf{k_{j+1}} = \mathsf{g_{x_{j+1}}}(\mathsf{r_j})] \left| \to \cdots \to [g_{x_n}(k_{n-1}) = F_k(x)] \right. \\ \qquad\qquad\qquad\qquad\quad \updownarrow \qquad\qquad\qquad\qquad\quad \left| \qquad\qquad\qquad\qquad\qquad \updownarrow \right. \\ \qquad\qquad\qquad\quad \ell(r_{j-1}, g_{x_j}(r_{j-1})) \qquad\quad \mathsf{y_i} \qquad \left| \quad \cdots \quad \ell(k_{n-1}, g_{x_n}(k_{n-1})) \right. \end{cases}$$

Notice that replacing $\ell(r_j, g_{x_{j+1}}(r_j))$ by an input instance $y_i = \ell(k_{i-1}, g_{x_i}(k_{i-1}))$ is possible, if x_{j+1} exactly selects the given g_{x_i} (happens with probability $\Pr_{\{0,1\}^b} [x_{j+1} = x_i] = 2^{-b}$), and that this substitution does not change the distribution of $(H_j, L_j), (H_{j+1}, L_{j+1})$, as we could equivalently and coincidentally have chosen $r_j = k_{i-1}$ as the pre-image to the leakage in first place already. Now, Inv asks $\mathsf{A}^{2^b\mathsf{PRG},\mathsf{P}}$ to distinguish (H_{j+1}, L_{j+1}') from (H_j, L_j'), which by construction succeeds with probability $> \varepsilon/(n+1)$ in time $\tau + (q+1) \cdot t_{PRF}$ (considering the additional effort for simulation of the hybrids), and in the course of which at least one of the leakages must have been inverted along no more than p queries to $2^b\mathsf{PRG}$ (as argued above). The algorithm Inv then simply records all queries submitted by $\mathsf{A}^{2\mathsf{PRG},\mathsf{P}}$, and returns a randomly chosen one as the final output. Hence, the overall probability for a particular input $\ell(k_i, g_{x_{i+1}}(k_i))$ to have become inverted then comes to $\geq \varepsilon/(2^b \cdot p \cdot (n+1))$, which contradicts the assumed hardness of inversion. ∎

5.3   Hardware Implementations with Super-Exponential Leakages

In this section, we recap the cryptographic and implementation criteria of efficient pseudorandom functions with the aim of illustrating how hardness of inversion can be preserved at implementation level by careful design choices (we refer to [3,21] for concrete hardware implementations and side-channel analysis). To this end, we consider block-ciphers efficient GGMbased PRFimplementations [3,28,32,33]. This particular choice turns a PRF into a block-cipher mode of operation which can be seen as the traversal of the GGM b-ary tree construction, as formally described in Definition 7. Note that in general a PRF would map into $\{0,1\}^n$ for some fixed n, which we set equal to the security parameter κ in our treatment. This is especially convenient in cases where the PRF is constructed from block-ciphers in the form BC : $\{0,1\}^\kappa \times \{0,1\}^\kappa \to \{0,1\}^\kappa$ as they allow for more regular hardware design.

Definition 7 (Block-cipher based GGM-tree PRF). *Let κ be a security para-meter and* BC : $\{0,1\}^\kappa \times \{0,1\}^\kappa \to \{0,1\}^\kappa$ *be a SP-based block-cipher with b-bit S-boxes. Let $x \in \{0,1\}^m$ be an input message, partitioned into b-bit blocks $x_1, \ldots, x_n$ ($m = nb$) and $k \in \{0,1\}^\kappa$ a secret value. A PRF* F: $\{0,1\}^\kappa \times \{0,1\}^m \to \{0,1\}^\kappa$ *is constructed as follows, using the efficient GGM-tree construction:*

$$
\begin{cases}
y_0 & = k & \text{[Initialization]} \\
y_i & = \mathsf{BC}y_{i-1}(x_i^1||x_i^2||...||x_i^{\kappa/b}) & i = 1, \ldots, n & \text{[Iteration]} \\
\mathsf{F}_k(x) & := y_n & & \text{[Output]}
\end{cases}
\tag{9}
$$

It is easy to see that the hardness of inversion to some extent relies on (implies) high complexity: given that a function f_i is bijective and deterministic, it is necessary for f_i to have a large image (or, equivalently pre-image) space, so as to avoid brute-force pre-image search. For example, if the pre-image space has a cardinality that is polynomial in the security parameter, then a plain search by trial-and-error will – in polynomial time – dig up a pre-image. This requirement practically translates into demanding large intermediate values, say $\kappa = 128$, as well as large datapath in the corresponding hardware architecture (ideally, this is steered by κ as well). Indeed, it is easy to verify that serialized hardware architectures, which only process a small amount of data at a time (say, 8-bit), would lead to easy invertible computations in the initial and final computations, if no additional resis-tance is ensured by hard-to-invert physical functions. On the other side, it can be noted that the requirement of having a large image alone is still insufficient to achieve hardness of inversion in practice. Namely, if the intermediate functions $f_i : \{0,1\}^\kappa \times \{0,1\}^\kappa \to \{0,1\}^\kappa$ can be viewed as the concatenation of smaller (independent) functions, then the adversary can address those individual parts independently and break the "large image" requirement. This is practically the case of standard SP-based block-ciphers when e.g., the initial or the final state are processed by small-size independent additions and S-box functions. In this case, the adversary can take a divide-and-conquer approach and would require only a slightly increased effort to get rid of the so called algorithmic noise induced by the excluded functions [18]. However, the implementation of leakage-resilient PRF, as provided in Definition 7, ensures the dependency of such functions by splitting the inputs into words of size b, which are then *replicated* to fit the block cipher input length κ. This careful choice of the inputs (which limits the data complexity to 2^b by construction) together with our realistic modeling assumptions given in Sect. 3 lead now to *super-exponential* leakages for the adversary. Indeed, in the worst case, the adversary has to enumerate all the $(\kappa/b)!$ permutations of κ/b keywords to fully recover the secret state back, once the κ/b individual keywords are (in the best case) successfully recovered by classical divide-and-conquer side-channel means [21].

6   Conclusion

In this work, we made an attempt to provide a constructive *critique* to secure hardware design aimed at bridging the gap between theoretical leakage-resilient constructions and practical side-channel resistant implementations of efficient GGM-tree PRFs. We put forward how hard-to-invert leakages can provide a unified approach to embrace security in presence of leakages from both theoretical and practical sides. In fact, if from one side it has been often observed that standard cell libraries and layout rules typically lead to physical functions with very limited complexity, like the Hamming-weight or the Hamming-distance, on the other side, it seems natural to focus the attention on the design and implementation of algorithmic steps to achieve hard-to-invert leakages. This fact outlines how theoreticians, cryptographers and hardware designers can collaborate at different levels of abstraction, but on a common ground with clearly defined goals, to devise leakage-resilient constructions, cryptographic algorithms and implementation criteria to ultimately build physically secure cryptographic devices.

Acknowledgment. The authors would like to thank the reviewers for the constructive and helpful comments.

References

1. Abdalla, M., Belaïd, S., Fouque, P.-A.: Leakage-resilient symmetric encryption via re-keying. In: Bertoni, G., Coron, J.-S. (eds.) CHES 2013. LNCS, vol. 8086, pp. 471–488. Springer, Heidelberg (2013)
2. Abdalla, M., Bellare, M.: Increasing the lifetime of a key: a comparative analysis of the security of re-keying techniques. In: Okamoto, T. (ed.) ASIACRYPT 2000. LNCS, vol. 1976, p. 546. Springer, Heidelberg (2000)
3. Belaid, S., De Santis, F., Heyszl, J., Mangard, S., Medwed, M., Schmidt, J.M., Standaert, F.X., Tillich, S.: Towards fresh re-keying with leakage-resilient PRFs: cipher design principles and analysis. J. Cryptogr. Eng. **4**, 1–15 (2014)
4. Bellare, M., Desai, A., Jokipii, E., Rogaway, P.: A concrete security treatment of symmetric encryption. In: Proceedings of the 38th Annual Symposium on Foundations of Computer Science, p. 394. FOCS, IEEE Computer Society, Washington, DC, USA (1997)

5. Brier, E., Clavier, C., Olivier, F.: Correlation power analysis with a leakage model. In: Joye, M., Quisquater, J.-J. (eds.) CHES 2004. LNCS, vol. 3156, pp. 16–29. Springer, Heidelberg (2004)
6. Chari, S., Rao, J.R., Rohatgi, P.: Template attacks. In: Kaliski, B.S., Koç, Ç.K., Paar, C. (eds.) Cryptographic Hardware and Embedded Systems - CHES 2002. LNCS, vol. 2523, pp. 13–28. Springer, Heidelberg (2003)
7. Coron, J.-S., Goubin, L.: On boolean and arithmetic masking against differential power analysis. In: Paar, C., Koç, Ç.K. (eds.) CHES 2000. LNCS, vol. 1965, p. 231. Springer, Heidelberg (2000)
8. Dodis, Y., Kalai, Y.T., Lovett, S.: On cryptography with auxiliary input. In: Proceedings of the 41st Annual ACM Symposium on Theory of Computing, pp. 621–630. STOC 2009. ACM, New York, NY, USA (2009)
9. Dziembowski, S., Pietrzak, K.: Leakage-resilient cryptography. In: 49th Annual IEEE Symposium on Foundations of Computer Science, FOCS 2008, pp. 293–302. IEEE Computer Society, Philadelphia, PA, USA (2008)
10. Faust, S., Pietrzak, K., Schipper, J.: Practical leakage-resilient symmetric cryptography. In: Prouff, E., Schaumont, P. (eds.) CHES 2012. LNCS, vol. 7428, pp. 213–232. Springer, Heidelberg (2012)
11. Faust, S., Pietrzak, K., Schipper, J.: Practical leakage-resilient symmetric cryptography. In: Prouff, E., Schaumont, P. (eds.) CHES 2012. LNCS, vol. 7428, pp. 213–232. Springer, Heidelberg (2012)
12. Gierlichs, B., Batina, L., Tuyls, P., Preneel, B.: Mutual information analysis. In: Oswald, E., Rohatgi, P. (eds.) CHES 2008. LNCS, vol. 5154, pp. 426–442. Springer, Heidelberg (2008)
13. Goldreich, O.: Foundations of Cryptography: Basic Applications, vol. 2. Cambridge University Press, New York (2004)
14. Goldreich, O., Goldwasser, S., Micali, S.: How to construct random functions. J. ACM **33**(4), 792–807 (1986)
15. Goubin, L.: A sound method for switching between boolean and arithmetic masking. In: Koç, Ç.K., Naccache, D., Paar, C. (eds.) CHES 2001. LNCS, vol. 2162, p. 3. Springer, Heidelberg (2001)
16. Heyszl, J., Mangard, S., Heinz, B., Stumpf, F., Sigl, G.: Localized electromagnetic analysis of cryptographic implementations. In: Dunkelman, O. (ed.) CT-RSA 2012. LNCS, vol. 7178, pp. 231–244. Springer, Heidelberg (2012)
17. Kirschbaum, M.: Power analysis resistant logic styles - design, implementation, and evaluation. Ph.D. thesis (2011)
18. Mangard, S.: Hardware countermeasures against DPA – a statistical analysis of their effectiveness. In: Okamoto, T. (ed.) CT-RSA 2004. LNCS, vol. 2964, pp. 222–235. Springer, Heidelberg (2004)
19. Mangard, S., Oswald, M.E., Popp, T.: Power Analysis Attacks - Revealing the Secrets of Smart Cards. Springer, Heidelberg (2007)
20. Medwed, M., Standaert, F.-X., Großschädl, J., Regazzoni, F.: Fresh re-keying: security against side-channel and fault attacks for low-cost devices. In: Bernstein, D.J., Lange, T. (eds.) AFRICACRYPT 2010. LNCS, vol. 6055, pp. 279–296. Springer, Heidelberg (2010)
21. Medwed, M., Standaert, F.-X., Joux, A.: Towards super-exponential side-channel security with efficient leakage-resilient PRFs. In: Prouff, E., Schaumont, P. (eds.) CHES 2012. LNCS, vol. 7428, pp. 193–212. Springer, Heidelberg (2012)
22. Messerges, T.S.: Using second-order power analysis to attack DPA resistant software. In: Paar, C., Koç, Ç.K. (eds.) CHES 2000. LNCS, vol. 1965, p. 238. Springer, Heidelberg (2000)

23. Messerges, T.S.: Securing the AES finalists against power analysis attacks. In: Schneier, B. (ed.) FSE 2000. LNCS, vol. 1978, p. 150. Springer, Heidelberg (2001)
24. Micali, S., Reyzin, L.: Physically observable cryptography. In: Naor, M. (ed.) TCC 2004. LNCS, vol. 2951, pp. 278–296. Springer, Heidelberg (2004)
25. Moradi, A., Kirschbaum, M., Eisenbarth, T., Paar, C.: Masked dual-rail precharge logic encounters state-of-the-art power analysis methods. IEEE Trans. Very Large Scale Integr. (VLSI) Syst. **20**, 1578–1589 (2012)
26. Nikova, S., Rijmen, V., Schläffer, M.: Secure hardware implementation of non-linear functions in the presence of glitches. In: Lee, P.J., Cheon, J.H. (eds.) ICISC 2008. LNCS, vol. 5461, pp. 218–234. Springer, Heidelberg (2009)
27. Peeters, E., Standaert, F.-X., Donckers, N., Quisquater, J.-J.: Improved higher-order side-channel attacks with FPGA experiments. In: Rao, J.R., Sunar, B. (eds.) CHES 2005. LNCS, vol. 3659, pp. 309–323. Springer, Heidelberg (2005)
28. Pietrzak, K.: A leakage-resilient mode of operation. In: Joux, A. (ed.) EURO-CRYPT 2009. LNCS, vol. 5479, pp. 462–482. Springer, Heidelberg (2009)
29. Popp, T., Kirschbaum, M., Zefferer, T., Mangard, S.: Evaluation of the masked logic style MDPL on a prototype chip. In: Paillier, P., Verbauwhede, I. (eds.) CHES 2007. LNCS, vol. 4727, pp. 81–94. Springer, Heidelberg (2007)
30. Schindler, W., Lemke, K., Paar, C.: A stochastic model for differential side channel cryptanalysis. In: Rao, J.R., Sunar, B. (eds.) CHES 2005. LNCS, vol. 3659, pp. 30–46. Springer, Heidelberg (2005)
31. Standaert, F.-X.: How leaky is an extractor? In: Abdalla, M., Barreto, P.S.L.M. (eds.) LATINCRYPT 2010. LNCS, vol. 6212, pp. 294–304. Springer, Heidelberg (2010)
32. Standaert, F.-X., Pereira, O., Yu, Y.: Leakage-resilient symmetric cryptography under empirically verifiable assumptions. In: Canetti, R., Garay, J.A. (eds.) CRYPTO 2013, Part I. LNCS, vol. 8042, pp. 335–352. Springer, Heidelberg (2013)
33. Standaert, F.X., Pereira, O., Yu, Y., Quisquater, J.J., Yung, M., Oswald, E.: Leakage resilient cryptography in practice. In: Sadeghi, A.R., Naccache, D. (eds.) Towards Hardware-Intrinsic Security. Information Security and Cryptography, pp. 99–134. Springer, Heidelberg (2010)
34. Vaudenay, S.: Decorrelation: a theory for block cipher security. J. Cryptol. **16**(4), 249–286 (2003)
35. Yu, Y., Standaert, F.X., Pereira, O., Yung, M.: Practical leakage-resilient pseudo-random generators. In: Proceedings of the 17th ACM Conference on Computer and Communications Security, CCS 2010, pp. 141–151. ACM, New York, NY, USA (2010)

RSA and Elliptic Curve
Least Significant Bit Security

Dionathan Nakamura and Routo Terada[(⊠)]

Computer Science Department, University of São Paulo, São Paulo, Brazil
{nakamura,rt}@ime.usp.br

Abstract. The security of the least significant bit (LSB) of the secret key in the Elliptic Curve Diffie-Hellman protocol (and of the message in the RSA) is related to the security of the whole key (and of the whole message, respectively). Algorithms to invert these cryptographic algorithms, making use of oracles that predict the LSB, have been published. We implement two of these algorithms, identify critical parameters, and modify the sampling to achieve a significant improvement in running times.

Keywords: RSA · Elliptic curve · Diffie-Hellman · Least significant bit · Oracle · Integer factorization · Discrete log problem

1 Introduction

Cryptographic algorithms are based on some computational problems that are considered to be hard, i.e., there is no known polynomial algorithm to solve them. For example, the Discrete Logarithm Problem (DLP) and the Integer Factorization Problem (IFP). The relation among cryptographic algorithms and their problems are frequent subjects of research, such as the security levels between two algorithms and their related problems.

In cryptography there is a constant need for pseudo-random number generators, and many efforts have been made to ensure that these pseudo-random number generators (PRNGs) are cryptographically secure.

Cryptographic algorithms and their corresponding problems are related to cryptographically secure pseudo-random number generators [5]. We can mention the Blum, Blum, Shub [4] pseudo-random number generator as an example of a cryptographically secure PRNG based on the Rabin encryption [20]. It iteratively uses several Rabin encryptions and uses the LSB of each encryption to compose the generated number. A similar PRNG based on RSA [21] is described in [19, Algorithm 5.35].

Many interesting studies relate the security of cryptographic schemes with the bits of the secret key, or with the bits of the ciphertext, especially with the LSB [1,3,6,7,11,13–16,22].

D. Nakamura—Partially funded by Coordenadoria de Aperfeiçoamento de Pessoal de Nível Superior (CAPES).

R. Terada—Partially funded by Fundação de Amparo à Pesquisa (FAPESP) grant 2011/50761-2.

© Springer International Publishing Switzerland 2015
D.F. Aranha and A. Menezes (Eds.): LATINCRYPT 2014, LNCS 8895, pp. 146–161, 2015.
DOI: 10.1007/978-3-319-16295-9_8

These studies analyze the RSA scheme and the elliptic curve Diffie-Hellman (ECDH) scheme [12]. For RSA the interest is in the *plaintext LSB*, and for ECDH the interest is in *the exchanged private key LSB*. Since RSA and Rabin schemes are very similar, their studies are conducted almost interchangeably. To verify the LSB security is equivalent to verifying the security of pseudo-random numbers generated by the previously mentioned PRNGs.

Alexi *et al.* [1] showed it is possible to invert RSA using an LSB oracle with probability $(1/2 + \varepsilon)$, such that $\varepsilon > 0$ is small, but non-negligible. This paper is referred to as ACGS.

Boneh *et al.* [6] showed the LSB of the exchanged point's abscissa is unpredictable, i.e., the existence of an LSB oracle would imply breaking ECDH. This paper is referred to as BS.

Fischlin *et al.* [13] showed how the running times of those algorithms can be very high, sometimes even exceeding the brute force complexity. This fact motivated this paper. The published papers on this subject take a theoretical approach, so the behaviour in software implementations and experiments was the goal to achieve. Despite the fact that their algorithms showed a theoretical convergence, our goal was to verify if they still converged with our existing PRNG, and also to verify if the running times were reasonable, even with restricted computational resources.

We implement ACGS for the RSA and BS for ECDH. During the implementations we identified the parameters that significantly impact the running times., and found where the errors were overestimated in previous papers. The number of oracle queries and running times were minimized.

The main contributions of this paper are:

- An implementation of the RSA LSB attack;
- An implementation of the ECDH LSB attack;
- Identification of the most relevant parameters for the running time;
- A reduction of the parity function bound;
- A significant reduction in the number of oracle queries;
- ACGS became faster than IFP for practical values.

We verified empirically that, if an adversary has access to an LSB oracle, there is an imbalance between the running times of the RSA LSB and ECDH LSB attacks. We give evidences of how large this imbalance is in Sect. 4. For example, given access to an LSB oracle, to invert RSA-1024 with the implementations in this work, 138.4×10^3 years (see Table 3) are needed, while to invert ECC-160, 80×10^6 years are needed (see Table 5).

Theoretical analysis given in previous published papers compare security levels of RSA and ECC, but *without* an LSB oracle. For example, Bos *et al.* [8] compare RSA-1024 and ECC-160 and Lenstra *et al.* [18] compare RSA-1825 and ECC-160.

Paper Organization. In Sect. 2 we present preliminary concepts. In Sect. 3 we describe the algorithms and the techniques for their implementations. Our results are shown and explained in Sect. 4. In Sect. 5 we present concluding remarks and suggestions for future research.

2   Preliminaries

We consider elliptic curves E over a finite field $\mathbb{F}$ of prime characteristic $\mathrm{char}(\mathbb{F}) = p$, $p \neq 2$, $p \neq 3$, where p is relatively large. We represent an elliptic curve $E/\mathbb{F}$ in the form of a simplified Weierstrass equation $E : y^2 = x^3 + ax + b$, where $a, b \in \mathbb{F}$, ∞ is the point at infinity, and the discriminant $\Delta = 4a^3 + 27b^2 \neq 0$.

Suppose we have ciphertext $y \equiv x^e \mod N = RSA(x)$ and we wish to obtain $z \equiv (cx)^e \mod N = RSA(cx)$ for some constant c. The RSA multiplicative (homomorphic) property is:

$$RSA(cx) = RSA(c).RSA(x)$$

Note that if c and $RSA(x)$ are known, it is possible to obtain the value of $RSA(cx)$ without knowing the original message x. Likewise, it is possible to obtain $RSA(2^{-1}cx)$. Another important fact: the division by 2 in the modular arithmetic is just the product of the multiplicative inverse of 2 modulus N. The multiplicative property and the division by 2 are used in ACGS. Similar computations are made with points of elliptic curves.

The modulus N is denoted by an upper case letter N and its length in bits $n = \lfloor \lg N \rfloor + 1$ is denoted by lower case letter n. If x is an integer, then the remainder of x modulus N is denoted by $[x]_N$, that is, $[x]_N = x \mod N$.

2.1   BKGCD Algorithm

A variation of the GCD (Greatest Common Divisor) algorithm called "binary GCD" is detailed in Sect. 4.5.2 (Algorithm B), page 321 of [17].

It uses the following properties by Stein [24]:

- if $|b|$ and $|c|$ are both even, then $\mathrm{GCD}(b, c) = 2 \cdot \mathrm{GCD}(\frac{b}{2}, \frac{c}{2})$;
- if $|b|$ is even and $|c|$ odd, then $\mathrm{GCD}(b, c) = \mathrm{GCD}(\frac{b}{2}, c)$;
- if $|b|$ and $|c|$ are both odd, then $\mathrm{GCD}(b, c) = 2 \cdot \mathrm{GCD}(\frac{b+c}{2}, \frac{b-c}{2})$;
- in the last case, either $\frac{b+c}{2}$ or $\frac{b-c}{2}$ is divisible by 2.

The above properties of the binary GCD are used to develop a modified binary GCD in [3], and a similar algorithm by Brent and Kung [9] is used in Algorithm 3.1 on page 152. From now on, this binary GCD is called BKGCD.

2.2 $PAR()$ Function

The $PAR()$ function is a parity function. For example, $PAR(b, y)$ returns the parity of $[bx]_N$ ($y = x^e mod N$). It is shown as Algorithm 2.1, on page 149.

$PAR()$ queries an oracle $\mathcal{O}_N$ with advantage $\varepsilon > 0$ that guesses the LSB of the RSA *plaintext* x with modulus N. Suppose this advantage is equal to $\varepsilon : 0 < \varepsilon \leq \frac{1}{2}$, where ε is small, but non-negligible. Then, the probability of the oracle to guess the $LSB(x)$ is, formally, as follows: $\mathrm{Adv}_{e,N}^x(\mathcal{O}_N) = \left| \mathrm{Pr}_{e,N}[\mathcal{O}_N(e, N) = \mathrm{LSB}(x)] - \frac{1}{2} \right| > \varepsilon > 0$.

Algorithm 2.1. Parity function $PAR(d, y)$.

Input: factor $d \in \mathbb{Z}_N$ (In Algorithm 3.1 $d = a$ or $d = b$ or $d = (a+b)/2$),
        $y = \mathrm{RSA}(x)$.
Output: the $\mathrm{LSB}(\mathrm{abs}_N(dx))$.

1   // Definition of $abs_N()$ on page 149.
2   **begin**
3      $counter_0 \leftarrow 0$;
4      $counter_1 \leftarrow 0$;
5   **for** $i \leftarrow 1$ **to** m **do**
6   choose randomly $r_i \in_R \mathbb{Z}_N$;
7   **if** $\mathcal{O}_N(\mathrm{RSA}(r_i x)) = \mathcal{O}_N(\mathrm{RSA}(r_i x + dx))$ **then**
8             $counter_0 \leftarrow counter_0 + 1$;
9   **else**
10             $counter_1 \leftarrow counter_1 + 1$;
11   **if** $counter_0 > counter_1$ **then**
12   **return** *0*;
13   **else return** *1* ;

The $PAR()$ function is a voting scheme that draws m samples to compute the parity of dx. The amount of votes are added to the variables $counter_0$ and $counter_1$, representing even and odd numbers, respectively. The result, parity bit 0 or 1, is determined by the majority of votes: $counter_0$ versus $counter_1$.

Each parity vote is obtained by verifying if $\mathrm{LSB}(rx) = \mathrm{LSB}(rx + dx)$. The equality holds iff dx is even, because an even number doesn't change the parity of another number by addition.

A modular reduction (also called *overlapping* or *wraparound*) may happen to the sum $rx + dx$ and, since N is odd, the vote is inverted. To avoid overlapping, the choices of a and b should lead to small $[ax]_N$ and $[bx]_N$. Here, $[ax]_N$ small means $\mathrm{abs}_N(ax) < \frac{\varepsilon}{2} N$, where:

$$\mathrm{abs}_N(x) \overset{def}{=} \begin{cases} [x]_N & \text{if } [x]_N < N/2 \\ N - [x]_N & \text{otherwise} \end{cases}$$

This way, if $[dx]_N$ is small then the probability of a modular reduction to occur in $rx + dx$ is also small, namely, $(\frac{\varepsilon}{2})$.

Next, we describe how, with the voting scheme of the $PAR()$ function, it is possible to build an almost perfect parity function, for sufficiently large m (the sample size) to achieve successful convergences.

In Line 7 of Algorithm 2.1 two oracle queries are made. To avoid one of the two queries, ACGS uses the parity $PAR()$ function in a way that the oracle is only queried for the right side of equality, $(rx + dx)$. On the left side, (rx), the LSB is computed by dividing the set of integers modulus N in intervals, and applying an hypothesis technique on them, as described next.

1. choose $k, l \in \mathbb{Z}_N$
2. compute m values $[r_i x]_N$ such that $[r_i x]_N \equiv [(k + il)x]_N$ for $i = 1, 2, ...m$
3. let $t = [kx]_N$ and $z = [lx]_N$, so that $[r_i x]_N \equiv t + iz \bmod N$
4. for $\varepsilon > 0$, $0 \leq j < 4/\varepsilon$, assume t in one of the following $4\varepsilon^{-1}$ intervals is known:

$$\left[j \frac{\varepsilon N}{4}, (j + 1) \frac{\varepsilon N}{4} \right]$$

5. for $0 \leq j' < \frac{4m}{\varepsilon}$, assume z in one of the following $4m\varepsilon^{-1}$ intervals is known:

$$\left[j' \frac{\varepsilon N}{4m}, (j' + 1) \frac{\varepsilon N}{4m} \right]$$

6. assume $LSB(t)$ and $LSB(z)$ are known
7. with the knowledge from (4), (5), and (6), it is easy to compute $LSB(r_i x)$

Note that by the hypothesis technique described above, we have $2 \cdot 4\varepsilon^{-1} \cdot 2 \cdot 4m\varepsilon^{-1} = 2^6 m\varepsilon^{-2}$ intervals.

We run Algorithm 2.1 with each of these alternatives and in only one of them the $LSB(r_i x)$ is computed correctly, and we call this instance **the correct alternative**.

3   Methods and Implementations

Two cryptographic schemes are analyzed, RSA and ECDH. In this section we present methods that use oracles to try to (probabilistically) invert these schemes. We also give details about the implementation of such methods.

3.1   RSA Algorithm

In this section we summarize the RSA LSB attack of Alexi et al. [1].

Let $RSA(x) = x^e \bmod N$. Next, we describe how the $PAR()$ function and BKGCD are used in ACGS to (probabilistically) invert RSA. The detailed description is shown as Algorithm 3.1 on page 152, that is briefly explained in the following paragraphs.

Part 1 of Algorithm 3.1 (Line 2), randomly chooses $a, b \in \mathbb{Z}_N$ with uniform distribution and independently.

Part 2 (Lines 3–27) computes two random values $[ax]_N$ and $[bx]_N$, it makes sure that $[ax]_N$ is odd[1] and then computes the $GCD([ax]_N, [bx]_N)$ using the Brent and Kung's GCD algorithm [9], as explained in Sect. 2.1. The $PAR()$ function (defined in Sect. 2.2) verifies if the parity of the two arguments is the same or not. Note that as the while-loops in Part 2 evolves, the intermediate values $[ax]_N$ and $[bx]_N$ remain small, because of the way the $PAR()$ function is defined.

---

[1] This requirement is NOT explicitly written in [1].

Part 3 (Lines 28–30) In Line 31, $RSA(ax)$ is computed using the RSA homomorphism mentioned in Sect. 2. In Line 32, the chosen $[ax]_N$ and $[bx]_N$ are expected to be coprime, thus $GCD([ax]_N, [bx]_N) = 1$. Since $RSA(1)$ is always 1, $GCD(RSA([ax]_N), RSA([bx]_N)) = RSA([cx]_N) = 1$ for some constant c, and $[cx]_N = 1$. Line 32 is such that $\text{RSA}(ax) = \pm 1$, so that $c \equiv a \bmod N$ and $x \equiv a^{-1} \bmod N$. If $\text{RSA}(ax) \neq \pm 1$, the algorithm returns to Line 2 and chooses a and b again expecting the chosen values $[ax]_N$ and $[bx]_N$ to be coprime.

From a theorem by Dirichlet [17, Sect. 4.5.2 Theorem D p. 324], the probability that two random integers in the interval $[-K, K]$ are coprime converges to $\frac{6}{\pi^2}$ as $K \to \infty$. This is around $60,8\,\%$. So with only two random attempts of a and b, $[ax]_N$ and $[bx]_N$ are expected to be coprime.

On the Bounds m (sample size) and *GCDlimit*. A condition that makes Algorithm 3.1 return to Line 2 is the variable *GCDlimit*, a bound to BKGCD (Lines 10,11,16,17,25,26 of Algorithm 3.1). This bound is for the case the $PAR()$ function (Lines 7, 13, 21) fails and the algorithm gets into an infinite loop.

In the correct alternative, as defined in Sect. 2.2, ACGS recovers the original message.

Alexi *et al.* [1] used the Chebyshev's inequality to estimate the error probability of $PAR()$ function to be $\Pr(\text{PAR to err}) = \frac{4}{m\varepsilon^2}$. From this estimation, a sufficiently large sample size m for the parity function to be almost perfect can be derived. So, $m \stackrel{def}{=} 64n\varepsilon^{-2}$ was chosen in [1] (so that $\Pr(\text{PAR to err}) \leq \frac{1}{16n}$), and the $PAR()$ function converged successfully. In our experiments we confirmed that with an error probability of $\frac{1}{16n}$, the $PAR()$ function, on average, runs without any error in all function calls made by BKGCD (with a maximum of *GCDlimit* $= 6n + 3$).

We run additional experiments and noticed that, on average, the bound *GCDlimit* for BKGCD can be less than $6n + 3$. We observed that even half of it, $(6n + 3)/2$, is enough to assure convergence. Details of this observation are described in Sect. 4.1.

The ACGS running time corresponds to ε^{-2} attempts to obtain small ax and bx, times two attempts to guarantee that they are coprime, (since by Dirichlet's theorem, $\frac{\pi^2}{6}$ that is < 2), times the running time of $2^6 m\varepsilon^{-2}$ alternatives, $6n + 3$ calls to the $PAR()$ function, and m samples. Altogether we have the following total running time:

$$\varepsilon^{-2} \cdot 2 \cdot 2^6 m\varepsilon^{-2} \cdot (6n + 3) \cdot m \approx 3 \cdot 2^8 \varepsilon^{-4} nm^2 \approx 3 \cdot 2^8 \varepsilon^{-4} n \left(\frac{64n}{\varepsilon^2}\right)^2 \approx 3 \cdot 2^{20} \varepsilon^{-8} n^3$$

$$= O(\varepsilon^{-8} n^3)$$

Experimental results are presented in Sect. 4.1.

3.2   ECDH Algorithm

In this section we summarize the ECDH LSB attack of Boneh *et al.* [6].

The BS algorithm makes an adaptation of ACGS to be used for ECDH. Consider a cyclic group of prime order curve $E(\mathbb{F}_p)$ with a generator G of prime

Algorithm 3.1. ACGS Invertion of RSA. (Alexi *et al.* [1])

Input: $y = \text{RSA}(x) = x^e \bmod N$, modulus N and the encryption exponent e.
Output: original message x.

1 **begin**
    // Part 1: Randomization
2   Choose $a, b \in_R \mathbb{Z}_N$ uniformly and independently;
    // Part 2: BKGCD - Brent-Kung's GCD of $[ax]_N$ and $[bx]_N$
3    $\alpha \leftarrow n$;
4    $\beta \leftarrow n$;
5   GCDlimit $\leftarrow 0$;
6   // GCDlimit and PAR() are explained in Sections 33 and 33
7   **while** $\text{PAR}(a, y) = 0$ **do**
8      $a \leftarrow [a/2]_N$;
9      $\alpha \leftarrow \alpha - 1$;
10   GCDlimit $\leftarrow$ GCDlimit $+ 1$;
11   **if** GCDlimit $> 6n + 3$ **then** GoTo(*2*);
12   ;
13   **repeat**
14   **while** $\text{PAR}(b, y) = 0$ **do**
15       $b \leftarrow [b/2]_N$;
16       $\beta \leftarrow \beta - 1$;
17   GCDlimit $\leftarrow$ GCDlimit $+ 1$;
18   **if** GCDlimit $> 6n + 3$ **then** GoTo(*2*);
19   ;
20   **if** $\beta \leq \alpha$ **then**
21   Swap(a,b);
22   Swap(α,β);
23   **if** $\text{PAR}(\frac{a+b}{2}, y) = 0$ **then**
24       $b \leftarrow [(a+b)/2]_N$;
25   **else**
26       $b \leftarrow [(a-b)/2]_N$;
27   GCDlimit $\leftarrow$ GCDlimit $+ 1$;
28   **if** GCDlimit $> 6n + 3$ **then** GoTo(*2*);
29   ;
30   **until** $b = 0$;
    // Part 3: Inversion
31   **if** $\text{RSA}(ax) \neq \pm 1$ **then** GoTo(*2*);
32   // $GCD([ax]_N, [bx]_N) = 1 = [ax]_N$, $x = a^{-1} \bmod N$ $c \leftarrow [\pm a^{-1}]_N$;
33   **return** c ;   /* c is the original message x */

---

order q and two private keys $a, b \in [1, q-1]$. We define the secret of Diffie-Hellman function $DH_{E,G}(aG, bG) = abG$ as the x coordinate of the point abG, $[abG]_x$. In this section we suppose the existence of an oracle $\mathcal{O}_p$ for $[abG]_x$, such that for a, b uniformly distributed in $[1, q-1]$, $\mathcal{O}_p$ is an oracle with advantage $\varepsilon > 0$

of guessing correctly the LSB of $[abG]_x$ of the Diffie-Hellman function. Then, we denote $\mathrm{Adv}^x_{E,G}(\mathcal{O}_p) = \left|\mathrm{Pr}_{a,b}[\mathcal{O}_p(E, G, aG, bG) = \mathrm{LSB}([abG]_x)] - \frac{1}{2}\right| > \varepsilon > 0$.

Let $E(\mathbb{F}_p)$ be an elliptic curve over $\mathbb{F}_p$ by the Weierstrass equation $y^2 = x^3 + Ax + B$. For a $\lambda \in \mathbb{F}_p^*$, define $\phi_\lambda(E)$ to be the elliptic curve $Y^2 = X^3 + A\lambda^4 X + B\lambda^6$, called *twisted elliptic curve*. Since $4(A\lambda^4)^3 + 27(B\lambda^6)^2 = (4A^3 + 27B^2)\lambda^{12} \neq 0$, $\phi_\lambda(E)$ is an elliptic curve for any $\lambda \in \mathbb{F}_p^*$. It is easy to verify that these *twists* are in fact isomorphisms. So, for points $Q, R, S \in E$, with $Q = (x, y)$, we have $(Q)_\lambda = Q_\lambda = (\lambda^2 x, \lambda^3 y) \in \phi_\lambda(E)$ and for $Q + R = S$, we have $Q_\lambda + R_\lambda = S_\lambda$ (homomorphic property). Yet, with these *twists*, we define, for an initial curve E_0, a family of isomorphic curves $\phi_\lambda(E_0)_{\lambda \in \mathbb{F}_p^*}$.

Informally, the BS algorithm is as follows. Let K_{ab} be the agreed upon point of ECDH over the curve E_0 with the public keys PK_a and PK_b. The x coordinate of K_{ab} is the secret key. The same way we did with ACGS, we choose a' and b' (we denote a', b' in order to avoid confusing them with the private keys a, b) expecting $[a'x]_p$ and $[b'x]_p$ to be small. Similarly to BKGCD, a' and b' generate values d, that we call λ, and a query to the parity function is made with $\mathrm{PAR}(\lambda, PK_a, PK_b)$. The queries for the LSB is different. The $PAR()$ function is queried for the LSB of $\lambda^2 x$, because $(K_{ab})_\lambda = (\lambda^2 x, \lambda^3 y)$. So, the query is made to a point that is on another curve, different from E_0.

Boneh *et al.* [6] consider an oracle $\mathcal{O}_p$ with advantage $\varepsilon > 0$ in predicting the LSB in the curve E_0. That means that $\mathcal{O}_p$ doesn't have advantage in every curve of the family $\phi_\lambda(E_0)_{\lambda \in \mathbb{F}_p^*}$. By definition, $\mathcal{O}_p$ keeps this advantage for at least a fraction δ of the curves in the isomorphic family.

A problem to overcome is that BS queries the $PAR()$ function with many λ values. To solve this problem a new oracle $\mathcal{B}_p$ is constructed in [6] such that its success probability inside and outside the fraction δ of the curves is known, equal to $\varepsilon\delta/8$.

Next a brief description of the BS algorithm is given.

1. Input is $\langle E, G, PK_a, PK_b \rangle$, where $PK_a = aG$ and $PK_b = bG$ and G of prime order q. The goal is to compute the point $C = abG$;
2. As a and b are fixed and unknown, we randomize the process by defining $PK_{ra} = a_r aG$ and $PK_{rb} = b_r bG$, for $a_r, b_r \in [1, q-1]$, hoping that the values $a_r a$ and $b_r b$ lead to a case that the oracle $\mathcal{B}_p$ has non-negligible advantage;
3. Let $DH_{E,G}(PK_{ra}, PK_{rb}) = D$. The goal is to compute D to obtain C, because $C = c_r \cdot D$, where $c_r \equiv (a_r b_r)^{-1} \mod q$;
4. Now, an algorithm similar to ACGS is executed with the $\mathcal{B}_p$ oracle, as described informally above. Please see the BS Algorithm below.
5. To make sure that the desired values $a_r a$ and $b_r b$ are found, it might be necessary to repeat the process $\frac{8}{\varepsilon\delta}$ times.

By the end of the BS execution, a list of candidates C for the point abG is obtained. But, unlike ACGS, we cannot automatically identify the correct alternative. This fact forces us to execute all alternatives and try to identify the correct one at the end. This identification is possible thanks to an algorithm by Shoup [23, Theorem 7]. Informally, it is shown by Shoup [23] that an algorithm

Algorithm 3.2. BS Algorithm (Boneh *et al.* [6]).

Input: $E(\mathbb{F}_p)$, G, PK_a, PK_b, where $PK_a = aG, PK_b = bG$.

Output: x-coordinate of $C = abG$.

```
1  begin
       // Part 1: Randomization
2      Choose a_r, b_r ∈_R [1, q − 1] uniformly and independently;
       // Part 2: Brent-Kung's GCD
3      α ← length(q);
4      β ← length(q);
5      GCDlimit ← 0;
6      while PAR((a_r)², PK_a, PK_b) = 0 do
7          a_r ← a_r/2;
8          α ← α − 1;
9          GCDlimit ← GCDlimit + 1;
10         if GCDlimit > 6n + 3 then GoTo(2);
11         ;

12     repeat
13         while PAR((b_r)², PK_a, PK_b) = 0 do
14             b_r ← b_r/2;
15             β ← β − 1;
16             GCDlimit ← GCDlimit + 1;
17             if GCDlimit > 6n + 3 then GoTo(2);
18             ;
19         if β ≤ α then
20             Swap(a_r,b_r);
21             Swap(α,β);

22         if PAR( (a_r+b_r)²/4 , PK_a, PK_b) = 0 then
23             b_r ← (a_r + b_r)/2;
24         else
25             b_r ← (a_r − b_r)/2;
26         GCDlimit ← GCDlimit + 1;
27         if GCDlimit > 6n + 3 then GoTo(2);
28         ;
29     until b_r = 0;
       // Part 3: Inversion of x-coordinate
30     if a_r PK_a = (x_a, y_a) ≠ (±1, y_a) then GoTo(2);
31     // here, x-coordinate of C is ±a_r^{-1} mod(q − 1)
32     C_x ← ±a_r^{-1} mod(q − 1);
33     return C_x ;        /* x coordinate of C = abG is the desired goal */
```

that outputs a list of candidates for the Diffie-Hellman function can be easily converted into an algorithm that computes the Diffie-Hellman function.

Experimental results are presented in Sect. 4.2.

3.3   Relic Cryptographic Library

To implement the ACGS and BS algorithms, we used the C programming language and the **Relic Toolkit** version 0.3.0 [2] cryptographic library.

We used SECG_P160 and NIST_P224 curves of the library for ECDH. For RSA, we used 1024, 2048 and 5000 bits modulus, and also a 128 bits theoretical modulus, to register the most demanding running times. We chose 1024 and 5000 bits because of [13], and 2048 bits because of the current NIST recommendation[2]. Moreover, because of the equivalence of the security level [8], we chose 160 and 224 bits for elliptic curves.

Regarding the oracle implementations, the answer for an ECDH query is quick, requiring just a few computations. For RSA, a decryption is required for each and every query, resulting in a much greater running time.

The PRNG built inside Relic is the FIPS 185-2 based on SHA1. Relic library already has timing functions (*benchmark*) to record the running times, and we used the most accurate version of these timing functions (HPROC). For the simulation of the algorithms and the running time data collection, we used an Intel core 2 Duo T5450 of 1,66 Ghz with 2 GB RAM.

4   Results

In this section we present our results of ACGS for RSA and BS for ECDH.

4.1   ACGS Results

The success of the ACGS algorithm depends on either a and b resulting in a small ax and bx ($abs_N(ax) < \frac{\varepsilon N}{2}$), and also a and b being coprime. For reasonable running times, we implemented the algorithm in a way that those conditions are almost always met, as we describe next.

With these initial conditions met, and the algorithm always running with the correct input alternative (as defined in Subsect. 2.2), we registered the running times of ACGS. However, running and observing the output of ACGS for 1024 bits takes a very long time, and further longer for small advantages, $\varepsilon < 0.05$. So we decided to compute the running time for the correct input alternative of RSA-128.

Figure 1 shows the running times of a 128 bits RSA with m samples. Note that as the sample size doubles, the running time also doubles.

Arithmetic operations in the algorithm have a variable time, that is related to the number of bits of the computed numbers. The exponentiation time with 256 bits numbers is greater than with 128 bits numbers, and the exponentiation time of 512 bits is greater than with 256 bits numbers, and so on. Figure 2 shows the running time of each call to the $PAR()$ function with respect to the size of the operands; the $PAR()$ function makes 16 oracle queries. Note that the

---

[2] http://csrc.nist.gov/groups/ST/toolkit/key_management.html.

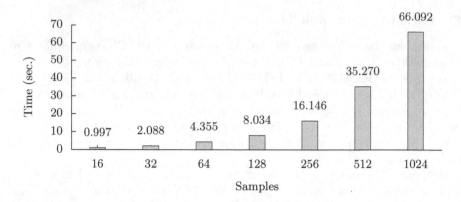

Fig. 1. ACGS running time and the number of samples for RSA-128

running time for 5000 bits is only indicated by an arrow in order not to interfere with the graphic proportions.

On the bounds m (sample size) and $GCDlimit$. Now we describe how we obtained more efficient running times for the ACGS algorithm.

The sampling number m is a critical parameter. In the original ACGS it is $m \stackrel{def}{=} \frac{64n}{\varepsilon^2}$ by definition. That way, the value of m depends only on the length of the modulus N and the oracle advantage ε. In Table 1 we show some examples of sample size and distinct pairs n, ε.

We noticed that even for values of m smaller than adopted by the original ACGS algorithm, it is successful in many cases. Our experiments showed that the value $m \stackrel{def}{=} \frac{4}{\varepsilon^2}$ does not affect successful convergences. Consequently the running time was reduced substantially. An interesting observation is that if we

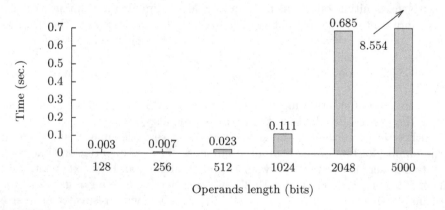

Fig. 2. Running time of the $PAR()$ function in ACGS and operands length ($m = 16$).

Table 1. Sample size and oracle advantage

	128 bits	1024 bits	5000 bits
0.4	51,200	409,600	2,000,000
0.3	91,023	728,178	3,555,556
0.2	204,800	1,638,400	8,000,000
0.1	819,200	6,553,600	32,000,000
0.05	3,276,800	26,214,400	128,000,000
0.01	81,920,000	655,360,000	3,200,000,000

replace $m \overset{def}{=} \frac{4}{\varepsilon^2}$ in the theoretical upperbound $\Pr(\text{PAR to err}) < \frac{4}{m\varepsilon^2}$, we get

$$\Pr(\text{PAR to err}) < \frac{4}{\frac{4}{\varepsilon^2}\varepsilon^2} = 1$$

that indicates this upperbound is not tight.

Figure 3 shows the examples for our new sampling method. This figure shows only the advantage, since in this new method the modulus length does not interfere with the sample size. It is worth comparing $4/\varepsilon^2$ with the values in Table 1: they are much smaller. E.g., for 128 bits and $\varepsilon = 0.1$, it is around 2,000 times smaller.

Using this new sampling, we ran ACGS for RSA-128. The algorithm was successful in successive executions. We observed the following results: $\varepsilon = 0.4$: ran 66 times and then stopped; $\varepsilon = 0.3$: ran 33 times and then stopped; $\varepsilon = 0.2$: ran 53 times and then stopped; $\varepsilon = 0.1$: ran 44 times and then stopped; $\varepsilon = 0.05$: ran 29 times and then stopped; $\varepsilon = 0.01$: ran 2 times and then continued.

For $\varepsilon = 0.01$, it ran successfully twice, but we had to abort since each test took about 40 min to complete. For $\varepsilon = 0.2$ and $\varepsilon = 0.05$, with $m = 100$ and $m = 1600$ respectively, we had some problems, since the tests were failing very shortly.

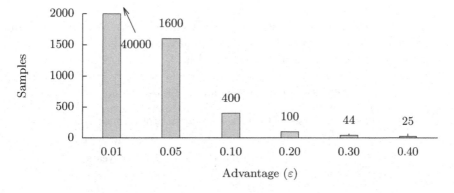

Fig. 3. Oracle advantage and sample size, for RSA-128

However, by adding just one unit in the sampling the results were achieved ($m = 101, m = 1601$).

We observed a particular effect running ACGS with the correct input alternative (as defined in Subsect. 2.2). For a 128 bits RSA, in 200 runs, the BKGCD function made between 324 and 368 calls to the $PAR()$ function, i.e., between $2.53n$ and $2.88n$. No run with more than $3n$ calls to the $PAR()$ function with the correct input alternative happened.

The BKGCD function bound is defined as $GCDlimit = 6n+3$, but according to our observations, this value could be lowered to $GCDlimit = 3n$ without affecting the convergence success. Note that this reduction affects not only the correct alternative, but also all other alternatives, offering a reduction of the ACGS running time by half.

That way, the new running time includes the initial conditions time ($\varepsilon^{-2} \cdot 2$), every alternatives time ($2^6 m \varepsilon^{-2}$), the BKGCD function time ($3n$), the $PAR()$ function time (m) and the oracle time (1, we consider it to be constant): (ε^{-2})(2) $(2^6 m \varepsilon^{-2})(3n)(m)(1) = 3 \cdot 2^{11} \varepsilon^{-8} n$.

Table 2 shows the new complexity measures together with the original estimates. We also added the running time of IFP algorithm [10] as in [13]. Note that ACGS is now faster than IFP.

Table 2. Comparison of existing complexities

	Estimate	1024 bits	2048 bits	5000 bits
IFP	$\exp(1{,}9(\ln N)^{\frac{1}{3}}(\ln\ln N)^{\frac{2}{3}})$	$6.409 \cdot 10^{25}$	$5.817 \cdot 10^{34}$	$3.755 \cdot 10^{50}$
Orig. ACGS	$3 \cdot 2^{20} \varepsilon^{-8} n^3$	$3.378 \cdot 10^{31}$	$2.702 \cdot 10^{32}$	$3.932 \cdot 10^{33}$
New ACGS	$3 \cdot 2^{11} \varepsilon^{-8} n$	$6.292 \cdot 10^{22}$	$1.258 \cdot 10^{23}$	$3.072 \cdot 10^{23}$

Now we can use the running times of the $PAR()$ function in Fig. 2 to obtain the time estimates. In this figure, we have only 16 samples, but since we are using $\varepsilon = 0.1$, we have $m = 400$. Let tPAR be the running time of the $PAR()$ function in Fig. 2, the proportional estimate is $\frac{400}{16}$ tPAR $\cdot 3n \cdot 2^6 m \varepsilon^{-2} \cdot \varepsilon^{-2} \cdot 2 = 3{,}84 \cdot 10^{10} \cdot$ tPAR $\cdot n$.

Table 3 shows the measured running times to invert RSA. We have a column for the time on the correct input alternative, and a column for the total time of the algorithm, and also the time tPAR from Fig. 2.

Table 3. Estimated time for ACGS execution

n (bits)	tPAR (sec.)	Correct alternative	Total time
1024	0.111	19.7 days	$138.4 \cdot 10^3$ years
2048	0.685	243.5 days	$1.7 \cdot 10^6$ years
5000	8.554	20.3 years	$52.1 \cdot 10^6$ years

4.2   BS Results

Now we describe how we obtained more efficient running times for the BS algorithm.

We applied the same approach as we did for ACGS: use a new sampling method.

Regarding the BKGCD function, there is no problem with the bound equal to $3 \lg p$. However, regarding the sampling, we have to consider that two oracles are used, the oracle $\mathcal{O}_p$, and the oracle $\mathcal{B}_p$. The oracle $\mathcal{B}_p$ is driven by δ that defines the proportion of curves which $\mathcal{O}_p$ has advantage ε. The ideal case is when $\delta = 1$, that makes the $\mathcal{O}_p$ oracle applicable to every curve of the family.

In the case $\delta = 1$ we have that BS is similar to ACGS. Based on this value, we can obtain the running times for the $PAR()$ function with operands of 160 bits and 224 bits for ECDH, as shown in Fig. 4.

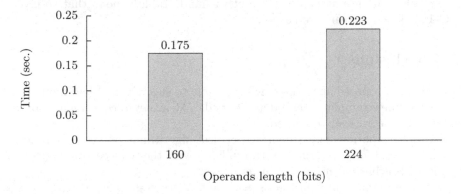

Fig. 4. The $PAR()$ function running time and operands length ($m = 40000$).

We chose the $m \stackrel{def}{=} \frac{8}{\varepsilon^2}$ sampling size and we executed experiments of the BS for ECDH with 160 bits. The algorithm was successful in several successive executions until it failed or was aborted by the time bound. We had the following results: $\varepsilon = 0.4$: ran 368 and then stopped; $\varepsilon = 0.3$: ran 247 and then stopped; $\varepsilon = 0.2$: ran 183 and then stopped; $\varepsilon = 0.1$: ran 862 and then continued; $\varepsilon = 0.05$: ran 35 and then continued; $\varepsilon = 0.01$: ran once and then continued.

We estimated the running times for $\varepsilon = 0.1$ and $m = 800$, with the tPAR time in Fig. 4. We have a total time of $\frac{800}{40000}$tPAR $\cdot 3 \lg p \cdot 2^6 m\varepsilon^{-2} \cdot \varepsilon^{-2} \cdot 2 = 6.144 \cdot 10^7 \cdot$ tPAR $\cdot \lg p$. From this equation, we built Table 4 with estimated running times for BS.

We analyzed the worst case, where $\delta < 1$ and close to zero. We fixed $\delta = 0.1$, so that in only 10 % of the curves the $\mathcal{O}_p$ oracle has advantage inside the isomorphic family. So we use the $\mathcal{B}_p$ oracle that implies a new sampling $m \stackrel{def}{=} \frac{8}{\varepsilon \delta^2}$ that is related to the $\mathcal{O}_p$ oracle queries by the $\mathcal{B}_p$ oracle.

This new sampling is obtained using the same methodology as before. Table 5 shows the estimated times to find the ECDH keys using the $\mathcal{B}_p$ oracle.

Table 4. Estimated running time for the BS algorithm.

$\lg p$ (bits)	tPAR (s)	Correct altern	Total time
160	0.175	5.6 min	54.6 years
224	0.223	9.9 min	97.3 years

Table 5. Estimated time for the BS algorithm with the $\mathcal{B}_p$ oracle.

$\lg p$ (bits)	Correct altern	Total time
160	15.6 years	$80 \cdot 10^6$ years
224	44.7 years	$229 \cdot 10^6$ years

Comparison. Table 3 shows that to invert RSA-1024 with our implementations, 138.4×10^3 years are needed. On the other hand, Table 5 shows that to invert ECC-160, 80×10^6 years are needed.

5   Conclusions

Based on the proposed algorithms, we were able to successfully build implementations with new sampling method and bounds. We achieved convergence speeds higher than theoretically expected.

With our experiments we identified critical algorithmic parameters in order to significantly reduce the running times. Even with the reduced running times, the methods achieved successful convergences.

Our results show how fast it is to invert a cryptographic scheme with a minimum knowledge of the LSB. In particular, for RSA with 1024 bits ACGS required much less time to invert than the IFP time. This evidence was not known from previous published analyses.

Future work. There is an ACGS version [1] for the Rabin scheme, which could also be implemented for comparison. An idea to prove the validity of our methods is to create a faulty RSA implementation (revealing the LSB through a secondary channel) and then adapt it as an oracle for ACGS. Such work would show practical implications of our methods.

References

1. Alexi, W., Chor, B., Goldreich, O., Schnorr, C.-P.: RSA and rabin functions: certain parts are as hard as the whole. SIAM J. Comput. **17**(2), 194–209 (1988)
2. Aranha, D.F., Gouvêa, C.P.L.: RELIC is an Efficient LIbrary for Cryptography. http://code.google.com/p/relic-toolkit/
3. Ben-Or, M., Chor, B., Shamir, A.: On the cryptographic security of single RSA bits. In: ACM Symposium on Theory of Computing (STOC 1883), pp. 421–430. ACM Press, Baltimore, April 1983

4. Blum, L., Blum, M., Shub, M.: A simple unpredictable pseudo-random number generator. SICOMP: SIAM J. Comput. **15**, 364–383 (1986)
5. Blum, M., Micali, S.: How to generate cryptographically strong sequence of pseudo-random bits. SIAM J. Comput. **13**, 850–864 (1984)
6. Boneh, D., Shparlinski, I.E.: On the unpredictability of bits of the elliptic curve Diffie–Hellman scheme. In: Kilian, J. (ed.) CRYPTO 2001. LNCS, vol. 2139, p. 201. Springer, Heidelberg (2001)
7. Boneh, D., Venkatesan, R.: Hardness of computing the most significant bits of secret keys in Diffie-Hellman and related schemes. In: Koblitz, N. (ed.) CRYPTO 1996. LNCS, vol. 1109, pp. 129–142. Springer, Heidelberg (1996)
8. Bos, J.W., Kaihara, M.E., Kleinjung, T., Lenstra, A.K., Montgomery, P.L.: On the security of 1024-bit RSA and 160-bit elliptic curve cryptography. http://eprint. iacr.org/2009/389 (2009)
9. Brent, R.P., Kung, H.T.: Systolic VLSI arrays for polynomial GCD computation. IEEE Trans. Comput. **33**, 731–736 (1984)
10. Buhler, J.P., Lenstra, H.W., Pomerance, C.: Factoring integers with the number field sieve. In: Lenstra, A.K., Lenstra, H.W. (eds.) The Development of the Number Field Sieve. LNM, vol. 1554, pp. 50–94. Springer, Heidelberg (1993)
11. Chevalier, C., Fouque, P.-A., Pointcheval, D., Zimmer, S.: Optimal randomness extraction from a Diffie-Hellman element. In: Joux, A. (ed.) EUROCRYPT 2009. LNCS, vol. 5479, pp. 572–589. Springer, Heidelberg (2009)
12. Diffie, W., Hellman, M.: New directions in cryptography. IEEE Trans. Inf. Theory **22**(6), 644–654 (1976)
13. Fischlin, R., Schnorr, C.-P.: Stronger security proofs for RSA and rabin bits. J. Cryptol. **13**(2), 221–244 (2000)
14. Goldwasser, S., Micali, S., Tong, P.: Why and how to establish a private code on a public network (extended abstract). In: FOCS, pp. 134–144. IEEE, Chicago, Illinois, November 1982
15. Hofheinz, D., Kiltz, E.: Practical chosen ciphertext secure encryption from factoring. In: Joux, A. (ed.) EUROCRYPT 2009. LNCS, vol. 5479, pp. 313–332. Springer, Heidelberg (2009)
16. Jetchev, D., Venkatesan, R.: Bits security of the elliptic curve Diffie–Hellman secret keys. In: Wagner, D. (ed.) CRYPTO 2008. LNCS, vol. 5157, pp. 75–92. Springer, Heidelberg (2008)
17. Knuth, D.E.: The Art of Computer Programming. Seminumerical Algorithms, vol. 2, 2nd edn. Addison-Wesley, Reading (1981)
18. Lenstra, A.K., Verheuil, E.R.: Selecting cryptographic key sizes. J. Cryptol. **14**, 255–293 (1999)
19. Menezes, A.J., Vanstone, S.A., Van Oorschot, P.C.: Handbook of Applied Cryptography. CRC Press, Boca Raton (1996)
20. Rabin, M.: Digitalized signatures as intractable as factorization. Technical report MIT/LCS/TR-212, MIT Laboratory for Computer Science, January 1979
21. Rivest, R.L., Shamir, A., Adleman, L.M.: A method for obtaining digital signatures and public key cryptosystems. Commun. ACM **21**(2), 120–126 (1978)
22. Roh, D., Hahn, S.G.: On the bit security of the weak Diffie-Hellman problem. Inf. Process. Lett. **110**, 799–802 (2010)
23. Shoup, V.: Lower bounds for discrete logarithms and related problems. In: Fumy, W. (ed.) EUROCRYPT 1997. LNCS, vol. 1233, pp. 256–266. Springer, Heidelberg (1997)
24. Stein, J.: Computational problems associated with Racah algebra. J. Comput. Phys. **1**(3), 397–405 (1967)

Isogeny Volcanoes of Elliptic Curves and Sylow Subgroups

Mireille Fouquet[1], Josep M. Miret[2], and Javier Valera[2]([✉])

[1] Institut de Mathématiques de Jussieu, Université Paris Diderot - Paris 7,
Paris, France
fouquet@math.univ-paris-diderot.fr
[2] Dept. de Matemàtica, Universitat de Lleida, Lleida, Spain
{miret,jvalera}@matematica.udl.cat

Abstract. Given an ordinary elliptic curve over a finite field located in the floor of its volcano of ℓ-isogenies, we present an efficient procedure to take an ascending path from the floor to the level of stability and back to the floor. As an application for regular volcanoes, we give an algorithm to compute all the vertices of their craters. In order to do this, we make use of the structure and generators of the ℓ-Sylow subgroups of the elliptic curves in the volcanoes.

Keywords: Elliptic curves · Isogeny volcanoes · Sylow subgroups · Finite fields

1 Introduction

In the last decades, the usage of elliptic curves over finite fields in the design of secure cryptography protocols has grown significantly. Nevertheless, not all elliptic curves are useful in cryptography based on the discrete logarithm problem, since they must satisfy certain requirements related to their group orders or their embedding degrees. Concerning their group orders, they must be of the form $f \cdot q$ with q prime and f a small integer, otherwise the curves are vulnerable to the Pohlig-Hellman attack [17]. Regarding their embedding degrees, they must be ≥ 6 for curves of 160 bits, otherwise the curves are vulnerable to the MOV attack [12].

Isogenies between elliptic curves over finite fields, in particular, prime degree isogeny chains, have long been a subject of study with different approaches, since they play a central role in the SEA algorithm (see [3,18]) to compute the group order of an elliptic curve. The basic idea of this algorithm is the computation of the trace of the Frobenius endomorphism of a curve modulo different suitably chosen small primes ℓ.

Given two ordinary elliptic curves E and E' over a finite field $\mathbb{F}_q$ with endomorphism rings $\mathcal{O}$ and $\mathcal{O}'$, respectively, and an isogeny $\mathcal{I} : E \to E'$ of degree a prime ℓ such that $\ell \nmid q$, Fouquet and Morain [6] introduced, from the Kohel's Ph.D. thesis [9], the notion of direction of an ℓ-isogeny. It is *ascending, horizontal* or *descending* whether the index $[\mathcal{O}' : \mathcal{O}]$ is ℓ, 1 or $1/\ell$ respectively. With this

© Springer International Publishing Switzerland 2015
D.F. Aranha and A. Menezes (Eds.): LATINCRYPT 2014, LNCS 8895, pp. 162–175, 2015.
DOI: 10.1007/978-3-319-16295-9_9

notion of direction for the ℓ-isogenies, the set of isomorphism classes of ordinary elliptic curves over $\mathbb{F}_q$ with group order $N = q + 1 - t$, $|t| \leq 2\sqrt{q}$, can be represented as a directed graph, whose vertices are the isomorphism classes and its arcs represent the ℓ-isogenies between curves in two vertices. It is worth remarking that if two vertices are connected by an arc, the corresponding dual ℓ-isogeny is represented as an arc in the other direction.

Each connected component of this graph is called *volcano of ℓ-isogenies* due to its peculiar shape. Indeed, it consists of a cycle that can be reduced to one point, called *crater*, where from its vertices hang $\ell + 1 - m$ complete ℓ-ary trees being m the number of horizontal ℓ-isogenies. Then, the vertices can be stratified into levels in such a way that the curves in each level have the same endomorphism ring. The bottom level is called the *floor* of the volcano.

Knowing the cardinality of an elliptic curve, Kohel [9] and recently Bisson and Sutherland [1] describe algorithms to determine its endomorphism ring taking advantage of the relationship between the levels of its volcano and the endomorphism rings at those levels. When the cardinality is unknown, Fouquet and Morain [6] give an algorithm to determine the *height* (or depth) of a volcano using exhaustive search over several paths on the volcano to detect the crater and the floor levels. As a consequence, they obtain computational simplifications for the SEA algorithm, since they extend the moduli ℓ in the algorithm to prime powers ℓ^s.

In [15], Miret et al. showed the relationship between the levels of a volcano of ℓ-isogenies and the ℓ-Sylow subgroups of the curves. All curves in a fixed level have the same ℓ-Sylow subgroup. At the floor, the ℓ-Sylow subgroup is cyclic. When ascending by the *volcanoside*, that is, by the levels which are between the floor and the crater, the ℓ-Sylow subgroup structure is becoming balanced. The first level, if it exists, where the ℓ-Sylow subgroup is balanced, is called *stability level*. If this level does not exist, the stability level is the crater of the volcano. Recently, Ionica and Joux [7] have developed a method to decide whether the isogeny with kernel a subgroup generated by a point of order ℓ is an ascending, horizontal or descending ℓ-isogeny using a symmetric pairing over the ℓ-Sylow subgroup of a curve [8].

Volcanoes of ℓ-isogenies have also been used by Sutherland [20] to compute the Hilbert class polynomials. Another application has been provided by Bröker, Lauter and Sutherland [2] in order to compute modular polynomials. To reach these goals, in both works, it is necessary to determine the vertices of the craters of the volcanoes. On the other hand, some specific side channel attacks, the so-called Zero Value Point attacks, can be avoided using isogenies or more precisely volcanoes of ℓ-isogenies [16].

In this paper, given an ordinary elliptic curve $E/\mathbb{F}_q$, the structure $\mathbb{Z}/\ell^r\mathbb{Z} \times \mathbb{Z}/\ell^s\mathbb{Z}$, $r > s$, of its ℓ-Sylow subgroup and a point P_r of order ℓ^r, we construct a chain of ℓ-isogenies starting from E and ending at a curve at the floor of the volcano. This chain first is ascending, then horizontal and finally descending. When $h \geq 1$ and $2h < r + s$, being h the height of the volcano, all the vertices of its crater can be obtained by using repeatedly this sort of chains. Therefore we present an algorithm to perform this task.

In the following, we consider ordinary elliptic curves defined over a finite field $\mathbb{F}_q$, with cardinality unknown. We assume that the characteristic p of $\mathbb{F}_q$ is different from 2 and 3. We denote by ℓ a prime that does not divide q. Furthermore, in Sects. 3 and 4 we assume that the ℓ-Sylow subgroup of the considered curve is not trivial.

2   Preliminaries

In this section we introduce some notations that are used in the sequel concerning ℓ-isogenies, volcanoes of ℓ-isogenies and ℓ-Sylow subgroups of elliptic curves.

We denote by $E/\mathbb{F}_q$ an elliptic curve defined over the finite field $\mathbb{F}_q$, by $E(\mathbb{F}_q)$ its group of rational points with O_E its neutral element and by $j(E)$ its j-invariant.

Given an ordinary elliptic curve $E/\mathbb{F}_q$ with group order $N = q + 1 - t$, where t is the trace of the Frobenius endomorphism of $E/\mathbb{F}_q$, its endomorphism ring $\mathcal{O} = \text{End}(E)$ can be identified with an order of the imaginary quadratic field $\mathcal{K} = \mathbb{Q}(\sqrt{t^2 - 4q})$ (see [19]). The order $\mathcal{O}$ satisfies [4]

$$\mathbb{Z}[\pi] \subseteq \mathcal{O} \subseteq \mathcal{O}_\mathcal{K},$$

where $\mathcal{O}_\mathcal{K}$ is the ring of integers of $\mathcal{K}$ and π is the Frobenius endomorphism of $E/\mathbb{F}_q$. Writing $t^2 - 4q = g^2 D_\mathcal{K}$, where $D_\mathcal{K}$ is the discriminant of $\mathcal{K}$, it turns out that g is the conductor of the order $\mathbb{Z}[\pi]$ in the maximal order $\mathcal{O}_\mathcal{K}$. Then the conductor f of $\mathcal{O}$ divides g.

A volcano of ℓ-isogenies [6] is a directed graph whose vertices are isomorphism classes of ordinary elliptic curves over a finite field $\mathbb{F}_q$ and where the arcs represent ℓ-isogenies among them. These graphs consist of a unique cycle (with one, two or more vertices) at the top level, called crater, and from each vertex of the cycle hang $\ell + 1$, ℓ or $\ell - 1$ (depending of the number of horizontal ℓ-isogenies) ℓ-ary isomorphic complete trees, except in the case where the volcano is reduced to the crater. The vertices at the bottom level, called floor of the volcano, have only one ascending outgoing arc. In the other cases each vertex has $\ell + 1$ outgoing arcs: for the vertices in the volcanoside, one is ascending and ℓ are descending, while for the vertices on the crater it depends on its typology (and it can be easily explained for each case). The case where we encounter a vertex with j-invariant $j = 0$ or $j = 1728$ is slightly different and is not treated in this paper. We denote by $V_\ell(E/\mathbb{F}_q)$ the volcano of ℓ-isogenies where $E/\mathbb{F}_q$ belongs. We remark that if $E'/\mathbb{F}_q$ is another curve on the volcano, $V_\ell(E'/\mathbb{F}_q) = V_\ell(E/\mathbb{F}_q)$.

Lenstra [11] proved that $E(\mathbb{F}_q) \simeq \mathcal{O}/(\pi - 1)$ as $\mathcal{O}$-modules, from where one can deduce that $E(\mathbb{F}_q) \simeq \mathbb{Z}/n_1\mathbb{Z} \times \mathbb{Z}/n_2\mathbb{Z}$. By writing $\pi = a + g\omega$ with

$$a = \begin{cases} (t-g)/2 \\ t/2 \end{cases} \quad \text{and} \quad \omega = \begin{cases} \frac{1+\sqrt{D_\mathcal{K}}}{2} & \text{if } D_\mathcal{K} \equiv 1 \pmod 4 \\ \sqrt{D_\mathcal{K}} & \text{if } D_\mathcal{K} \equiv 2,3 \pmod 4 \end{cases}$$

we obtain that $n_2 = \gcd(a - 1, g/f)$, $n_2 \mid n_1$, $n_2 \mid q - 1$ and $\#E(\mathbb{F}_q) = n_1 n_2$. This implies that on a volcano of ℓ-isogenies the group structure of all the curves with same endomorphism ring, i.e. at the same level, is identical.

From this classification of the elliptic curves, the relationship between the structure of the ℓ-Sylow subgroup $E[\ell^\infty](\mathbb{F}_q)$ of an elliptic curve $E/\mathbb{F}_q$ and its location in the volcano of ℓ-isogenies $V_\ell(E/\mathbb{F}_q)$ is deduced.

Proposition 1. [15] Let $E/\mathbb{F}_q$ be an elliptic curve whose ℓ-Sylow subgroup is isomorphic to $\mathbb{Z}/\ell^r\mathbb{Z} \times \mathbb{Z}/\ell^s\mathbb{Z}$, $r \geq s \geq 0$, $r + s \geq 1$.

- If $s < r$ then E is at level s in the volcano with respect to the floor;
- If $s = r$ then E is at least at level s with respect to the floor.

As said in the introduction, we call stability level the level where from this one down to the floor, the structure of the ℓ-Sylow subgroup is different at each level (we therefore allow the stability level to be the crater). Ionica and Joux [7] call it the first level of stability. The curves located above the stability level (including this one) until the crater, if they exist, have ℓ-Sylow subgroup isomorphic to $\mathbb{Z}/\ell^{\frac{n}{2}}\mathbb{Z} \times \mathbb{Z}/\ell^{\frac{n}{2}}\mathbb{Z}$, being $n = v_\ell(N)$, n even and N the cardinality of the curves (see [15]). A volcano whose crater is equal to the stability level is called a *regular* volcano. Otherwise it is called an *irregular* volcano. Notice that if n is odd, then the volcano is regular. If n is even, it can be regular or irregular.

The height h of a volcano of ℓ-isogenies coincides with the ℓ-valuation of the conductor g of $\mathbb{Z}[\pi]$. This value, assuming n is kwown, can be completely determined in most cases (see [15]).

Concerning the ℓ-Sylow subgroup $E[\ell^\infty](\mathbb{F}_q)$ of an elliptic curve $E/\mathbb{F}_q$, Miret et al. [14] gave a general algorithm to determine its structure $\mathbb{Z}/\ell^r\mathbb{Z} \times \mathbb{Z}/\ell^s\mathbb{Z}$, $r \geq s \geq 0$, together with generators P_r and Q_s, without knowing the cardinality of the curve. Their method starts computing either one point of order ℓ of $E(\mathbb{F}_q)$, if the ℓ-Sylow subgroup is cyclic, or two independent points of order ℓ, otherwise. Then, in an inductive way, the algorithm proceeds computing one point of order ℓ^{k+1} for one or two points of order ℓ^k until reaching those of maximum order. If the cardinality of the curve is known, Ionica and Joux [7] give a probabilistic algorithm to compute the ℓ-Sylow structure more efficiently than the preceeding one.

Finally, we say that a point $Q \in E(\mathbb{F}_q)$ is ℓ-divisible or ℓ-divides if there exists another point $P \in E(\mathbb{F}_q)$ such that $\ell P = Q$. We say, as well, that P is an ℓ-divisor of Q.

3   A Particular Chain of ℓ-Isogenies

Given an elliptic curve $E/\mathbb{F}_q$, which is on the floor of the volcano $V_\ell(E/\mathbb{F}_q)$, we determine a chain of ℓ-isogenies in the volcano from the floor to the stability level and back to the floor. More precisely, if $h \geq 1$ and the ℓ-Sylow subgroup of $E/\mathbb{F}_q$ is isomorphic to $\mathbb{Z}/\ell^n\mathbb{Z}$, then we give a chain of length n starting at the floor to the stability level and descending back to the floor.

3.1   Behaviour of the ℓ-Sylow Subgroup Through Particular ℓ-Isogenies

In this subsection, we study the changes in the ℓ-Sylow subgroup when we consider isogenies defined by the quotient of subgroups of order ℓ.

Lemma 2. *Let $\mathbb{Z}/\ell^r\mathbb{Z} \times \mathbb{Z}/\ell^s\mathbb{Z}$ with $r > s > 0$ be the group isomorphic to the ℓ-Sylow subgroup of an elliptic curve $E/\mathbb{F}_q$. Let $P_r \in E(\mathbb{F}_q)$ and $Q_s \in E(\mathbb{F}_q)$ be two linearly independent points whose orders are respectively ℓ^r and ℓ^s. Denote $P_1 = \ell^{r-1}P_r$ and $Q_1 = \ell^{s-1}Q_s$.*

(i) Either the isogenous curve $E' \simeq E/\langle P_1 \rangle$ has ℓ-Sylow subgroup isomorphic to $\mathbb{Z}/\ell^{r-1}\mathbb{Z} \times \mathbb{Z}/\ell^{s+1}\mathbb{Z}$ and the ℓ-isogeny of kernel $\langle P_1 \rangle$ is ascending or the isogenous curve $E' \simeq E/\langle P_1 \rangle$ has ℓ-Sylow subgroup isomorphic to $\mathbb{Z}/\ell^r\mathbb{Z} \times \mathbb{Z}/\ell^s\mathbb{Z}$ and the ℓ-isogeny of kernel $\langle P_1 \rangle$ is horizontal.

(ii) Either the isogenous curve $E'' \simeq E/\langle Q_1 \rangle$ has ℓ-Sylow subgroup isomorphic to $\mathbb{Z}/\ell^{r+1}\mathbb{Z} \times \mathbb{Z}/\ell^{s-1}\mathbb{Z}$ and the ℓ-isogeny of kernel $\langle Q_1 \rangle$ is descending or the isogenous curve $E'' \simeq E/\langle Q_1 \rangle$ has ℓ-Sylow subgroup isomorphic to $\mathbb{Z}/\ell^r\mathbb{Z} \times \mathbb{Z}/\ell^s\mathbb{Z}$ and the ℓ-isogeny of kernel $\langle Q_1 \rangle$ is horizontal.

(iii) In the case that $E/\mathbb{F}_q$ is on the crater of the volcano, then the ℓ-Sylow subgroup of $E'/\mathbb{F}_q$ is isomorphic to $\mathbb{Z}/\ell^r\mathbb{Z} \times \mathbb{Z}/\ell^s\mathbb{Z}$, that is, the ℓ-isogeny of kernel $\langle P_1 \rangle$ is horizontal.

Proof. By [15] the action of an ℓ-isogeny over the ℓ-Sylow subgroup of an elliptic curve $E/\mathbb{F}_q$ is, if $E[\ell^\infty](\mathbb{F}_q) \simeq \mathbb{Z}/\ell^r\mathbb{Z} \times \mathbb{Z}/\ell^s\mathbb{Z}$ with $r > s > 0$, of the form $\mathbb{Z}/\ell^r\mathbb{Z} \times \mathbb{Z}/\ell^s\mathbb{Z}$, $\mathbb{Z}/\ell^{r+1}\mathbb{Z} \times \mathbb{Z}/\ell^{s-1}\mathbb{Z}$ or $\mathbb{Z}/\ell^{r-1}\mathbb{Z} \times \mathbb{Z}/\ell^{s+1}\mathbb{Z}$ depending on the direction of the ℓ-isogeny. By looking at the orders of the images of P_r and Q_s with the considered ℓ-isogeny, we can conclude. ∎

Lemma 3. *Let $\mathbb{Z}/\ell^r\mathbb{Z} \times \mathbb{Z}/\ell^s\mathbb{Z}$ with $r > s > 0$ be the group isomorphic to the ℓ-Sylow subgroup of an elliptic curve $E/\mathbb{F}_q$. Let $P_r \in E(\mathbb{F}_q)$ and let $Q_s \in E(\mathbb{F}_q)$ be two linearly independent points whose orders are respectively ℓ^r and ℓ^s. Let $\mathcal{I}$ denote the isogeny from E to E' of degree ℓ such that $\ker\mathcal{I} = \langle \ell^{r-1}P_r \rangle$. Then there exists exactly one point R of the form*

$$Q_s \quad \text{or} \quad P_r + kQ_s, \quad 0 \leq k < \ell,$$

which does not ℓ-divide in $E(\mathbb{F}_q)$ and $\mathcal{I}(R)$ ℓ-divides in $E'(\mathbb{F}_q)$, but does not ℓ^2-divide.

Proof. First of all, the set of points R in the ℓ-Sylow subgroup $\langle P_r, Q_s \rangle$ which do not ℓ-divide are, up to multiples, of one of the following forms

$$P_r + k_1\ell P_r + k_2 Q_s, \qquad 0 \leq k_1 < \ell^{r-1}, \; 0 \leq k_2 < \ell^s \tag{1}$$

$$Q_s + k_1\ell^{r-s}P_r + k_2\ell Q_s, \qquad 0 \leq k_1 < \ell^s, \; 0 \leq k_2 < \ell^{s-1} \tag{2}$$

$$Q_s + k_1\ell^{r-s-i}P_r + k_2\ell Q_s, \qquad \begin{array}{c} 0 \leq k_1 < \ell^{s+i}, \; 0 \leq k_2 < \ell^{s-1}, \\ 0 < i < r - s \end{array} \tag{3}$$

Since all the points of the form $k_1\ell P_r$ and $k_2\ell Q_s$ ℓ-divide in $E(\mathbb{F}_q)$, if some point of the form (1), (2) or (3) has an image point under $\mathcal{I}$ that ℓ-divides in $E'(\mathbb{F}_q)$, then at least one of the points Q_s or $P_r + kQ_s$, $0 \leq k < \ell$, has an image point under $\mathcal{I}$ that also ℓ-divides.

We denote $\hat{\mathcal{I}}$ the dual of $\mathcal{I}$. We denote by $Q'_s = \mathcal{I}(Q_s)$. We have seen in our first lemma that $|Q'_s| = \ell^s$. Suppose there exists $Q'_{s+x} \in E'(\mathbb{F}_q)$ such that $\ell^x Q'_{s+x} = Q'_s$ with $x > 1$. We have

$$\ell Q_s = \hat{\mathcal{I}}(\mathcal{I}(Q_s)) = \hat{\mathcal{I}}(Q'_s) \quad \text{and} \quad \ell^x \hat{\mathcal{I}}(Q'_{s+x}) = \hat{\mathcal{I}}(\ell^x Q'_{s+x}) = \hat{\mathcal{I}}(Q'_s) = \ell Q_s.$$

Therefore $|\hat{\mathcal{I}}(Q'_{s+x})| = \ell^{x+s-1}$. Since $x > 1$, we get $x + s - 1 > s$ and this is not possible since the elements of $\langle Q_s \rangle$ have at most order equal to ℓ^s. The same argument holds to prove that the image of $P_r + kQ_s$ at most ℓ-divides in $E'(\mathbb{F}_q)$.

There is at least one point that does not ℓ-divide in $E(\mathbb{F}_q)$ whose image by $\mathcal{I}$ ℓ-divides since the order of the ℓ-Sylow subgroup is invariant by isogeny. The same argument shows that there is only one point, up to multiples, that does not ℓ-divide in $E(\mathbb{F}_q)$ whose image by $\mathcal{I}$ ℓ-divides.

Let us remark that the points R of the form (1) are of order ℓ^r, the ones of the form (2) are of order ℓ^s and the ones of the form (3) are of order ℓ^{s+i}. This consideration shows us that if the ℓ-isogeny is ascending the unique R that does not ℓ-divide whose image ℓ-divides in $E'(\mathbb{F}_q)$ is of the form (2), while if the ℓ-isogeny is horizontal the unique R is of the form (3) with order ℓ^{r-1}.

Proposition 4. *Let E be an elliptic curve defined over $\mathbb{F}_q$. Let $P_n \in E(\mathbb{F}_q)$ be a point of order ℓ^n which does not ℓ-divide. Denote by R a point of order ℓ of $E(\overline{\mathbb{F}_q})$ which generates a Galois invariant subgroup G of $E(\mathbb{F}_q)$. Let P_{n+1} be a point of E in some extension $\mathbb{F}_{q^k}$ of $\mathbb{F}_q$ such that $\ell P_{n+1} = P_n$. Let $\mathcal{I} : E \to E'$ the isogeny of kernel G. Then, the abscissa of the point $\mathcal{I}(P_{n+1})$ is rational, that is $x(\mathcal{I}(P_{n+1})) \in \mathbb{F}_q$, if and only if*

$$f_{\mathcal{I}}(x) = (x - x(P_{n+1}))(x - x(P_{n+1} + R)) \cdots (x - x(P_{n+1} + (\ell-1)R)) \in \mathbb{F}_q[x].$$

Proof. The coefficients of the polynomial $f_{\mathcal{I}}(x)$ are the elementary symmetric polynomials $\mathbf{S}_r$, $1 \leq r \leq \ell$, in the abscissas of the points in $P_{n+1} + \langle R \rangle$. In [13], these elementary symmetric polynomials are given in terms of the so called generalized Vélu parameters w_i of the curve,

$$w_i = (2i + 3)S^{(i+2)} + \frac{(i+1)b_2}{2}S^{(i+1)} + \frac{(2i+1)b_4}{2}S^{(i)} + \frac{ib_6}{2}S^{(i-1)}, \quad (4)$$

where $S^{(j)}$ indicates the j-th power sum of the abscissas of the points in $\langle R \rangle \setminus \{O_E\}$. Therefore, the r-th elementary symmetric polynomial in the abscissas of the points in $P_{n+1} + \langle R \rangle$ is given by

$$\mathbf{S}_r = S_{r-1}X + S_r + \sum_{i=0}^{r-2}(-1)^i w_i S_{r-i-2},$$

where X is the abscissa of the isogenous point $\mathcal{I}(P_{n+1})$ and S_j is the j-th elementary symmetric polynomial in the abscissas of points in $\langle R \rangle \setminus \{O_E\}$. Therefore, $X = x(\mathcal{I}(P_{n+1})) \in \mathbb{F}_q$ if and only if $\mathbf{S}_r \in \mathbb{F}_q, \forall r \in \{1, \ldots, \ell\}$.

Lemma 5. *Let E' be an elliptic curve defined over $\mathbb{F}_q$. We suppose that the ℓ-torsion subgroup of $E'/\mathbb{F}_q$ is generated by two points P' and Q' linearly independent. Let $\mathcal{I} : E' \to E$ be the isogeny of kernel $\langle P' \rangle$. We denote by Q the image of Q' by $\mathcal{I}$. Then the dual isogeny $\hat{\mathcal{I}}$ is the isogeny from E with kernel equal to $\langle Q \rangle$.*

Proof. Let $\mathcal{I}'$ be the isogeny from E with kernel equal to $\langle Q \rangle$. The kernel of the isogeny $\mathcal{I}' \circ \mathcal{I}$ is the ℓ-torsion subgroup of E'. Therefore, the composition $\mathcal{I}' \circ \mathcal{I}$ is equal to the multiplication by $[\ell]$ over the curve E', and hence $\mathcal{I}' = \hat{\mathcal{I}}$. Therefore $(\hat{\mathcal{I}} \circ \mathcal{I})(P') = O_{E'}$ and $(\hat{\mathcal{I}} \circ \mathcal{I})(Q') = O_{E'}$. By definition of $\mathcal{I}$, we have $\mathcal{I}(P') = O_E$ and $\mathcal{I}(Q') = Q$. Hence $\hat{\mathcal{I}}(Q) = O_{E'}$ and the subgroup generated by Q is in the kernel of $\mathcal{I}'$. But $\hat{\mathcal{I}}$ is an isogeny of degree ℓ and since Q is a point of order ℓ over E, $\ker \hat{\mathcal{I}} = \langle Q \rangle$.

We now show how we can obtain a chain of points on isogenous curves that do not ℓ-divide. This chain of non ℓ-divisible points gives us the key of our chain of ℓ-isogenies.

Proposition 6. *Let E be an elliptic curve defined over $\mathbb{F}_q$ with ℓ-Sylow subgroup isomorphic to $\mathbb{Z}/\ell^r\mathbb{Z} \times \mathbb{Z}/\ell^s\mathbb{Z}$ with $r \geq s \geq 0$ and $r \geq 2$. Let $P_k \in E(\mathbb{F}_q)$ of order ℓ^k, $k \geq 2$, such that P_k is not ℓ-divisible. Consider the isogeny $\mathcal{I}_1 : E \to E^{(1)}$ of kernel $\langle P_1 \rangle$, where $P_1 = \ell^{k-1}P_k$, and the isogeny $\mathcal{I}_2 : E^{(1)} \to E^{(2)}$ of kernel $\langle \mathcal{I}_1(P_2) \rangle$, where $P_2 = \ell^{k-2}P_k$.*
Suppose that the point $\mathcal{I}_1(P_k)$ in $E^{(1)}(\mathbb{F}_q)$ does not ℓ-divide.
  We, then, have two different cases depending on the value of k.

- *Case $k > 2$: the point $\mathcal{I}_2(\mathcal{I}_1(P_k))$ in $E^{(2)}(\mathbb{F}_q)$ does not ℓ-divide.*
- *Case $k = 2$: the point $\mathcal{I}_2(\mathcal{I}_1(P_k))$ is $O_{E^{(2)}}$ and the ℓ-torsion subgroup of $E^{(2)}/\mathbb{F}_q$ is cyclic.*

Proof. In the case $k > 2$, in order to prove that, under the isogeny $\mathcal{I}_2 : E^{(1)} \to E^{(2)}$, the point $\mathcal{I}_2(\mathcal{I}_1(P_k))$ does not ℓ-divide, let $P_{k+1} \in E(\overline{\mathbb{F}_q})$ such that $\ell P_{k+1} = P_k$. Assume $\mathcal{I}_2(\mathcal{I}_1(P_{k+1})) \in E^{(2)}(\mathbb{F}_q)$. From Proposition 4, if $x(\mathcal{I}_2(\mathcal{I}_1(P_{k+1}))) \in \mathbb{F}_q$, the polynomial

$$\prod_{m=0}^{\ell-1} (x - x(\mathcal{I}_1(P_{k+1}) + m\mathcal{I}_1(P_2)))$$

would have all its coefficients in $\mathbb{F}_q$. Nevertheless, if we consider the dual isogeny $\hat{\mathcal{I}}_1 : E^{(1)} \to E$ with kernel $\langle R \rangle$, where $R \in E^{(1)}(\overline{\mathbb{F}_q})$ and $\langle R \rangle \neq \langle \mathcal{I}_1(P_2) \rangle$ by Lemma 5, it turns out that $\hat{\mathcal{I}}_1(\mathcal{I}_1(P_{k+1})) = \ell P_{k+1} = P_k$. Hence the abscissa $x(\hat{\mathcal{I}}_1(\mathcal{I}_1(P_{k+1}))) \in \mathbb{F}_q$ and again from Proposition 4, the coefficients of the polynomial

$$\prod_{m=0}^{\ell-1} (x - x(\mathcal{I}_1(P_{k+1}) + mR))$$

belong to $\mathbb{F}_q$. Therefore, since the greatest common divisor of these two polynomials is the linear factor $x - x(\mathcal{I}_1(P_{k+1}))$, we get $x(\mathcal{I}_1(P_{k+1})) \in \mathbb{F}_q$. Besides, if the ordinate of the point $\mathcal{I}_2(\mathcal{I}_1(P_{k+1}))$ belongs to $\mathbb{F}_q$ as well, then the ordinate of $\mathcal{I}_1(P_{k+1}) \in E^{(1)}(\mathbb{F}_q)$, which is a contradiction. The relationship between these ordinates can be derived from the formula which expresses the ordinate of the image of a point P under an isogeny of kernel G in terms of the coordinates of P and the elementary symmetric polynomials in the abscissas of points of G (see [10]). If $k = 2$, we can see that $E_1[\ell](\mathbb{F}_q)$ is a non cyclic subgroup generated by $\mathcal{I}_1(P_2)$ and another point Q. Assume $E_2[\ell](\mathbb{F}_q)$ is as well a non cyclic group. Then $E_2[\ell](\mathbb{F}_q)$ is generated by $\mathcal{I}_2(Q)$ and another point P. Therefore there exists a point $P_3 \in E(\overline{\mathbb{F}_q})$ such that $\ell P_3 = P_2$ and $\mathcal{I}_2(\mathcal{I}_1(P_3)) = P$. By using the same argument as for the case $k > 2$, we get $P_3 \in E(\mathbb{F}_q)$, which is a contradiction.

Corollary 7. *Let E be an elliptic curve defined over $\mathbb{F}_q$ with ℓ-Sylow subgroup isomorphic to $\mathbb{Z}/\ell^r\mathbb{Z} \times \mathbb{Z}/\ell^s\mathbb{Z}$ with $r > s \geq 0$ and $r \geq 2$ such that E is under the crater of its volcano of ℓ-isogenies. Let $P_r \in E(\mathbb{F}_q)$ such that P_r is of order ℓ^r. We denote by $E^{(1)}$ (resp. $E^{(2)}$, $E^{(3)}$, ..., $E^{(r)}$) the quotient of the curve E (resp. $E^{(1)}$, $E^{(2)}$, ..., $E^{(r-1)}$) by the subgroup generated by P_1 (resp. the images of P_2, P_3, ..., P_r). Then the successive images of P_r in $E^{(1)}$, $E^{(2)}$, ..., $E^{(r-1)}$ never ℓ-divide unless in $E^{(r)}$ where the image of P_r is $O_{E^{(r)}}$ and the ℓ-torsion subgroup of $E^{(r)}/\mathbb{F}_q$ is cyclic.*

Proof. Since the curve is below the crater, by *i)* of Lemma 2, the first ℓ-isogeny is ascending and therefore the ℓ-Sylow subgroup of $E^{(1)}/\mathbb{F}_q$ is isomorphic to $\mathbb{Z}/\ell^{r-1}\mathbb{Z} \times \mathbb{Z}/\ell^{s+1}\mathbb{Z}$ and hence the image of P_r does not ℓ-divide. By induction of Proposition 6, the result follows.

3.2   From Floor to Stability Level and Back to Floor

The preceding results lead us to consider the chain of ℓ-isogenies defined by the successive quotients of subgroups of order ℓ determined from a point of the initial curve whose order is the maximum power of ℓ.

Theorem 8. *Let E be an elliptic curve defined over $\mathbb{F}_q$ with ℓ-Sylow subgroup isomorphic to $\mathbb{Z}/\ell^n\mathbb{Z}$. Let P_n be a generator of this subgroup and, for all $k \in \mathbb{N}$, $k < n$, we denote by P_k the point $\ell^{n-k}P_n$. We suppose that the height h of the volcano $V_\ell(E/\mathbb{F}_q)$ is ≥ 1.*

(i) If the curves of the crater of the volcano $V_\ell(E/\mathbb{F}_q)$ have ℓ-Sylow subgroup isomorphic to $\mathbb{Z}/\ell^{\frac{n}{2}}\mathbb{Z} \times \mathbb{Z}/\ell^{\frac{n}{2}}\mathbb{Z}$, then the chain of ℓ-isogenous successive curves E, $E^{(1)}$, $E^{(2)}$, ..., $E^{(n-1)}$ given by the subgroups generated by P_1, resp. the images of P_2, P_3, ..., P_n consists of $n/2$ ascending ℓ-isogenies until reaching the stability level and $n/2$ descending ℓ-isogenies.

(ii) If the curves of the crater of the volcano $V_\ell(E/\mathbb{F}_q)$ have ℓ-Sylow subgroups isomorphic to $\mathbb{Z}/\ell^r\mathbb{Z} \times \mathbb{Z}/\ell^s\mathbb{Z}$ with $r > s = h$, then the chain of ℓ-isogenous successive curves E, $E^{(1)}$, $E^{(2)}$, ..., $E^{(n-1)}$ given by the subgroups generated by P_1, resp. the images of P_2, P_3, ..., P_n consists of h ascending ℓ-isogenies until reaching the crater, $n - 2h$ horizontal ℓ-isogenies and finally h descending ℓ-isogenies.

Proof. Consider the successive isogenies $\mathcal{I}_i : E^{(i-1)} \to E^{(i)}$, $i = 1, \ldots, n$, where $E^{(0)} = E$, whose kernels are the subgroups generated by the successive images of the points $P_i = \ell^{n-i} P_n$ under the previous isogenies. Since the ℓ-Sylow subgroup of E is cyclic, E is at the floor of the volcano and it has a unique isogeny $\mathcal{I}_1 : E \to E^{(1)}$ which is ascending. The following isogenies of the sequence, from Lemma 2, must be ascending or horizontal until reaching either a curve, if it exists, with a balanced ℓ-Sylow subgroup isomorphic to $\mathbb{Z}/\ell^{\frac{n}{2}}\mathbb{Z} \times \mathbb{Z}/\ell^{\frac{n}{2}}\mathbb{Z}$ or a curve on the crater. Thus, the isogenies of the sequence are ascending from the floor to the stability level.

In the case *(i)*, n is even and the curve $E^{(\frac{n}{2})}$ has ℓ-Sylow subgroup isomorphic to $\mathbb{Z}/\ell^{\frac{n}{2}}\mathbb{Z} \times \mathbb{Z}/\ell^{\frac{n}{2}}\mathbb{Z}$. From Corollary 7, the image of the point P_n under the isogeny $\mathcal{I}_{\frac{n}{2}}$ does not ℓ-divide in the curve $E^{(\frac{n}{2})}$. This implies that the ℓ-Sylow subgroup of the curve $E^{(\frac{n}{2}+1)}$ cannot be of the form $\mathbb{Z}/\ell^{\frac{n}{2}}\mathbb{Z} \times \mathbb{Z}/\ell^{\frac{n}{2}}\mathbb{Z}$ since the point P_n does not ℓ-divide. Hence, the isogeny $\mathcal{I}_{\frac{n}{2}+1} : E^{(\frac{n}{2})} \to E^{(\frac{n}{2}+1)}$ is descending. By Lemma 2, the following isogenies of the sequence are descending.

In the case *(ii)*, by Lemma 2, we might encounter a sequence of horizontal isogenies and then the rest of the isogenies will be descending.

We will first treat the case $r > s + 1$. We reach the crater with the curve $E^{(s)}$. Its ℓ-Sylow subgroup is generated by $P_n^{(s)}$ the successive image of P_n of order ℓ^r, $r = n - s$, and a point $Q_s^{(s)}$ of order ℓ^s. The isogeny $\mathcal{I}_{s+1}$ is the quotient of $E^{(s)}$ by $\langle \ell^{r-1} P_n^{(s)} \rangle$. Therefore $\mathcal{I}_{s+1}(P_n^{(s)})$ is of order ℓ^{r-1} and $\mathcal{I}_{s+1}(Q_s^{(s)})$ is of order ℓ^s. Since the isogeny cannot be ascending, it has to be horizontal and therefore a point of the form $\mathcal{I}_{s+1}(P_n^{(s)} + kQ_s^{(s)})$ ℓ-divides in $E^{(s+1)}$ by Lemma 3. By Corollary 7, we have that $1 \le k < \ell$. This point $P_n^{(s)} + kQ_s^{(s)}$ is an ℓ^s-divisor of $P_r^{(s)}$ but not an ℓ^{s-1}-divisor of $P_{r+1}^{(s)}$. This argument can be repeated until we reach the isogeny $\mathcal{I}_{n-s}$ defined by the quotient by $\langle P_{n-s}^{(n-s-1)} \rangle$. In the curve $E^{(n-s)}$, the point $P_{n-s+1}^{(n-s)} = \ell^{s-1} P_n^{(n-s)}$ does not have ℓ^i-divisors with $i \ge s$. Therefore the point $P_n^{(n-s)}$ is now a generator of order ℓ^s of the ℓ-Sylow subgroup of $E^{(n-s)}$. A $\mathbb{Z}/\ell^r\mathbb{Z}$ component of the ℓ-Sylow subgroup is obtained with the ℓ^i-divisors of the point $P_n^{(n-s)} + kQ_s^{(n-s)}$. By Lemma 2, the isogeny $\mathcal{I}_{n-s+1}$ defined by the quotient by $\langle P_{n-s+1}^{(n-s)} \rangle$ is either horizontal or descending and by Corollary 7 the isogeny is descending. By Lemma 2, the following isogenies are descending and since we have $s - 1$ left, the last curve is at the floor of the volcano.

At last, we treat the case $r = s + 1$. The isogeny $\mathcal{I}_{s+1}$ is the quotient of $E^{(s)}$ by $\langle \ell^{r-1} P_n^{(s)} \rangle$. Therefore $\mathcal{I}_{s+1}(P_n^{(s)})$ is of order ℓ^{r-1}, $r - 1 = s$, and $\mathcal{I}_{s+1}(Q_s^{(s)})$ is of order ℓ^s. Here, we can have either, like the precedent case, a point $\mathcal{I}_{s+1}(P_n^{(s)} + kQ_s^{(s)})$ that ℓ-divides in $E^{(s+1)}$ or the point $\mathcal{I}_{s+1}(Q_s^{(s)})$ that ℓ-divides in $E^{(s+1)}$. By a similar argument as the previous one, the following isogenies are descending until the floor of the volcano.

The same method given in Theorem 8 works when considering an elliptic curve $E/\mathbb{F}_q$ located in a level higher than the floor and lower than the stability level, in the sense that the ℓ-isogeny chain obtained is ascending from $E/\mathbb{F}_q$ to the stability level and descending to the floor.

3.3   An Example

Now we show an example of ℓ-isogeny chain starting from a curve at the floor of the volcano determined by the kernels of the successive images of the points in the ℓ-Sylow subgroup of the initial curve.

Let us consider the curve over the field $\mathbb{F}_p$, $p = 10009$, given by the equation

$$E/\mathbb{F}_p : y^2 = x^3 + 8569x + 2880,$$

whose 3-Sylow subgroup is cyclic isomorphic to $\mathbb{Z}/3^5\mathbb{Z}$ generated by the point $P_5 = (9137, 1237)$. Then, the chain of 3-isogenies determined by this point is given by

$$\begin{array}{cccccccccc}
996 & \to & 8798 & \to & 8077 & \to & 2631 & \to & 3527 & \to & 8123 \\
(5,0) & & (4,1) & & (3,2) & & (3,2) & & (4,1) & & (5,0)
\end{array}$$

where we give the curves by their j-invariants ($j(E) = 996$) and we put in brackets the integers (r, s) which determine the structure $\mathbb{Z}/3^r\mathbb{Z} \times \mathbb{Z}/3^s\mathbb{Z}$ of the 3-Sylow subgroups of the curves.

The corresponding sequence of the generators $\langle P, Q \rangle$ of the 3-Sylow subgroups, together with the integers (r, s) of the structure $\mathbb{Z}/3^r\mathbb{Z} \times \mathbb{Z}/3^s\mathbb{Z}$ and the point determining the kernel of the isogeny is:

$$\begin{array}{ccccccc}
\langle P_5 \rangle & \to & \langle P_5^{(1)}, Q_1^{(1)} \rangle & \to & \langle P_5^{(2)}, Q_2^{(2)} \rangle & \to \\
(5,0) \ 3^4 P_5 & & (4,1) \quad 3^3 P_5^{(1)} & & (3,2) \quad 3^2 P_5^{(2)} &
\end{array}$$

$$\begin{array}{ccccccc}
\langle Q_3^{(3)}, P_5^{(3)} \rangle & \to & \langle Q_4^{(4)}, P_5^{(4)} \rangle & \to & \langle Q_5^{(5)} \rangle \\
(3,2) \quad 3P_5^{(3)} & & (4,1) \quad P_5^{(4)} & & (5,0)
\end{array}$$

4   Going Around the Crater

In this section we give an application of the ℓ-isogeny chains introduced in the previous section. More precisely, given a regular volcano of ℓ-isogenies $V_\ell(E/\mathbb{F}_q)$ with height $h \geq 1$ satisfying $2h < v_\ell(\#E(\mathbb{F}_q))$ and whose crater has length $c > 2$, we present an algorithm to walk around the vertices of its crater. In order to do this we make use of the horizontal ℓ-isogenies of our particular chains. Throughout this section we suppose that the craters of the volcanoes have lengths > 2.

Proposition 9. *Let $E/\mathbb{F}_q$ be an elliptic curve whose ℓ-Sylow subgroup is isomorphic to $\mathbb{Z}/\ell^r\mathbb{Z} \times \mathbb{Z}/\ell^s\mathbb{Z}$ with $r > s > 0$ located in the crater of $V_\ell(E/\mathbb{F}_q)$. Let $E[\ell^\infty](\mathbb{F}_q) = \langle P, Q \rangle$ with $|P| = \ell^r$ and $|Q| = \ell^s$. Let $\mathcal{I}_1 \colon E \to E'$ be the ℓ-isogeny of kernel $\langle \ell^{r-1}P \rangle$ which is horizontal from Lemma 2(iii). Let $\mathcal{I}_2 \colon E \to E''$ be the other horizontal ℓ-isogeny of E. Then the dual ℓ-isogeny $\hat{\mathcal{I}}_2 \colon E'' \to E$ has kernel $\langle \ell^{r-1}\mathcal{I}_2(P) \rangle$ with $|\mathcal{I}_2(P)| = \ell^r$.*

Proof. The kernel of $\mathcal{I}_2$ is $\langle \ell^{s-1}Q + k\ell^{r-1}P \rangle$ for some $k \in \{0, \ldots, \ell-1\}$. From Lemma 5 the kernel of $\hat{\mathcal{I}}_2$ is $\langle \ell^{r-1}\mathcal{I}_2(P) \rangle$. Note that $\mathcal{I}_2(P)$ has order ℓ^r. Indeed, if $|\mathcal{I}_2(P)| < \ell^r$, then $\mathcal{I}_2(\ell^{r-1}P) = O_{E''}$. Hence $\ell^{r-1}P \in \ker \mathcal{I}_2 = \langle \ell^{s-1}Q + k\ell^{r-1}P \rangle$, which is a contradiction.

Corollary 10. Let $E_0 \xrightarrow{\mathcal{I}_0} E_1 \xrightarrow{\mathcal{I}_1} \cdots \xrightarrow{\mathcal{I}_{c-2}} E_{c-1} \xrightarrow{\mathcal{I}_{c-1}} E_0$ be the cycle of ℓ-isogenies of the crater of $V_\ell(E_0/\mathbb{F}_q)$. For all $i \in \{0, 1, \ldots, c-1\}$, let $E_i[\ell^\infty](\mathbb{F}_q) = \langle P_i, Q_i \rangle$ such that $|P_i| = \ell^r$ and $|Q_i| = \ell^s$ with $r > s > 0$. Then either, $\forall i \in \{0, 1, \ldots, c-1\}$, $\langle \ell^{r-1} P_i \rangle$ is the kernel of $\mathcal{I}_i$ or, $\forall i \in \{0, 1, \ldots, c-1\}$, $\langle \ell^{r-1} P_i \rangle$ is the kernel of $\hat{\mathcal{I}}_{(i-1) \bmod c}$.

As a consequence of Corollary 10 we can obtain, by using successive ℓ-isogeny chains, all vertices of the crater of $V_\ell(E/\mathbb{F}_q)$, since the horizontal ℓ-isogenies of the chains all go in the same direction (see Fig. 1). This idea is implemented in Algorithm 1.

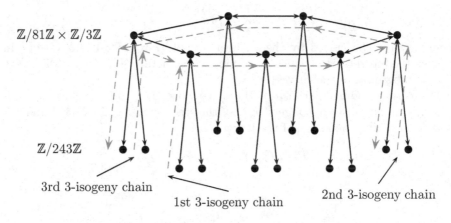

Fig. 1. Going around the crater by using 3-isogeny chains.

In order to study the cost of Algorithm 1 we need first to know the suitable number of chains to go around all the vertices in the crater. Knowing the parameters of the Algorithm 1 and assuming the length of the crater is c, the number of ℓ-isogeny chains required is $k = \left\lceil \frac{c}{n-2h} \right\rceil$. Indeed, since by the second part of Theorem 8 each of our ℓ-isogeny chains has $n - 2h$ horizontal ℓ-isogenies, we can go around all the curves on the crater by using k ℓ-isogeny chains. More precisely, starting in a curve on the floor of the volcano we ascend up to the crater and we walk through $n - 2h$ curves of the crater to descend again to the floor and we repeat the same process. From Corollary 10 we always take the same direction.

Thus, the cost of Algorithm 1 is given by $k(C_1 + n(C_2 + C_3))$ where C_1 is the cost to find a point of order ℓ^n, C_2 is the cost to compute an ℓ-isogeny using Vélu's formulae [21], and C_3 is the cost to compute the image of a given point under an ℓ-isogeny.

The cost C_1 of finding a point of order ℓ^n, assuming $\ell \ll \log q$ and the Extended Riemann Hypothesis, is $O(nM(\ell) \log q)$ with $M(\ell) = \ell \log \ell \log \log \ell$. Indeed, according to [14] it corresponds to compute a root of a polynomial of degree ℓ in $\mathbb{F}_q[x]$, which has cost $O(M(\ell) \log q)$, a total of $2n$ times. If we suppose known the cardinality of the elliptic curve, using the Algorithm 1 of [7], we have

---

Algorithm 1. CRATER$(E, \ell, n, h) \longrightarrow S$

Input: An ordinary elliptic curve E over $\mathbb{F}_q$ such that $E[\ell^\infty](\mathbb{F}_q) \simeq \mathbb{Z}/\ell^n\mathbb{Z}$ and the height h of $V_\ell(E/\mathbb{F}_q)$ is greater than 0 and $2h < n$.

Output: A sequence S containing an elliptic curve of each vertex of the crater of $V_\ell(E/\mathbb{F}_q)$.

$S \leftarrow [\,]$;

Compute a point $P \in E(\mathbb{F}_q)$ of order ℓ^n;

Compute the ℓ^h-isogeny $\mathcal{I}: E \to E'$ of kernel $\langle \ell^{n-h}P \rangle$;

$E \leftarrow E'$; $P \leftarrow \mathcal{I}(P)$; $E_{final} \leftarrow E$;

repeat

    $i \leftarrow n - h$;

    repeat

        $i \leftarrow i - 1$;

        Compute the ℓ-isogeny $\mathcal{I}: E \to E'$ of kernel $\langle \ell^i P \rangle$;

        $E \leftarrow E'$; $P \leftarrow \mathcal{I}(P)$; $S[\#S + 1] \leftarrow E$;

        $final \leftarrow E \simeq E_{final}$;

    until $final \vee i = h$;

    if $\neg final$ **then**

        Compute the ℓ^h-isogeny $\mathcal{I}: E \to E'$ of kernel $\langle P \rangle$;

        Compute a point $P' \in E'(\mathbb{F}_q)$ of order ℓ^n;

        Compute the ℓ^h-isogeny $\mathcal{I}': E' \to E''$ of kernel $\langle \ell^{n-h}P' \rangle$;

        $E \leftarrow E''$; $P \leftarrow \mathcal{I}'(P')$;

    end if

until $final$;

return S;

$C_1 = O(\log q)$. The cost C_2 of computing an ℓ-isogeny using Vélu's formalae is $O(\ell)$. Finally, by [5], the cost C_3 of evaluating a given point under an ℓ-isogeny is $O(\ell)$. Therefore, the total cost is either $O(knM(\ell)\log q)$ or $O(k\log q)$ whether the cardinality is unknown or not.

In Table 1 we give the costs to go around all the vertices in the crater of a volcano of ℓ-isogenies using the proposed procedure by Ionica and Joux [7] and using our Algorithm 1 assuming the cardinality is known. Notice that while Ionica-Joux's algorithm computes ℓ-Sylow subgroups for each curve on the crater, our proposal computes $k = \left\lceil \frac{c}{n-2h} \right\rceil$ ℓ-Sylow subgroups of curves on the floor.

The Algorithm 1 has been implemented with MAGMA V2.10-8. It has been tested with several elliptic curves over $\mathbb{F}_q$ over an Intel Pentium M with 1.73 GHz. In Table 2 we give a sample of them including information about their

Table 1. Different costs with known cardinality.

Case	Ionica-Joux	Our proposal
Regular: $2h < n$	$O(c\log q)$	$O\left(\left\lceil \frac{c}{n-2h} \right\rceil \log q\right)$
Regular: $2h = n$	$O(c\log q)$	—

Table 2. Some timings about several volcanoes of ℓ-isogenies.

$q = p$	a	b	ℓ	n	h	c	t_1	t_2
15559	4188	7183	3	4	1	40	0.07	0.02
10000000141	7034565020	8371535734	3	6	1	5612	64.99	13.83
1000000001773	464414175298	982044907463	3	7	2	37906	1955.42	979.54
10000000061	5760822374	8478355374	5	4	1	4982	196.90	15.54
10000000061	4382731032	4661390138	5	5	1	5153	134.63	13.35
1000000011151	875978249672	248043522958	5	6	2	11310	506.98	104.69
1000000000063	676232083726	397006774798	7	5	1	3486	151.98	6.61
100000231	58130720	83739022	11	5	1	190	0.83	0.09

volcanoes of ℓ-isogenies and timings. In the second and third columns of the table we have denoted by a and b the coefficients of the elliptic curve with equation $y^2 = x^3 + ax + b$. In the eighth and ninth columns we provide the timings t_1 and t_2 (in seconds) corresponding to our implementation of Algorithm 1 assuming the cardinality is known or not.

Acknowledgments. The authors thank the reviewers for their valuable comments and specially Sorina Ionica for her suggestions which have improved this article. Research of the second and third authors was supported in part by grants MTM2013-46949-P (Spanish MINECO) and 2014 SGR1666 (Generalitat de Catalunya).

References

1. Bisson, G., Sutherland, A.V.: Computing the endomorphism ring of an ordinary elliptic curve over a finite field. J. Number Theory **131**(5), 815–831 (2011)
2. Bröker, R., Lauter, K., Sutherland, A.V.: Modular polynomials via isogeny volcanoes. Math. Comput. **81**(278), 1201–1231 (2012)
3. Couveignes, J.-M., Morain, F.: Schoof's algorithm and isogeny cycles. In: Huang, M.-D.A., Adleman, L.M. (eds.) ANTS 1994. LNCS, vol. 877, pp. 43–58. Springer, Heidelberg (1994)
4. Cox, D.A.: Primes of the Form $x^2 + ny^2$. Wiley-Interscience, New York (1989)
5. Doche, C., Icart, T., Kohel, D.R.: Efficient scalar multiplication by isogeny decompositions. In: Yung, M., Dodis, Y., Kiayias, A., Malkin, T. (eds.) PKC 2006. LNCS, vol. 3958, pp. 191–206. Springer, Heidelberg (2006)
6. Fouquet, M., Morain, F.: Isogeny volcanoes and the SEA algorithm. In: Fieker, C., Kohel, D.R. (eds.) ANTS 2002. LNCS, vol. 2369, pp. 276–291. Springer, Heidelberg (2002)
7. Ionica, S., Joux, A.: Pairing the volcano. Math. Comput. **82**(281), 581–603 (2013)
8. Joux, A., Nguyen, K.: Separating decision Diffie-Hellman from computational Diffie-Hellman in cryptographic groups. J. Cryptol. **16**(4), 239–247 (2003)
9. Kohel, D.: Endomorphism rings of elliptic curves over finite fields. Ph.D. thesis, University of California, Berkeley (1996)
10. Lercier, R.: Algorithmique des courbes elliptiques dans les corps finis. Ph.D. thesis, École Polytechnique, Paris (1997)

11. Lenstra Jr., H.W.: Complex multiplication structure of elliptic curves. J. Number Theory **56**(2), 227–241 (1996)
12. Menezes, A., Okamoto, T., Vanstone, S.: Reducing elliptic curve logarithms to logarithms in a finite field. IEEE Trans. Inf. Theory **39**, 1639–1646 (1993)
13. Miret, J., Moreno, R., Rio, A.: Generalization of Vélu's formulae for isogenies between elliptic curves. In: Proceedings of the Primeras Jornadas de Teoría de Números Publicacions Matemàtiques, vol. Extra, pp. 147–163 (2007)
14. Miret, J., Moreno, R., Rio, A., Valls, M.: Computing the ℓ-power torsion of an elliptic curve over a finite field. Math. Comput. **78**(267), 1767–1786 (2009)
15. Miret, J., Moreno, R., Sadornil, D., Tena, J., Valls, M.: Computing the height of volcanoes of ℓ-isogenies of elliptic curves over finite fields. Appl. Math. Comput. **196**(1), 67–76 (2008)
16. Miret, J., Sadornil, D., Tena, J., Tomàs, R., Valls, M.: On avoiding ZVP-attacks using isogeny volcanoes. In: Chung, K.-I., Sohn, K., Yung, M. (eds.) WISA 2008. LNCS, vol. 5379, pp. 266–277. Springer, Heidelberg (2009)
17. Pohlig, S., Hellman, M.: An improved algorithm for computing algorithms over $GF(p)$ and its cryptographyc significance. IEEE Trans. Inf. Theory **24**, 106–110 (1978)
18. Schoof, R.: Counting points on elliptic curves over finite fields. J. Théor. Nombres Bordeaux **7**(1), 219–254 (1995)
19. Silverman, J.H.: The Arithmetic of Elliptic Curves. Graduate Texts in Mathemathics. Springer-Verlag, New York (1986)
20. Sutherland, A.V.: Computing Hilbert class polynomials with the Chinese remainder theorem. Math. Comput. **80**(273), 501–538 (2011)
21. Vélu, J.: Isogenies entre courbes elliptiques. Comptes Rendus De L'Academie Des Sciences Paris, Serie I-Mathematique, Serie A **273**, 238–241 (1971)

Privacy

Beating the Birthday Paradox in Dining Cryptographer Networks

Pablo García[1], Jeroen van de Graaf[2], Alejandro Hevia[3]([✉]), and Alfredo Viola[4]

[1] Universidad Nacional de San Luis, San Luis, Argentina
[2] Depto. de Ciência da Computação, Universidade Federal de Minas Gerais,
Belo Horizonte, Brazil
[3] Department of Computer Science, University of Chile, Santiago, Chile
ahevia@dcc.uchile.cl
[4] Instituto de Computación, Universidad de la República, Montevideo, Uruguay

Abstract. A Dining Cryptographer Network (DC-Net) allows multiple players to broadcast messages without disclosing the identity of the sender. However, due to their probabilistic nature, message collisions can occur, meaning that two or more messages sent by different participants end up occupying the same slot, causing these messages to be lost. In this work, we evaluate two different strategies to deal with collisions. When repeating a DC-net sequentially, honest parties who see that their message did not collide can switch to sending a null message, effectively decreasing the collision probability in subsequent rounds. When repeating a DC-net in parallel, no feedback exists, and there will always remain a non-zero probability that one message collides in every round. We analyze both strategies with respect to the number of parties, the number of slots, the number of repetitions and the probability of success. We obtain exact but rather convoluted combinatorial formulas for both cases, together with more tractable approximations, the correctness of which has been demonstrated by simulations.

1 Introduction

Dining Cryptographers and Collisions. A well-known protocol providing anonymous message broadcast is the Dining Cryptographers protocol, developed by Chaum [4]. A DC-net is a randomized protocol which allows a set of parties to send messages anonymously. Typically, a DC-net protocol is parameterized by the number of participants, n, and its *bandwidth*, m. In the simplest variant of a DC-net, each party must randomly choose a *slot number* between 1 and m in order to send her message.

A serious problem of DC nets, already recognized in the original paper, is the fact that two or more parties can, by coincidence, submit their message to the same slot, leading to a collision. Here we focus on *random collisions* which occur because of users choosing the same slot. One way to deal with collisions is to run a slot reservation protocol before the actual message is sent; another

© Springer International Publishing Switzerland 2015
D.F. Aranha and A. Menezes (Eds.): LATINCRYPT 2014, LNCS 8895, pp. 179–198, 2015.
DOI: 10.1007/978-3-319-16295-9_10

is to run a mix-net first to deal with the slot reservations [6]. Here we explore another direction.

Because of these collisions, we can consider the original DC-net as an unreliable communication channel. We say that a DC-net is reliable if, for each participant, the probability that she does not succeed in publishing her message (anonymously) is *negligible* in the security parameter κ. Theoretically this should be interpreted as exponentially small in κ, and practically as below a very small parameter such as 10^{-10} or smaller.

Summary of Our Results. In this work, we explore new ways to build collision-free (or reliable) DC-net protocols. Our focus is not the low-level design of new DC-net protocols, but instead a precise evaluation of the efficiency of at least two simple ways to build collision-free DC-nets: sequential repetition and parallel repetition.

Sequential Repetition. A first way to obtain a reliable DC-net is to consider r sequential repetitions of an unreliable DC-net, for some fixed parameter $r > 1$. If the message for a given party is successfully sent (has not collided with some other message) then that party starts sending the null message in any subsequent repetition of the basic DC-net. After r rounds, all parties stop.

Parallel Repetition. A second way to obtain a reliable DC-net is to consider k parallel repetitions of the basic DC-net, for some fixed parameter $k \geq 1$. Each party sends a copy of the message in each parallel instance, henceforth called *channel*, of the DC-net protocol. In this case we say a message was successfully sent if the message did not collide in at least one channel.

For both scenarios, we develop a quantitative analysis allowing a trade-off between the probability of success, the number of participants n, the total number of available position m, and either the number of sequential repetitions r or the number of parallel repetitions k.

We remark that both Chaum's DC-net as well as Golle-Juels' DC-net do not really deal with collisions. In their papers [4,17] the authors seem to suggest the strategy that the protocol needs to be repeated until each message is transmitted collision-free – but no concrete algorithm is given.

Interestingly, as a byproduct of our analysis of the collision probability of these iterated constructions, we show that collisions in Golle-Juels' DC-net are not random under the standard static adversary that corrupts $t < n$ players. In particular, the collision probability under such adversary for some honest user' message is not t/n, but instead depends on the *value* of the message. Obtaining a DC-net with random collisions can be easily done at the price of an extra round. See Sect. 8 for details.

Related Work. An important motivation for our research is the observation that anonymous broadcast can only be achieved in two ways: either by using DC-nets or by using Mixnets, also invented by David Chaum [3]. However, a main difference between these two alternatives is that DC-nets provide unconditional anonymity (at the cost of a significant message overhead), whereas in Mixnets the anonymity is inherently computational. So at some time in the future, when the NSA has broken RSA or ElGamal, it can retro-actively trace messages sent through Tor back to their originators. To eliminate this drawback, we believe that practical prototypes of unconditional DC-nets should be implemented, and this paper should be regarded as a theoretical contribution towards this effort.

Note that the Golle-Juels variant of DC-nets and its practical implementations, Dissent [5] and Verdict [6] do not satisfy this requirement, because in that approach unconditional privacy is traded away in favor of efficiency: noninteractive DHKE and pseudo-random number generators, implying computational anonymity only. We are not convinced this is the right trade-off and believe that more effort should be spent to preserve unconditional privacy.

As a strategy to deal with disrupters we advocate the pro-active use of cryptography (bit commitments, proofs of knowledge etc.), just as proposed in [17,21]. The original proposal by Chaum suggested the use of traps, see Waidner and Pfitzmann [22] for details. But we are somewhat sceptical about them. First, traps assume sequential repetition, which in some cases is not available or not desirable. Second, traps represent a re-active solution: disrupters need to be identified and excluded after they have started misbehaving. This means that a set of sleeping Sybils can mount a collaborative denial-of-service attack and bring down the network for a significant period of time. We therefore prefer pro-active approaches.

Herbivore [15] uses a different topological model which reduces the impact of disrupters. It also uses a slot reservation phase to avoid (or at least, reduce) collisions which is in spirit similar to our sequential approach. However, their performance metric is not the number of rounds, but the number of bits transmitted per node in order to send a packet anonymously and successfully.

The idea of using parallel Dining Cryptographers was first suggested in [20] (in German), but as far as we know no quantitative results have ever been published. Preliminary results of our research were published recently in [11,12] (in Spanish). Very recently yet another strategy was proposed to deal with collisions which makes no a priori assumption about a distribution on the messages submitted [9], using techniques from networking. An optimal solution offering unconditional privacy is presented in [10].

2 Preliminaries

Protocols, Messages, and Adversaries: For any integer $s > 0$, let $[s]$ denote the set $\{1, \ldots, s\}$. We model a DC-net as a multi-sender broadcast anonymous channel as follows. We consider a setting with $n + 1$ parties (n is some fixed constant) where sender parties $P_1, P_2, \ldots, P_n$ attempt to send messages $M_1, M_2, \ldots, M_n$ (resp.) to a receiver party R, in an anonymous and reliable way. The communication between parties can be observed and even be influenced by a (possibly computationally unbounded) adversary A which can statically corrupt at most

t senders. In this work, since we start from secure DC-nets, we do not fix a particular model of interaction between the adversary and the parties. Instead, we describe the protocols in terms of the messages input to the protocol and the messages received by R. If M is a tuple of n messages for the sender parties we write $\mathsf{Exec}_\Pi(P(M), A)(\kappa)$ for the output of party R when the parties $P = (P_1, P_2, \ldots, P_n)$ and R execute the protocol in the presence of adversary A for security parameter κ. We assume that this output is a set of messages. Notice that the execution is randomized, so once M is fixed, $\mathsf{Exec}(P(M), A)(\kappa)$ is a random variable. Also, we assume that any initialization that the protocol requires (i.e. a public key infrastructure) are performed at the beginning of the execution.

Commitments: In this work, we use commitment schemes [19] to implement the parallel iterated DC-net. In its basic form, commitment schemes are a tuple $(\mathcal{K}, \mathsf{Comm}, \mathsf{Open})$, where the key generation algorithm $\mathcal{K}$ produce the commitment parameters μ, after being executed during setup time by a trusted player. Then, randomized algorithm Comm takes μ and a message M, and produces a pair commitment value and opening $(\mathcal{C}, op) \leftarrow \mathsf{Comm}_\mu(M)$. Algorithm Open allows opening the commitment using M and op. Yet, the security requirements we assumed on the commitment schemes are stronger, including non-malleability. Due to space constraints we refer the interested reader to the work of Abdalla et al. [1].

3 Sequential Repetition of DC-Nets

Perhaps the most well-known strategy to fix the collision problem is to sequentially repeat the protocol – so much it is pervasively mentioned in many places, including [4,17]. The strategy is simple: given an anonymous broadcast protocol Γ, fix a number of rounds $r \geq 1$ and simply instruct parties to execute Γ in sequence r times. In each execution, all senders attempt to transmit. For each sender, if there is no collision (her message goes through) then that party transmits the null message (denoted $\perp$) from that point forward.

The main benefit of this strategy is simplicity. The resulting reliable DC-net protocol is very easy to implement in a *black-box* manner, given access to Γ.

We assume each player P_i can detect if a message (not necessarily sent by P_i) was involved in a collision or not (see [2] for a way to implement this assumption).

Sequential Repetition of DC-Net: Let $\Gamma(M)$ denote a single instance of an unreliable DC-net protocol executed by players $P_1, \ldots, P_n$ under input $M = (M_1, \ldots, M_n)$ where M_j is the input of player j. We let $\overline{\Gamma}^{(r)} = \Gamma_1, \Gamma_2, \cdots, \Gamma_r$ be the protocol consisting of r sequential repetitions of protocol Γ, so $\Gamma_j = \Gamma$ for $j = 1, \ldots, k$. All parties participate in every sequential iteration of the protocol. In the description below, we let $c_i = (c_{i,j})_{j \in [r]}$ be the vector of slots randomly chosen by player P_i. In the j-th iteration of Γ, player P_i transmits her message in slot $c_{i,j} \in [m]$. Protocol $\overline{\Gamma}^{(r)}$ takes input M and proceeds as follows:

1. **Main Loop:** All players do the following, from $j = 1$ up to $j = r$:
    (i) **Commit to a slot:** Each player P_i chooses a value $c_{i,j}$ uniformly at random in $[m]$ as the transmission slot for the current execution of the DC-net. Player P_i then computes and broadcast the associated commitment $\mathcal{C}_{i,j} = \text{Com}(c_{i,j})$.
    (ii) **Computing and broadcasting the pads:** Each party P_i participates in the execution of protocol $\Gamma_j = \Gamma$ using M_i as input and $c_{i,j}$ as the transmission slot. In Γ_j, any non-interactive zero-knowledge proof of correctness[1] is modified so P_i can prove that the slot $c_{i,j}$ used to prepare the pad is consistent with the value committed in the previously broadcasted $\mathcal{C}_{i,j}$. Let $d_j = (c_{i,j})_{i \in [n]}$ be the vector of slots used by all parties in round j.
    (iii) **Compute the transmitted values:** Let $S_j \leftarrow \text{Exec}_{\Gamma_j}(P(M; d_j), A)(\kappa)$ denote the final output of Γ_j. If message $M_i \in S$ then P_i sets $M_i \leftarrow \perp$ in order to transmit the null message in every subsequent round.
2. **Compute the final set of transmitted values:** The output of $\overline{\Gamma}^{(r)}$ is then the union (as sets) of S_j over $j \in [r]$.

Notice that a message is successful if there exists a round $j \in [r]$ such that the message is successfully broadcast under protocol Γ_j.

4   Analysis of Sequential Repetition of DC-Nets

In this section we analyze the collision probability of sequential approach in terms of the number of players n, the number of slots per channel m, and the number of rounds r. The sequential DC protocol can be modelled as a problem of balls-in-bins, where balls correspond to messages and bins to slots (positions).

Following the symbolic method presented in [8], we let $G_m(x, z) = (e^{z/m} + (x - 1)z/m)^m$. In this setting the coefficient $n![x^k z^n]G_m(x, z)$ is the probability that when throwing n balls in m bins we have k bins with one ball. These bins are filled by the messages that go through, and as a consequence do not have to be sent again in the next round.

Theorem 1. *With the convention that $y_0 = z$, let*

$$\hat{G}_m(x, y_0, y_1, \ldots, y_r) = \prod_{i=1}^{n} G_m(x \, y_i, y_{i-1}/y_i). \tag{1}$$

Then $n![x^n, z^n, y_1^0, \ldots, y_r^0]$ is the probability generating function that all messages go through in less than or equal r round when n messages are sent in m slots, using the sequential algorithm.

Sketch of the proof: The variable y_i "marks" the number of balls that go through in round i. Intuitively y_{i-1}/y_i takes care of the fact that the messages that go through round $i - 1$ do not have to be sent in round i.   □

---

[1] For example, the term $\sigma''_{i,\theta}$ in the Golle-Juels protocol, see Appendix A.

Even though this is an exact approach, it is far from trivial to extract coefficients and to find the probability that more than r rounds are needed.

By the Poisson Law which governs the asymptotic behavior of the problem in the case $n = \lambda m$ with $\lambda > 0$ a constant ([8], pag. 177), the *expected* number of bins with one ball when $m, n \to \infty$ and $n = \lambda m$ is $ne^{-\lambda}$. It can be seen that this value is concentrated around its mean.

Let $\lambda_0 = n/m < 1$. Then, the *expected* number of balls that fail in this round is $n - ne^{-\lambda_0}$, and as a consequence after the first round, the expected value of the proportional constant λ_1 is $\lambda_0(1 - e^{-\lambda_0})$.

If we iterate this process, then after iteration i, the expected number of messages that has to be sent is $n_{i+1} = \lambda_{i+1}m$ with $\lambda_{i+1} = \lambda_i(1 - e^{-\lambda_i})$. The process is finished at the iteration f such that $n_f < 1$, that is $\lambda_f < 1/m = \lambda_0/n$.

Theorem 2. *Let $0 < \delta < 1$ and $0 < \lambda_0 < 1$ with $(1+\delta)\lambda_0 < 1$, and let f be the number of iterations made in the sequential algorithm when n balls are initially thrown in $m = n/\lambda_0$ bins. Moreover, let*

$$low(n, \lambda_0, \delta) := \left\lfloor \lg\left(\log\left(\frac{n}{\lambda_0}\right)\right) - \lg\left(-\log(((1-\delta)(1 - \lambda_0/2)\lambda_0))\right) \right\rfloor$$

and

$$high(n, \lambda_0, \delta) := \left\lceil \lg\left(\log\left(\frac{n}{\lambda_0}\right)\right) - \lg\left(-\log((1+\delta)\lambda_0)\right) \right\rceil .$$

Then,

$$\Pr\left[low(n, \lambda_0, \delta) \leq f \leq high(n, \lambda_0, \delta)\right] > 1 - e^{-\frac{\delta^2(1-e^{-\lambda_0})}{3}n} .$$

where $\lg$ *denotes the logarithm in base 2.*

Sketch of the Proof. This is a proof by induction based on Chernoff bounds. After the first iteration we have an average of $n_1 = n_0(1 - e^{-\lambda_0})$ balls that had collisions, where $n = n_0 = \lambda_0 m$.

By Chernoff bounds, we have

$$\Pr\left[\lambda_1 > (1+\delta)\lambda_0(1 - e^{-\lambda_0})\right] = \Pr\left[n_1 > (1+\delta)n_0(1 - e^{-\lambda_0})\right]$$
$$< e^{-\frac{\delta^2(1-e^{-\lambda_0})}{3}n},$$
$$\Pr\left[\lambda_1 < (1-\delta)\lambda_0(1 - e^{-\lambda_0})\right] = \Pr\left[n_1 < (1-\delta)n_0(1 - e^{-\lambda_0})\right]$$
$$< e^{-\frac{\delta^2(1-e^{-\lambda_0})}{2}n}.$$

Moreover, since $u - u^2/2 \leq 1 - e^u \leq u$, then

$$\Pr\left[\lambda_1 > (1+\delta)\lambda_0^2\right] < e^{-\frac{\delta^2(1-e^{-\lambda_0})}{3}n},$$
$$\Pr\left[\lambda_1 < (1-\delta)\left(1 - \lambda_0/2\right)\lambda_0\right] < e^{-\frac{\delta^2(1-e^{-\lambda_0})}{2}n}.$$

In general,

$$Pr[\lambda_{i+1} > (1+\delta)\lambda_i^2 \geq (1+\delta)\lambda_i(1 - e^{-\lambda_i})] < e^{-\frac{\delta^2(1-e^{-\lambda_i})}{3}n},$$
$$Pr[\lambda_{i+1} < (1-\delta)(1-\lambda_0/2)\lambda_i \leq (1-\delta)(1-\lambda_i/2)\lambda_i$$
$$\leq (1-\delta)\lambda_i(1 - e^{-\lambda_i})] < e^{-\frac{\delta^2(1-e^{-\lambda_i})}{2}n}.$$

The sequential process finishes when $n_f = 1$ (or equivalently when $\lambda_f = 1/m = n/\lambda_0$). Furthermore, since

$$\prod_{i=0}^{f-1} \Pr\left[\lambda_{i+1} > (1+\delta)\lambda_i^2\right] < \Pr\left[\lambda_1 > (1+\delta)\lambda_0^2\right],$$

$$\prod_{i=0}^{f-1} \Pr\left[\lambda_{i+1} < (1-\delta)(1-\lambda_0/2)\lambda_i\right] < \Pr\left[\lambda_1 < (1-\delta)(1-\lambda_0/2)\lambda_0\right],$$

then, by chaining all the inequalities,

$$\Pr\left[\lambda_0/n = \lambda_f > ((1+\delta)\lambda_0)^{2^f} \geq \frac{((1+\delta)\lambda_0)^{2^f}}{(1+\delta)}\right] < e^{-\frac{\delta^2(1-e^{-\lambda_0})}{3}n}, \qquad (2)$$

$$\Pr\left[\lambda_0/n = \lambda_f < (C\lambda_0)^{2^f} \leq \frac{(C\lambda_0)^{2^f}}{C}\right] < e^{-\frac{\delta^2(1-e^{-\lambda_0})}{2}n}, \qquad (3)$$

with $C = (1-\delta)(1-\lambda_0/2)\lambda_0$.

The theorem then follows by finding f from Eqs. (2) and (3), and taking the complement of these events. $\qquad\square$

It is important to note that this leads to $O(\log\log(n))$ rounds. In terms of space, if $m = n$, this means we have a $O(n\log\log(n))$ space algorithm, improving the $O(n^2)$ cost given by the birthday paradox.

Experimental Results. For the experiments we have considered $\delta = 0.1$ and $\lambda_0 \in \{0.3, 0.5, 0.7, 0.9\}$. Moreover, for each of these values, and for $n \in \{10^4, 10^5, 10^6\}$ we have calculated $low(n, \lambda_0, \delta)$, $high(n, \lambda_0, \delta)$, calculated the probability of f being in the estimated range, run 100 experiments, and seen how f falls in the estimated range. The probability is estimated up to 10 decimal digits (so, errors less than 10^{-10}, will lead to probability equal to 1).

The results are shown in Fig. 1. It can be seen that not only the upper and lower bounds are very good but that all the experiments fall in the estimated range of values.

n	λ_0	Prob	low	$\lg\left(\log\left(\frac{n}{\lambda_0}\right)\right)$	high	# low	# low + 1	# low + 2
10^4	0.3	0.9996	2	4	4	0	76	24
10^4	0.5	0.9999	3	4	5	0	97	3
10^4	0.7	0.9999	3	4	6	0	95	5
10^4	0.9	0.9999	3	4	10	0	97	3
10^5	0.3	1	3	4	4	10	90	0
10^5	0.5	1	3	4	5	0	86	14
10^5	0.7	1	3	4	6	0	88	12
10^5	0.9	1	3	4	11	0	85	15
10^6	0.3	1	3	4	4	0	100	0
10^6	0.5	1	3	4	5	0	29	71
10^6	0.7	1	3	4	6	0	28	72
10^6	0.9	1	4	4	11	24	76	0

Fig. 1. Experimental results of the sequential protocol.

5 Parallel Repetition of DC-Nets

Section 3 presented a variant that relies on a reactive strategy on behalf of the players. This implies that all players have to be online during the execution of the DC-net protocol, in particular, during $r = \mathcal{O}(n \log(\log(n)))$ iterations. In some situations this is not possible or even desirable.

In this section we present a solution for a round-optimal reliable version of the DC-net protocol (just $r + 1$ rounds if the underlying unreliable DC-net takes r rounds). In particular, it is *almost* non-interactive (just 2 rounds) if the underlying DC-net is non-interactive (e.g. Golle-Juels). Like the original versions, it consists of three phases: (1) a preliminary phase, in which all pairs of players exchange random keypads; (2) a two round phase, consisting of a commitment round followed by a broadcast round, in which each party publishes **one** contribution; (3) a message computation phase, in which the actual messages are computed.

In the strategy described in Sect. 3 phase 2 was repeated and parties adapted their new inputs as a function of the results of phase 3. We change the model in this section by assuming that repeating phases 2 and 3 is not possible. We assume each party can only broadcast **one** contribution (message) in phase 2, and that is her only chance. So if a collision occurs, it means that no posterior correction is possible. This implies that the probability of a collision should be made to be negligible.

At the cost of extra bandwidth, this is not difficult to achieve: we assume the existence of several (say k) copies of the same DC-net. We refer to each DC-net as a *channel*, (reserving the word *net* for the whole construction consisting of k channels). Each party chooses randomly a different slot in each channel and sends the *same* contribution to each of these channels. Now we say that a message transmission *fails* if the message collides in all the channels; it *succeeds* if in at least one channel the message is alone in the slot.

Parallel Repetition of DC-Nets: As before, let $\Gamma(\boldsymbol{M})$ denote a single instance of an unreliable DC-net protocol executed by players $P_1, \ldots, P_n$ under input $\boldsymbol{M} = (M_1, \ldots, M_n)$, where M_j is the input of player j.

We let $\hat{\Gamma}^{(k)} = \Gamma_1 \| \Gamma_2 \| \cdots \| \Gamma_k$ be the protocol consisting of k parallel repetitions of the same protocol $\Gamma_\ell = \Gamma$, for $\ell = 1, \ldots, k$. All players participate in each parallel protocol Γ_ℓ, $\ell = 1, \ldots, k$. In the description below, we denote by $\boldsymbol{c_i} = (c_{i,\ell})_{\ell \in [k]}$ the vector of slots randomly chosen by player P_i so $c_{i,\ell} \in [m]$ is the slot used in the ℓ-th parallel channel Γ. Protocol $\hat{\Gamma}^{(k)}$ takes input $\boldsymbol{M} = (M_1, \ldots, M_n)$ and proceeds as follows:

1. **First round, committing to the slots:** Each player P_i chooses a k-dimensional vector of slots $\boldsymbol{c_i} = (c_{i,\ell})_{\ell \in [k]}$ by choosing each $c_{i,\ell}$ uniformly at random in $[m]$. Player P_i then computes and broadcasts the corresponding vector of commitments $\boldsymbol{C_i} = \mathrm{Com}(c_{i,\ell})_{\ell \in [k]}$.
2. **Second round, computing the pads:** Each player P_i prepares to runs k copies of Γ, namely Γ_1 through Γ_k, using the same input M_i in all parallel executions Γ_ℓ. That is, if $M_{\ell,i}$ denotes the message used by player P_i in execution Γ_ℓ, then $M_{\ell,i} = M_i$ and the slot used by player P_i is $c_{i,\ell}$.
3. **Second round, Computing the Proof of Unique Message:** Each player P_i then computes a zero-knowledge proof ρ_i that the same message is used in each parallel copy of Γ.
4. **Second round, broadcast of proofs and anonymous broadcast of messages:** Each player P_i broadcasts ρ_i and then runs runs k parallel copies of protocol Γ, where any non-interactive zero-knowledge proof of correctness is modified to also prove that the slot used ($c_{i,\ell}$) is consistent with the previously broadcast commitment $\mathcal{C}_i$.
5. **Second round, computing received messages in each parallel channel:** Each player first checks that ρ_j is a valid proof for player P_j, $j \in [n]$. If not and if there is a reconstruction procedure for Γ, this is triggered with respect to player P_j in each parallel run. If there is no reconstruction procedure, the whole parallel protocol aborts and starts again, excluding the misbehaving party. Let $\mathsf{Exec}_{\Gamma_\ell}(P(\boldsymbol{M}; \{\boldsymbol{c_i}\}_{i \in [n]}), A)(\kappa)$ denote the final output of protocol Γ_ℓ. The output of $\hat{\Gamma}^{(k)}$ is then the union (as sets) over $\ell \in [k]$ of $\mathsf{Exec}_{\Gamma_\ell}(P(\boldsymbol{M}; \{\boldsymbol{c_i}\}_{i \in [n]}), A)(\kappa)$.

The following section presents the analysis of the collision probability of this approach in terms of the number of players n, the number of slots per channel m, and the number of parallel channels k.

6   Analysis of Parallel Repetition of DC-Nets

The parallel DC protocol can also be modelled as a problem of balls-in-bins, where as before balls correspond to messages and bins to slots (positions).

In this setting we have a total of S bins, n balls and k parallel channels. In each channel, each of the n balls chooses a position among $m = S/k$ bins.

The ball i ($1 \leq i \leq n$) is successful if in at least one of the channels falls in a bin with only one ball.

We want to calculate $P_{m,n,k}$, the probability that all balls are successful if k parallel channels are used. Moreover, the goal, for given values of n and $S = km$ is to find the value of k that maximizes $P_{m,n,k}$.

These kind of problems of balls-in-bins are known as the occupancy problem; see for instance [7,18]. However, this specific problem does not seem to have been analyzed.

We have obtained an exact expression for $P_{m,n,k}$ which can be described as a convolution of $2k - 1$ nested sums. Even though this expression is exact, it is very difficult to handle and far from trivial. We present it here, but then we use an approximate model to predict the probability of success.

In the sequel $\left\{ {a \atop b} \right\}$ denotes the Stirling numbers of the second kind (number of ways of placing a elements in b non-empty sets). Moreover the falling factorials are defined as $m^{\underline{j}} = m(m-1)(m-2)\ldots(m-j+1)$.

The proof of the following Fact can be found in the standard literature of the problem (for example in [8]).

Fact 3. *Let $C_{n,m}$ be the number of ways of placing n balls in m bins in such a way that no bin has one ball, and $T_{m,n,r} = \sum_{j \geq 0} m^{\underline{j+r}} C_{n-r,j}$. Moreover, let $N_{m,n,r}$ be the number of ways of throwing n balls in m bins in such a way that r of the bins contain 1 ball. Then*

1. $C_{n,m} = \sum_{i \geq 0} (-1)^i \binom{n}{i} \left\{ {n-i \atop m-i} \right\}$.
2. $N_{m,n,r} = \binom{n}{r} T_{m,n,r}$.
3. $P_{m,n,1} = \frac{N_{m,n,n}}{m^n} = \frac{m^{\underline{n}}}{m^n}$.

Sketch of the proof: Using symbolic techniques (as in [8]), $N_{m,n,r} = n![u^r z^n]$ $(e^z + (u-1)z)^m$ and $C_{N,M} = N![z^n](e^z - 1 - z)^M$.

A more direct approach will be useful for the generalization.

– Choose r balls to go to the r bins with one ball. This gives the factor $\binom{n}{r}$.
– Choose r bins and place the remaining $n - r$ balls in the other $m - r$ bins in such a way that no bin has 1 ball. This gives the factor $T_{m,n,r}$. It can be achieved as follows
  1. Choose r bins to place 1 ball at each one ($m^{\underline{r}}$).
  2. Choose j bins from the remaining $m - r$ bins to place the $n - r$ balls that give collisions. There are $(m - r)^{\underline{j}}$ ways of doing this. Together with the previous selection, this gives a factor $m^{\underline{r+j}}$.
  3. Place the $(n - r)$ balls in the j bins in such a way that none of these j bins are empty or have one ball. This contributes with the factor $C_{n-r,j}$.

□

Fact 3 clearly divides the calculation of $N_{m,n,r}$ in two parts. The first one, $\binom{n}{r}$, is related with the selection of successful balls and the second one, $T_{m,n,r}$, with the selection of the bins and placement of the unsuccessful balls. This division is key to derive a general results for all k. Since we were unable to use symbolic techniques for the problem, we present a direct approach.

Theorem 4. *Let* $I_k = \{0 \le j_1 = s_1 \le n-t\} \cup \{0 \le j_i \le s_i \le n-t, \ 2 \le i \le k\}$

be a set of indices, $\delta_i = s_i - j_i$ *and* $J_k = \sum_{i=1}^{k} j_i$. *Let* $Q_{m,n,t}^{(k)}$ *be the probability of having* t *failures when throwing* n *balls in* k *parallel channels with sets of* m *bins. Then*

1. $Q_{m,n,t}^{(k)} = \frac{R_{m,n,t}^{(k)}}{n^{km}}$, *with* $R_{m,n,t}^{(k)} = \binom{n}{t} \sum_{I_k} c_{n,t}^{(k)} \prod_{i=1}^{k} N_{m,n,s_i}$ *and.*

$$c_{n,t}^{(1)} = 1 \text{ which equals } \binom{J_0}{n-t-s_1} \text{ when } s_1 = n-t,$$

$$c_{n,t}^{(k)} = \frac{c_{n,t}^{(k-1)}}{\binom{J_{k-2}}{n-t-s_{k-1}}} \binom{n-t-J_{k-2}}{j_{k-1}} \binom{J_{k-2}}{\delta_{k-1}} \binom{J_{k-1}}{n-t-s_k}, \quad 2 \le k.$$

2. $P_{m,n,k} = Q_{m,n,0}^{(k)}.$

Sketch of the proof: The factor $\binom{n}{t}$ gives the number of ways to choose the t balls that fail in all the channels. As a consequence the other $n-t$ balls should fall in an bin with one ball in at least one of the channels. To analyze the algorithm it is better to think it as a sequential process.

In this setting, s_i is the number successful balls in channel i, j_i is the number of balls whose *first* successful try is channel i and J_i is the *total* number of successful balls up to channel i, with $1 \le i \le k$. Then $s_1 = j_1$ since this is the first successful channel for these balls, and $j_k = n - t - J_{k-1}$ since at the end $n - t$ balls are successful ($J_k = n - t$).

Moreover, for $1 \le i < k$, the j_i balls whose first successful try is channel i should be taken from the ones that have failed in all previous channels ($n-t-J_{i-i}$ of them), giving the factor $\binom{n-t-J_{i-1}}{j_i}$. Furthermore, the other $\delta_i = s_i - j_i$ balls should be chosen among the one already successful, giving the factor $\binom{J_{i-1}}{\delta_i}$. Since s_k are successful in the last channel, we have to choose the other $n - t - s_k$ unsuccessful balls (but *successful* ones in the k channels) among the already successful J_{k-1} balls (giving the factor $\binom{J_{k-1}}{n-t-s_k}$).

The recursive definition of $c_{n,t}^{(k)}$ then follows by induction on k (the number of channels). Notice that when $k = 1$, $c_{n,t}^{(1)} = \binom{J_0}{n-t-s_1}$. Since $J_0 = 0$ this coefficient is 0, unless $s_1 = n - t$, when it takes the value of 1. This is actually the case, since all the $n - t$ balls should be successful in the first try.

The proof is then completed by noticing that $\prod_{i=1}^{k} N_{m,n,s_i}$ counts the number of ways to place all the $n - s_i$ balls in $m - s_i$ bins in such a way that no bin has one ball.   □

Since this exact expression is very difficult to handle with large numbers, and asymptotic results are difficult to achieve, an alternate, approximate approach will be given to find an analytic expression to predict the results found by our simulations.

As it is presented in Sect. 4, if n balls are thrown in m bins, then the expected number of bins with 1 ball is

$$E[\# \; bins \; with \; 1 \; ball] = n\left(1 - \frac{1}{m}\right)^{n-1}. \tag{4}$$

If $n, m \to \infty$ this value can be approached by $ne^{-n/m}$. Its variance is

$$Var[\# \; bins \; with \; 1 \; ball] = n(n-1)\left(1 - \frac{2}{m}\right)^{n-2}\left(1 - \frac{1}{m}\right)$$
$$+ n\left(1 - \frac{1}{m}\right)^{n-1} - n^2\left(1 - \frac{1}{m}\right)^{2n-2}. \tag{5}$$

Moreover, since when $n, m \to \infty$ this variance is approached by $ne^{-n/m}(1 - e^{-n/m})$, the number of bins with one ball is concentrated around its mean.

In the approximate model, let $p = (1 - 1/m)^{n-1}$. So the expected number of bins with one ball is np. Then, p can be interpreted as the probability that a given ball is successful when n balls are thrown in m bins. Furthermore $q = 1 - p$ can be interpreted as the probability that a given ball fails, since it falls in an bin with two or more balls.

In this setting, we may consider a Bernoulli process where the probability that a given ball fails in all the k channels is q^k. Hence, with probability $(1 - q^k)$ a given ball succeeds in at least one of the channels. As a consequence, the probability that all the n balls succeed in at least one channel is $(1 - q^k)^n$. Then, we have the following approximation, with $S = km$.

Theorem 5. *Let, in the approximate model, $p = (1 - 1/m)^{n-1}$ and $S = km$. Moreover, let $\hat{P}_{S,n,k}$ be the probability (in this model) that when n balls are thrown in k parallel channels each with m bins, all the balls are successful. Then*

1.

$$\hat{P}_{S,n,k} = \left(1 - \left(1 - \left(1 - \frac{k}{S}\right)^{n-1}\right)^k\right)^n.$$

2. *When $S, n \to \infty$ and $k = o(S)$, then $p \approx e^{\frac{-nk}{S}}$, and then*

$$\hat{P}_{S,n,k} \approx \left(1 - \left(1 - e^{\frac{-nk}{S}}\right)^k\right)^n.$$

3. *In this setting, for fixed S and n, the optimal number of channels (the value of k that maximizes $\hat{P}_{S,n,k}$) is $k = \lfloor \frac{Sx_0}{n} \rfloor$ with $x_0 = log(2)$.*
4. *The optimal probability is*

$$\hat{P}^*_{S,n} \approx \left(1 - \frac{1}{2^{\frac{Sx_0}{n}}}\right)^n.$$

5. *For given n, the minimal value of S to achieve an exponentially small error $1/2^b$ (with b a constant like 80, for example) is*

$$S = \frac{n}{x_0}(b + lg(n)),$$

where lg is the base 2 logarithm.

Sketch of the proof: Parts 1 and 2 are straightforward.

For parts 3 and 4, to find the optimal value of k, let us call $x = kn/S$ and so

$$\frac{\partial}{\partial k}\hat{P}_{S,n,k} \approx -\frac{\hat{P}_{S,n,k}}{1-(1-e^{-x})^k} \frac{n(1-e^{-x})^k}{1}\left(log(1-e^{-x}) + \frac{xe^{-x}}{1-e^{-x}}\right).$$

The maximum is at the value x_0 such that $\left(log(1-e^{-x_0}) + \frac{x_0 e^{-x_0}}{1-e^{-x_0}}\right) = 0$, giving $x_0 = log(2)$. As a consequence, the optimal number of channels is $k = \lfloor\frac{Sx_0}{n}\rfloor$ (note that since $S = \Omega(n)$, then $k = o(S)$ as requested), and so, the optimal probability in this setting satisfies:

$$\hat{P}_{S,n}^* \approx \left(1 - \frac{1}{2^{\frac{Sx_0}{n}}}\right)^n. \tag{6}$$

Moreover, for part 5, given n and a constant b (like $b = 80$), if we want an exponentially small error in b ($\hat{P}_{S,n}^* = 1 - 1/2^b$, with $b > 0$), then from (6) and the approximation $(1 - \frac{1}{2^b})^{\frac{1}{n}} = 1 - \frac{1}{n2^b} + O(\frac{1}{n^2})$ then S should verify:

$$S = \frac{n}{x_0}(b + lg(n)), \tag{7}$$

where lg is the base 2 logarithm. □

Experimental Results. When $S = o(n^2)$ (e.g. $n^{2-\epsilon}$, for a constant $\epsilon > 0$) the probability of success tends to 0 with n (e.g. e^{-n^ϵ}). It is then reasonable to assume for our experiments that $S = O(n^2)$.

It is very hard to compute the exact probabilities for large values of k and n using Theorem 4; these values are calculated only up to $k = 4$ and small values of n. We use this to check against the approximations calculated in parts 1 and 2 of Theorem 5. For $k = 1$ the approximation is very bad but we use the exact value of $m^{\underline{n}}/m^n$ ($S = m$ in this case).

Each experiment consists of running 100.001 trials and counting how many successful trials are found. Underlined and in bold are the maximal probabilities (in this case $k = 10$). The table shows that even for small values of n, the approximate and asymptotic formulae agree very well with the experimental results (and with the exact probabilities when they could be computed) (Fig 2).

The next table tests part 5 of Theorem 5. In this setting given b and n the formula gives the minimum value of S such that, when using the optimal number k_{opt} of channels the probability of error is less than 2^{-b}. We normalize

S	n	k	exact	approx1	approx2	experimental
360	23	1	0.4877	-	-	0.4861
360	23	2	0.7566	0.7348	0.7165	0.7587
360	23	3	0.8992	0.8961	0.8848	0.8967
360	23	4	0.9500	0.9493	0.9421	0.9497
360	23	5	-	0.9704	0.9654	0.9717
360	23	6	-	0.9801	0.9763	0.9802
360	23	8	-	0.9877	0.9849	0.9878
360	23	9	-	0.9891	0.9866	0.9891
360	23	10	-	**0.9898**	**0.9874**	**0.9901**
360	23	12	-	0.9898	0.9873	0.9897
360	23	15	-	0.9869	0.9838	0.9865
360	23	18	-	0.9799	0.9759	0.9803
360	23	20	-	0.9718	0.9670	0.9717
360	23	24	-	0.9410	0.9348	0.9406
360	23	30	-	0.8246	0.8226	0.8248

Fig. 2. Experimental results for the probability of success for a total of $S = 360$ and $k = 23$ messages, slightly adjusted values from the birthday paradox.

the probability of success with the estimated value of error, and the quotient should be close to one. The table shows that even for small values of n the formulae are very good. For values of b up to 15, the test is run 200001 times (with an average of 8 failures) (Fig. 3).

n	b	S_{opt}	k_{opt}	experimental/$(1 - 2^{-b})$
23	12	549	16	1.000134
23	15	648	19	1.000015
30	12	732	16	1.000119
30	15	862	19	1.000005
40	12	1000	17	1.000034
40	15	1173	20	0.999995
50	12	1273	17	1.000024
50	15	1490	20	1.000010

Fig. 3. Experimental results for the number of slots needed to guarantee a probability of failure less than 2^{-b}. Ideally the last should be 1

7   Instantiating the Two Approaches

The two composition strategies analyzed in this paper can be instantiated with several DC-nets yielding different overall properties. Using an information-theoretical DC-net like the one proposed by van de Graaf [21] (based on the DC-net proposed by Chaum [4]) we can obtain privacy against computationally unbounded adversaries. If, instead, we use Golle and Juels' (short) DC-net [17], then privacy against polynomial-time adversaries can be achieved. Moreover, the reconstruction procedure of Golle-Juels' DC-net translates into a collision-free iterated DC-net with stronger robustness, where misbehaving parties can not only be detected, but their contribution removed from the protocol result.

In any case, the analysis of the collision probability of Sects. 4 and 6 only assumed two conditions over the underlying DC-net channel: (a) the adversary can disrupt at most one *slot* per compromised party in each broadcast, where the disrupted slot *cannot* depend on the slot selection by honest parties; and (b) for the parallel construction, each party must transmit the same message in all parallel channels. Condition (b) can be easily achieved by using (possibly information-theoretically hiding) commitments and non-interactive zero-knowledge proofs. Condition (a), on the other hand, it is trickier to achieve. In fact, contrarily to what it was claimed, the well-known DC-net protocol by Golle and Juels [17] does not achieve this property. Section 8 explains the attack and presents a simple modification that fixes the protocol.

We believe the two conditions mentioned above suffice to prove security in the standalone model under a reasonable definition of secure, single-channel DC-net. Due to space constraints a formal definition of security as well as the security proof for the constructions are left to the full version of this paper.

Cost Analysis. In order to evaluate the overhead of the two composition strategies, we compare them with a single execution of a DC-net achieving a similar collision probability. Given the assumptions mentioned above, the cost associated to a single DC-net execution with S slots and n senders (each one sending a single message) is at least $O(n)$ commitments for the slots (one per sender), plus the total numbers of bits $C_{S,n}$ transmitted during a single DC-net with S slots and n players. Furthermore, we make the reasonable assumption that $C_{S,n}$ is linear in S. The computation costs are simply those associated to computing the commitments and verifying the committed slots are consistent with the actual slots used.[2]

In terms of communication costs, our sequential repetition approach with n senders, $m = n$ slots and $r = \log \log(n)$ rounds requires sending $O(n \log \log(n))$ commitments (one per sender per round), plus $O(C_{n,n} \log \log(n))$ bits. In terms of computation costs, it is easy to see that the costs are r times the cost of executing a single DC-net.

For the parallel repetition approach with n senders, S slots and $k = \lfloor \frac{S \log(2)}{n} \rfloor$ parallel executions, the total communication costs are now $O(nk)$ commitments (one per sender per parallel channel), the total length of the proof of message equality among parallel channels (namely nT_k bits if the cost of such proof is T_k per sender) plus $kC_{S/k,n}$ bits, the communication costs of k parallel executions of a S/k-slot DC-net. The computation costs are k times the cost of running a single DC-net plus the costs associated to computing the commitments and the proofs, and then verifying the committed slots are consistent with the actual slots used.

---

[2] The exact computation costs depends on the DC-net used.

In order to achieve a high enough probability of success (a probability that does not goes to 0 when n tends to infinity), the standard single-channel DC-net must use $\Omega(n^2)$ slots. But, even assuming $O(n^2)$ slots for the single-channel DC-net, the communication overhead of our sequential repetition construction is *much less* since it is $O(C_{n,n} \log\log(n))$.

For the parallel repetition construction, the communicational overhead is $O(nT_k)$ bits from the proofs of message equality plus the extra commitments ($O(nk) = O(S)$ compared with $O(n)$). The computational overhead is simply the costs of verifying the proofs of message equality. This follows from the fact that running $k = S/n$ copies of a single n-slot DC-net is comparable to running a single S-slot DC-net both in communication as well as computational costs. To the contrary of a standard DC-net, our construction *guarantees* an exponentially small probability of failure (part 5 of Theorem 5 for $b = n$) with only $O(n^2)$ space.

8   Jamming Attack on Golle-Juels' DC-Net

In the abstract of their paper [17], Golle and Juels argue that their construction achieves "high-probability detection and identification of cheating players". It turns out that this is not quite true if the messages sent are not encrypted (the standard case when DC-nets are used as broadcast channels), as the malicious players can still disrupt honest players' transmissions by causing collisions with probability 1 – well above the collision probability of honest users – without being detected. Details follow.

The Golle-Juels' short DC-Net protocol (see Appendix A for a more detailed description) follows Chaum's standard DC-net paradigm where each player P_i initially prepares a m-dimensional vector (or "pad"). This pad satisfies a neat property: if no player sends messages, the pair-wise multiplication of all player's pads results in the all-ones m-dimensional vector. Player P_i can use this property to transmit a message $M_i \neq 1$ by simply multiplying M_i into one of the components of his pad. The index of this component, called P_i's *communication slot* and denoted c_i, is chosen uniformly at random in $[m]$.

To prevent that a malicious player transmits several messages simply multiplying them into different component of the pad, a distinguished characteristic of Golle-Juels' protocol is that it uses proof of knowledge techniques [16] to force each player to select at most one slot. Then, since a corrupted player $\hat{P}_j$ can choose *at most* a single slot to transmit a value, it may seem that $\hat{P}_j$ cannot do more harm that an honest player: both honest players and corrupted players would have the same collision probability as the other players.

Unfortunately, this reasoning is flawed if the adversary is *rushing* [13]. A rushing adversary is allowed to "speak last" in every communication round, even in synchronous networks. Consider now a single player $\hat{P}_j$ corrupted under such an adversary. This player can wait until all honest players have submitted their pads and proofs, and then

1. internally and privately compute the resulting output by using an honestly-produced pad as its own ($\hat{P}_j$'s pad). This allows the adversary to not only obtain the messages sent by all honest parties, but also identify all the positions where those messages were sent (it does NOT allow the adversary to know *who* sent it, though). And then,
2. the adversary can "choose" $\hat{P}_j$'s slot c_j to be equal to any of the slots where honest parties have sent their messages. Clearly, the adversary can also choose the value of $\hat{P}_j$'s message M_j; this message can be chosen so the "collision" with the honest player's message in slot c_j produces *an arbitrary message* of the adversary's choice. Then, $\hat{P}_j$ outputs his (honestly computed with slot c_j) pad and proof.

The resulting message in slot c_j will be completely controlled by the adversary: it can be made look like a collision or not[3]. Furthermore, this strategy gives free range for the adversary to select her target slot: for example, the adversary could select a position based on the value of the honest player's message. Clearly, the outcome is a collision with probability 1.

Fixing the problem: We can prevent the attack by simply forcing each player to commit to her slot using non-malleable commitments [1]. Consider the description of the protocol given in Appendix A. To prevent the degenerate case where a corrupted player copies everything (commitments, pads, proofs) together, one could modify these proofs to assure that the secret exponent x_i used in the computation of the pad $W_i^{(\theta,\ell)}$, for each $\theta \in [m]$, is the same exponent on P_i's public key y_i. This worst case can be dealt, however, much more easily by rejecting repeated proofs under some fix ordering of the players.

Acknowledgements. The authors would like to thank the anonymous referees for their helpful comments and suggestions which significatively helped to improve the paper presentation. The third author also thanks the support of INRIA Chile.

A   The Golle-Juels' DC-Net

The DC-net protocol proposed by Golle and Juels [17] (called Short DC-net) directly extends the ideas from Chaum's DC-net [4]. In Chaum's original protocol, parties share secret keys (called *keypads*) in a pairwise fashion, which satisfies that the addition of all keypads effectively cancels out. To broadcast a message, a party P_i adds (xors) the message into one of her pairwise-shared keypads, so when keys are later combined, keys cancel out leaving her message as the result. Notice that the Chaum's original protocol outputs at most a single message per execution.

Golle and Juels' protocol dispenses with sharing private keys by having each pair of parties non-interactively compute the keypads as Diffie-Hellman

---

[3] See [2] for techniques to always detect this situation as a collision.

keys. These keypads satisfy the same property – they cancel out once all are combined – and enjoy specific algebraic properties (inherited from the pairing-based key agreement used) that allow public verification of correctness of each party's keypad. In consequence, misconstructed keypads can be identified. Moreover, since each party's private keys are initially threshold-shared among for all parties, reconstruction of incorrectly-computed keypads is possible.

The full protocol is detailed below.

COMMON PUBLIC INPUTS: $k \in \mathbb{N}$, the security parameter, $n \in \mathbb{N}$ the number of parties, $t \in \mathbb{N}$ a fixed integer, $z \in \mathbb{N}$ a counter, and $\ell \in \mathbb{N}$ the parallel index. Also, G_1, G_2, e, g_1, g, h, where G_1, G_2 are cyclic groups of prime order $p = \mathcal{O}(2^k)$, such that $e: G_1 \times G_1 \to G_2$ is an admissible pairing, g_1 is a generator of G_1, and g, h are independently-chosen generators of G_2, and $H: \{0,1\}^* \to G_1$ a collision-resistant hash function.

PUBLIC INPUTS: $y_1, \ldots, y_n$, where $y_i = g_1^{x_i}$ is the public key of P_i corresponding to private key x_i.

PRIVATE INPUTS: P_i has input $M_i \in G_2$, and secret key $x_i \in \mathbb{Z}_p$. Vía a one-time setup procedure, P_i also gets a private key gets a (n,t)-share $x_{j,i}$ of party P_j's the secret key x_j.

PUBLIC OUTPUT: Each honest P_i obtains a multiset $\mathcal{M} = \{M_1, \ldots, M_n\} \subset G_2$ consisting of (almost) all messages sent by parties.

(1) **Transmission Step:** In this round, each party P_i does as follows:

1. First, P_i creates a *shared pad* by computing, for each $\theta \in \{1, \ldots, n\}$
   $Q_{\theta,\ell} \xleftarrow{\$} H(z\|\theta\|\ell)$ and $W_i^{(\theta,\ell)} \leftarrow \Pi_{j \in \{1,\ldots,n\}, j \neq i} e(Q_{\theta,\ell}, y_j)^{\delta_{i,j} x_i}$
   where $\delta_{i,j} = 1$ if $i < j$ and -1 otherwise.

2. P_i then chooses $c_i \xleftarrow{\$} \{1, \ldots, n\}$. Then, P_i computes her transmission vector $V_i^{(\ell)} = (V_i^{(\theta,\ell)})_{\theta \in \{1,\ldots,n\}}$ as $V_i^{(\theta,\ell)} \leftarrow m_i \cdot W_i^{(\theta,\ell)}$ if $\theta = c_i$ and $V_i^{(\theta,\ell)} \leftarrow W_i^{(\theta,\ell)}$ if not.

3. P_i computes the verification information $\sigma_i = (\sigma_i', \{\sigma_{i,\theta}''\}_{\theta \in \{1,\ldots,n\}})$ as follows:

   (a) For each $\theta \in \{1, \ldots, n\}$, P_i does the following:

      i. P_i picks $r_{i,\theta} \xleftarrow{\$} \mathbb{Z}_p$, and sets $d_{i,\theta} \leftarrow h^{r_{i,\theta}}$ if $\theta \neq c_i$, and $d_{i,\theta} \leftarrow gh^{r_{i,\theta}}$ if $\theta = c_i$.

      ii. P_i computes $\sigma_{i,\theta}''$ as $PoK\{(x_i, r_{i,\theta}) : (V_i^{(\theta,\ell)} = e(\Pi_{j \in \{1,\ldots,n\}, j \neq i} y_j^{\delta_{i,j}}, Q_{\theta,\ell})^{x_i} \wedge d_\theta = h^{r_{i,\theta}}) \vee (d_\theta = gh^{r_{i,\theta}})\}$.

   (b) Using $r = \sum_{j \in \{1,\ldots,n\}} r_j$ as the witness, P_i computes σ_i' as $PoK\{r \in \mathbb{Z}_p : (\Pi_{\theta=1}^n (d_\theta))/g = h^r\}$.

4. P_i then broadcasts $(V_i^{(\ell)}, \sigma_i)$.

5. Let $(V_i^{(\ell)}, \sigma_i)$ be the values broadcast by P_i. Each P_j verifies the validity of proofs σ_i. Let $\Omega \subset \{1, \ldots, n\}$ be the subset of parties whose proofs fail. If $\Omega = \emptyset$, each P_j outputs the multiset $\mathcal{M}$ defined by $\mathcal{M} \leftarrow \{\Pi_{i \in \{1,\ldots,n\}} V_i^{(\theta,\ell)} : \theta \in [m]\}$. and finishes the protocol. Otherwise, set $\mathcal{M} \leftarrow \emptyset$ and execute the reconstruction phase of the protocol.

(1) **Correction Step:** If x_j is the secret key of party P_j, we let $x_{i,j}$ be the private share of x_j hold by P_i, and let $y_{i,j} = g_1^{x_{i,j}}$ be the corresponding public share. We assume the shares were computed using known techniques, eg. [14], so they satisfy $x_{i,j} = f_i(j)$ where $f_i(x)$ is a random polynomial of degree $(t-1)$ such that $f_i(0) = x_i$. Let $\lambda_{i,\Gamma} = \Pi_{j \in \Gamma, j \neq i} \frac{j}{j-i}$ be the corresponding Lagrange coefficients used to recover x_j from the shares $x_{j,i}$, that is, $x_j = \sum_{i \in \Gamma} x_{j,i} \lambda_{i,\Gamma}$.

In this round, each party P_i reconstructs the shared pad for each (misbehaving) party P_j in Ω as by working as follows:

1. First, P_i broadcasts $Z_{j,i} \leftarrow Q_{\theta,\ell}^{x_{j,i}}$ for each $\theta \in \{1, \ldots, n\}$.
2. Let $\Theta = \{Z_{u,v}\}$ be the set of all broadcast shares, where $Z_{u,v}$ was sent by P_v. Each party checks the validity of share $Z_{u,v}$ by checking that
$e(Z_{u,v}, g_1) \overset{?}{=} e(y_{u,v}, Q_{\theta,\ell})$. Let $\Gamma \subset \Theta$ be a subset of shares that satisfy the above validity condition, where $|\Gamma| = t$.
3. Then, P_i reconstructs the shared pad for each party $P_j \in \Omega$ as follows: for each $\theta \in \{1, \ldots, n\}$, $\hat{V}_j^{(\theta,\ell)} \leftarrow \Pi_{u \in \{1,\ldots,n\}, u \neq j} e(\Pi_{i \in \Gamma} Z_{j,i}^{\lambda_{i,\Gamma}}, y_u)^{\delta_{j,u}}$.
4. Finally, P_i outputs the multiset $\mathcal{M} \leftarrow \{ \Pi_{i \in \{1,\ldots,n\} \backslash \Omega} V_i^{(\theta,\ell)} \cdot \Pi_{j \in \Omega} \hat{V}_j^{(\theta,\ell)} : \theta \in [m] \}$ and finishes the protocol.

References

1. Abdalla, M., Benhamouda, F., Blazy, O., Chevalier, C., Pointcheval, D.: SPHF-friendly non-interactive commitments. In: Sako, K., Sarkar, P. (eds.) ASIACRYPT 2013, Part I. LNCS, vol. 8269, pp. 214–234. Springer, Heidelberg (2013)
2. Barthe, G., Hevia, A., Luo, Z., Rezk, T., Warinschi, B.: Robustness guarantees for anonymity. In: CSF, pp. 91–106. IEEE Computer Society (2010)
3. Chaum, D.: Untraceable electronic mail, return addresses, and digital pseudonyms. Commun. ACM **24**(2), 84–88 (1981)
4. Chaum, D.: The dining cryptographers problem: Unconditional sender and recipient untraceability. J. Cryptology **1**(1), 65–75 (1988)
5. Corrigan-Gibbs, H., Ford, B.: Dissent: accountable anonymous group messaging. In: Al-Shaer, E., Keromytis, A.D., Shmatikov, V. (eds.) Proceedings of the 17th ACM Conference on Computer and Communications Security, CCS 2010, pp. 340–350. ACM (2010)
6. Corrigan-Gibbs, H., Wolinsky, D.I., Ford, B.: Proactively accountable anonymous messaging in Verdict. In: King, S.T. (ed.) USENIX Security, pp. 147–162. USENIX Association (2013)
7. Feller, W.: An Introduction to Probability Theory and its Applications, 3rd edn. Wiley, New York (1968)
8. Flajolet, P., Sedgewick, R.: Analytic Combinatorics. Cambridge University Press, New York (2009)
9. Franck, C.: Dining cryptographers with 0.924 verifiable collision resolution. CoRR, abs/1402.1732 (2014). http://arxiv.org/abs/1402.1732
10. Franck, C., van de Graaf, J.: Dining cryptographers are practical (preliminary version). CoRR, abs/1402.2269 (2014). http://arxiv.org/abs/1402.2269
11. García, P.: Optimización de un protocolo noninteractive dining cryptographers. Master's thesis, Universidad Nacional de San Luiz, 2013. Universidad Nacional de San Luiz (2013)

12. García, P., van de Graaf, J., Montejano, G., Bast, S., Testa, O.: Implementación de canales paralelos en un protocolo non interactive dining cryptographers. In: 43 Jornadas Argentinas de Informática e Investigación Operativa (JAIIO 2014), Workshop de Seguridad Informática (WSegI 2014) (2014)

13. Gennaro, R., Jarecki, S., Krawczyk, H., Rabin, T.: Secure distributed key generation for discrete-log based cryptosystems. In: Stern, J. (ed.) EUROCRYPT 1999. LNCS, vol. 1592, pp. 295–310. Springer, Heidelberg (1999)

14. Gennaro, R., Jarecki, S., Krawczyk, H., Rabin, T.: Secure distributed key generation for discrete-log based cryptosystems. In: Stern, J. (ed.) EUROCRYPT 1999. LNCS, vol. 1592, pp. 295–310. Springer, Heidelberg (1999)

15. Goel, S., Robson, M., Polte, M., Sirer, E.G.: Herbivore: A scalable and efficient protocol for anonymous communication. Technical report TR2003-1890, Computing and Information Science, Cornell University (2003). http://www.cs.cornell.edu/people/egs/papers/herbivore-tr.pdf

16. Goldwasser, S., Micali, S., Rackoff, C.: The knowledge complexity of interactive proof systems. SIAM J. Comput. **18**(1), 186–208 (1989)

17. Golle, P., Juels, A.: Dining cryptographers revisited. In: Cachin, C., Camenisch, J.L. (eds.) EUROCRYPT 2004. LNCS, vol. 3027, pp. 456–473. Springer, Heidelberg (2004)

18. Kolchin, V., Sevastyanov, B., Chistyakov, V.P.: Random Allocations. Wiley, New York (1978)

19. Pedersen, T.P.: Non-interactive and information-theoretic secure verifiable secret sharing. In: Feigenbaum, J. (ed.) CRYPTO 1991. LNCS, vol. 576, pp. 129–140. Springer, Heidelberg (1992)

20. Pfitzmann, A.: Diensteintegrierende Kommunikationsnetze mit teilnehmerüberprüfbarem Datenschutz. Informatik-Fachberichte, vol. 234. Springer (1990)

21. van de Graaf, J.: Anonymous one-time broadcast using non-interactive dining cryptographer nets with applications to voting. In: Chaum, D., Jakobsson, M., Rivest, R.L., Ryan, P.Y.A., Benaloh, J., Kutylowski, M., Adida, B. (eds.) Towards Trustworthy Elections. LNCS, vol. 6000, pp. 231–241. Springer, Heidelberg (2010)

22. Waidner, M., Pfitzmann, B.: The dining cryptographers in the disco: unconditional sender and recipient untraceability with computationally secure serviceability (abstract). In: Quisquater, J.-J., Vandewalle, J. (eds.) EUROCRYPT 1989. LNCS, vol. 434, p. 690. Springer, Heidelberg (1990)

Private Asymmetric Fingerprinting: A Protocol with Optimal Traitor Tracing Using Tardos Codes

Caroline Fontaine[1,2], Sébastien Gambs[3,4], Julien Lolive[1,4(✉)], and Cristina Onete[3]

[1] UMR CNRS 6285 Lab-STICC, Institut TELECOM, TELECOM Bretagne, Plouzané, France
{caroline.fontaine,julien.lolive}@telecom-bretagne.eu
[2] UMR 6285 Lab-STICC, CNRS, Plouzané, France
[3] Université de Rennes 1, Rennes, France
{sgambs,maria-cristina.onete}@irisa.fr
[4] Inria Rennes Bretagne-Atlantique / IRISA, Rennes, France

Abstract. Active fingerprinting schemes were originally invented to deter malicious users from illegally releasing an item, such as a movie or an image. To achieve this, each time an item is released, a different fingerprint is embedded in it. If the fingerprint is created from an anti-collusion code, the fingerprinting scheme can trace colluding buyers who forge fake copies of the item using their own legitimate copies. Charpentier, Fontaine, Furon and Cox were the first to propose an asymmetric fingerprinting scheme based on Tardos codes – the most efficient anti-collusion codes known to this day. However, their work focuses on security but does not preserve the privacy of buyers. To address this issue, we introduce the first privacy-preserving asymmetric fingerprinting protocol based on Tardos codes. This protocol is optimal with respect traitor tracing. We also formally define the properties of correctness, anti-framing, traitor tracing, as well as buyer-unlinkability. Finally, we prove that our protocol achieves these properties and give exact bounds for each of them.

Keywords: Fingerprinting · Watermarking · Anti-collusion code · Tardos code · Privacy · Anonymity

1 Introduction

A huge amount of digital items, such as pictures, songs and movies, is downloaded on a daily basis, both from centralized content platforms and via peer-to-peer (P2P) networks. Some of these items have a commercial value (e.g., those

Caroline Fontaine—This work has received a French governmental support granted to the COMIN Labs excellence laboratory and managed by the National Research Agency in the "Investing for the Future" program under reference ANR-10-LABX-07-01.

© Springer International Publishing Switzerland 2015
D.F. Aranha and A. Menezes (Eds.): LATINCRYPT 2014, LNCS 8895, pp. 199–218, 2015.
DOI: 10.1007/978-3-319-16295-9_11

sold on VoD platforms), while others have rather a personal value (e.g., user-generated content). However, all digital material are associated with copyright laws that should be respected. Indeed, even legitimate holders of copyrighted material are typically not allowed to distribute it without an explicit permission. This type of infringement must be traceable in the sense that the culprits should be identifiable.

One way of addressing this issue is to personalize the delivered content through active fingerprinting techniques [9,24,26]. This process embeds a fingerprint (i.e., a sequence of symbols, usually bits) into an item every time it is sold or delivered. Such fingerprints are uniquely linked to single transactions, thus implicitly also to single buyers. Fingerprinting can be achieved by watermarking techniques, making the fingerprint imperceptible to human eyes and robust with respect to some transformations (e.g., item resizing or change of format).

A major challenge in fingerprinting is to thwart *collusion attacks*, in which several owners of an item combine their legitimate copies to create an illegitimate, untraceable copy before distributing it. *Anti-collusion codes*, like Tardos code [30], were specifically designed to counter this threat. Indeed, if the fingerprints embedded into items are codewords of an anti-collusion code, then the structure of the code allows the retrieval of the identity of at least one of the colluders.

Related Work on Fingerprinting. In this paper we consider two main types of participants: the *merchant*, who retails several items, and the *buyer*, who purchases (receives) an item. While we use this notation, which is standard in the fingerprinting literature, our protocol can be extended to non-monetary transactions. In particular, our protocol is specifically designed to be independent of the mechanism used for the payment[1]. The underlying fingerprinting scheme is usually encapsulated by running a *fingerprinting protocol* between the buyer and the merchant, which should prevent both parties from cheating. However, for most existing fingerprinting protocols, security comes at the cost of the buyer's privacy.

In particular, in *symmetric fingerprinting protocols* the merchant embeds a unique fingerprint in each distributed copy of an item. When the buyer downloads such a copy, the merchant links the buyer's identity to the unique fingerprint in that copy. Apart from *knowing* the identity of all the buyers, the merchant can also *frame* them by releasing a copy bound to them and accusing them of illegal redistribution. *Asymmetric fingerprinting protocols* [24,26] solve this weakness by generating the fingerprint through a secure bipartite protocol between the merchant and the buyer. At the end of the transaction, the merchant learns a part of the fingerprint, called the *halfword*, which is used to

---

[1] In particular, we need to provide the buyers with transaction-specific pseudonyms during the buying phase of our protocol, rather than assuming than these pseudonyms are inherited from the payment scheme. This property is achieved through the use of a group signature scheme. The only assumption that we make on the payment protocol is that it preserves the anonymity of the buyer.

trace a buyer's identity but is not enough to frame him. Indeed, the complete fingerprinted item is only delivered to the buyer.

However, in both the symmetric and the asymmetric cases, the merchant still learns the identity of all buyers. An *anonymous fingerprinting protocol* [22,23,25] goes a step further by ensuring the anonymity of buyers as long as they behave honestly (i.e., they do not illegally distribute items) [1,6,28]. For instance in [25], Pfitzmann and Waidner modified a previous construction [26] to obtain *revocable anonymity*, thus forbidding (honest) user profiling by the merchant. The anonymity is guaranteed by using a trusted third party, the *registration center* (RC), who knows the identity of buyers but not the items they bought. This protocol was extended in [22] by using e-cash, so that the RC provides the buyer with a digital coin at buyer registration. As illegal item redistribution amounts to coin replays, which can be detected as double-spending. Camenisch [6] extended the protocol in [25] by using group signatures. This is also the case in our approach but our method is fundamentally different as it is designed to be able to handle the structure of anti-collusion Tardos codes. Another anonymous fingerprinting protocol due to Abdul, Gaborit, and Carré [1] is based on DC-nets, private information retrieval and group signatures. Finally more recently in [28], the authors proposed an anonymous fingerprinting protocol using group signatures, homomorphic encryption and zero-knowledge proofs. They also provided a formal security framework for anonymous fingerprinting protocols. As our attack model is slightly different, we have chosen to build a fully-fledged syntax and security definitions from scratch.

Our Contributions. In this paper we combine the strong traitor-tracing capacity of Tardos codes-based fingerprinting, as first exploited in [9], with black-box cryptographic primitives providing strong buyer privacy. Our goal is to design the first asymmetric fingerprinting protocol based on Tardos codes providing strong privacy features. One of the main challenge here is to accommodate the complex structure of these anti-collusion codes and achieve provable security.

Our first contribution is to formalize a full security and privacy model for privacy-preserving fingerprinting protocols with traitor-tracing capacities (PFP-TT), including the properties of correctness, buyer unlinkability, anti-framing and traitor-tracing. For merchants dealing with strictly digital material, this will not affect logistics, while providing a real asset to buyer privacy. Our second contribution is the introduction of our protocol, the first PFP-TT provably attaining these properties. Our proposal benefits from the optimal traitor-tracing properties of Tardos codes as well as from the security and privacy features provided by other cryptographic primitives such as Non-Interactive Zero-Knowledge Proofs of Knowledge (NIZK-PK), group signatures, symmetric encryption and oblivious transfer. Finally, our third contribution is to give exact security bounds for each of these properties with respect to our protocol.

2   Tardos Codes

Tardos codes [30] are probabilistic anti-collusion codes particularly adapted to the context of fingerprinting protocols. In this setting, fingerprints are codewords[2] drawn at random according to a distribution **p**, which is *half-sparse* and *half-dense*, chosen secretly at setup by the merchant. Tardos codes are the shortest and most efficient anti-collusion codes available today. In our construction, we specifically use the improved version of binary Tardos codes due to Škorić, Katzenbeisser and Celik [32].

A Tardos code is defined by the following parameters: the size $|\chi|$ of the alphabet it uses, the maximum number n of buyers, the maximum size c for a collusion of malicious buyers tolerated by the code and maximum false alarm probability $\delta << 1$ (denoted in the literature as ϵ_1) for a given innocent buyer to be wrongly accused. The design of this code is so efficient that the real risk for a given buyer to be wrongly accused is much smaller than this loose theoretical bound [8], leading to a very small global false alarm probability[3] even when we set $\delta = 0.001$. The *length* m of the code is of the form $m = Ac^2\lceil\ln(1/\delta)\rceil$ while the *accusation threshold* is of the form $Z = Bc\lceil\ln(1/\delta)\rceil$. In the literature, one can find many attempts to minimize the constants A and B, mainly by improving the accusation process. In practice in this paper, we set $A = 2\pi^2$ and $B = 2\pi$ as in [32], which ensures a uniform efficient accusation, whatever the colluding buyers' strategy.

To our knowledge, Tardos codes have only been used in one other asymmetric fingerprinting protocol, [9], which we briefly describe below. This protocol begins with a *setup phase* in which, on input the parameters of the Tardos code, a probabilistic generation algorithm outputs a vector **p** of size m containing identical and independently distributed (IID) random probabilities p_j (more details about their distribution are provided later). These probabilities are only known by the merchant and kept secret from all other entities. Afterwards, a *quantization* is realized such that $p_j = \frac{L_j}{N}$ with $L_j \in [N-1]$, for an integer N. The quantization process consists in choosing a number L_j of bits equal to 1, and a number $N - L_j$ of bits equal to 0 for each j. These bits are used to generate an $N \times m$ WORM (*Write Once Read Many*) memory [21], containing encrypted watermarked bits. The merchant stores both the quantizations and the keys used for the encryptions.

During the *buying phase*, the buyer and the merchant generate the specific fingerprint embedded in the item. In particular, the buyer runs an oblivious transfer (OT) protocol with the merchant, resulting in the recovery of m keys (one per column of the WORM). These keys decrypt the content of m cells in the WORM, allowing for the retrieval of several item bits. The buyer and merchant then run a two-party protocol to embed this fingerprint into the item and let the merchant learn the halfword. The buyer only retrieves his own copy of the item. The halfword is stored by the merchant and used to trace misbehaving purchasers.

Whenever a forgery is detected (i.e., the merchant finds an illegal copy of the item), an *accusation phase* follows in [9], during which a score S_j is computed

---

[2] In both [9] and in our work, only binary Tardos codes are used. We therefore just describe the parameters used in the binary case.

[3] By "global false alarm" we refer to the probability that *any* innocent buyer (rather than merely a particular one) is falsely accused.

for each suspicious buyer, based on his halfword and the forged fingerprint Y. Computed as in [32], this score reflects the probability that Buyer B_j colluded to create the forgery. If S_j is greater than the accusation threshold for the halfword, the buyer is deemed *guilty*, while otherwise he is assumed to be *innocent*. Note that the merchant does not know the full fingerprint of the buyer, but only the halfword, which enables him to compute only partial accusation scores. To provide a full score and a final decision, a second score has to be computed based on the whole fingerprint. This second score is the one that truly reflects the probability for the buyer to be guilty. To proceed with this final computation, a *judge* forces the suspicious buyer to reveal his whole fingerprint (refusing to cooperate for the buyer will lead the judge to believe that he is guilty). Thus, in [9] the role of the judge is twofold: (1) he checks that the merchant has computed the partial scores properly (i.e., without cheating), and (2) he computes the second score that is needed to proceed to the final accusation.

3   Security Model

In this section, we first describe the system and adversary models before defining the security and privacy requirements for privacy-preserving asymmetric fingerprinting protocols.

3.1   System and Adversary Models

System Model. We consider a system composed of the following participants.

- A set $\mathcal{B}$ of n *buyers*, denoted $B_1, \ldots, B_n$, whose identities are confirmed by a certification authority. The buyers can receive digital content (hereafter called an *item*) after conducting a transaction with a *merchant*.
- A single *merchant* who owns several *items* $I_1, \ldots, I_\ell$, can create fingerprinted copies of these items and – upon a successful transaction with a buyer – allows the buyer to retrieve a fingerprinted copy of an item.
- A *certification authority* (CA) that registers buyers into the system. We implicitly assume that the credentials of the merchant are certified, but not necessarily by the same registration authority.
- An *opening authority* (OA) that may lift the anonymity of malicious buyers upon receiving proof of their misbehavior from the merchant. Sometimes the CA may also act as an OA, but a separation of these roles limits the possibility of abuse.

We assume that the the merchant and buyers communicate over an anonymous channel (e.g., across the TOR network). Furthermore, if a buyer has to pay for an item, we require the payment transaction to be anonymous. These assumptions are necessary, but not sufficient. As pointed out in [18], an anonymous channel can only *preserve* privacy, but not *create* it. Indeed, the message content must also ensure the sender- and receiver-anonymity. In our context, we are only concerned with protecting the privacy of buyers. We also require that buyers receive their credentials from the CA via a secure (i.e., authenticated and confidential) channel. We describe further assumptions on the watermarking and fingerprinting process in the appendix.

Adversary Model. Each of the merchant's items is divided in blocks of equal size (the size depends of the medium and the watermarking scheme W). In this work, we assume that the items all have m blocks, denoted $I^1, \ldots, I^m$. If an item is composed of less blocks, the merchant is assumed to "pad" the item.

We define the security and privacy properties of a Privacy-preserving Fingerprinting Protocol with Traitor-Tracing (PFP-TT) capacity (see also Appendix A.1) in a game-based manner. Intuitively, a PFP-TT consists of the following algorithms. The *Setup* algorithm outputs a set of public and private parameters. The *Buyer Registration* algorithm, in which a buyer registers to the CA, produces as output to the buyer his private group credentials. The *Item Preparation* algorithm, allows the merchant to output a prepared item matrix τ_I and a proof π_I that the item was correctly prepared. The *Buying* algorithm, enables the buyer to retrieve a set of keys and positions indicating the entries in the fingerprinting table τ_I from which the buyer will recover the item. The merchant gets as output a *halfword*, enabling him to detect misbehaving buyers. In addition, the PFP-TT also integrates a *Recovery* algorithm, allowing the buyer to retrieve an item with a unique fingerprint as well as an *Accusation* algorithm, use by the merchant to generate a list of buyers and a proof that these buyers have misbehaved. Finally, the *Opening* algorithm is run by the OA to lift the anonymity of misbehaving buyers based on the proof that he has received from the merchant.

For the security and privacy models, we consider an adversary $\mathcal{A}$ who is given access to oracles enabling him to interact with, and on behalf of, honest parties. These oracles are sketched below and presented in detail in Appendix A.2.

Buying. We describe three oracles related to the purchase of an item. The first oracle Buy* enables a malicious merchant to choose an arbitrary input input* and to deviate from the protocol in his transaction with a buyer. The second one, Execute, allows an honest-but-curious merchant to run a buying protocol with a chosen buyer for a chosen item. Finally the third oracle, BBuy, can be used by a buyer to run a transaction with the merchant for some chosen item.

Opening. The opening oracle allows a merchant to query the result of the Opening algorithm, without knowing the secret information of the OA.

Corrupt. The corruption oracle returns the private key of a chosen buyer.

Collude. This oracle takes as input a set of fingerprinted copies and an arbitrary collusion strategy, and outputs a forged fingerprinted copy. The only restriction applying to this oracle is the marking assumption, which states that if all the input fingerprinted copies have for a particular block the same fingerprint $f_I^{i,j}$, then the fingerprinted block of the resulting forgery *cannot* correspond to another valid fingerprinted block $f_I^i \neq f_I^{i,j}$.

Accuse. This oracle simulates the Accuse algorithm.

In the following section, we define the security game for each property in terms of an additional oracle called Test, which changes at each game to reflect the required security or privacy property.

3.2   Security and Privacy Requirements

In this section, we outline the security and privacy requirements that our **PFP-TT** should guarantee. In each security game, the adversary is given access to a subset of the algorithms and oracles presented in the previous section as well as to the Test oracle. Due to space restrictions, we only informally describe the Test oracles here and leave the full description to the Appendix.

Correctness. In this game, the challenger runs the **Setup** algorithm, which outputs the secret and public parameters spar and ppar, then proceeds to prepare all items I using the **IPrep** algorithm. The adversary is given ppar and the corresponding proofs of preparation tuples (τ_I, π_I) for each item I. The adversary can adaptively query the $\mathsf{Test}^{\mathsf{Corr}}$ oracle, which runs an honest buyer-merchant execution, and then calls the recovery algorithm. The oracle returns 0 if the honest execution is wrongly formatted (i.e., the inputted item or buyer identifiers – see Appendix A – correspond to non-existing values) or if it fails to recover the correct set of fingerprinted blocks, while otherwise the oracle returns 1. The adversary wins if at least one $\mathsf{Test}^{\mathsf{Corr}}$ query returns 0. We define the *advantage* of the adversary $\mathcal{A}$ as $\mathsf{Adv}_{\mathcal{A}}^{\mathsf{correct}} = \mathbb{P}[\mathcal{A} \text{ wins}]$.

Definition 1. *A PFP-TT is $(N_{\mathsf{Test}}, \epsilon)$-correct if any polynomial-time adversary $\mathcal{A}$ against the correctness of PFP-TT making at most N_{Test} queries to the $\mathsf{Test}^{\mathsf{Corr}}$ oracle wins with advantage $\mathsf{Adv}_{\mathcal{A}}^{\mathsf{correct}} \leq \epsilon$. Asymptotically, the protocol is N_{Test}-correct if any adversary $\mathcal{A}$ has negligible probability to win with at most N_{Test} queries.*

Buyer Unlinkability. Following the approaches of [12] and [17], this game considers the adversary to be a merchant, who can be honest-but-curious or malicious. This behaviour is denoted by a flag $\mathsf{flag} \in \{\mathsf{hbc}, \mathsf{mal}\}$. Thus, $\mathcal{A}$ receives the public parameters and the merchant's share of spar, before running **IPrep** at will. If $\mathsf{flag} = \mathsf{hbc}$, the adversary may query the oracles Execute and Corrupt, while if $\mathsf{flag} = \mathsf{mal}$, $\mathcal{A}$ may query Buy^* and Corrupt. The adaptive $\mathsf{Test}^{\mathsf{BUnlink}}$ oracle takes as input two buyers B_i and B_j, choosing consistently either the first or the second, depending on a hidden bit b. The adversary can run Execute (respectively Buy^*) at will with the buyer, then eventually free both buyers. We say that the adversary $\mathcal{A}$ wins if he can guess the bit b. We consider the same adversary classes as defined by Vaudenay [31]: *weak* adversaries cannot corrupt, *forward* adversaries only follow corruption queries by further corruption queries, and *strong* adversaries can cheat arbitrarily without any restrictions. We define the advantage of the adversary is this game $\mathcal{A}$ as $\mathsf{Adv}_{\mathcal{A}}^{\mathsf{ID-priv}} := \mathbb{P}[\mathcal{A} \text{ wins}] - \frac{1}{2}$.

Definition 2. *A PFP-TT is $(N_{\mathsf{Test}}, \epsilon)$-buyer-unlinkable with respect to a flag-merchant (in which $\mathsf{flag} \in \{\mathsf{hbc}, \mathsf{mal}\}$) if any polynomial-time adversary $\mathcal{A}$ against* *the buyer unlinkability of* PFP-TT *making at most* N_{Test} *queries to the* $\text{Test}^{BUnlink}$ *oracle wins with advantage* $\text{Adv}_{\mathcal{A}}^{\text{ID}-\text{priv}} \leq \epsilon$. *Asymptotically, the protocol is* N_{Test}-*buyer-unlinkable if it is* $(N_{\text{Test}}, \nu(1^\lambda))$-*buyer-unlinkable.*

Anti-framing. This property guarantees both anti-framing (i.e., a buyer cannot be framed by a malicious merchant) and exculpability (i.e., a buyer cannot be framed by a collusion of malicious buyers) by considering a generic collusion between a malicious merchant and a set of malicious buyers. The adversary uses the secret information of the merchant and Corrupt queries to collude with buyers. He may also query Collude (for any of the corrupted buyers) and a testing oracle, which allows the merchant to simulate the opening algorithm from the OA. The objective of $\text{Test}^{\text{NoFrame}}$ is to verify whether $\mathcal{A}$ can produce a convincing proof allowing to lift the anonymity of a buyer who is still honest (i.e., uncorrupted) when the oracle is queried. The adversary wins if at least one $\text{Test}^{\text{NoFrame}}$ query returns 1. We define the advantage of the adversary as $\text{Adv}_{\mathcal{A}}^{\text{no}-\text{frame}} = \mathbb{P}[\mathcal{A} \text{ wins}]$.

Definition 3. *A* PFP-TT *is* $(N_{\text{Test}}, \epsilon)$-*unframeable if any polynomial-time adversary* $\mathcal{A}$ *against the anti-framing of* PFP-TT *making at most* N_{Test} *queries to the* $\text{Test}^{\text{NoFrame}}$ *oracle wins with advantage* $\text{Adv}_{\mathcal{A}}^{\text{no}-\text{frame}} \leq \epsilon$. *Asymptotically, the protocol is* N_{Test}-*unframeable if it is* $(N_{\text{Test}}, \nu(1^\lambda))$-*unframeable.*

Traitor-Tracing. In traitor-tracing, an adversary can use the public parameters and all the private keys of the buyers he controls to simulate transactions and obtain legitimate fingerprinted copies $\text{FI}_{\text{B},\text{I}}$. The Test^{TT} oracle runs the Collude oracle on any subset of the legitimate copies, before calling the Accuse and Open oracles (thus simulating the attempt of a merchant to trace forgeries), and wins if the opening reveals no identity associated with one of legitimate copies in the original input subset. The advantage of the adversary is $\text{Adv}_{\mathcal{A}}^{TT} = \mathbb{P}[\mathcal{A} \text{ wins}]$.

Definition 4. *A* PFP-TT *is* $(N_{\text{Test}}, c, \epsilon)$-*traitor-tracing if any polynomial-time adversary* $\mathcal{A}$ *against the traitor-tracing of* PFP-TT *making at most* N_{Test} *queries to the* Test^{TT} *oracle for at most* c *colluders, wins with advantage* $\text{Adv}_{\mathcal{A}}^{TT} \leq \epsilon$. *Asymptotically, the protocol is* N_{Test}-*traitor-tracing if it is* $(N_{\text{Test}}, \nu(1^\lambda))$-*traitor-tracing.*

4 Privacy-Preserving Asymmetric Fingerprinting Using Tardos Codes

In order to attain the strong security and privacy requirements outlined in Sect. 3, we introduce a new privacy-preserving fingerprinting protocol based on Tardos codes. Our solution extends the asymmetric fingerprinting protocol based on Tardos code due to [9]. Our protocol relies on cryptographic primitives such as group signatures, oblivious transfer and NIZK-PK to achieve buyer unlinkability. Recall that we also assume that merchants and buyers communicate over an anonymous channel.

4.1 Protocol Description

Intuitively, our protocol consists of six phases: Setup, Registration, Item preparation, Buying, Item Recovery and Accusation. These phases are also depicted in Figs. 1, 2 and 3. We proceed by describing each phase, first informally and then formally.

Setup. The *Setup* phase provides relevant keys to, respectively, the Certification and the Opening Authorities. More precisely, during the Setup procedure, the algorithm GSKGen is run to output the master key for user registration sk_{CA} and a trapdoor key for opening signatures sk_{OA}. These keys are given respectively to the CA and the OA. The CA also generates the public group signature key pk_G.

Registration. Buyers are then *registered* with the CA receiving secret keys for a group signature scheme[4]. Formally, we assume that the CA keeps track of fraudulent buyers and that he communicates with buyers via secure channels. During the BReg procedure, when given as input the identity of a B_i, the CA checks whether this buyer is black-listed (in this case, CA outputs a special symbol $\perp$). Otherwise, the CA transmits a private group-signature key sk_{B_i} via the secure channel.

Item Preparation. Independently of the buyer registration process, the merchant runs the *Preparation* phase. Each item is assumed to have a uniform length of m so-called blocks (shorter items are padded to this length). During item preparation, the merchant generates the parameters for the Tardos codes (namely the false alarm rate δ – which directly gives the bound for the maximum number c of detected colluders). The value δ must be smaller than a maximum false-alarm rate certified by the CA. Subsequently, for each item, the merchant generates a Write-Once-Read-Many matrix (WORM), in which each entry (i, j) corresponds to the encryption of a fingerprinted copy of the i-th block of the item. More precisely the block is fingerprinted with either a 0 or a 1 symbol, depending on the identically and independently at random probabilities p_i. Finally, the merchant proves that the preparation process was properly done by computing a NIZK-PK [3,15]. Finally, the WORM and the NIZK-PK are published as depicted in Fig. 1.

Formally, during the item preparation procedure IPrep, the merchant first generates the parameters of the Tardos code: c, Z and δ. These parameters depend on the maximum item length m. Afterwards, the merchant draws the vector of probabilities $\mathbf{p} := p_1, \ldots, p_m$. Then, for each item I_t the merchant proceeds by: (1) generating the bits $f_{I_t}^{i,j}$ for $i \in \{1, \ldots, m\}$ and $j \in \{1, \ldots, N\}$, which are stored in an $N \times m$ matrix $\mathcal{F}_{I_t}$, (2) generating $N \times m$ symmetric-encryption keys $k_{I_t}^{i,j}$ of an IND-CPA-secure symmetric encryption scheme, storing them in a matrix κ_{I_t} for item I_t, (3) creating the WORM τ_{I_t} whose entries $fb_{I_t}^{i,j} =$

---

[4] Group signatures [2,4,10] provide anonymity to buyers for each transaction *regardless* of whether or not our protocol is coupled with an anonymous electronic payment mean. Furthermore, they enable the OA to trace signatures back to the signers during the Accusation phase.

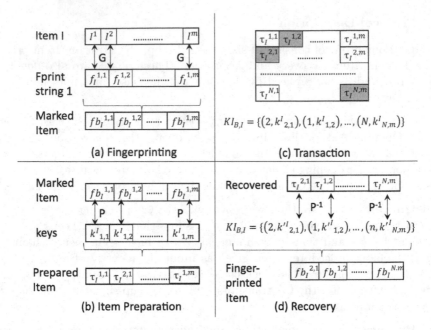

Fig. 1. The basic `PFP-TT` mechanics, including (a) item fingerprinting, (b) item preparation, (c) the transaction itself between the merchant and the buyer and (d) item recovery.

$\mathsf{Enc}(\mathsf{k}_{I_t}^{i,j}, \mathsf{W}(I_t^i, f_{I_t}^{i,j}))$ are the encryption, under one of the generated keys, of a fingerprinted block of the item with a specific fingerprint bit, and (4) generating a NIZK-PK[5], proving that the Tardos codes parameters are generated genuinely, that $\delta < \delta_{\max}$, and that the WORM is correctly set-up. This proof is denoted as π_{I_t}. For each item, the WORM and the proofs are published. We depict the item fingerprinting, item preparation, item transaction and item recovery in Fig. 1.

Buying. The buying process is one of the most fundamental phases of our protocol and it consists of two parts. The first part consists in the generation of the transaction information while during the second part, the buyer and merchant alternatively exchange the roles of *sender* and *receiver* as they run several sequential rounds of oblivious transfer (OT) [4,5,7,11,13,14,19,20,27] (for a total of m rounds). More precisely, the buyer retrieves a sequence of m tuples of keys and WORM positions for an item while the merchant gets to learn half of the indices recovered by the buyer, which make up the halfword. To prevent attacks in which the buyer inserts a false key into the OT, the merchant will link

---

[5] This NIZK is a proof of correctness in which the witness consists of the maximum number c of detected colluders, the probability δ that an innocent is wrongly accused, the accusation threshold Z, and probabilities p_i for $i \in \{1, \ldots, m\}$. The statement proved by the NIZK-PK consists of the following conditions: $m = 2\pi c^2 [\ln \frac{1}{\delta}]$ and that $p_i = (\sin r_i)^2$ for some random r_i uniformly picked in a specific interval (see [9]).

at each transaction, the value of the key to a randomly generated nonce, which will be recovered together with the halfword.

In our protocol, these two objectives are achieved by *interleaving*, respectively, rounds of 1-out-of-N and 1-out-of-2 oblivious transfer as follows. First, the buyer retrieves, in a 1-out-of-N OT round, a tuple key/nonce, the first input being from the corresponding column of the WORM while the second is generated at random by the merchant at each transaction (here the merchant plays the role of the sender in the OT while the buyer is the receiver). At the end of this OT round, we interleave a round of 1-out-of-2 OT, in which the merchant recovers either the index, the key and the random nonce retrieved by the buyer as well as a group signature on the key and the transaction number, or a bogus tuple of values of the same format. These inputs are randomized by the buyer before being inputted to the OT protocol. Thus, the merchant will not learn until he receives the input whether it was genuine (in which case the key/nonce values will be the correct ones) or not. This process continues by interleaving 1-out-of-N and 1-out-of-2 OT rounds until respectively all the m index-key tuples and the halfword are recovered.

Formally, during the IBuy procedure, the merchant and buyers communicate via an anonymous and one-sided-authenticated channel[6]. Whenever a buyer B_i wants to buy an item I_t from the merchant, the latter sends the current timestamp and a randomly chosen transaction number N_T. The buyer signs the message consisting of those parameters and the requested item, by using his private group key sk_{B_i}. He sends the item name and this signature $\sigma_{I_t}^{B_i}$ to the merchant, who verifies it. If the verification fails, the merchant aborts by outputting the error symbol $\bot$.

Then the merchant generates $N * m$ random numbers $\{R^{i,j}\}$, for $i \in \{1, \ldots, N\}$ and $j \in \{1, \ldots, m\}$. The buyer first runs a 1-out-of-N OT protocol, retrieving a key $\mathsf{k}_{I_\ell}^{i,1}$ and the corresponding random value $R^{i,1}$. Afterwards the merchant runs a 1-out-of-2 OT with the buyer, retrieving either (1) a tuple $(\mathsf{k}_{I_\ell}^{i,1}, R^{i,1}, \sigma_i)$, such that σ_i is a group signature on the tuple $(\mathsf{k}_{I_\ell}^{i,1}, N_T)$, in which N_T was the transaction number signed by the buyer before, or (2) a random value r respecting the appropriate format. The merchant verifies the signature, before checking that the tuple $(\mathsf{k}_{I_\ell}^{i,1}, R^{i,1})$ is consistent with his input for the previous OT round. If both checks succeeds, the index i is added to the halfword Hw. Otherwise, the halfword Hw remains unchanged. This process is repeated until m keys have been retrieved by the buyer. The merchant checks at regular intervals during the retrieval process that he has retrieved about half the indices of the fingerprint blocks. If this condition is not satisfied, he aborts the transaction. The frequency of these checks depends on m. Since m is presumably large, it is reasonable to expect that the number of recovered bits in the halfword is about a half of the total number of keys the buyer has recovered. At the end of

---

[6] For the purpose of attaining the exact bounds of the Theorem in Sect. 4.2, we additionally assume that buyers only have black-box access to the protocols during the buying process. For a detailed discussion of this assumption, see the remark on privacy versus traitor-tracing.

this transaction, the buyer has recovered a set $\mathbf{KI}_{B,I}$ of size m whose elements are tuples of the form $(i, k_{I_I}^{i,j})_{j=1}^m$ while the merchant has retrieved a set Hw of indices.

Recovery. The buyer will use the indices and keys he has recovered in order to obtain a fingerprinted copy of the item by using the WORM published for that item. Specifically, the keys are used to decrypt m cells of the WORM, with exactly one entry on each column. Formally, given the keys and the index sets, the buyer can recover a fingerprinted item $\mathsf{FI}_{B,I}$.

Accusation and Opening. If the merchant suspects that a copy of a specific item is a forgery, he computes accusation scores for each transaction of that item. If the score exceeds the accusation threshold, this indicates that the signer of the transaction signature has colluded to forge the item. The transaction signatures, together with the signatures retrieved with the halfword Hw, for *each* such score, together with a NIZK-PK that the scores were well computed, are forwarded to the OA. If the NIZK-PK verifies, the OA *opens* the forwarded group signatures, retrieving the identities of the signers. In our protocol, the use of the NIZK-PK effectively replaces the first role of the judge in the work of Charpentier and co-authors [9], mentioned in Sect. 2, as it ensures that the merchant cannot cheat during the computation of this score or arbitrary lift the anonymity of buyers. Note that the second role of the judge, which consists in computing the full scores, remains and is accomplished by the OA in our case (but this could also be delegated to another trusted entity).

Formally, given a forged item FI_{B^*,I_t}^* of an item I_t, during the Accuse phase, the merchant computes the Tardos scores S_k for each transaction k made for I_t, comparing them with the threshold for the halfword. For each score above this threshold, the merchant adds to a list $\mathcal{L}$ the signature $\sigma_{I_t}^{B^*}$ received during that transaction. The merchant proves using a NIZK-PK the correctness of the accusation scores[7]. Finally, the merchant forwards the index of the forged item I_t, the list of signatures $\mathcal{L}$, and the proof π_M to the OA.

Then, during the Open phase, if the proof verifies, then the OA uses the master opening key sk_OA to lift the anonymity of the signatures contained the list $\mathcal{L}$. If the signed message opens to an item that is not the same as the item forwarded by the merchant, and for which the computation was performed, then the OA does not return the output. Depending on the practical deployment of our protocol, the merchant may also be blacklisted if he misbehaves.

Remark: Privacy Versus Traitor Tracing. The probabilistic way in which we run the 1-out-of-2 OT protocol is designed to preserve both the buyer's privacy (and ensuring anti-framing) – by not revealing more than the strictly necessary indices of a buyer's copy of an item – *and* the traitor-tracing capacity

---

[7] The witness for this NIZK-PK consists of the transaction transcripts for the guilty parties (including the group signatures for the 1-out-of-2 OT rounds), their scores (computed as in [32]) and the threshold. The NIZK-PK statement is that the scores are correctly computed, that they are higher than the threshold, and that the signed messages sent along with the proof are indeed the ones associated to the transactions.

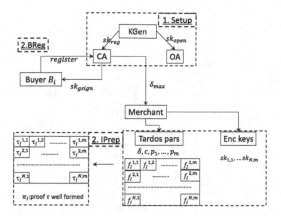

Fig. 2. System interactions: Setup, Buyer registration and Item preparation.

of Tardos codes – by giving the halfword to the merchant. However, the fact that the merchant does not learn whether the received input (i.e., the index and verifiable proof) will be the real or the simulated input until he verifies them means that with a non-negligible probability, the merchant will recover less than exactly half of the indices. However, the bounds we give in Sect. 4.2 only hold if the merchant can recover at least half of these positions. In practice, this could be ensured by allowing the buyer to only access in a black-box manner to the buying process (i.e., using a trusted but obfuscated application), and additionally authorizing the merchant to *request* further OT rounds by proving to the CA that he has recovered less than the required amount of information. Another way to ensuring this is by having the trusted application running a different type of protocol, in which instead of 1-out-of-2 OT, the merchant simply specifies $m/2$ positions for which the indices are returned in black-box manner, without revealing the request or the output to the buyer. In addition, note that if the merchant receives slightly less than the halfword, the traitor-tracing bound is only marginally decreased. However, for the statements in Sect. 4.2 we make the assumption that the merchant always recovers the entire halfword (abstracting away from the method ensuring this).

4.2   Security and Privacy Properties

Due to space restrictions, we only give here the theorem including the security properties of our protocol, leaving the formal proofs for the full version of the paper.

Theorem 1. *Let our protocol be implemented with a group signature scheme* GSScheme= *(*GSKGen, Join, Sign, Vf, Open, Revoke*), a symmetric encryption scheme* EScheme= *(*EKGen, Enc, Dec*), a certification scheme* Cert = *(*CSign, CVf*), and a* 1*-out-of-*N *OT protocol. We also assume the existence of: a one-side authenticated and one-side anonymous channel between the merchant and each buyer, a*

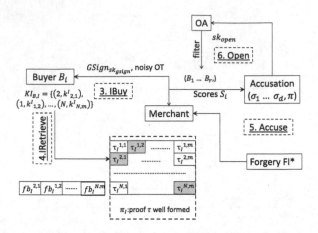

Fig. 3. System interactions: item transaction, item recovery, accusation and opening.

secure channel between the CA and the buyers, and of two NIZK-PKs NIZK-PK$_1$, and NIZK-PK$_2$ (for proving respectively (1) that prepared item matrices are well-formed and (2) the correctness of the accusation-score computation). The following properties hold:

Correctness. For every $(N_{\text{Test}}, \epsilon)$-correctness adversary $\mathcal{A}$ there exist: an adversary $\mathcal{A}_1$ against the correctness of GSScheme, an adversary $\mathcal{A}_2$ against the correctness of Cert for the CA, an adversary $\mathcal{A}_3$ against the correctness of the symmetric encryption scheme, an adversary $\mathcal{A}_4$ against the correctness of the OT scheme, and adversaries $\mathcal{A}_5$ against the correctness of the NIZK-PK$_1$, such that:

$$\epsilon \leq \sum_{i=1}^{5} \mathsf{Adv}_{\mathcal{A}_i}^{\text{correctness}}.$$

Buyer Unlinkability. For every $(N_{\text{Test}}, \epsilon)$-strong-buyer-unlinkability adversary $\mathcal{A}$ against our protocol, there exists: an adversary $\mathcal{A}_1$ against the full anonymity of GSScheme, an adversary $\mathcal{A}_2$ against the anonymity of the merchant-buyer communication channel, an adversary $\mathcal{A}_3$ against the security of the CA-buyer channel, and adversary $\mathcal{A}_4$ against the soundness of the NIZK-PK$_1$, such that:

$$\epsilon \leq N_{\text{Test}}\mathsf{Adv}_{\mathcal{A}_1}^{\text{full-anon}} + N_{\text{Test}}n(\mathsf{Adv}_{\mathcal{A}_2}^{\text{anon}} + \mathsf{Adv}_{\mathcal{A}_3}^{\text{sec}}) + \ell\mathsf{Adv}_{\mathcal{A}_4}^{\text{snd}}.$$

Anti-Framing. For any $(N_{\text{Test}}, \epsilon)$-anti-framing adversary $\mathcal{A}$ against our protocol, there exist: an adversary $\mathcal{A}_1$ against the full-traceability of GSScheme, and adversaries $\mathcal{A}_2$ and $\mathcal{A}_3$ against the soundness of respectively NIZK-PK$_1$ and NIZK-PK$_2$, such that:

$$\epsilon \leq \mathsf{Adv}_{\mathcal{A}_1}^{\text{full-trace}} + \ell(\mathsf{Adv}_{\mathcal{A}_2}^{\text{Snd}} + \mathsf{Adv}_{\mathcal{A}_3}^{\text{Snd}}) + N_{\text{Test}}n\delta.$$

Traitor-Tracing. For any $(N_{\mathsf{Test}}, c, \epsilon)$-*traitor-tracing adversary* $\mathcal{A}$ *against our protocol there exist: an adversary* $\mathcal{A}_1$ *against the full-traceability of* GSScheme *and an adversary* $\mathcal{A}_2$ *against the soundness of* NIZK-PK$_2$ *such that:*

$$\epsilon \leq \mathsf{Adv}_{\mathcal{A}_1}^{\mathsf{full-trace}} + N_{\mathsf{Test}}\mathsf{Adv}_{\mathcal{A}_2}^{\mathsf{Snd}}.$$

5 Conclusion

The work presented in this paper is the first step towards integrating strong privacy features into fingerprinting protocols. Our solution extends the original asymmetric fingerprinting protocol of Charpentier, Fontaine, Furon and Cox by adding buyer-unlinkability, while preserving a high traitor-tracing probability by using Tardos codes. Furthermore, we formally define the properties of correctness, buyer-unlinkability, as well as anti-framing and traitor-tracing for PFP-TT, finally giving exact bounds for the security and privacy properties of our protocol. We are currently implementing our protocol in order to evaluate its practical efficiency as well as its scalability.

References

1. Abdul, W., Gaborit, P., Carré, P.: Private anonymous fingerprinting for color images in the wavelet domain. In: Proceedings of SPIE Multimedia on Mobile Devices, vol. 7542 (2010)
2. Bellare, M., Micciancio, D., Warinschi, B.: Foundations of group signatures: formal definitions, simplified requirements, and a construction based on general assumptions. In: Biham, E. (ed.) Advances of Cryptology EUROCRYPT 2003. LNCS, vol. 2656, pp. 614–629. Springer-Verlag, Heidelberg (2003)
3. Blum, M., Feldman, P., Micali, S.: Non-interactive zero-knowledge and its applications. In: Proceedings of the Annual Symposium on the Theory of Computing (STOC), pp. 103–112 (1988)
4. Boneh, D., Boyen, X.: Short signatures without random oracles. In: Cachin, C., Camenisch, J.L. (eds.) EUROCRYPT 2004. LNCS, vol. 3027, pp. 56–73. Springer, Heidelberg (2004)
5. Brassard, G., Crépeau, C., Robert, J.M.: All-or-nothing disclosure of secrets. In: Odlyzko, A.M. (ed.) CRYPTO 1986. LNCS, vol. 263, pp. 234–238. Springer, Heidelberg (1987)
6. Camenisch, J.L.: Efficient anonymous fingerprinting with group signatures. In: Okamoto, T. (ed.) ASIACRYPT 2000. LNCS, vol. 1976, pp. 415–428. Springer, Heidelberg (2000)
7. Camenisch, J.L., Neven, G., Shelat, A.: Simulatable adaptive oblivious transfer. In: Naor, M. (ed.) EUROCRYPT 2007. LNCS, vol. 4515, pp. 573–590. Springer, Heidelberg (2007)
8. Cérou, F., Furon, T., Guyader, A.: Experimental assessment of the reliability for watermarking and fingerprinting schemes. EURASIP J. Inf. Secur. **2008**, 12 (2008). Article ID 414962
9. Charpentier, A., Fontaine, C., Furon, T., Cox, I.: An asymmetric fingerprinting scheme based on tardos codes. In: Filler, T., Pevný, T., Craver, S., Ker, A. (eds.) IH 2011. LNCS, vol. 6958, pp. 43–58. Springer, Heidelberg (2011)

10. Chaum, D., van Heyst, E.: Group signatures. In: Davies, D.W. (ed.) EUROCRYPT 1991. LNCS, vol. 547, pp. 257–265. Springer, Heidelberg (1991)
11. Chu, C.-K., Tzeng, W.-G.: Efficient k-out-of-n oblivious transfer schemes with adaptive and non-adaptive queries. In: Vaudenay, S. (ed.) PKC 2005. LNCS, vol. 3386, pp. 172–183. Springer, Heidelberg (2005)
12. Gambs, S., Onete, C., Robert, J.: Prover anonymous and deniable distance-bounding authentication. In: Proceedings of ACM AsiaCCS 2014, Accepted for publication. ACM Press (2014)
13. Green, M., Hohenberger, S.: Blind identity-based encryption and simulatable oblivious transfer. In: Kurosawa, K. (ed.) ASIACRYPT 2007. LNCS, vol. 4833, pp. 265–282. Springer, Heidelberg (2007)
14. Green, M., Hohenberger, S.: Universally composable adaptive oblivious transfer. In: Pieprzyk, J. (ed.) ASIACRYPT 2008. LNCS, vol. 5350, pp. 179–197. Springer, Heidelberg (2008)
15. Groth, J., Ostrovsky, R., Sahai, A.: Perfect non-interactive zero knowledge for NP. In: Vaudenay, S. (ed.) EUROCRYPT 2006. LNCS, vol. 4004, pp. 339–358. Springer, Heidelberg (2006)
16. Groth, J., Sahai, A.: Efficient non-interactive proof systems for bilinear groups. In: Smart, N.P. (ed.) EUROCRYPT 2008. LNCS, vol. 4965, pp. 415–432. Springer, Heidelberg (2008)
17. Hermans, J., Pashalidis, A., Vercauteren, F., Preneel, B.: A new RFID privacy model. In: Atluri, V., Diaz, C. (eds.) ESORICS 2011. LNCS, vol. 6879, pp. 568–587. Springer, Heidelberg (2011)
18. Kohlweiss, M., Maurer, U., Onete, C., Tackmann, B., Venturi, D.: Anonymity-preserving public-key encryption: a constructive approach. In: De Cristofaro, E., Wright, M. (eds.) PETS 2013. LNCS, vol. 7981, pp. 19–39. Springer, Heidelberg (2013)
19. Lindell, A.Y.: Efficient fully-simulatable oblivious transfer. In: Malkin, T. (ed.) CT-RSA 2008. LNCS, vol. 4964, pp. 52–70. Springer, Heidelberg (2008)
20. Naor, M., Pinkas, B.: Efficient oblivious transfer protocols. In: Proceedings of the 12-th ACM-SIAM Symposium on Discrete Algorithms (SODA 2001), pp. 448–457. SIAM (2001)
21. Oprea, A., Bowers, K.D.: Authentic time-stamps for archival storage. In: Backes, M., Ning, P. (eds.) ESORICS 2009. LNCS, vol. 5789, pp. 136–151. Springer, Heidelberg (2009)
22. Pfitzmann, B., Sadeghi, A.-R.: Coin-based anonymous fingerprinting. In: Stern, J. (ed.) EUROCRYPT 1999. LNCS, vol. 1592, pp. 150–164. Springer, Heidelberg (1999)
23. Pfitzmann, B., Sadeghi, A.-R.: Anonymous fingerprinting with direct non-repudiation. In: Okamoto, T. (ed.) ASIACRYPT 2000. LNCS, vol. 1976, pp. 401–414. Springer, Heidelberg (2000)
24. Pfitzmann, B., Schunter, M.: Asymmetric fingerprinting. In: Maurer, U.M. (ed.) EUROCRYPT 1996. LNCS, vol. 1070, pp. 84–95. Springer, Heidelberg (1996)
25. Pfitzmann, B., Waidner, M.: Anonymous fingerprinting. In: Fumy, W. (ed.) EUROCRYPT 1997. LNCS, vol. 1233, pp. 88–102. Springer, Heidelberg (1997)
26. Pfitzmann, B., Waidner, M.: Asymmetric fingerprinting for larger collusions. In: Proceedings of the 4-th ACM conference on Computer and Communications Security (ACM CCS 1997), pp. 151–160. ACM Press (1997)
27. Rabin, M.: How to exchange secrets with oblivious transfer. Harvard University Technical Report and IACR Eprint archive, report 187/2005 (1981). http://eprint.iacr.org/2005/187

28. Rial, A., Deng, M., Bianchi, T., Piva, A., Preneel, B.: A provably secure anonymous buyer-seller watermarking protocol. IEEE Trans. Inf. Forensics Secur. **5**, 920–9310 (2010). IEEE
29. Stern, J.P.: A new and efficient all-or-nothing disclosure of secrets protocol. In: Ohta, K., Pei, D. (eds.) ASIACRYPT 1998. LNCS, vol. 1514, pp. 357–371. Springer, Heidelberg (1998)
30. Tardos, G.: Optimal probabilistic fingerprint codes. In: Proceedings of the 35-th ACM Symposium on Theory of Computing (STOC 2003), pp. 116–125. ACM Press (2003)
31. Vaudenay, S.: On privacy models for RFID. In: Kurosawa, K. (ed.) Advances in Cryptology–ASIACRYPT 2007. LNCS, vol. 4833, pp. 68–87. Springer, Heidelberg (2007)
32. Škorić, B., Katzenbeisser, S., Celik, M.: Symmetric tardos fingerprinting codes for arbitrary alphabet sizes. Des. Codes Crypt. **46**, 137–166 (2008). Springer-Verlag

A   The Full Security Model

A.1   Watermarking and Fingerprinting Assumptions

Watermarking and fingerprinting assumptions. In our context, a protocol is run between a buyer and the merchant each time the former wants to recover a specific item. At the end of the protocol, the buyer can retrieve the item such that each block of the item is fingerprinted with exactly one bit. Thus, each buyer's version of the item is personalized with a unique fingerprint. We assume that the fingerprint is embedded in the item by means of a watermarking technique, which is imperceptible to humans and robust with respect to certain attacks. More specifically:

1. The watermarking function, denoted by W, does not allow an adversary to recover even a single bit of the fingerprint.
2. The watermarking technique is robust with respect to *signal attacks*, such as compressing, printing, scanning, resizing or cropping of the digital medium representing the item.
3. A collusion of malicious users can combine parts of their copies to create a forged item. However, they are restricted by the fact that if they have the same fingerprint block recurring at the same position in *all* their water-marked copies, they cannot output a copy in which the fingerprint bit at that specific position is *different* than the one they have in all their copies. This well-known assumption is called the *marking assumption* in the literature. Note that, if the collusion only involves a single buyer, this assumption precludes this buyer from producing a different (forged) fingerprint of the item.

In a *collusion attack*, several buyers combine parts of the legitimate fingerprinted items they own in order to forge an illegitimate copy. More precisely, they may combine the bits of their watermarks in an arbitrary manner, even adding erasures or errors, under the sole restriction of the marking assumption (see above). Examples of collusion attacks include the majority and minority rules, as well as the random choice. In the majority rule, the colluders choose for each block the most frequent fingerprint block in their copies (without necessarily knowing the value of this block) while in the minority rule, the less frequent fingerprinted block is chosen. Finally, in the random choice strategy, a random fingerprint is chosen amongst the available ones. A quite different strategy is a fusion of blocks. In this attack, the marking assumption has for consequence that for some blocks the fingerprint generated will be an error or an erasure, but never a valid fingerprinted block. The objective of the Tardos code is precisely to guarantee that in case of a collusion, with high probability at least one of its member will be traced.

A.2   A Formal Description of PFP-TT Schemes

Definition. A Privacy-preserving Fingerprinted Protocol with Traitor-Tracing capacities is a tuple of algorithms PFP-TT = (Setup, BReg, IPrep, IBuy, IRecover, Open) such that:

Setup: when given as input a security parameter 1^λ, this algorithm returns secret parameters spar (to be divided between the CA and the OA) and the public parameters ppar that are available to all parties. We assume that the remaining algorithms all implicitly take as input the public parameters ppar.

BReg: when given as input spar and a buyer's identity B, the buyer registration algorithm outputs either a secret key sk_B for B, or $\perp$.

IPrep: when given as input an item I, the item preparation algorithm outputs the prepared item τ_I (in our case, a Write-Once-Read-Many WORM table [21]), a proof π_I that the item has been correctly formed, a matrix of keys κ_{I_I} and a matrix $\mathcal{F}_I$ of fingerprints used for the preparation.

IBuy: the interactive buyer-merchant algorithm takes as input an item I, the secret key sk_B of a buyer, and a key-matrix κ_{I_I} generated at item preparation. The output is a set $\mathbf{KI}_{B,I}$ and some auxiliary information $aux_{B,I}$ (in our case a halfword).

IRecover: when given as input the key set $\mathbf{KI}_{B,I}$, the prepared item τ_I and the proof π_I, the recovery algorithm returns a fingerprinted item $FI_{B,I}$ or the symbol $\perp$.

Open: when given as input a (merchant-generated) proof π_M and spar, this algorithm outputs a set of buyer identities, denoted $\{B_i\}_{i=1}^d$ or an error symbol $\perp$. The value d is at most equals to the number of buyers c running a collusion attack.

Accuse: the accusation algorithm takes as input a fingerprinted copy FI and the set of all auxiliary information $aux_{B,I}$ obtained from honest transactions, and outputs a proof π.

Formal Oracles. Adversary interaction is captured by the following oracles:

Buy*(I, B, input^*): This oracle allows an adversary (in particular a malicious merchant) to deviate from protocol and execute the IBuy algorithm for item I and buyer B with malicious input input*. It returns the full output of the IBuy algorithm and the transcript of the transaction.

Execute(I, B): This oracle takes as input an item identifier I and a buyer identifier B, and simulates the execution of the IBuy algorithm for buyer B and item I for an honest merchant input. The oracle outputs the two values produced by the buying algorithm: the keys $\mathbf{KI}_{B,I}$ and the auxiliary information $\text{aux}_{B,I}$, as well as the transcript of the transaction.

BBuy(I, sk_B): This oracle takes as input a buyer's secret key sk_B and an item I and runs the IBuy algorithm, returning $\mathbf{KI}_{B,I}$ and the full transcript.

Open(π_M): This oracle takes as input a proof π_M and runs the opening algorithm Open on input π_M and the secret parameters spar, outputting a set of identities $\{B_i\}_{i=1}^d$. The oracle Open returns this set of identities.

Corrupt(B): This oracle takes as input a buyer identifier B and outputs the buyer's secret key sk_B.

Collude$(\{\text{FI}_{B_i,I}\}_{i=1}^k, \text{strategy})$: This oracle takes as inputs a set of at most $k \leq c$ legitimately-bought fingerprinted copies $\{\text{FI}_{B_i,I}\}_{i=1}^k$, and a strategy strategy outputs a forged fingerprinted copy, $\text{FI}_{\tilde{B},\tilde{I}}$. The strategy can be arbitrary with the following restriction: if for some block i of the item the recovered fingerprinted block $\text{fb}_I^{i,j}$ of *all* the colluding users embeds the γ-bit fingerprint $f_I^{i,j}$, then the corresponding fingerprinted block of the forged item FI_{B^*,I^*} *must* embed the fingerprint $f_I^{i,j}$ (this is a consequence of the marking assumption).

Accuse$(\text{FI}_{,})$: This oracle runs Accuse on input the fingerprinted copy $\text{FI}_{,,}$ a matrix of keys κ_{I_I}, and a matrix of fingerprints $\mathcal{F}_I$, outputting the proof π.

The Test oracles We proceed by listing the formal Test oracles for each property.

Correctness: when given as input a product identifier I and a buyer identifier B, Test^{Corr} runs Execute(I, B), outputting the keys $\mathbf{KI}_{B,I} = \{(j, \text{k}_I^{\tilde{i},j}\}_{j \in \{1,\dots,N\}}$, for consecutive values of i (if the values are not consecutive or have the wrong format, Test^{Corr} returns 0). The algorithm IRecover is subsequently run on input the keys $\mathbf{KI}_{B,I}$, the table τ_I, and the proof π_I, outputting the series of blocks $\text{FI}_{B,I}$ (else, if $\perp$ is output, the oracle Test^{Corr} returns 0). The oracle tests if for each entry $[\text{FI}_{B,I}]_{i,j}$, it holds that $[\tau_I]_{i,j} = \text{P}([\text{FI}_{B,I}]_{i,j}, [\kappa_{I_I}]_{i,j})$ for some one-way trapdoor preparation function P. If this last check fails, the oracle outputs 0 while otherwise it outputs 1.

Buyer-unlinkability: when given as input two buyer identities B_i and B_j, and a text parameter text $\in \{\text{draw}, \text{free}\}$, the Test$_b^{\text{BUnlink}}$ oracle, which keeps an internal database $\mathcal{D}_{\text{Test}^{\text{BUnlink}}}$, consistently associates either the first or the second input buyer identities with a handle handle depending on an input bit b. In this mode, once the Test$^{\text{BUnlink}}(\cdot, \cdot, \text{draw})$ query is run, the adversary may interact with the anonymized buyer by means of the Execute and respectively Buy* oracles (we modify these oracles to take as input the handle handle instead of the identifier of the buyer). The adversary may also choose to interact with other buyers or corrupt them. Finally, the adversary will free the two buyers by means of a $\mathsf{Test}^{\mathsf{BUnlink}}(\cdot, \cdot, \mathsf{free})$ query. If the adversary queries the Test oracle with text input draw while the current handle has not been released, this oracle returns $\perp$. Similarly, trying to free a handle while no handle is currently associated to any buyer will yield the output $\perp$.

Anti-framing: when given as input a proof π_M, $\mathsf{Test}^{\mathsf{NoFrame}}$ runs the Open oracle as a black box, receiving the set of identities $\{B_i\}_{i=1}^d$. The oracle checks if at least one identity output by Open is uncorrupted at the time of the $\mathsf{Test}^{\mathsf{NoFrame}}$ query. If this statement is true, the oracle outputs 1 while otherwise it returns 0.

Traitor-tracing: when given as input a set of honest fingerprinted copies $\{\mathsf{FI}_{B_i,\mathsf{I}}\}_{i=1}^k$ and a strategy strategy, $\mathsf{Test}^{\mathsf{TT}}$ internally runs Collude, outputting a forged copy $\mathsf{FI}_{,}$. Subsequently, it runs Accuse on input $\mathsf{FI}_{,}$, receiving the proof π. This proof is given as input to the Open oracle, which returns a set of identities $\{B_j\}_{j=1}^d$. If there exists some buyer B^* such that one of the inputs was $\mathsf{FI}_{B^*,\mathsf{I}}$ *and* B^* is amongst the outputs of the Open query, then the oracle $\mathsf{Test}^{\mathsf{TT}}$ returns 1 and the proof π, while otherwise it returns 0.

Anonymous Authentication with Shared Secrets*

Joël Alwen[1], Martin Hirt[1], Ueli Maurer[1], Arpita Patra[2],
and Pavel Raykov[1(✉)]

[1] Department of Computer Science, ETH Zurich, Zürich, Switzerland
{alwenj,martin.hirt,ueli.maurer,pavel.raykov}@inf.ethz.ch
[2] Department of Computer Science and Automation,
Indian Institute of Science, Bangalore, India
arpitapatra10@gmail.com

Abstract. Anonymity and authenticity are both important yet often conflicting security goals in a wide range of applications. On the one hand for many applications (say for access control) it is crucial to be able to verify the identity of a given legitimate party (a.k.a. entity authentication). Alternatively an application might require that no one but a party can communicate on its behalf (a.k.a. message authentication). Yet, on the other hand privacy concerns also dictate that anonymity of a legitimate party should be preserved; that is no information concerning the identity of parties should be leaked to an outside entity eavesdropping on the communication. This conflict becomes even more acute when considering anonymity with respect to an active entity that may attempt to impersonate other parties in the system.

In this work we resolve this conflict in two steps. First we formalize what it means for a system to provide both authenticity and anonymity even in the presence of an active man-in-the-middle adversary for various specific applications such as message and entity authentication using the constructive cryptography framework of [Mau11,MR11]. Our approach inherits the composability statement of constructive cryptography and can therefore be directly used in any higher-level context. Next we demonstrate several simple protocols for realizing these systems, at times relying on a new type of (probabilistic) Message Authentication Code (MAC) called *key indistinguishable* (KI) MACs. Similar to the key hiding encryption schemes of [BBDP01] they guarantee that tags leak no discernible information about the keys used to generate them.

1 Introduction

1.1 Anonymous Authentication

Anonymity and authenticity are both important yet often conflicting security goals in a wide range of applications. On the one hand "entity authentication" is a core functionality needed for implementing access control both in physical

---

* The unabridged version of this paper appears in [AHM+14a].
A. Patra—Work done while the author was at ETH Zurich.

D.F. Aranha and A. Menezes (Eds.): LATINCRYPT 2014, LNCS 8895, pp. 219–236, 2015.
DOI: 10.1007/978-3-319-16295-9_12

and digital systems. Moreover for many applications we are also required to authenticate *what* is being said, a security goal more commonly referred to as "message authentication". In both cases an implicit assumption underlying the systems is that each user has some unique identifying information associated with them which they can use either to prove who they are or what they are saying.

On the other hand, in a world where privacy matters, providing identifying information over public channels leads to an inherent conflict between the desire for authenticity (the property that no one else can claim to be you) and anonymity (the guarantee that external parties learn nothing about your identity). The problem is especially acute in light of the fact that many authentication protocols for physical access control are implemented using RFID tokens, where the communication can easily be eavesdropped, and the tokens can often be accessed wirelessly from a significant distance and without the consent or even awareness of the owner. Moreover mobile phones, constantly communicating over the public radio spectrum, also make use of uniquely identifying information for authenticating their communication with the network. Even on the internet when using a proxy or onion routing service to hide one's IP address a user interacts with a service requiring some form of authentication (say a VPN) may still not enjoy anonymity if the service does not make use of an anonymous authentication protocol.

In particular we stress that using cryptographic tools (to achieve secrecy and/or authenticity) over an anonymous channel generally destroys the anonymity of the channel. For example, if a challenge-response protocol based on a MAC and shared secret keys is used for client authentication, then the MAC values (aka tags) may leak partial information about the key, which means that an adversary can recognize that the same client is involved in different sessions, i.e., one loses unlinkability and hence also anonymity.

The goal of this work is to resolve this conflict allowing for the design of systems which provably guarantee both properties regardless of the greater context in which they are used.

1.2   Our Contributions

On the highest level we achieve our stated goal via two phases. First we cleanly, formally and composably capture what it means for a system to provide both authenticity and anonymity even in the presence of an active man-in-the-middle adversary for an array of specific applications such as message and entity authentication. Next we prove the security of several simple protocols for realizing these systems, at times relying on a new type of Message Authentication Code (MAC) introduced in [AHM+14b].

Formalizing Anonymous Authentication. In more detail, the first contribution of this work is to intuitively model a variety of resources providing anonymity. We do this in the constructive cryptography framework of [Mau11, MR11][1],

---

[1] One could also give an equivalent formulation in the UC framework.

inheriting its general composability guarantees. Concretely we define anonymous variants of insecure channels ($\mathcal{F}_{A\text{-}IC}$), authenticated channels ($\mathcal{F}_{A\text{-}AC}$), secure channels ($\mathcal{F}_{A\text{-}SC}$) and entity authentication ($\mathcal{F}_{A\text{-}EA}$). Each primitive is modelled as an ideal resource which explicitly shows the abilities and limits of an active man-in-the-middle adversary. For example, the ideal resource of an anonymous secure channel allows a sender to send a message to a receiver such that the adversary learns only the length of the message. The only actions permitted to the adversary are to cause delivery of any message previously sent by a sender (but without learning either the contents of the message or the identity of its sender).

Constructions. The second contribution is to prove security for various constructions of stronger anonymous primitives from weaker ones (see Fig. 1 for the overview). In particular, we build anonymous variants of authenticated channels from insecure channels (and a pairwise shared-key setup denoted with $\mathcal{K}$), entity authentication from authenticated channels (and a type of insecure broadcast channel denoted with $\mathcal{F}_{IB}$) and secure channels from authenticated channels. While most of the constructions are relatively immediate the proofs are often significantly more involved and it is these we consider to be the second contribution of this work. Some of the information theoretic constructions are decidedly unpractical (and should be viewed more as feasibility results) but we also provide several optimizations decreasing both communication and computational complexity. Combined with the pseudo-random function (PRF) based MAC of [AHM+14b] these give rise to practically interesting protocols for constructing anonymous message authentication and entity authentication from anonymous insecure channels and insecure broadcast in the shared key setting.

All computational constructions in this work rely on a novel primitive called *key indistinguishable* (KI) message authentication codes (MAC). Similar to the notion of a key hiding public key encryption scheme given in [BBDP01] these are (probabilistic) MACs such that tags generated with *different* keys cannot be distinguished from tags generated with the *same* key.[2] This notion was introduced in [AHM+14b] where a variety of constructions are given based on both blackbox primitives (such as PRFs, weak PRFs and Hash Proof Systems [HPS]) as well as concrete number theoretic assumptions (such as Decisional Diffie-Hellman [DDH], Learning with Errors [LWE] and Learning Parity with Noise [LPN]).

Deterministic Devices. We remark that while most of our protocols require parties to be probabilistic (which may be a problem for extremely light-weight computing devices) they can easily and generically be translated into stateful but deterministic parties by using a PRG.[3]

---

[2] Or more generally using the same or different states.

[3] In particular the security proof for the probabilistic setting then automatically carries over (at least in a computational sense) by preceding the proof with a hybrid argument replacing the output of each call to the PRG with fresh random numbers.

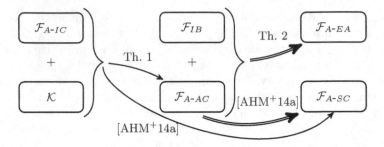

Fig. 1. The overview of resources and relations among them. The single arrow ($\longrightarrow$) denotes computational constructions, while double arrow ($\Longrightarrow$) denotes information-theoretic constructions.

Robustness to Side-Channels. An important consequence of how we define these resources is that we capture various types of information which may potentially be available to an adversary through side-channels. Technically this is done by providing extra capabilities and information to the distinguisher[4] D whose goal it is to tell the construction and ideal resource apart. In particular while no information concerning the identities of anonymous parties leaks on the adversarial interface of a given resource, D can trivially obtain such information from say the receivers interface. Thus D can make use of such information to aid in its task. In other words even an adversary equipped with such a side-channel learns no more from its interface to the real communication resource used to run the protocol than it does from the side-channel and the adversarial interface to the ideal resource. It is this property which is central to the intuitive claim of composability for all our constructions.[5]

Besides side-channels leaking information related to identities, we model another important type of side-channel that concerns the relative order in which parties respond. In particular, responses are always initiated at the behest of D. This is of particular interest in a setting with mobile phones of differing computational power or RFID tags positioned at differing distances from the adversary.

Exact Security. Similar to [AHM+14b], all reductions we give come with an exact security analysis (as opposed to asymptotic ones). We see at least two advantages in taking this approach. First, such results greatly facilitate comparing the quality/efficiency trade-off obtained via different constructions especially when based on the same underlying cryptographic assumptions. A somewhat less common but equally relevant advantage is that such statements make explicit the benefits obtained by enforcing constraints on the adversary through implementation choices. Take for example a protocol whose security degrades say in $q/|\mathcal{M}|$: the number of times an adversary can interact with a client divided by the size of the messages space supported by a MAC. Normally such a protocol

---

[4] called the "environment" in the language of UC.
[5] And more abstractly, this property plays an important role in the composition theorem of [Mau11].

would require a MAC with at least 160-bit messages to be considered secure. However, if implemented on hardware which guarantees failure after a limited number of interactions, say $q \leq 2^{10}$ (a common assumption in the RFID setting) the MAC now needs only to support 100-bit messages potentially reducing the hardware costs of the resulting implementation significantly.

1.3   Related Work

We provide an overview of the related literature which can, roughly speaking, be divided up into those concerned with generic entity authentication, anonymous entity authentication for RFIDs and anonymous authentication for mobile phone networks.

Entity Authentication. A large body of work going back almost 30 years has considered a range of security notions for unilateral entity authentication which, broadly speaking, consist of an "information collection" phase followed by an "attack" phase. An elegant overview elucidating their relationships can be found in [MT12]. Borrowing the language of [MT12] we can informally characterize a given security notion using three types of oracles C, S and T, namely the honest client (prover), server (verifier) and a transcript oracle[6] respectively. A particular security notion is then defined via two subsets of these oracles indicating the resources available to the adversary during the two respective phases. For example $(\{C\}, \{S\})$-auth security, traditionally referred to as *active* security, allows the adversary oracle access to the client in the first phase while only allowing access to the server in the attack phase. Another common example is $(\{T\}, \{S\})$-auth which is traditionally referred to as *passive* security.

While classical entity authentication protocols [FS86,GQ88,Sch89,Oka92] focus on the public key setting satisfying weaker security notions (i.e. *passive* and *active* security) more recent works (especially in the context of RFID identification protocols) are set in the shared-key model and achieve increasingly strong types of security culminating in several variants of man-in-the-middle (MiM) security. In this work we consider $(\{C, S\}, \{S\})$-auth which [MT12] show to be strictly weaker than $(\{\}, \{C, S\})$-auth (also used in [BR93, Vau10] for example). We justify this choice by arguing that the later MiM variant is in fact stronger than needed in real world applications. In particular it is unavoidable that an adversary, with online access to C during the attack phase, can convince S to accept (e.g. by blindly forwarding messages). Thus, in contrast to $(\{\}, \{C, S\})$-auth, we have opted for a security notion which does not rule out adversaries convincing S using an *online* C through more involved means such as by modifying C's messages.[7]

---

[6] Upon each invocation the transcript oracle outputs a freshly sampled transcript between the honest server and client.

[7] As is done for example in the separating example between the two notions in [MT12].

Anonymous Authentication for RFIDs. Anonymous authentication has primarily been studied in the context of entity authentication especially in the context of RFID systems, where anonymity is of particular interest. Unfortunately these results are tailored to this specific application and do not put forward a general framework for anonymous authentication as in this work.

Furthermore, most of the suggested schemes are proven secure using game-based notions [Vau10, HPVP11, DLYZ11], where security of the real world system is guaranteed only within the particular (often complex) context described by the game. Therefore it often remains unclear within which greater context such protocols can be used. For example, in perhaps the most popular such notion of [Vau10] and its derivatives (ex. [HPVP11]), the anonymity of a new session is defined only against adversaries which do not learn the identities of clients involved in previous (successful) sessions. In particular this means such protocols can not, a priori, be used in a greater context where an adversary may have access to say the full output of an RFID reader at any point in time. However ideally one would hope for a protocol which is again anonymous once the adversary loses such side-channel access to the card reader.

Composable security (UC) is considered in [ACdM05], but this solution assumes purely passive tags, which an adversary can easily overwrite, hence security cannot be achieved against a MiM adversary. Recently, [BLdMT09] and [BM11] provide composable (UC) security, but the clients must be stateful.

Note that the term "anonymous authentication" is at times also used for group authentication, where the *server* must not learn which particular client is authenticating. This is not the scope of this paper.

Anonymous Authentication for Mobile Phone Networks. Besides the RFID setting the models and definitions in this work also apply to authentication requirements in other radio communication networks such as mobile and satellite phone systems. For example, in the specification of the 3rd (and later) generation mobile phone systems by the 3GPP, a crucial key agreement (KA) phase between end-users and the network operator is described. While this phase involves communication over the public radio spectrum (an insecure, but potentially anonymous communication channel) the specification explicitly lists both end-user anonymity (preservation) and authenticity as key design goals for this phase [rGPP12]. To achieve this, a unique long-term secret is shared between each end-user device, called a Universal Subscriber Identification Module (USIM), and the services provider. Crucially, because the actual mobile devices (e.g. cellphones, tablets) are not trusted, all secret key operations on the user-end are performed on the USIM; a very light-weight computing device. This in turn has led to the use of light-weight cryptographic primitives in the form of symmetric key algorithms such as those used in this work.

Existing solutions leave much to be desired both in terms of anonymity [KAC08, AMRR11, AMR+12] and authenticity [TM12] and a large body of work with ad-hoc security arguments focuses on improving the status quo (e.g. [BR05, KO05, GVI06, SAJ07, CRS11, CRS12]). Notable exceptions are the works of [AMRR11, TM12, AMR+12] which make use of more formal symbolic analysis to discover some vulnerabilities in existing solutions. Finally in the concurrent and independent work of [LSWW13] a complex game-based security model, targeted specifically at the setting of UMTS/LTE client-network communication (including the KA), is presented. Moreover existing solutions are shown to satisfy a limited notion of anonymity and (roughly speaking) $(\{C, S\}, \{S\})$-authenticity under some novel yet plausible assumptions concerning key component functions used in the protocols.

In light of these results we view our work as making significant progress towards developing a formal yet intuitively tractable and secure model, compatible with the language and tools of modern cryptography, for capturing and analyzing the stated design goals of mobile phone networks together with examples of rigorously analyzed solutions. While we by no means claim the model (nor protocols) to be directly applicable in this arena we do believe them to represent an important step forward towards developing more satisfactory security guarantees for this extensively deployed application.

Anonymous Public-Key Encryption. Kohlweiss et al. [KMO+13] initiated the study of anonymity in the constructive cryptography framework. They considered a single sender multi-receiver setting where the sender and receivers communicate via a receiver-anonymous insecure channel. In such a setting they apply a public-key encryption in order to achieve a receiver-anonymous *confidential* channel. While achieving confidentiality with public-key encryption is straightforward, one needs to additionally assume that the employed scheme is *key-private* [BBDP01], i.e., which essentially means that for two given public keys and one ciphertext, one cannot decide for which public key this ciphertext is valid.

A related notion of the signer anonymity has been studied for signatures schemes [YWDW06]. Intuitively, these schemes guarantee that the signature for a secret random message hides the identity of the signer.

Multilateral-Anonymous Communication. Entity authentication providing both sender and receiver anonymity has been studied in [AF04, JKT08]. In [AF04] the authors define the notion of anonymity allowing each client to reveal and prove its identity to certain other clients (chosen according to some policy), and hide it from the remaining clients. Then, [AF04] gives two protocols implementing the proposed anonymity notion and analyzes them in the applied pi calculus security model [AF01]. In [JKT08] the authors present an authenticated key-exchange protocol that provides anonymity for the parties exchanging a key.

1.4   Outline

In Sect. 2 we briefly review the constructive cryptography framework of [Mau11, MR11] (with support for making exact security statements), and review two security notions for MAC schemes.

In Sect. 3 we present the anonymous authentication resources for various applications and provide information theoretic and computational constructions.

We also describe some more efficient variants thereof. In particular we describe a protocol providing a trade-off allowing a potentially much more efficient server protocol for realizing entity authentication at the cost of maintaining a short but mutable state on the client side.

2   Definitions

We review the constructive cryptography framework which we use to define anonymous authentication protocols. Next, in this section we define several security notions for MACs including key-indistinguishability as well as a variety of unforgeability definitions.

2.1   Constructive Cryptography and Exact Security

The primary goal of this work is to build anonymous authentication protocols under various assumptions such that the resulting protocols can be used as a building block within *any* greater system. This gives rise to two defining characteristics of our security notions.

On the one hand we require an *arbitrarily* composable security guarantee. For this we use the constructive cryptography [Mau11,MR11] (CC) framework which allows for real/ideal type definitions supporting very strong "general" composability. On the other hand in practice concrete security matters. So departing somewhat from the asymptotic statements used in many such definitions we provide precise security claims detailing the exact security as a function of the properties of the underlying assumptions. In doing so we provide a method to evaluate the practicality and efficiency of using different constructions and basing security on different concrete assumptions.

Recalling the CC Framework. We recall the main ideas of the CC framework and refer to [Mau11,MR11] for further details. Similar to the Universal Composability (UC) framework of Canetti [Can01] the CC framework makes use of two types of basic computational systems. The first are called *resources* which are equipped with a set of interfaces.[8] We generally denote resources using calligraphic capital letters such as $\mathcal{R}$ and $\mathcal{F}$. Each interface captures the capabilities of a particular party which interacts with the system. The second type of systems are called *converters* (such as a protocol π or simulator σ). Converters have an internal interface through which they are connected to the available resources and an external interface through which the composed system is accessed by the context in which it is being used. In contrast to the UC framework we do not assume the presence of a network of insecure channels and instead explicitly define all communication resources we use in any given statement.

The CC framework allows for the presence of several resources for which the $\|$ operator is used. For example the system where say resources modeling a

---

[8] In the language of UC we speak of ideal functionalities and of ITM communication tapes in the language of ITMs.

setup $\mathcal{R}$ and an insecure channel $\mathcal{F}$ are present is denoted by $(\mathcal{R}\|\mathcal{F})$. Moreover resources can be composed with (possibly multiple) converters giving rise to a network of computational systems which we refer to as a *composed system*. More concretely a broadcast channel (resource) together with several protocols (converters) forms a composed system. To denote this composed system obtained by attaching say a protocol π to resource $\mathcal{R}$ on interface I we write $\pi^I \mathcal{R}$.[9] As usual a security definition in the CC framework involves equating two composed systems; intuitively modeling the "real" and "ideal" worlds respectively.

We wish to capture the intuition of providing "security against an attacker". For this we model the capabilities of an adversary using a given resource via an *adversarial interface*. Generally a resource $\mathcal{R}$ modeling the real world provides non-trivial capabilities on the adversarial interface (say full man-in-the-middle (MiM) access) while the ideal resource $\mathcal{F}$ only provides very restricted capabilities on that interface. The desired intuition is then captured by detailing a *simulator* converter which attaches to adversarial interface of $\mathcal{F}$ and produces an on-line translation (a.k.a. a simulation) making the interface look like the adversarial interface of $\mathcal{R}$. Unlike UC, the CC framework does not model the adversary as an (arbitrary) separate entity (converter). Instead CC's approach could be thought of as UC in the special case of the dummy adversary, which acts as a transparent conduit between its inner and outer interfaces.[10]

Finally the CC framework allows for security definitions which support "general composition". That is the real and ideal composed systems are interchangeable *regardless* of the context in which they are used.[11] For this CC like UC uses an online adaptive distinguisher D (i.e. the "environment" in UC) whose goal it is to tell the two systems apart. Intuitively D models the arbitrary context in which the systems might be used and relative to which they should be interchangeable. Technically D is given access to all interfaces of either the real or ideal system and, after arbitrary interaction with the system, D outputs a bit indicating whether it believes this was the real or the ideal system.

Exact Security. We develop the notation for making exact security statements in the CC framework. Take a fixed pair of systems $\mathcal{R}$ and $\mathcal{F}$ with k interfaces. We consider a parametrized class of $(t, x_1, \ldots, x_k)$-distinguishers D running in time t and querying the ith interface at most x_i times during the interaction

---

[9] We note that resources and composed systems are actually computational objects of the same type and so at times we also use calligraphic capital letters to denote a composed system.

[10] Indeed, as shown in the so called "Dummy Lemma" for various UC type frameworks, this restriction results in no loss of generality while making security proofs far more tractable.

[11] This stands in contrast to say game based definitions which instead guarantee certain properties of a real world system only within the particular context captured by the game. For example the anonymity of the authentication protocols defined in [Vau10, HPVP11] holds only with respect to adversaries which remain oblivious to which parties have previously authenticated themselves during the life of the system (even for the "wide adversary" variants).

with a system ($\mathcal{R}$ or $\mathcal{F}$). The advantage of a specific D from this class $\Delta^{\mathsf{D}}(\mathcal{R}, \mathcal{F})$ is defined to be $|\Pr[\mathsf{D}(\mathcal{R}) \to 1] - \Pr[\mathsf{D}(\mathcal{F}) \to 1]|$. We will write $\mathcal{R} \approx_\alpha \mathcal{F}$ (where $\alpha = (t, x_1, \dots, x_k, \epsilon)$) to say that all distinguishers from this class have advantage at most ϵ.[12] In this work all systems are also parameterized by a implicit security parameter λ. It can be understood to be fixed to an arbitrary value once and for all and then shared across all systems in any given theorem.

2.2   Message Authentication Codes (MAC)

In this subsection we introduce the syntax and security properties for message authentication codes. Note that the following presentation is adapted from [AHM+14b].

Syntax. A message authentication code MAC = {KG, TAG, VRFY} is a triple of algorithms with associated key space $\mathcal{K}$, message space $\mathcal{M}$, and tag space $\mathcal{T}$.

- **Key Generation.** The probabilistic key generation algorithm $k \leftarrow \mathsf{KG}(1^\lambda)$ takes as input a security parameter $\lambda \in \mathbb{N}$ (in unary) and outputs a secret key $k \in \mathcal{K}$.
- **Tagging.** The probabilistic authentication algorithm $\tau \leftarrow \mathsf{TAG}_k(m)$ takes as input a secret key $k \in \mathcal{K}$ and a message $m \in \mathcal{M}$ and outputs an authentication tag $\tau \in \mathcal{T}$.
- **Verification.** The deterministic verification algorithm $\mathsf{VRFY}_k(m, \tau)$ takes as input a secret key $k \in \mathcal{K}$, a message $m \in \mathcal{M}$ and a tag $\tau \in \mathcal{T}$ and outputs an element of the set {Accept, Reject}.

Next we define some useful properties such a triple of algorithms can have such as completeness and unforgeability. We also discuss a less common security notion for MACs, called key indistinguishability [AHM+14b] which can only be achieved by *randomized* MACs. Each of the following definitions depend on a security parameter λ. However, in line with the above discussion on the treatment of the security parameter in our constructive statements, we omit λ from our notation. Instead, to avoid clutter, we assume that all security properties in any given statement share the same fixed value of λ.

Completeness. We say that MAC has completeness error η if for all $m \in \mathcal{M}$,

$$\Pr[\mathsf{VRFY}_k(m, \tau) = \texttt{Reject} :\quad k \leftarrow \mathsf{KG}(1^\lambda), \tau \leftarrow \mathsf{TAG}_k(m)] \leq \eta.$$

Unforgeability. We recall the standard notion security for (randomized) MACs; namely unforgeability under chosen message (and chosen verification) query

---

[12] More specifically in this work the underlying cryptographic assumptions used give rise to the properties of the real world resource $\mathcal{R}$ while the implementation choices can allow for bounding properties of D. The final distinguishing advantage of the real and ideal systems is usually a function of both types of properties.

attack (**uf-cmva**). We denote by $\mathbf{Adv}_{\mathsf{MAC}}^{\mathsf{uf\text{-}cmva}}(\mathsf{A}, \lambda)$, the *advantage* of the adversary A in forging the message for a random key $k \leftarrow \mathsf{KG}(1^\lambda)$. Formally it is the probability that the following experiment outputs 1.

Experiment. $\mathbf{Exp}_{\mathsf{MAC}}^{\mathsf{uf\text{-}cmva}}(\mathsf{A}, \lambda)$

- $k \leftarrow \mathsf{KG}(1^\lambda)$
- *Invoke* $\mathsf{A}^{\mathsf{TAG}_k(\cdot), \mathsf{VRFY}_k(\cdot, \cdot)}$.
- *Output* 1 *if* A *queried* (m^*, τ^*) *to* $\mathsf{VRFY}_k(\cdot, \cdot)$ *s.t.* $\mathsf{VRFY}_k(m^*, \tau^*) = \mathsf{Accept}$ *and* A *did not receive* τ^* *by querying* m^* *to* $\mathsf{TAG}_k(\cdot)$.

The above experiment can be weakened by relaxing the winning condition of the experiment $\mathbf{Exp}_{\mathsf{MAC}}^{\mathsf{uf\text{-}cmva}}$ to require that m^* has not previously been queried to $\mathsf{TAG}_k(\cdot)$. We refer to the resulting notion as *weakly* unforgeable while referring to the more stringent security notions as *strongly* unforgeable. In general in this work unless stated explicitly otherwise we always mean the strong variants. Finally we can remove the adversary's access to the verification oracle in which case we refer the experiment as **cma** rather than **cmva**.

We refer to an efficient (i.e. PPT) adversary A playing a **cmva** type experiments as a (t, q_t, q_v)-adversary if it runs in time at most t, and for any pair of oracles with a fixed key A makes at most q_t tag and q_v verification queries.

Definition 1 (Unforgeability of MACs). *A message authentication scheme* MAC *is* (t, q_t, q_v, ϵ)-**uf-cmva** *secure if for any* (t, q_t, q_v)-*adversary* A *we have:*

$$\mathbf{Adv}_{\mathsf{MAC}}^{\mathsf{uf\text{-}cmva}}(\mathsf{A}, \lambda) := \Pr[\mathbf{Exp}_{\mathsf{MAC}}^{\mathsf{uf\text{-}cmva}}(\mathsf{A}, \lambda) \to 1] \leq \epsilon.$$

Key Indistinguishability. The notion of key indistinguishability (KI) guarantees that tags leak no information about the underlying key (or state). This allows us to use such a scheme to implement authentication anonymously. We note that such a property is not implied by even the strongest of unforgeability notions defined above.[13] The intuition we capture for KI is that an adversary can not tell a single pair of tag and verify oracles from two pairs of such oracles with different states (including secret keys). In other words if an adversary has access to 4 oracles (2 tag and 2 verify oracles) it can not tell if the tag (and verify) oracles actually use the same state or not.

To formalize this we introduce some notation. For keys $k_0, k_1 \in \mathcal{K}$ we write $[k_0, k_1]$ to denote the 4-tuple of oracles $(\mathsf{TAG}_{k_0}, \mathsf{VRFY}_{k_0}, \mathsf{TAG}_{k_1}, \mathsf{VRFY}_{k_1})$. Moreover we write $[k_0, k_0]$ to denote a similar 4-tuple but where the TAG oracles share their entire internal state including secret key (and similarly for the VRFY oracles).

---

[13] Indeed this is not difficult to see. For example we can modify any (say **uf-cmva**) unforgeable scheme as follows such that it is clearly not key indistinguishable. Double the key size, use the first half of the key in conjunction with the original TAG algorithm to tag the message and then append the second half of the key to the resulting tag. Clearly the scheme remains unforgeable however it is trivial to tell tags issued under different keys apart.

In other words calls to the first and third oracle of $[k_0, k_0]$ are answered by essentially the same oracle (and similarly for the second and fourth oracle).[14]

Experiment. $\mathbf{Exp}_{\mathsf{MAC}}^{\mathsf{ki-cmva}}(A, \lambda)$

- $k_0, k_1 \leftarrow \mathsf{KG}(1^\lambda)$, $c \leftarrow \{0, 1\}$
- *Sample output* $c' \leftarrow A^{[k_0, k_c]}$.
- *If a tag obtained from the left oracle (namely TAG_{k_0}) was verified using the right verification oracle (namely VRFY_{k_c}) or vice versa, then output a uniform random bit.*
- *Otherwise if $c = c'$ output 1 and 0 otherwise.*

As usual, in the above experiment we have made a non-triviality constraint; namely that A is not allowed to make a query (m, τ) to oracle VRFY_{k_c} if τ was obtained from TAG_{k_0} for message m (and vice versa).

As before in the following definition we say that an adversary A is a (t, q_t, q_v)-adversary if it runs in time at most t and for *each* pair of oracles with a given key makes at most q_t tag and q_v verification queries. So in total such an adversary can make up to $2q_t$ tag queries namely by making q_t queries to TAG_{k_0} and TAG_{k_c}.

Definition 2 (Key Indistinguishability). *Let λ be an (implicit) fixed security parameter. A message authentication scheme* MAC *is* (t, q_t, q_v, ϵ)-**ki-cmva** *secure if for any* (t, q_t, q_v)-*adversary* A *we have*

$$\mathbf{Adv}_{\mathsf{MAC}}^{\mathsf{ki-cmva}}(A, \lambda) := 2\left|\Pr[\mathbf{Exp}_{\mathsf{MAC}}^{\mathsf{ki-cmva}}(A, \lambda) \rightarrow 1] - \frac{1}{2}\right| \leq \epsilon.$$

Moreover if MAC *is* $(t, q_t, 0, \epsilon)$-**ki-cmva** *then we call it* (t, q_t, ϵ)-**ki-cma** *secure. In particular in the* **ki-cma** *experiment we simply omit all verification oracles.*

3   Anonymous Authentication as Real/Ideal Transformations

In this section we define a range of anonymous resources together with various computational and information theoretic protocols for constructing them. We also describe several optimizations including two practically relevant protocols.

---

[14] For stateful MACs it is important that the full state (and not just the secret key) be shared between matching oracles in $[k_0, k_0]$. Suppose we have a secure MAC which hides all information about the secret keys. We can modify the TAG algorithm to keep a counter which it appends to each tag τ it outputs. Clearly the scheme still hides all information about the secret key. However it is unclear how such a scheme might be used to achieve anonymity. Indeed it is trivial to tell say the 10th tag issued for key k_0 from the 3rd tag issued for different key k_1.

3.1 Anonymous Message Authentication

We begin by focusing on anonymous message authentication. We prove that using a KI and unforgeable MAC one can construct anonymous variants of authenticated channels from insecure channels in the shared key setting thereby reducing the problem of anonymous message authentication to building such MACs. In [AHM+14a] we also give an optimization which provides the trade-off of improving receiver efficiency (given an optimistic but realistic assumption) at the cost of requiring senders to be stateful.[15]

We define an anonymous insecure chan-
nel $\mathcal{F}_{A\text{-}IC}$ (intuitively depicted on the right
side) which captures the minimal communica-
tion resource we require for achieving any type
of anonymous authentication. Intuitively this is a
multi-sender/single-receiver channel which provides the guarantee that the identity of the sender remains hidden on the adversary's interface. The scheduling and content of messages being sent is externally driven (technically they are provided by the distinguisher D) and we model an active adversary with full control over message delivery and content. Finally once the adversary chooses to deliver a message to the receiver, the receiver learns the content of that message but not (a priori) the identity of the original sender. Indeed this identity may not even be well defined as the adversary may have mauled an original sent message or even invented a completely new message for delivery.

Next we define a multi-sender/single-receiver
anonymous *authenticated* channel $\mathcal{F}_{A\text{-}AC}$, (intuitively depicted on the right side and described
formally in Fig. 2). The difference to $\mathcal{F}_{A\text{-}IC}$ are
two-fold. First the adversary is now restricted to
delivering only messages m which were originally
sent by one of the senders and second upon deliv-
ery of m the receiver additionally learns the identity of the original sender.

Finally we construct $\mathcal{F}_{A\text{-}AC}$ from $\mathcal{F}_{A\text{-}IC}$ in a shared key setting modeled via n key-distribution resources $\mathcal{K} = \{\mathcal{K}_i \mid i \in [n]\}$[16] where, upon initialization each such resource $\mathcal{K}_i$ samples a fresh key and outputs it both to the corresponding sender and to the receiver. Formally, $\mathcal{K}$ is a 2-interface resource which upon initialization samples $k_i \leftarrow_R \mathcal{K}$ and outputs it on both interfaces. The protocol for realizing $\mathcal{F}_{A\text{-}AC}$ from $\mathcal{F}_{A\text{-}IC}$ and $\mathcal{K}_i$ uses a MAC scheme $\mathsf{MAC} = (\mathsf{TAG}, \mathsf{VRFY})$ with message space $\mathcal{M}$. In particular to send a message $m \in \mathcal{M}$ the sender obtains shared key k_i from $\mathcal{K}_i$, computes a tag $\tau = \mathsf{TAG}_{k_i}(m)$ and outputs (m, τ) to $\mathcal{F}_{A\text{-}IC}$ on interface S_i. When the receiver obtains a message of the form (m, τ) from interface R of $\mathcal{F}_{A\text{-}IC}$ it looks for a key k_i (obtained from $\mathcal{K}_i$) such that $\mathsf{VRFY}_{k_i}(m, \tau) = \mathbf{true}$. If such a key is found output (m, i) and otherwise output $\perp$ (on the external interface).

---

[15] For some applications (such as entity authentication for light-weight devices) this reflects a design choice for senders already common in practice.

[16] We use the standard notation $[n]$ to denote the set $\{1, \ldots, n\}$.

INIT: $M \leftarrow \emptyset$, *counter* $\leftarrow 0$
ON INTERFACE S_i:
    CASE (m): *counter* $\leftarrow$ *counter* $+ 1$, $M \leftarrow M \cup \{(counter, i, m)\}$; output
    $(counter, m)$ on interface A
ON INTERFACE A:
    CASE $(j \in \mathbb{N} \cup \{\perp\})$: If $\exists (j, i, m) \in M$ then output (i, m) on interface R (and
    otherwise output $\perp$).

Fig. 2. The anonymous authenticated channel $\mathcal{F}_{A\text{-}AC}$

We prove the construction secure using a somewhat involved sequence of hybrid systems as summarized in the following theorem and the proof can be found in [AHM+14a]. While the result is not surprising the proof reveals a subtlety arising from the somewhat non-standard use of unforgeability in a multi-user setting. As a consequence, in terms of exact security the construction loses double the expected unforgebaility term ϵ' per sender. The details can be found in [AHM+14a].

Theorem 1. *The trivial protocol* $\pi = (\pi^{S_1}, \cdots, \pi^{S_n}, \rho^R)$ *described above realizes* $\mathcal{F}_{A\text{-}AC}$ *from* $\mathcal{F}_{A\text{-}IC}$ *and* $\mathcal{K}$. *More precisely, there exists a simulator* σ *such for any* $t, q_t, q'_t, q'_v \in \mathbb{N}$ *and* $\epsilon, \epsilon' > 0$, *distinguisher* D *and MAC scheme* MAC *with message space* $\mathcal{M}$ *such that:*

– MAC *is* (t, q_t, ϵ)-**ki-cma** *secure,* $(t, q'_t, q'_v, \epsilon')$-**uf-cmva** *secure and has* η *completeness error.*
– D *runs in time* t, *sends* q'_v *messages through* A, $\min(q'_t, \frac{q'_v}{n}, \frac{q_t}{n})$ *messages through* S_i *(for all* $i \in [n]$*).*

we get that $\Delta^D[\pi(\mathcal{F}_{A\text{-}IC}||\mathcal{K}), \sigma^A(\mathcal{F}_{A\text{-}AC})] \leq 2n\epsilon' + q'_v\eta + n\epsilon$.

3.2 Anonymous Entity Authentication

We describe a multi-session and multi-user anonymous entity authentication resource $\mathcal{F}_{A\text{-}EA}$ in such a way that we can prove that a standard challenge response protocol indeed constructs it with statistical security.

Resource $\mathcal{F}_{A\text{-}EA}$ models multiple (sequential) authentication sessions initiated via the server interface. Clients respond to the most recent pending authentication challenge whenever prompted to do so via their interface. Each session results either in the server accepting a particular identity or else failing (denoted with a special output $\perp$). The adversary, assumed to be controlling the scheduling of the underlying communication channel, is given control over forwarding challenges from the server to the client (via the QUERY command). However, it learns nothing more than the relative order of responses generated by clients thereby capturing the intuitive goal of anonymity. In particular, if the relative order is random, it is impossible for the adversary to link clients' responses in different sessions.[17] Further the adversary can, at any point, forward a client's response on to the server.

To capture the intuitive goal of entity authentication we equip $\mathcal{F}_{A\text{-}EA}$ with an internal set *Responded* which keeps track of the set of clients which have forwarded their response for the *current* authentication session. In particular, *Responded* is cleared whenever a new authentication session is initiated and, crucially, for any given session the adversary can only cause identities contained in *Responded* to be output on the server's interface. In other words, the only identities ever accepted at the end of a session are those which respond *during* the session regardless of all previous actions taken on any interface.[18] A formal description capturing this behavior can be found in Fig. 3.

INIT: *InSession* ← **false**, *counter* ← 0, *Responded* ← ∅, $\forall i \in [n]$ msg_i ← **false**
ON INTERFACE C_i:
    CASE (**RESPOND**):
        If (msg_i = **true**) then
            msg_i ← **false**, *counter* ← *counter* + 1
            If (*InSession* = **true**) then *Responded* ← *Responded* ∪ {(*counter*, i)}
            Output *counter* on interface A
ON INTERFACE S:
    CASE (**GO**): *InSession* ← **true**, *Responded* ← ∅; output **GO** on interface A
ON INTERFACE A:
    CASE (**QUERY**): $\forall i \in [n]$ msg_i ← **true**
    CASE ($j \in \mathbb{N} \cup \{\bot\}$):
        If (*InSession* = **true**) then
            InSession ← **false**
            If $\exists (j, i) \in$ *Responded* then output i on interface S, otherwise output $\bot$

Fig. 3. The ideal resource of anonymous entity authentication $\mathcal{F}_{A\text{-}EA}$

To verify that $\mathcal{F}_{A\text{-}EA}$ captures our intended intuition we show that a very simple challenge-response protocol indeed constructs $\mathcal{F}_{A\text{-}EA}$ from $\mathcal{F}_{A\text{-}AC}$ as expected. Subsequently we describe several optimizations of interest for a more practical scenario.

In order to send the challenge from server to clients we assume the presence of a type of single-sender/multi-receiver insecure broadcast channel $\mathcal{F}_{IB}$. Put simply any message input by the sender is output to the adversary and any message input by the adversary is delivered to all receivers.[19] The server protocol

---

[17] In case the relative order of clients' responses in different sessions is known to be correlated (e.g., by one client possessing a faster hardware than the others and being always the first to respond), the unlinkability of sessions is not guaranteed.

[18] As described in the introduction, in the language of [TM12] this corresponds precisely to ({C, S}, {S})-authenticity.

[19] A formal description can be found in [AHM+14a].

ρ to realize $\mathcal{F}_{A\text{-}EA}$ using $\mathcal{F}_{A\text{-}AC}$ and $\mathcal{F}_{IB}$ is extremely simple. For each new authentication session it chooses a fresh random challenge $r \leftarrow_R \mathcal{M}$ and broadcasts it using $\mathcal{F}_{IB}$. When it receives a response (i, r') from $\mathcal{F}_{A\text{-}AC}$ it outputs identity i if $r' = r$ and otherwise $\perp$. The ith client protocol π^{C_i} is equally simple; it is equipped with a message buffer which stores the most recent message received from $\mathcal{F}_{IB}$. Whenever π receives the command to respond it checks if its message buffer is full and if so forwards the content to interface S_i of $\mathcal{F}_{A\text{-}AC}$. A formal description of this protocol and the proof that it constructs $\mathcal{F}_{A\text{-}EA}$ for $(\mathcal{F}_{A\text{-}AC}\|\mathcal{F}_{IB})$ as stated in the following theorem can be found in [AHM+14a].

Theorem 2. *The protocol* $\pi = (\pi^{C_1}, \cdots, \pi^{C_n}, \rho^S)$ *described above realizes* $\mathcal{F}_{A\text{-}EA}$ *from* $\mathcal{F}_{A\text{-}AC}$ *and* $\mathcal{F}_{IB}$. *More precisely, there exists a simulator* σ *such for any* $t, q_s, q_v \in \mathbb{N}$, *distinguisher* D *sending* q_v *messages through interface* A *and starting* q_s *sessions, and a challenge set* $\mathcal{M}$ *we get that* $\Delta^D[\pi(\mathcal{F}_{A\text{-}AC}\|\mathcal{F}_{IB}), \sigma^A (\mathcal{F}_{A\text{-}EA})] \leq \frac{q_s(q_s + q_v)}{|\mathcal{M}|}$.

We briefly remark on some variants of this result. Similar to the optimization for building $\mathcal{F}_{A\text{-}SC}$ from $(\mathcal{F}_{A\text{-}IC}\|\mathcal{K})$ here too when using $(\mathcal{F}_{A\text{-}IC}\|\mathcal{K})$ in place of $\mathcal{F}_{A\text{-}AC}$ the response from the clients need not include the random challenge r. Moreover the same trade-off for the "optimistic setting" described in [AHM+14a] can also be applied here to improve server efficiency using stateful clients. Finally, when using KI MACs over $(\mathcal{F}_{A\text{-}IC}\|\mathcal{K})$ underneath the challenge-response protocol we observe that it suffices to use only universally unforgeable MACs[20] instead of **uf-cmva** ones. Intuitively, this is because the only messages for which producing a fresh tag could impersonate a client are the random challenges chosen by the server protocol. However for given (t, q_t, q_v, ϵ)-**uf-cmva** secure MAC the exact distinguishing advantage between the real and ideal systems is smaller (by an additive factor of $(n\epsilon - 1)q_s$) than if the MAC is only (t, q_t, q_v, ϵ)-secure against universal forgeries.[21]

This observation can be interpreted in two ways. On the one hand for a given MAC based challenge-response authentication protocol we can weaken the assumptions on the MAC for obtaining secure entity authentication. On the other hand we can make use of potentially more efficient (but slightly more forgeable) MAC schemes for constructing $\mathcal{F}_{A\text{-}EA}$.

References

[ACdM05] Ateniese, G., Camenisch, J., de Medeiros, B.: Untraceable RFID tags via insubvertible encryption. In: ACM Conference on Computer and Communications Security, pp. 92–101 (2005)

---

[20] Universal unforgeability is a relaxed security notion for MACs where the adversary only wins by producing a fresh (valid) tag for a uniform random message chosen by the challenger.

[21] The security loss arises because in addition to having to guess for which client an impersonation attack will arise (see [AHM+14b]) the reduction to universal unforgeability must also guess during which of the q_s sessions the attack occurs so as to properly plant its random challenge message from the universal unforgeability game.

[AF01]   Abadi, M., Fournet, C.: Mobile values, new names, and secure communication. SIGPLAN Not. **36**(3), 104–115 (2001)

[AF04]   Abadi, M., Fournet, C.: Private authentication. Theor. Comput. Sci. **322**(3), 427–476 (2004)

[AHM+14a] Alwen, J., Hirt, M., Maurer, U., Patra, A., Raykov, P.: Anonymous authentication with shared secrets. Cryptology ePrint Archive, Report 2014/073 (2014). http://eprint.iacr.org/

[AHM+14b] Alwen, J., Hirt, M., Maurer, U., Patra, A., Raykov, P.: Key-indistinguishable message authentication codes.Cryptology ePrint Archive, Report 2014/107 (2014 to appear in SCN 2014)

[AMR+12] Arapinis, M., Mancini, L.I., Ritter, E., Ryan, M., Golde, N., Redon, K., Borgaonkar, R.: New privacy issues in mobile telephony: fix and verification. In: ACM CCS, pp. 205–216. ACM (2012)

[AMRR11] Arapinis, M., Mancini, L.I., Ritter, E., Ryan, M.: Formal analysis of UMTS privacy. CoRR, abs/1109.2066 (2011)

[BBDP01] Bellare, M., Boldyreva, A., Desai, A., Pointcheval, D.: Key-privacy in public-key encryption. In: Boyd, C. (ed.) ASIACRYPT 2001. LNCS, vol. 2248, pp. 566–582. Springer, Heidelberg (2001)

[BLdMT09] Burmester, M., Le, T.V., de Medeiros, B., Tsudik, G.: Universally composable RFID identification and authentication protocols. ACM Trans. Inf. Syst. Secur. **12**(4), 1–33 (2009)

[BM11]   Burmester, M., Munilla, J.: Lightweight RFID authentication with forward and backward security. ACM Trans. Inf. Syst. Secur. **14**(1), 11 (2011)

[BR93]   Bellare, M., Rogaway, P.: Entity authentication and key distribution. In: Stinson, D.R. (ed.) CRYPTO 1993. LNCS, vol. 773, pp. 232–249. Springer, Heidelberg (1994)

[BR05]   Barbeau, M., Robert, J.-M.: Perfect identity concealment in UMTS over radio access links. In: WiMob (2), pp. 72–77. IEEE (2005)

[Can01]   Canetti, R.: Universally composable security: A new paradigm for cryptographic protocols. In: FOCS, pp. 136–145 (2001)

[CRS11]   Choudhury, H., Roychoudhury, B., Saikia, D.K.: UMTS user identity confidentiality: An end-to-end solution. In: WOCN, pp. 1–6. IEEE (2011)

[CRS12]   Choudhury, H., Roychoudhury, B., Saikia, D.K.: Enhancing user identity privacy in LTE. In: TrustCom, pp. 949–957. IEEE C. Soc. (2012)

[DLYZ11] Deng, R.H., Li, Y., Yung, M., Zhao, Y.: A zero-knowledge based framework for RFID privacy. J. Comp. Sec. **19**(6), 1109–1146 (2011)

[FS86]   Fiat, A., Shamir, A.: How to prove yourself: practical solutions to identification and signature problems. In: Odlyzko, A.M. (ed.) CRYPTO 1986. LNCS, vol. 263, pp. 186–194. Springer, Heidelberg (1987)

[GQ88]   Guillou, L.C., Quisquater, J.-J.: A practical zero-knowledge protocol fitted to security microprocessor minimizing both transmission and memory. In: Günther, C.G. (ed.) EUROCRYPT 1988. LNCS, vol. 330, pp. 123–128. Springer, Heidelberg (1988)

[GVI06]   Gódor, G., Varadi, B., Imre, S.: Novel authentication algorithm of future networks. In: ICN/ICONS/MCL, p. 80. IEEE Computer Society (2006)

[HPVP11] Hermans, J., Pashalidis, A., Vercauteren, F., Preneel, B.: A new RFID privacy model. In: Atluri, V., Diaz, C. (eds.) ESORICS 2011. LNCS, vol. 6879, pp. 568–587. Springer, Heidelberg (2011)

[JKT08]   Jarecki, S., Kim, J., Tsudik, G.: Beyond secret handshakes: affiliation-hiding authenticated key exchange. In: Malkin, T. (ed.) CT-RSA 2008. LNCS, vol. 4964, pp. 352–369. Springer, Heidelberg (2008)

[KAC08]   Khan, M., Ahmed, A., Cheema, A.R.: Vulnerabilities of UMTS access domain security architecture. In: SNPD, pp. 350–355 (2008)

[KMO+13] Kohlweiss, M., Maurer, U., Onete, C., Tackmann, B., Venturi, D.: Anonymity-preserving public-key encryption: a constructive approach. In: De Cristofaro, E., Wright, M. (eds.) PETS 2013. LNCS, vol. 7981, pp. 19–39. Springer, Heidelberg (2013)

[KO05]   Køien, G.M., Oleshchuk, V.A.: Location privacy for cellular systems; analysis and solution. In: Danezis, G., Martin, D. (eds.) PET 2005. LNCS, vol. 3856, pp. 40–58. Springer, Heidelberg (2006)

[LSWW13] Lee, M.-F., Smart, N.P., Warinschi, B., Watson, G.: Anonymity guarantees of the UMTS/LTE authentication and connection protocol. Cryptology ePrint Archive, Report 2013/027 (2013). http://eprint.iacr.org/

[Mau11]   Maurer, U.: Constructive cryptography – a new paradigm for security definitions and proofs. In: Mödersheim, S., Palamidessi, C. (eds.) TOSCA 2011. LNCS, vol. 6993, pp. 33–56. Springer, Heidelberg (2012)

[MR11]   Maurer, U., Renner, R.: Abstract cryptography. In: ICS, pp. 1–21. Tsinghua University Press (2011)

[MT12]   Mol, P., Tessaro, S.: Secret-key authentication beyond the challenge-response paradigm: Definitional issues and new protocols. Manuscript, December 2012

[Oka92]   Okamoto, T.: Provably secure and practical identification schemes and corresponding signature schemes. In: Brickell, E.F. (ed.) CRYPTO 1992. LNCS, vol. 740, pp. 31–53. Springer, Heidelberg (1993)

[rGPP12] 3rd Generation Partnership Project. TS 33.102 - 3G security; Security architecture V11.5.0 (2012)

[SAJ07]   Sattarzadeh, B., Asadpour, M., Jalili, R.: Improved user identity confidentiality for UMTS mobile networks. In: ECUMN, pp. 401–409. IEEE Computer Society (2007)

[Sch89]   Schnorr, C.-P.: Efficient identification and signatures for smart cards. In: Brassard, G. (ed.) CRYPTO 1989. LNCS, vol. 435, pp. 239–252. Springer, Heidelberg (1990)

[TM12]   Tsay, J.-K., Mjølsnes, S.F.: A vulnerability in the UMTS and LTE authentication and key agreement protocols. In: Kotenko, I., Skormin, V. (eds.) MMM-ACNS 2012. LNCS, vol. 7531, pp. 65–76. Springer, Heidelberg (2012)

[Vau10]   Vaudenay, S.: Privacy models for RFID schemes. In: Ors Yalcin, S.B. (ed.) RFIDSec 2010. LNCS, vol. 6370, pp. 65–65. Springer, Heidelberg (2010)

[YWDW06] Yang, G., Wong, D.S., Deng, X., Wang, H.: Anonymous Signature Schemes. In: Yung, M., Dodis, Y., Kiayias, A., Malkin, T. (eds.) PKC 2006. LNCS, vol. 3958, pp. 347–363. Springer, Heidelberg (2006)

Cryptanalysis

On Key Recovery Attacks Against Existing Somewhat Homomorphic Encryption Schemes

Massimo Chenal$^{(\boxtimes)}$ and Qiang Tang

APSIA Group, SnT, University of Luxembourg, 6, rue Richard Coudenhove-Kalergi,
1359 Luxembourg, Luxembourg
{massimo.chenal,qiang.tang}@uni.lu

Abstract. In his seminal paper at STOC 2009, Gentry left it as a future
work to investigate (somewhat) homomorphic encryption schemes with
IND-CCA1 security. At SAC 2011, Loftus et al. showed an IND-CCA1
attack against the somewhat homomorphic encryption scheme presented
by Gentry and Halevi at Eurocrypt 2011. At ISPEC 2012, Zhang, Plantard and Susilo showed an IND-CCA1 attack against the somewhat
homomorphic encryption scheme developed by van Dijk et al. at Eurocrypt 2010.

In this paper, we continue this line of research and show that most
existing somewhat homomorphic encryption schemes are not IND-CCA1
secure. In fact, we show that these schemes suffer from key recovery
attacks (stronger than a typical IND-CCA1 attack), which allow an
adversary to recover the private keys through a number of decryption oracle queries. The schemes that we study in detail, include those by Brakerski and Vaikuntanathan at Crypto 2011 and FOCS 2011, and that by
Gentry, Sahai and Waters at Crypto 2013. We also develop a key recovery
attack that applies to the somewhat homomorphic encryption scheme by
van Dijk et al., and our attack is more efficient and conceptually simpler than the one developed by Zhang et al.. Our key recovery attacks
also apply to the scheme by Brakerski, Gentry and Vaikuntanathan at
ITCS 2012, and we also describe a key recovery attack for the scheme
developed by Brakerski at Crypto 2012.

Keywords: Somewhat homomorphic encryption · Key recovery attack ·
IND-CCA1 security

1 Introduction

In 1978, Rivest, Adleman and Dertouzos [24] introduced the concept of privacy
homomorphism and asked whether it is possible to perform arbitrary operations
on encrypted ciphertexts. 30 years later, Gentry [11] gave a positive answer by
proposing an ingenious approach to construct fully homomorphic encryption
(FHE) schemes. With this approach, we can start with a somewhat homomorphic
encryption (SHE) scheme that can perform only limited number of operations
on ciphertexts (i.e. it can evaluate only low-degree polynomials). Then, through

© Springer International Publishing Switzerland 2015
D.F. Aranha and A. Menezes (Eds.): LATINCRYPT 2014, LNCS 8895, pp. 239–258, 2015.
DOI: 10.1007/978-3-319-16295-9_13

the so-called bootstrapping step, we can turn this SHE scheme into an FHE scheme. Even though SHE schemes are less powerful than FHE schemes, they can already be used in many useful real-world applications, such as medical and financial applications [22]. Note that researchers have proposed the concept of leveled FHE schemes (e.g. [2,17]), which allow third parties to evaluate any circuits up to a certain depth. In the following discussion, we treat these schemes as SHE.

1.1 Related Work

After Gentry's work, many SHE and FHE schemes have been proposed. Based on the underlying hardness assumptions, these schemes can be categorized as follows.

(1) The first category starts with Gentry [10,11]. A number of variations, optimizations and implementations appear in [14,25]. The security of these schemes are based on hard problems on lattices.
(2) The second category starts with van Dijk et al. [27]. More variants, implementation and optimizations appear in [6–8]. The security of these schemes rely on the approximate greatest common divisor (AGCD) problem and some variants. It is worth mentioning that Ding and Tao [9] claim to have found an algorithm to solve the AGCD problem with some special parameters in polynomial time. However, the AGCD problem and its variants are still believed to be hard.
(3) The third category starts with Brakerski and Vaikuntanathan [3,4]. More variants, implementations and optimizations appear in [1,2,16,17,22]. The security of these schemes are based on the learning with errors (LWE) and on the ring-learning with errors (RLWE) problems.

See Fig. 1 for a graphical visualization of the main families.

Recently, Nuida [23] proposed a new framework for noise-free FHE, based on finite non-commutative groups. This is completely different from everything appeared in literature so far, since the ciphertext in all known schemes carry some noise. Nevertheless, a secure instantiation has yet to be found.

There exists another family of schemes, based on the NTRU encryption scheme [18]. In [20] it is shown how to obtain a homomorphic encryption scheme in a multi-user setting, introducing the notion of multi-key homomorphic encryption where it is possible to compute any function on plaintexts encrypted under multiple public keys. The multi-key FHE of [20] is based on the NTRU scheme [18] and on ideas introduced in [2]. We will not focus on NTRU-based multi-key homomorphic encryption schemes.

All known SHE and FHE schemes have been developed with the aim of being IND-CPA secure (resistant against a chosen-plaintext attack). In [11], Gentry left it as a future work to investigate SHE schemes with IND-CCA1 security (i.e. secure against a non-adaptive chosen-ciphertext attack). At this moment, we have the following results.

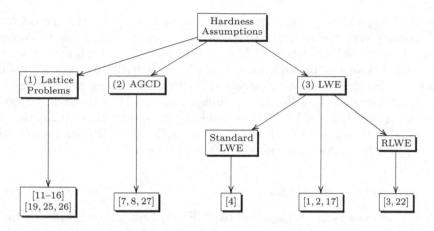

Fig. 1. Hardness assumptions and relevant papers

- No SHE and FHE scheme can be IND-CCA2 (secure against adaptive chosen-ciphertext attack). The reason is straightforward, based on the fact that the adversary is allowed to manipulate the challenged ciphertext and submit it to the decryption oracle in an IND-CCA2 attack.
- With Gentry's approach, the resulted FHE scheme cannot be IND-CCA1 secure. The reason is also straightforward, based on the fact that the private key is encrypted and the adversary is able to submit the ciphertext to the decryption oracle.
- Loftus et al. [19] showed that Gentry's SHE scheme [11] is not IND-CCA1 secure and presented an IND-CCA1 attack against the variation proposed in [14]. They also showed that the same attack applies to the other variant by Smart and Vercauteren [25]. In fact, the attacks are both key recovery attacks. Moreover, they modified the SHE in [25] and proved its IND-CCA1 security based on a new assumption. Zhang et al. [28] presented an IND-CCA1 attack against the SHE scheme in [27], which can recover the secret key with $O(\lambda^2)$ queries where λ is the security parameter.

In theory, IND-CPA security may be enough for us to construct cryptographic protocols, in particular if we assume semi-honest attackers. However, key recovery attacks will pose serious threat for practical usage of SHE and FHE. If a malicious attacker submits manipulated ciphertexts and observes the behavior (side channel leakage) of the decryptor, then it may be able to recover all plaintexts in the system. Therefore, it is very desirable to design SHE and FHE with IND-CCA1 security, or at least to avoid key recovery attacks.

1.2   Our Contributions

In this paper, we continue the line of work of [19,28] to present key recovery attacks for the SHE schemes [4] (which, following the literature, we will refer to as the BV11b SHE scheme), [3] (the BV11a SHE scheme), [17] (the GSW13 SHE scheme), and [1] (the Bra12 SHE scheme). Our attacks can also be applied to the SHE scheme in [2] (the BGV12 SHE scheme). We also develop a new key recovery attack against the SHE scheme in [27] (the vDGHV10 scheme), and our attack is more efficient and conceptually simpler than that from [28]. Our results essentially show that the SHE schemes underlying the FHE schemes in category (3) above are not IND-CCA1 secure. Combining the results from [19,28], we can conclude that most existing SHE schemes, except that from [19], suffer from key recovery attacks so that they are not IND-CCA1 secure.

1.3 Structure of the Paper

In Sect. 2, we recall some background on SHE and FHE schemes. Starting from Sect. 4, we are going to develop key recovery attacks against the aforementioned SHE schemes. In Sect. 7, we conclude the paper. In the full version of this paper, [5], we give the algorithmic description of our key recovery attacks, as well as the efficiency analysis.

2 Preliminaries

Let $\mathbb{N}$ be the set of natural numbers, $\mathbb{Z}$ the ring of integers, $\mathbb{Q}$ the field of rational numbers, and $\mathbb{F}_q$ a finite field with q elements, where q is a power of a prime p. In particular, we will consider often $\mathbb{F}_p = \mathbb{Z}/p\mathbb{Z} = \mathbb{Z}_p$. If $r \in \mathbb{Z}_q$, we indicate as r^{-1} its inverse in $\mathbb{Z}_q$, i.e. that value such that $r^{-1} \cdot r = 1 \bmod q$. For a ring R and a (two-sided) ideal I of R, we consider the quotient ring R/I. For given vectors $\mathbf{v} = (v_1, \ldots, v_n)$, $\mathbf{w} = (w_1, \ldots, w_n) \in \mathbb{Z}_q^n$, we let $\langle \mathbf{v}, \mathbf{w} \rangle = \sum_i v_i w_i$ the dot product of $\mathbf{v}, \mathbf{w}$. For a given rational number $x \in \mathbb{Q}$, we let $\lfloor x \rceil$, $\lfloor x \rfloor$ and $\lceil x \rceil$ be respectively the rounding function, the floor function and the ceiling function. For a given integer $n \in \mathbb{N}$, $\lfloor n + 1/2 \rceil = n + 1$. To indicate that an element a is chosen uniformly at random from a set A we use notation $a \xleftarrow{\$} A$. For a set A, we let its cardinality be $|A|$. We consider also the standard basis $\{\mathbf{e}_i\}_{i=1}^n$ of $\mathbb{R}^n$, where the coefficients of $\mathbf{e}_i$ are all 0 except for the i-th coefficient, which is 1.

Unless otherwise specified, λ will always denote the security parameter of the encryption schemes. In the asymmetric schemes we are going to discuss, the secret key will be denoted as sk, and the public key will be pk.

2.1 Homomorphic Encryption

The following definitions are adapted from [11]. We only assume bit-by-bit public-key encryption, i.e. we only consider encryption schemes that are homomorphic with respect to boolean circuits consisting of gates for addition and multiplication mod 2. Extensions to bigger plaintext spaces and symmetric-key setting are straightforward; we skip the details.

Definition 1 (Homomorphic Encryption). *A public key homomorphic encryption (HE) scheme is a set* $\mathcal{E} = (\mathsf{KeyGen}_{\mathcal{E}}, \mathsf{Encrypt}_{\mathcal{E}}, \mathsf{Decrypt}_{\mathcal{E}}, \mathsf{Evaluate}_{\mathcal{E}})$ *of four algorithms, all of which must run in polynomial time. When the context is clear, we will often omit the index* $\mathcal{E}$.

$\mathsf{KeyGen}(\lambda) = (\mathsf{sk}, \mathsf{pk})$

    – *input:* λ

    – *output:* sk; pk

$\mathsf{Encrypt}(\mathsf{pk}, m) = c$

    – *input:* pk *and plaintext* $m \in \mathbb{F}_2$

    – *output: ciphertext* c

$\mathsf{Decrypt}(\mathsf{sk}, c) = m'$

    – *input:* sk *and ciphertext* c

    – *output:* $m' \in \mathbb{F}_2$

$\mathsf{Evaluate}(\mathsf{pk}, C, (c_1, \dots, c_r)) = c_e$

    – *input:* pk, *a circuit* C, *ciphertexts* $c_1, \dots, c_r$, *with* $c_i =$ $\mathsf{Encrypt}(\mathsf{pk}, m_i)$

    – *output: ciphertext* c_e

Definition 2 (Correct Homomorphic Decryption). *The public key homomorphic encryption scheme* $\mathcal{E} = (\mathsf{KeyGen}, \mathsf{Encrypt}, \mathsf{Decrypt}, \mathsf{Evaluate})$ *is correct for a given t-input circuit C if, for any key-pair* (sk, pk) *output by* $\mathsf{KeyGen}(\lambda)$, *any t plaintext bits* $m_1, \dots, m_t$, *and any ciphertexts* $\bar{c} = (c_1, \dots, c_t)$ *with* $c_i \leftarrow \mathsf{Encrypt}_{\mathcal{E}}(\mathsf{pk}, m_i)$, *we have that* $\mathsf{Decrypt}(\mathsf{sk}, \mathsf{Evaluate}(\mathsf{pk}, C, \bar{c})) = C(m_1, \dots, m_t)$.

Definition 3 (Homomorphic Encryption). *The public key homomorphic encryption scheme* $\mathcal{E} = (\mathsf{KeyGen}, \mathsf{Encrypt}, \mathsf{Decrypt}, \mathsf{Evaluate})$ *is homomorphic for a class $\mathcal{C}$ of circuits if it is correct for all circuits $C \in \mathcal{C}$. We say that $\mathcal{E}$ is a fully homomorphic encryption (FHE) scheme if it is correct for all boolean circuits.*

Informally, a homomorphic encryption scheme that can perform only a limited number of operations is called a somewhat homomorphic encryption (SHE) scheme.

2.2   Security Definitions

The following definitions are taken from [19]. The security of a public-key encryption scheme in terms of indistinguishability is normally presented as a game between a challenger and an adversary $\mathcal{A} = (\mathcal{A}_1, \mathcal{A}_2)$. The scheme is considered secure if no adversary can win the game with significantly greater probability than an adversary who must guess randomly. The game runs in two stages:

– $(\mathsf{pk}, \mathsf{sk}) \leftarrow \mathsf{KeyGen}(1^\lambda)$
– $(m_0, m_1) \leftarrow \mathcal{A}_1^{(\cdot)}(\mathsf{pk})$ /* Stage 1 */
– $b \leftarrow \{0, 1\}$
– $c^* \leftarrow \mathsf{Encrypt}(m_b, \mathsf{pk})$
– $b' \leftarrow \mathcal{A}_2^{(\cdot)}(c^*)$ /* Stage 2 */

The adversary is said to win the game if $b = b'$, with the advantage of the adversary winning the game being defined by

$$\mathrm{Adv}_{\mathcal{A}, \mathcal{E}, \lambda}^{\mathrm{IND\text{-}atk}} = |\Pr(b = b') - 1/2|$$

A scheme is said to be IND-atk secure if no polynomial time adversary $\mathcal{A}$ can win the above game with non-negligible advantage in the security parameter λ. The precise security notion one obtains depends on the oracle access one gives the adversary in its different stages:

- If $\mathcal{A}$ has access to no oracles in either stage then atk=CPA (indistinguishability under chosen plaintext attack).
- If $\mathcal{A}$ has access to a decryption oracle in stage one then atk=CCA1 (indistinguishability under non-adaptive chosen ciphertext attack).
- If $\mathcal{A}$ has access to a decryption oracle in both stages then atk=CCA2, often now denoted simply CCA (indistinguishability under adaptive chosen ciphertext attack).
- If $\mathcal{A}$ has access to a ciphertext validity oracle in both stages, which on input of a ciphertext determines whether it would output $\perp$ or not on decryption, then atk=CVA.

According to the definition, in order to show that a scheme is not IND-CCA1 secure, we only need to show that an adversary can guess the bit b with a non-negligible advantage given access to the decryption oracle in Stage 1. Formally, in a *key recovery attack*, an adversary can output the private key given access to the decryption oracle in Stage 1. In comparison, a key recovery attack is stronger than a typical IND-CCA1 attack.

3   Key Recovery Attack Against the vDGHV10 Scheme

In [28], Zhang, Plantard and Susilo presented a key recovery attack against the SHE scheme from [27]. Given $O(\lambda^2)$ decryption oracle queries, an attacker can recover the private key. Let η be the bit-length of the secret key p, $O(\lambda^2) = 3(\eta + 3)$ in the best case.

We describe here a more efficient and conceptually simpler key recovery attack. Our attack is optimal in the sense that it recovers directly the secret key with at most η oracle queries. Note that the decryption oracle outputs one bit at a time.

We start by presenting the (asymmetric) SHE scheme as developed in [27]. The message space is $\mathcal{M} = \mathbb{Z}_2$. The scheme is parametrized by γ (bit-length of the integers in the public key), η (bit-length of the secret key), ρ (bit-length of the noise), and τ (the number of integers in the public key). We also consider a secondary noise parameter $\rho' = \rho + \omega(\log\lambda)$. For a specific ($\eta$-bit) odd positive integer p, consider the following distribution over γ-bit integers:

$$\mathcal{D}_{\gamma,\rho}(p) = \{\text{choose } q \xleftarrow{\$} \mathbb{Z} \cap [0, 2^\gamma/p), r \xleftarrow{\$} \mathbb{Z} \cap (-2^\rho, 2^\rho) : \text{output } x = pq + r\}$$

The algorithms of the vDGHV10 SHE scheme are defined as follows:

KeyGen(λ)

   – sk: odd η-bit integer
$$p \overset{\$}{\leftarrow} (2\mathbb{Z}+1) \cap [2^{\eta-1}, 2^{\eta}).$$

   – restart unless x_0 odd, $r_p(x_0)$ even
$$(r_p(x) = x - \lfloor x/p \rceil \cdot p \in (-p/2, p/2])$$

   – pk $= (x_0, x_1, \ldots, x_\tau)$.

   – sample $x_i \overset{\$}{\leftarrow} \mathcal{D}_{\gamma,\rho}(p)$ for $i = 0, \ldots, \tau$

   – relabel so that x_0 is the largest

Encrypt(pk, $m \in \mathcal{M}$)

   – choose a random subset $S \subseteq \{1, 2, \ldots, \tau\}$

   – choose a random integer r in $(-2^{\rho'}, 2^{\rho'})$

   – output $c = [m + 2r + 2\sum_{i \in S} x_i]_{x_0}$

Decrypt(sk, c)

   – output $m' = (c \bmod p) \bmod 2$

Since $\eta = \#\text{bits}(p)$, we immediately obtain odd lower and upper bounds l_p and u_p, respectively, for p:

$$l_p = 2^{\eta-1} + 1 \leq p \leq u_p = 2^{\eta} - 1$$

Notice explicitly that p can only assume the odd values $2^{\eta-1} + 1, 2^{\eta-1} + 3, \ldots,$ $2^{\eta} - 3, 2^{\eta} - 1$. In particular, between $2^{\eta-1}$ and 2^{η} there are $2^{\eta-2}$ candidate values for p. We can also argue that between l_p and u_p there are $(u_p - l_p)/2 = 2^{\eta-2} - 1$ even integers. Let $H_{(l_p, u_p)} = \{0, 1, \ldots, 2^{\eta-2} - 2\}$, these integers can be denoted as $l_p + 2h + 1$ for $h \in H_{(l_p, u_p)}$.

Now, the idea of the key recovery attack is as follows: consider the 'ciphertext' $c = l_p + 2h + 1$ for a given $h \in H_{(l_p, u_p)}$. Submit c to the decryption oracle $\mathcal{O}_D$; we will obtain a bit $b \leftarrow \mathcal{O}_D(c) = (c \bmod p) \bmod 2$. There are two cases to distinguish:

b = 0 'Decryption is correct' (since c is even); hence $p > c$, i.e. $p \geq l_p + 2h + 2$.
      Update $l_p \leftarrow l_p + 2h + 2$.

b = 1 'Decryption is not correct'; hence $p < c$, i.e. $p \leq l_p + 2h$.
      Update $u_p \leftarrow l_p + 2h$.

Next, we repeat the decryption query with the updated values for l_p, u_p and with another even 'ciphertext' $c \in [l_p + 1, u_p - 1]$, and we stop when $u_p = l_p$. In particular, for efficiency we always choose c as the even integer in the middle of the interval $[l_p + 1, u_p - 1]$. It is easy to see that this attack leads to a full recovery of the secret key p with at most $\log(2^{\eta-2} - 2) \approx \eta$ oracle queries.

4 Key Recovery Attack Against the BV11b Scheme

In this section, we describe a key recovery attack against the SHE scheme from [4].

4.1   The BV11b SHE Scheme

The message space is $\mathcal{M} = \mathbb{Z}_2$. Let f be a polynomial in λ, i.e. $f(\lambda) = \text{poly}(\lambda)$. Consider $n = f(\lambda) \in \mathbb{N}$ and let $\epsilon \in (0, 1) \cap \mathbb{R}$. Assume an odd integer $q \in \mathbb{N}$ such that $q \in [2^{n^\epsilon}, 2 \cdot 2^{n^\epsilon})$, and an integer $m \geq n\log q + 2\lambda$. Let χ be a noise distribution over $\mathbb{Z}_q$ (it produces small samples, all of magnitude not greater than n). Finally, let $L \in \mathbb{N}$ be an upper bound on the maximal multiplicative depth that the scheme can homomorphically evaluate, say $L \approx \epsilon\log n$.

KeyGen(λ)

- pick $\mathbf{s_0}, \ldots, \mathbf{s_L} \overset{\$}{\leftarrow} \mathbb{Z}_q^n$
- pick a matrix $A \overset{\$}{\leftarrow} \mathbb{Z}_q^{m \times n}$
- pick a vector $\mathbf{e} \leftarrow \chi^m$
- compute $\mathbf{b} = A\mathbf{s_0} + 2\mathbf{e}$
- sk $= \mathbf{s_L}$
- pk $= (A, \mathbf{b})$

Encrypt(pk, $\mu \in \mathcal{M}$)

- pick $\mathbf{r} \overset{\$}{\leftarrow} \{0, 1\}^m$
- set $\mathbf{v} = A^{\mathrm{T}}\mathbf{r} \in \mathbb{Z}_q^n$
- set $w = \mathbf{b}^{\mathrm{T}}\mathbf{r} + \mu \in \mathbb{Z}_q$
- ciphertext $c = ((\mathbf{v}, w), l)$.

Decrypt(sk, $c = ((\mathbf{v}, w), L)$)

$$\mu = (w - \langle \mathbf{v}, \mathbf{s_L} \rangle \mod q) \mod 2$$

Notice that the vectors $\mathbf{s_1}, \ldots, \mathbf{s_{L-1}}$ are used in order to compute the evaluation key, which we omit here. We remark that during the homomorphic evaluation, the scheme generates ciphertexts of the form $c = ((\mathbf{v}, w), l)$, where the tag l indicates the multiplicative level at which the ciphertext has been generated (fresh ciphertexts are tagged with $l = 0$). Note that it always holds that $l \leq L$ due to the bound on the multiplicative depth, and that the output of the homomorphic evaluation of the entire circuit is expected to have $l = L$. As described in [4], the SHE scheme is only required to decrypt ciphertexts that are output by the evaluation step (which we omit here), and those will always have level tag L. Therefore, we always expect a ciphertext of the form $c = ((\mathbf{v}, w), L)$ and decryption is correctly performed using the secret key $\mathbf{s_L}$.

Apparently, we cannot decrypt level l ciphertexts $c = ((\mathbf{v}, w), l)$, for $1 \leq l < L$, since we are only allowed to decrypt level L ciphertexts. However, we can compute $L - l$ fresh encryptions of 1, namely $c_1, \ldots, c_{L-l}$. Then, we compute $c^* = \text{Evaluate}(\text{pk}, MUL, c, c_1, \ldots, c_{L-l})$ based on the homomorphic property, where MUL is the multiplication circuit. The resulting ciphertext c^* will encrypt the same message as c does, and with a tag level L. In particular, we can decrypt fresh ciphertexts.

4.2   Our Key Recovery Attack

We are going to recover the secret key $\mathbf{s_L} \in \mathbb{Z}_q^n$ component-wise, and bit by bit. For ease of notation, we will write $\mathbf{s}$ instead of $\mathbf{s_L}$. More precisely, we write $\mathbf{s} = (s_1, \ldots, s_n) \in \mathbb{Z}_q^n$. For every $1 \leq j \leq n$, we have $s_j \in \mathbb{Z}_q$ and therefore s_j can be written with a maximum number N of bits, where $N = \lfloor \log_2(q-1) \rfloor + 1$. We are going to recover the i-th bit of s_j, for all $1 \leq i \leq N$ and for all $1 \leq j \leq n$.

Intuitively, our attack works as follows. We start by finding the first bit of s_j for every $1 \leq j \leq n$; then we will recover the second bit of s_j for every $1 \leq j \leq n$;

and we stop until we reach the N-th bit. In order to do so, we have to choose a 'ciphertext' c to be submitted to the decryption oracle. Instead of submitting $c = (\mathbf{v}, w)$ for honestly-generated $\mathbf{v} \in \mathbb{Z}_q^n$ and $w \in \mathbb{Z}_q$, we submit $c^* = (\mathbf{x}, y)$ for some specifically-picked $\mathbf{x} \in \mathbb{Z}_q^n$ and $y \in \mathbb{Z}_q$. We omit to write the level tag since we can always obtain a level tag L from any $l \leq L$.

For any $1 \leq j \leq n$, let $(s_j)_2 := a_{j,N} a_{j,N-1} \cdots a_{j,1}$ be the binary representation of s_j (bits ordered from most significant to least significant). We have $a_{j,i} \in \{0,1\}$, for all $1 \leq i \leq N$.

Recovering $a_{j,1}$

We have to choose $\mathbf{x} \in \mathbb{Z}_q^n$ and $y \in \mathbb{Z}_q$ in such a way that $y - \langle \mathbf{x}, \mathbf{s} \rangle \bmod q = s_j$. To do so, pick $y = 0$ and $\mathbf{x} = (0, \ldots, 0, -1, 0, \ldots, 0)$ (where -1 is in position j). Then, we have $0 - (-1)s_j \bmod q = s_j \bmod q = s_j$. As a result, by modding out with 2, this will return the last bit $a_{j,1}$ of s_j.

Recovering $a_{j,2}$

Now that we know the last bit $a_{j,1}$ of s_j, we want to obtain $s_j^{(1)} := (s_j - a_{j,1})/2 \in \mathbb{Z}_q$ whose bit decomposition is the same as the bit decomposition of s_j, but with the last bit removed from it. Then, modding out by 2, we will get the desired bit. This translates to the following condition: find $\mathbf{x} \in \mathbb{Z}_q^n$ and $y \in \mathbb{Z}_q$ such that $y - \langle \mathbf{x}, \mathbf{s} \rangle \bmod q = (s_j - a_{j,1})/2$. Let $\mathbf{x} = (0, \ldots, 0, x_j, 0, \ldots, 0)$ (with x_j in j-th position). We have to find y and x_j such that $2y - s_j(2x_j + 1) = -a_{j,1} \bmod q$. Clearly, the solution is given by $x_j = -2^{-1} \bmod q$ and $y = -2^{-1}a_{j,1} \bmod q$. By querying the decryption oracle with the 'ciphertext' $c^* := (\mathbf{x}, y)$, we obtain the second-to-last bit $a_{j,2}$ of s_j.

Recovering $a_{j,m}$, for $1 \leq m \leq N$

Based on the above two cases, we generalize the procedure. Suppose we have found all bits $a_{j,i}$, for $1 \leq i \leq m - 1$. In order to recover the bit $a_{j,m}$, we choose $\mathbf{x} := (0, \ldots, 0, x_j, 0, \ldots, 0) \in \mathbb{Z}_q^n$ and $y \in \mathbb{Z}_q$ as follows: $x_j = -(2^{m-1})^{-1} \bmod q$ and $y = -(2^{m-1})^{-1}(\sum_{i=1}^{m-1} 2^{i-1}a_{j,i})$.

5   Key Recovery Attack Against the BV11a Scheme

In this section, we describe a key recovery attack against the symmetric-key SHE scheme from [3]. The attack also applies to the asymmetric-key SHE scheme.

5.1   The BV11a SHE Scheme

Consider primes $q = \text{poly}(\lambda) \in \mathbb{N}$, $t = \text{poly}(\lambda) \in \mathbb{Z}_q^*$. Let $n = \text{poly}(\lambda) \in \mathbb{N}$ and consider a polynomial $f(x) \in \mathbb{Z}[x]$ with $\deg(f) = n + 1$. The message space is $\mathcal{M} = R_t = \mathbb{Z}[x]/(f(x))$. Namely, a message is encoded as a degree n polynomial with coefficients in $\mathbb{Z}_t$. Let χ be an error distribution over the ring $R_q := \mathbb{Z}_q[x]/(f(x))$ and let $D \in \mathbb{N}$, which is related to the maximal degree of homomorphism allowed (and to the maximal ciphertext length). Parameters n, f, q, χ are public.

Keygen(λ)
- sample $s \leftarrow \chi$
- $\mathbf{s} = (1, s, s^2, \dots, s^D) \in R_q^{D+1}$
- sk $= s$

Decrypt(sk, $\mathbf{c} = (c_0, \dots, c_D) \in R_q^{D+1}$)

$$\mu = (\langle \mathbf{c}, \mathbf{s} \rangle \bmod q) \bmod t$$

Encrypt(sk, $\mu \in \mathcal{M}$)
- sample $a \xleftarrow{\$} R_q$ and $e \leftarrow \chi$
- compute $(a, b := as + te) \in R_q^2$
- compute $c_0 := b + \mu \in R_q$,
  $c_1 := -a$
- output $\mathbf{c} = (c_0, c_1) \in R_q^2$

We remark that while the encryption algorithm only generates ciphertexts $\mathbf{c} \in R_q^2$, homomorphic operations (as described in the evaluation algorithm which we omit here) might add more elements to the ciphertext. Thus, the most generic form of a decryptable ciphertext in this scheme is $\mathbf{c} = (c_0, \dots, c_d) \in R_q^{d+1}$, for $d \leq D$. Notice that 'padding with zeros' does not affect the ciphertext. Namely, $(c_0, \dots, c_d) \in R_q^{d+1}$ and $(c_0, \dots, c_d, 0, \dots, 0) \in R_q^{D+1}$ encrypt the same message $\mu \in R_t$.

5.2   Our Key Recovery Attack

We can write $s = s_0 + s_1 x + \cdots + s_n x^n \in \mathbb{Z}_q[x]/(f(x))$ with coefficients $s_j \in \mathbb{Z}_q$, $\forall 0 \leq j \leq n$. We will recover each coefficient s_j separately. Now, each s_j has at most $N := \lfloor \log_2(q-1) \rfloor + 1$ bits; therefore #bits(s) $\leq (n+1) \times N = (n+1) \times (\lfloor \log_2(q-1) \rfloor + 1)$ and each query to the oracle decryption will reveal a polynomial $\mu(x) = \mu_0 + \mu_1 x + \cdots + \mu_n x^n \in \mathbb{Z}_t[x]/(f(x))$; we have #bits($\mu$) $\leq (n+1) \times (\lfloor \log_2(t-1) \rfloor + 1)$. Therefore, the minimum number of oracle queries needed is given by

$$\left\lceil \frac{\#(\mathrm{bits}(s))}{\#(\mathrm{bits~revealed~by~an~oracle~query})} \right\rceil = \left\lceil \frac{\lfloor \log_2(q-1) \rfloor + 1}{\lfloor \log_2(t-1) \rfloor + 1} \right\rceil$$

We are going to query the decryption oracle with 'ciphertexts' of the form $\mathbf{c}_i^* := (h_i, y_i, 0, \dots, 0) \in R_q^{D+1}$ for some $h_i, y_i \in R_q$. We will describe in detail our attack in the case $t = 2$. An easy generalization for $t \geq 2$ is discussed later.

An easy case: $t = 2$.

We expect to query the decryption oracle at least $\lfloor \log_2(q-1) \rfloor + 1$ times and recover s_j, for all $0 \leq j \leq n$, bit by bit. Let $N = \#\mathrm{bits}(s_j) = \lfloor \log_2(q-1) \rfloor + 1$, $\forall 0 \leq j \leq n$; and let $(s_j)_2 = a_{j,N} a_{j,N-1} \cdots a_{j,1}$ be the binary representation of s_j, $\forall 0 \leq j \leq n$ (i.e., $a_{j,i} \in \{0, 1\}, \forall 1 \leq i \leq N$ and bits ordered most significant to least significant). For ease of notation, we write $\mathbf{c}^* = (h, y)$ instead of $\mathbf{c}^* = (h, y, 0, \dots, 0)$.

Recovering $a_{j,1}$, for all $0 \leq j \leq n$

For a submitted 'ciphertext' $\mathbf{c}^* = (h, y)$, decryption is given by $\langle \mathbf{c}^*, \mathbf{s} \rangle$ mod

$2 = h + ys \bmod 2$. We choose $h = \sum_{j=0}^{n} 0x^j = 0 \in R_q$ and $y = 1 + \sum_{j=1}^{n} 0x^j = 1 \in R_q$. The decryption oracle outputs

$$s \bmod 2 = (s_0 \bmod 2) + (s_1 \bmod 2)x + \cdots + (s_n \bmod 2)x^n$$
$$= a_{0,1} + a_{1,1}x + \cdots + a_{n,1}x^n$$

Therefore, we obtain the last bits $a_{j,1}$ for all $1 \le j \le n$, which are n bits of s.

Recovering $a_{j,2}$, $\forall 0 \le j \le n$

With $a_{j,1}$ for all $0 \le j \le n$, we are going to recover $a_{j,2}$, $\forall 0 \le j \le n$, as follows. We want to obtain

$$s^{(1)} := \frac{s - (a_{0,1} + a_{1,1}x + \cdots + a_{n,1}x^n)}{2}$$

$$= s_0^{(1)} + s_1^{(1)}x + \cdots + s_n^{(1)}x^n \in \frac{\mathbb{Z}_q[x]}{(f(x))}$$

for which the bit decomposition of the coefficients $s_j^{(1)}$ is the same as the bit decomposition of s_j, but with the last bit removed from it, for all $0 \le j \le n$. Then, by modding out with 2, we will get the desired bits. This translates to the following condition: find $\mathbf{c}^* = (h, y) = (h, y, 0, \dots, 0) \in R_q^{D+1}$ such that $\langle \mathbf{c}^*, \mathbf{s} \rangle = s^{(1)} := \frac{s - (a_{0,1} + a_{1,1}x + \cdots + a_{n,1}x^n)}{2}$, from which we obtain $2h + s(2y - 1) = -(a_{0,1} + a_{1,1}x + \cdots + a_{n,1}x^n)$. A solution is given by $y = 2^{-1} \in R_q$ and $h = -2^{-1}(a_{0,1} + a_{1,1}x + \cdots + a_{n,1}x^n) \in R_q$. Then, by modding out with 2 the 'decrypted ciphertext' $\mu = \langle \mathbf{c}^*, \mathbf{s} \rangle$, we recover the second-to-last bits $a_{j,2}$, for all $0 \le j \le n$.

Recovering $a_{j,m}$, **for** $1 \le m \le N$, $0 \le j \le n$

Suppose we have found all bits $a_{j,i}$, $\forall 1 \le i \le m - 1$ and $\forall 0 \le j \le n$. We want to recover $a_{j,m}$, $\forall 0 \le j \le n$. By a recursive argument, we find that we have to submit a 'ciphertext' $\mathbf{c}^* = (h, y)$ such that $y = (2^{m-1})^{-1} \in R_q$ and $h = -(2^{m-1})^{-1} \left(\sum_{j=0}^{n} d_j x^j \right)$ with $d_j = \sum_{i=1}^{m-1} 2^{i-1} a_{j,i}$.

This concludes the attack for the case $t = 2$. Efficiency-wise, the total number of oracle queries is $N = \lfloor \log_2(q - 1) \rfloor + 1$, which is optimal.

The general case: $t \ge 2$.

We consider now the general case in which $t \ge 2$ is a prime number in $\mathbb{Z}_q^*$. We want to find $s = s_0 + s_1 x + \cdots + s_n x^n \in \mathbb{Z}_q[x]/(f(x))$ and expect to query the decryption oracle $\left\lceil \frac{\lfloor \log_2(q-1) \rfloor + 1}{\lfloor \log_2(t-1) \rfloor + 1} \right\rceil$ times. With each query to the decryption oracle, we are going to recover $M = \lfloor \log_2(t - 1) \rfloor + 1$ bits of s_j, $\forall 0 \le j \le n$. The idea is that we are going to recover s_j, for all $0 \le j \le n$. In its representation in base t, s_j can be represented with N figures $a_{j,i} \in \{0, 1, \dots, t - 1\}$: $(s_j)_t = a_{j,N} a_{j,N-1} \cdots a_{j,1}$ where $N = \lfloor \log_t(q - 1) \rfloor + 1$; each $a_{j,i}$ is bounded by $t - 1$, which explains the value $M = \lfloor \log_2(t - 1) \rfloor + 1$.

Recovering $a_{j,1}$, $\forall 0 \leq j \leq n$

For a submitted 'ciphertext' $\mathbf{c}^* = (h, y) = (h, y, 0, \ldots, 0) \in R_q^{D+1}$, decryption works as follows: $\langle \mathbf{c}^*, \mathbf{s} \rangle \bmod t = x + ys \bmod t$. We choose $h = 0 \in R_q$ and $y = 1 \in R_q$. Then, the decryption oracle outputs

$$s \bmod t = (s_0 \bmod t) + (s_1 \bmod t)x + \cdots + (s_n \bmod t)x^n$$
$$= a_{0,1} + a_{1,1}x + \cdots + a_{n,1}x^n$$

as we wanted.

Recovering $a_{j,m}$, $\forall 1 \leq m \leq N$, $\forall 0 \leq j \leq n$

Suppose we know $a_{j,i}$, $\forall 1 \leq i \leq m-1$, $\forall 0 \leq j \leq n$. We want to recover $a_{j,m}$, for all $0 \leq j \leq n$. To do so, we submit to the decryption oracle a 'ciphertext' $\mathbf{c}^* = (h, y)$ such that $y = (t^{m-1})^{-1} \in R_q$, $h = -(t^{m-1})^{-1} \left(\sum_{j=0}^{n} d_j x^j \right)$, $d_j = \sum_{i=1}^{m-1} t^{i-1} a_{j,i}$. It is straightforward to verify that it works and we skip the details here.

5.3   Attacks Against the BGV12 and Bra12 SHE Schemes

The SHE scheme from [2] is closely related to the SHE schemes from [3,4]. This implies that the attacks from Sects. 4.2 and 5.2 can be directly applied against the SHE scheme from [2]. The details appear in [5]. The SHE scheme from [1] is conceptually different, however, a key recovery attack can also be developed; the details appear in [5].

6   Key Recovery Attack Against the GSW13 SHE Scheme

In this section, we describe a key recovery attack against the SHE scheme from [17]. We first give some useful preliminary definitions. Let $q, k \in \mathbb{N}$. Let l be the bit-length of q, i.e. $l = \lfloor \log_2 q \rfloor + 1$, and let $N = k \cdot l$. Consider a vector $\mathbf{a} := (a_1, \ldots, a_k) \in \mathbb{Z}_q^k$, and let $(a_i)_2 := a_{i,0} a_{i,1} \ldots a_{i,l-1}$ be the binary decomposition of a_i (bit ordered least to most significant), for every $i = 1, \ldots, k$. We define

$$\mathsf{BitDecomp}(\mathbf{a}) := (a_{1,0}, \ldots, a_{1,l-1}, \ldots, a_{k,0}, \ldots, a_{k,l-1}) \in \mathbb{Z}_q^N$$

For a given $\mathbf{a}' := (a_{1,0}, \ldots, a_{1,l-1}, \ldots, a_{k,0}, \ldots, a_{k,l-1}) \in \mathbb{Z}_q^N$, let

$$\mathsf{BitDecomp}^{-1}(\mathbf{a}') := (\sum_{j=0}^{l-1} 2^j \cdot a_{1,j}, \ldots, \sum_{j=0}^{l-1} 2^j \cdot a_{k,j}) \in \mathbb{Z}_q^k$$

We notice explicitly that $\mathbf{a}'$ does not necessarily lie in $\{0, 1\}^N$, but when it does then $\mathsf{BitDecomp}^{-1}$ is the inverse of $\mathsf{BitDecomp}$. For $\mathbf{a}' \in \mathbb{Z}_q^N$, we define

$$\mathsf{Flatten}(\mathbf{a}') := \mathsf{BitDecomp}(\mathsf{BitDecomp}^{-1}(\mathbf{a}')) \in \mathbb{Z}_q^N$$

When A is a matrix, let $\mathsf{BitDecomp}(A), \mathsf{BitDecomp}^{-1}(A), \mathsf{Flatten}(A)$ be the matrix formed by applying the operation to each row of A separately. Finally, for $\mathbf{b} := (b_1, \ldots, b_k) \in \mathbb{Z}_q$ let

$$\mathsf{PowersOf2}(\mathbf{b}) := (b_1, 2b_1, \ldots, 2^{l-1}b_1, \ldots, b_k, 2b_k, \ldots, 2^{l-1}b_k) \in \mathbb{Z}_q^N$$

It is easy to see that, for $\mathbf{a}, \mathbf{b} \in \mathbb{Z}_q^k$ and for $\mathbf{a}' \in \mathbb{Z}_q^N$,

- $\langle \mathsf{BitDecomp}(\mathbf{a}), \mathsf{Powersof2}(\mathbf{b}) \rangle = \langle \mathbf{a}, \mathbf{b} \rangle$
- $\langle \mathbf{a}', \mathsf{Powersof2}(\mathbf{b}) \rangle = \langle \mathsf{BitDecomp}^{-1}(\mathbf{a}'), \mathbf{b} \rangle = \langle \mathsf{Flatten}(\mathbf{a}'), \mathsf{Powersof2}(\mathbf{b}) \rangle$

6.1 The GSW13 SHE Scheme

The message space is $\mathcal{M} = \mathbb{Z}_q$ for a given modulus q with $\# \mathsf{bits}(q) = \kappa = \kappa(\lambda, L)$. Let $n = n(\lambda)$ be the lattice dimension and let $\chi = \chi(\lambda)$ be the error distribution over $\mathbb{Z}_q$ (chosen appropriately for LWE: it must achieve at least 2^λ security against known attacks). Choose $m = m(\lambda) = O(n \log q)$. So the parameters used in all algorithms are n, q, χ, m. We have that $l = \lfloor \log q \rfloor + 1$ is the number of bits of q, and we let $N = (n+1) \cdot l$.

Keygen(λ):
- $\mathbf{t} := (t_1, \ldots, t_n) \leftarrow \mathbb{Z}_q^n$
- $\mathsf{sk} := \mathbf{s} \leftarrow (1, -t_1, \ldots, -t_n) \subset \mathbb{Z}_q^{n+1}$
- $\mathbf{v} = \mathsf{Powersof2}(\mathbf{s}) \in \mathbb{Z}_q^N$; see[1]
- $B \xleftarrow{\$} \mathbb{Z}_q^{m \times n}$
- $\mathbf{e} \leftarrow \chi, \mathbf{e} \in \mathbb{Z}_q^m$
- $\mathbf{b} := B \cdot \mathbf{t} + \mathbf{e} =: (b_1, \ldots, b_m) \in \mathbb{Z}_q^m$.
- set A to be the $(n+1)$-column matrix consisting of $\mathbf{b}$ followed by the n columns of B

$$A = (\mathbf{b} \mid B) \in \mathbb{Z}_q^{m \times (n+1)}$$

- $\mathsf{pk} := A$.

We remark that $A \cdot \mathbf{s} = \mathbf{e}$.

Encrypt($\mathsf{pk}, \mu \in \mathcal{M}$):
- sample a matrix

$$R \xleftarrow{\$} \{0,1\}^{N \times m}$$

- output the ciphertext

$$C = \mathsf{Flatten}(\mu \cdot I_N + \\ + \mathsf{BitDecomp}(R \cdot A)) \in \mathbb{Z}_q^{N \times N}$$

Decrypt(sk, C):
- observe that the first l coefficients of $\mathbf{v}$ are $1, 2, \ldots, 2^{l-2}$
- among these coefficients, let $v_i = 2^i$ be in $(q/4, q/2]$
- let C_i be the i-th row of C
- compute $x_i := \langle C_i, \mathbf{v} \rangle$
- output $\mu' := \lfloor x_i / v_i \rceil$

The Decrypt algorithm can recover the message μ when it is in a 'small space' ($q = 2$, i.e. $\mathcal{M} = \mathbb{Z}_2$). For an algorithm that can recover any $\mu \in \mathbb{Z}_q$, we refer to the MPDec algorithm as described (as a special case) in [17] and in [21]. If the ciphertext is generated correctly, it is not difficult to show that $C \cdot \mathbf{v} = \mu \cdot \mathbf{v} + R \cdot A \cdot \mathbf{s} = \mu \cdot \mathbf{v} + R \cdot \mathbf{e} \in \mathbb{Z}_q^N$.

---

[1] $\mathbf{v} = \mathsf{Powersof2}(\mathbf{s}) = (s_1, 2s_1, \ldots, 2^{l-1}s_1, s_2, \ldots, 2^{l-1}s_2, \ldots, s_{n+1}, 2s_{n+1}, \ldots, 2^{l-1}s_{n+1})$
$= (1, 2, \ldots, 2^{l-1}, -t_1, -2t_1, \ldots, -2^{l-1}t_1, \ldots, -t_n, -2t_n, \ldots, -2^{l-1}t_n) \in \mathbb{Z}_q^{(n+1)l}$.

Now, the Decrypt algorithm uses only the i-th coefficient of the vector $C \cdot \mathbf{v} \in \mathbb{Z}_q^N$, i.e. $\langle C_i, \mathbf{v} \rangle = \mu \cdot v_i + \langle R_i, \mathbf{e} \rangle \in \mathbb{Z}_q$. Moreover, in the Decrypt step, i has to be such that $v_i := 2^i \in (q/4, q/2]$, with $i \in [1, 2, \ldots, 2^{l-1}]$. Now remember that $l = \lfloor \log q \rfloor + 1$ equals the number of bits of q. Hence we have

$$2^{l-3} \leq \frac{q}{4} < 2^{l-2} \leq \frac{q}{2} < 2^{l-1} \leq q < 2^l$$

Therefore the only possible value for $2^i \in (q/4, q/2]$ is 2^{l-2}. For this reason, Decrypt can be simply rewritten as

Decrypt(sk, C):
  - let C_{l-2} be the $(l-2)$-th row of C
  - compute $x_{l-2} := \langle C_{l-2}, \mathbf{v} \rangle$
  - output $\mu' := \lfloor x_{l-2}/2^{l-2} \rceil$

One could think of outputting as ciphertext only the $(l-2)$-th row C_{l-2} of the matrix C; this is actually not possible since the full matrix is still needed in order to perform the homomorphic operations (in particular, the multiplication of two ciphertexts). We will not discuss them here; see [17].

6.2   Our Key Recovery Attack

We are going to recover bit by bit each coefficient t_i of the secret vector $\mathbf{t} := (t_1, \ldots, t_n) \in \mathbb{Z}_q^n$. For every $1 \leq i \leq n$, let BitDecomp(t_i) $:= (t_{i,0}, t_{i,1}, \ldots, t_{i,l-1}) \in \mathbb{Z}_q^l$ bits ordered from least to most significant. We explicitly remark that $t_i = \sum_{j=0}^{l-1} 2^j t_{i,j}$. We will proceed as follows: start with $i = 1$ and recover, in this order, the bits from most to least significant. Then continue with $i = 2$, and so on until $i = n$. Let $x \in \mathbb{Z}_q$. Since #bits(q) $= l$, we have $x \leq q - 1 \leq 2^l - 2$. Moreover, we have #bits(x) $\leq \lfloor \log_2(q-1) \rfloor + 1 := l^*$. We have $l^* = l$ if q is not a power of 2, i.e. if $q \neq 2^h$, for any $h \in \{1, 2, \ldots, l-1\}$. Otherwise, $l^* = l - 1$. We will not distinguish between these two cases: just remark that if $l^* = l - 1$, then $t_{i,l-1} = 0$ for all $i \in \{1, 2, \ldots, n\}$.

Recovering BitDecomp(t_1)
We start by recovering BitDecomp(t_1). The trickiest part is to recover the most significant bit. We start by recovering $t_{1,l-1}, t_{1,l-2}, t_{1,l-3}$. We have to choose, and submit to the decryption oracle, a matrix $C \in \mathbb{Z}_q^{N \times N}$. Then the oracle will compute $x = \langle C_{l-2}, \mathbf{v} \rangle$ and will output the rounded value $\mu = \lfloor x/2^{l-2} \rceil$. Our attack works also, with a trivial modification, in the case we define the rounding function such that $\lfloor n + 1/2 \rceil := n$, for every $n \in \mathbb{N}$. Our strategy is to submit a matrix C whose entries are all 0 except for the $(l-2)$-th row C_{l-2}. Let $\mathbf{y} = (y_1, \ldots, y_N) \in \mathbb{Z}_q^N$ be the vector representing C_{l-2}.

We select $\mathbf{y} = (0, \ldots, 0, -1, 0, \ldots, 0) \in \mathbb{Z}_q^N$ where -1 is in $l+1$-th position, i.e.

$$y_i = \begin{cases} -1 & \text{if } i = l+1 \\ 0 & \text{otherwise} \end{cases}$$

Through the decryption oracle, we have $x = \langle \mathbf{y}, \mathbf{v} \rangle = -v_{l+1} = t_1 \in \mathbb{Z}_q$ and $\mu = \lfloor t_1/2^{l-2} \rfloor$. There are two cases.

1. $\mu = 0$. In this case, we have $0 \leq \frac{t_1}{2^{l-2}} < \frac{1}{2}$ i.e. $t_1 < 2^{l-3} = \sum_{j=0}^{l-4} 2^j + 1$. Then it must be $\boxed{t_{1,l-1} = t_{1,l-2} = t_{1,l-3} = 0}$.

2. $1 \leq \mu \leq 4$. In particular, $2^{l-3} \leq t_1 \leq 2^l - 2$. Then we have

$$(t_{1,l-1}, t_{1,l-2}, t_{1,l-3}) \in \{0,1\}^3 \backslash \{(0,0,0)\} \tag{1}$$

Next, query the decryption oracle with $\mathbf{y} = (0, \ldots, 0, -1, 0, 0, -1, 0, \ldots, 0) \in \mathbb{Z}_q^N$ with -1 in $(l-2)$-th and $(l+1)$-th positions:

$$y_i = \begin{cases} -1 & \text{if } i = l-2 \text{ or } i = l+1 \\ 0 & \text{otherwise} \end{cases}$$

Through the decryption oracle, we have $x = \langle \mathbf{y}, \mathbf{v} \rangle = t_1 - 2^{l-3} \geq 0$ and $\mu = \lfloor \frac{t_1 - 2^{l-3}}{2^{l-2}} \rfloor$. There are two cases:

2.1. $\mu = 0$. In this case, we have $0 \leq \frac{t_1 - 2^{l-3}}{2^{l-2}} < \frac{1}{2}$ i.e. $2^{l-3} \leq t_1 < 2^{l-2} = \sum_{j=0}^{l-3} 2^j + 1$. Then it must be $\boxed{t_{1,l-1} = t_{1,l-2} = 0}$. Condition (1) implies that $\boxed{t_{1,l-3} = 1}$.

2.2. $1 \leq \mu \leq 3$. In particular, $2^{l-2} \leq t_1 \leq 2^l - 2$. Then we have

$$(t_{1,l-1}, t_{1,l-2}) \in \{0,1\}^2 \backslash \{(0,0)\} \tag{2}$$

Next, query the decryption oracle with $\mathbf{y} = (0, \ldots, 0, -1, 0, -1, 0, \ldots, 0) \in \mathbb{Z}_q^N$, with -1 in $(l-1)$-th and $(l+1)$-th positions:

$$y_i = \begin{cases} -1 & \text{if } i = l-1 \text{ or } i = l+1 \\ 0 & \text{otherwise} \end{cases}$$

Through the decryption oracle, we have $x = \langle \mathbf{y}, \mathbf{v} \rangle = t_1 - 2^{l-2} \geq 0$ and $\mu = \lfloor \frac{t_1 - 2^{l-2}}{2^{l-2}} \rfloor$. There are two cases:

2.2.1. $\mu = 0$. In this case, we have $0 \leq \frac{t_1 - 2^{l-2}}{2^{l-2}} < \frac{1}{2}$ and $2^{l-2} \leq t_1 < 2^{l-2} + 2^{l-3} < 2^{l-1}$. This means that $\boxed{t_{1,l-1} = 0}$. Therefore, condition (2) implies that $\boxed{t_{1,l-2} = 1}$. Moreover, since we have $0 \leq t_1 - 2^{l-2} < 2^{l-3}$, we have that $\boxed{t_{1,l-3} = 0}$.

2.2.2. $1 \leq \mu \leq 2$. In particular, $2^{l-3} + 2^{l-2} \leq t_1$.
Next, choose $\mathbf{y} = (0, \ldots, 0, -1, -1, 0, -1, 0, \ldots, 0) \in \mathbb{Z}_q^N$, with -1 in $(l-2)$-th, $(l-1)$-th and $(l+1)$-th positions:

$$y_i = \begin{cases} -1 & \text{if } i = l-2, i = l-1 \text{ or } i = l+1 \\ 0 & \text{otherwise} \end{cases}$$

Query the decryption oracle with $\mathbf{y}$: we have $x = \langle \mathbf{y}, \mathbf{v} \rangle = t_1 - (2^{l-3} + 2^{l-2}) \geq 0$ and $\mu = \lfloor \frac{t_1 - (2^{l-3} + 2^{l-2})}{2^{l-2}} \rfloor$. There are two cases:

2.2.2.1. $\mu = 0$. In this case, we have $0 \le \frac{t_1 - 2^{l-3} - 2^{l-2}}{2^{l-2}} < \frac{1}{2}$,   i.e. $2^{l-3} +$ $2^{l-2} \le t_1 < 2^{l-1}$. This implies $\boxed{t_{1,l-1} = 0}$. Therefore, condition (2) gives $\boxed{t_{1,l-2} = 1}$. Moreover, we have $2^{l-3} \le t_1 - 2^{l-2} < 2^{l-2}$; hence $\boxed{t_{1,l-3} = 1}$.

2.2.2.2. $\mu = 1$. We have $2^{l-1} \le t_1 \le 2^l - 2$. This implies $\boxed{t_{1,l-1} = 1}$. We now have to recover $t_{1,l-2}, t_{1,l-3}$.
Query the decryption oracle with $\mathbf{y} = (0, \ldots, 0, -1, -1, 0, \ldots, 0) \in \mathbb{Z}_q^N$, with -1 in l-th and $(l+1)$-th positions:

$$y_i = \begin{cases} -1 & \text{if } i = l \text{ or } i = l+1 \\ 0 & \text{otherwise} \end{cases}$$

Through the decryption oracle, we have $x = \langle \mathbf{y}, \mathbf{v} \rangle = t_1 - 2^{l-1} \ge 0$ and $\mu = \left\lfloor \frac{t_1 - 2^{l-1}}{2^{l-2}} \right\rfloor$. There are two cases:

2.2.2.2.1. $\mu = 0$. In this case, we have $0 \le \frac{t_1 - 2^{l-1}}{2^{l-2}} < \frac{1}{2}$, i.e. $0 \le t_1 - 2^{l-1} < 2^{l-3} = \sum_{j=0}^{l-4} 2^j + 1$. This implies $\boxed{t_{1,l-2} = t_{1,l-3} = 0}$.

2.2.2.2.2. $1 \le \mu \le 3$. In particular, $2^{l-3} \le t_1 - 2^{l-1}$. Then we have

$$(t_{1,l-2}, t_{1,l-3}) \in \{0,1\}^2 \backslash \{(0,0)\} \tag{3}$$

Choose $\mathbf{y} = (0, \ldots, 0, -1, 0, -1, -1, 0, \ldots, 0) \in \mathbb{Z}_q^N$, with -1 in $(l-2)$-th, l-th and $(l+1)$-th positions:

$$y_i = \begin{cases} -1 & \text{if } i = l-2, i = l \text{ or } i = l+1 \\ 0 & \text{otherwise} \end{cases}$$

Query the decryption oracle with $\mathbf{y}$: we have $x = \langle \mathbf{y}, \mathbf{v} \rangle = t_1 - (2^{l-1} + 2^{l-3}) \ge 0$ and $\mu = \left\lfloor \frac{t_1 - 2^{l-1} - 2^{l-3}}{2^{l-2}} \right\rfloor$. There are two cases:

2.2.2.2.2.1. $\mu = 0$. In this case, we have $0 \le \frac{t_1 - 2^{l-1} - 2^{l-3}}{2^{l-2}} < \frac{1}{2}$, i.e. $2^{l-3} \le t_1 - 2^{l-1} < 2^{l-2}$. This means that $\boxed{t_{1,l-2} = 0}$. Condition (3) then implies $\boxed{t_{1,l-3} = 1}$.

2.2.2.2.2.2. $1 \le \mu \le 2$. In particular, $2^{l-2} \le t_1 - 2^{l-1} \le 2^l - 2 - 2^{l-1} = 2^{l-1} - 2$. Then, we have $\boxed{t_{1,l-2} = 1}$. We still have to find $t_{1,l-3}$. Next, query the decryption oracle with $\mathbf{y} = (0, \ldots, 0, -1, -1, -1, 0, \ldots, 0) \in \mathbb{Z}_q^N$, where -1 is in $(l-1)$-th, l-th and $(l+1)$-th positions:

$$y_i = \begin{cases} -1 & \text{if } i = l-1, i = l \text{ or } i = l+1 \\ 0 & \text{otherwise} \end{cases}$$

Through the decryption oracle, we have $x = \langle \mathbf{y}, \mathbf{v} \rangle = t_1 - (2^{l-1} + 2^{l-2}) \ge 0$ and $\mu = \left\lfloor \frac{t_1 - 2^{l-1} - 2^{l-2}}{2^{l-2}} \right\rfloor$. There are two cases:

2.2.2.2.2.2.1. $\mu = 0$. In this case,

$$0 \le \frac{t_1 - 2^{l-1} - 2^{l-2}}{2^{l-2}} < \frac{1}{2}, \text{ i.e. } 0 \le t_1 - 2^{l-1} - 2^{l-2} < 2^{l-3}$$

This implies that $\boxed{t_{1,l-3} = 0}$.

2.2.2.2.2.2.2. $\mu = 1$. Then $2^{l-3} \le t_1 - 2^{l-1} - 2^{l-2}$. This implies that $\boxed{t_{1,l-3} = 1}$.

At this point, we know the first three significant bits $t_{1,l-1}, t_{1,l-2}, t_{1,l-3}$ of t_1. Notice that we have recovered the first three most significant bits with at most 7 oracle queries. Next, we are going to recover $t_{1,l-4}$. Query the decryption oracle with

$$\mathbf{y} = (0, \ldots, 0, -t_{1,l-3}, -t_{1,l-2}, -t_{1,l-1}, -1, 0, \ldots, 0) \in \mathbb{Z}_q^N$$

where $-t_{1,i}$ is in $(i+1)$-th position. Then

$$x = \langle \mathbf{y}, \mathbf{v} \rangle = t_1 - (t_{1,l-1}2^{l-1} + t_{1,l-2}2^{l-2} + t_{1,l-3}2^{l-3})$$

Now, we have $0 \le x < 2^{l-3}$. Therefore, $\mu = \lfloor x/2^{l-2} \rceil = 0$, and so not useful at all to learn $t_{1,l-4}$. The idea is to 'shift' the bits 'to the left', i.e. towards the most significant. So, let us instead choose

$$\mathbf{y} = 2 \cdot (0, \ldots, 0, -t_{1,l-3}, -t_{1,l-2}, -t_{1,l-1}, -1, 0, \ldots, 0) \in \mathbb{Z}_q^N$$

So now $x = \langle \mathbf{y}, \mathbf{v} \rangle$ is such that $0 \le x < 2^{l-2}$. After submitting $\mathbf{y}$ to the decryption oracle, it will compute and output $\mu = \lfloor x/2^{l-2} \rceil$. Then $\boxed{t_{1,l-4} = \mu}$.

Now we can generalize and recover $t_{1,k}$, for all $k = l-4, l-5, \ldots, 1, 0$. This will complete the recovery of t_1. Suppose that, for a given k, we recovered already $t_{1,m}, \forall m \in [k+1, \ldots, l-1]$. We then recover $t_{1,k}$ by recurrence. Choose

$$\mathbf{y} = 2^{l-k-3}(0, \ldots, 0, -t_{1,k+1}, -t_{1,k+2}, \ldots, -t_{1,l-1}, -1, 0, \ldots, 0) \in \mathbb{Z}_q^N$$

with $-t_{1,i}$ in $(i+1)$-th position; i.e.

$$y_i = \begin{cases} -2^{l-k-3}t_{1,i-1} & \text{for } i \in [k+2, \ldots, l] \\ -2^{l-k-3} & \text{for } i = l+1 \\ 0 & \text{otherwise} \end{cases}$$

Then we have $x = \langle \mathbf{y}, \mathbf{v} \rangle = 2^{l-k-3}\left(t_1 - \sum_{j=k+1}^{l-1} t_{1,j}2^j\right)$ with $0 \le x < 2^{l-2}$. Then, $\boxed{t_{1,k} = \mu}$.

We recover completely t_1 after at most $7 + (l-3) = l+4$ oracle queries.

Recovering BitDecomp(t_r), **for every** $r \in [1, 2, \ldots, n]$
We can now generalize and recover BitDecomp(t_r), for every $r \in [1, 2, \ldots, n]$, in a way analogous to what has been done for the case $r = 1$. The only difference is that, when choosing $\mathbf{y} \in \mathbb{Z}_q^N$, we set -1 in position $rl + 1$. So, for a given $r \in [1, 2, \ldots, n]$, we have the following.

- Recovering the first three most significant bits $t_{r,l-1}, t_{r,l-2}, t_{r,l-3}$. This is done exactly as in the case of t_1, with the only modification $y_{l+1} = 0$ and $y_{rl+1} = -1$ always.
- Recovering $t_{r,k}$, for all $k = l - 4, l - 5, \ldots, 1, 0$. Suppose that, for a given k, we recovered already $t_{r,m}, \forall m \in [k+1, \ldots, l-1]$. We then recover $t_{r,k}$ by recurrence. Choose

$$\mathbf{y} = 2^{l-k-3}(0, \ldots, 0, -t_{r,k+1}, -t_{r,k+2}, \ldots, -t_{r,l-1}, 0, \ldots, 0, -1, 0, \ldots, 0) \in \mathbb{Z}_q^N$$

with $-t_{r,i}$ in $(i+1)$-th position and -1 in $(rl+1)$-th position; i.e.

$$y_i = \begin{cases} -2^{l-k-3}t_{r,i-1} & \text{for } i \in [k+2, \ldots, l] \\ -2^{l-k-3} & \text{for } i = rl+1 \\ 0 & \text{otherwise} \end{cases}$$

Then we have $x = \langle \mathbf{y}, \mathbf{v} \rangle = 2^{l-k-3}\left(t_r - \sum_{j=k+1}^{l-1} t_{r,j} 2^j\right)$ with $0 \le x < 2^{l-2}$. Then, $\boxed{t_{r,k} = \mu}$.

In summary, we can recover the secret key $\mathbf{t} \in \mathbb{Z}_q^n$ with at most $(l+4) \cdot n$ oracle queries.

7   Conclusion

In this paper, we have shown that the SHE schemes from [1–4,17] suffer from key recovery attacks when the attacker is given access to the decryption oracle. Combining the results from [19,28], we now know that most existing SHE schemes suffer from key recovery attacks, and so they are not IND-CCA1 secure. As such, a natural next step is to investigate whether it is possible to enhance these SHE schemes to avoid key recovery attacks and make them IND-CCA1 secure. One thing we should keep in mind is to preserve their homomorphic properties. Following the work of [19], one could think of tweaking the decryption step of a SHE scheme by including a ciphertext validity check in order to make sure that, with some high probability, the ciphertext is honestly generated by the attacker and not specifically chosen for the purpose of recovering a given bit (or bits) of the secret key. Unfortunately, we cannot directly apply the techniques from [19] due to the fact that the SHE scheme from [19] enjoys some particular algebraic properties which do not exist in other schemes. This means that we need to treat each SHE scheme individually.

Acknowledgments. Massimo Chenal is supported by an AFR PhD grant from the National Research Fund, Luxembourg. Qiang Tang is partially supported by a CORE (junior track) grant from the National Research Fund, Luxembourg.

References

1. Brakerski, Z.: Fully homomorphic encryption without modulus switching from classical GapSVP. In: Safavi-Naini, R., Canetti, R. (eds.) CRYPTO 2012. LNCS, vol. 7417, pp. 868–886. Springer, Heidelberg (2012)
2. Brakerski, Z., Gentry, C., Vaikuntanathan, V.: (Leveled) fully homomorphic encryption without bootstrapping. In: Proceedings of the 3rd Innovations in Theoretical Computer Science Conference, ITCS 2012, pp. 309–325. ACM (2012)
3. Brakerski, Z., Vaikuntanathan, V.: Fully homomorphic encryption from ring-LWE and security for key dependent messages. In: Rogaway, P. (ed.) CRYPTO 2011. LNCS, vol. 6841, pp. 505–524. Springer, Heidelberg (2011)
4. Brakerski, Z., Vaikuntanathan, V.: Efficient fully homomorphic encryption from (standard) LWE. In: Proceedings of the 2011 IEEE 52nd Annual Symposium on Foundations of Computer Science, FOCS 2011, pp. 97–106 (2011)
5. Chenal, M., Tang, Q.: On key recovery attacks against existing somewhat homomorphic encryption schemes. IACR Cryptology ePrint Archive, Report 2014/535 (2014)
6. Cheon, J.H., Coron, J.-S., Kim, J., Lee, M.S., Lepoint, T., Tibouchi, M., Yun, A.: Batch fully homomorphic encryption over the integers. In: Johansson, T., Nguyen, P.Q. (eds.) EUROCRYPT 2013. LNCS, vol. 7881, pp. 315–335. Springer, Heidelberg (2013)
7. Coron, J.-S., Mandal, A., Naccache, D., Tibouchi, M.: Fully homomorphic encryption over the integers with shorter public keys. In: Rogaway, P. (ed.) CRYPTO 2011. LNCS, vol. 6841, pp. 487–504. Springer, Heidelberg (2011)
8. Coron, J.-S., Naccache, D., Tibouchi, M.: Public key compression and modulus switching for fully homomorphic encryption over the integers. In: Pointcheval, D., Johansson, T. (eds.) EUROCRYPT 2012. LNCS, vol. 7237, pp. 446–464. Springer, Heidelberg (2012)
9. Ding, J., Tao, C.: A new algorithm for solving the approximate common divisor problem and cryptanalysis of the fhe based on gacd. IACR Cryptology ePrint Archive, Report 2014/042 (2014)
10. Gentry, C.: A Fully Homomorphic Encryption Scheme. Ph.D thesis, Stanford, CA, USA (2009)
11. Gentry, C.: Fully homomorphic encryption using ideal lattices. In: Proceedings of the Forty-First Annual ACM Symposium on Theory of Computing, STOC 2009, pp. 169–178. ACM (2009)
12. Gentry, C.: Computing arbitrary functions of encrypted data. Commun. ACM 53(3), 97–105 (2010)
13. Gentry, C., Halevi, S.: Fully homomorphic encryption without squashing using depth-3 arithmetic circuits. In: Proceedings of the 2011 IEEE 52nd Annual Symposium on Foundations of Computer Science, FOCS 2011, pp. 107–109 (2011)
14. Gentry, C., Halevi, S.: Implementing gentry's fully-homomorphic encryption scheme. In: Paterson, K.G. (ed.) EUROCRYPT 2011. LNCS, vol. 6632, pp. 129–148. Springer, Heidelberg (2011)
15. Gentry, C., Halevi, S., Smart, N.P.: Better bootstrapping in fully homomorphic encryption. In: Fischlin, M., Buchmann, J., Manulis, M. (eds.) PKC 2012. LNCS, vol. 7293, pp. 1–16. Springer, Heidelberg (2012)
16. Gentry, C., Halevi, S., Smart, N.P.: Fully homomorphic encryption with polylog overhead. In: Pointcheval, D., Johansson, T. (eds.) EUROCRYPT 2012. LNCS, vol. 7237, pp. 465–482. Springer, Heidelberg (2012)

258   M. Chenal and Q. Tang

258   M. Chenal and Q. Tang

17. Gentry, C., Sahai, A., Waters, B.: Homomorphic encryption from learning with errors: conceptually-simpler, asymptotically-faster, attribute-based. In: Canetti, R., Garay, J.A. (eds.) CRYPTO 2013, Part I. LNCS, vol. 8042, pp. 75–92. Springer, Heidelberg (2013)
18. Hoffstein, J., Pipher, J., Silverman, J.H.: NTRU: a ring-based public key cryptosystem. In: Buhler, J.P. (ed.) ANTS 1998. LNCS, vol. 1423, pp. 267–288. Springer, Heidelberg (1998)
19. Loftus, J., May, A., Smart, N.P., Vercauteren, F.: On CCA-secure somewhat homomorphic encryption. In: Miri, A., Vaudenay, S. (eds.) SAC 2011. LNCS, vol. 7118, pp. 55–72. Springer, Heidelberg (2012)
20. López-Alt, A., Tromer, E., Vaikuntanathan, V.: On-the-fly multiparty computation on the cloud via multikey fully homomorphic encryption. In: Proceedings of the Forty-Fourth Annual ACM Symposium on Theory of Computing, STOC 2012, pp. 1219–1234. ACM, New York (2012)
21. Micciancio, D., Peikert, C.: Trapdoors for lattices: simpler, tighter, faster, smaller. IACR Cryptology ePrint Archive, Report 2011/501 (2011)
22. Naehrig, M., Lauter, K., Vaikuntanathan, V.: Can homomorphic encryption be practical? In: Proceedings of the 3rd ACM Workshop on Cloud Computing Security Workshop, CCSW 2011, pp. 113–124 (2011)
23. Nuida, K.: A simple framework for noise-free construction of fully homomorphic encryption from a special class of non-commutative groups. IACR Cryptology ePrint Archive, Report 2014/097 (2014)
24. Rivest, R.L., Adleman, L., Dertouzos, M.L.: On data banks and privacy homomorphisms. In: De Millo, R.A., et al. (eds.) Foundations of Secure Computation, pp. 169–179. Academia Press, New York (1978)
25. Smart, N.P., Vercauteren, F.: Fully homomorphic encryption with relatively small key and ciphertext sizes. In: Nguyen, P.Q., Pointcheval, D. (eds.) PKC 2010. LNCS, vol. 6056, pp. 420–443. Springer, Heidelberg (2010)
26. Stehlé, D., Steinfeld, R.: Faster fully homomorphic encryption. In: Abe, M. (ed.) ASIACRYPT 2010. LNCS, vol. 6477, pp. 377–394. Springer, Heidelberg (2010)
27. van Dijk, M., Gentry, C., Halevi, S., Vaikuntanathan, V.: Fully homomorphic encryption over the integers. In: Gilbert, H. (ed.) EUROCRYPT 2010. LNCS, vol. 6110, pp. 24–43. Springer, Heidelberg (2010)
28. Zhang, Z., Plantard, T., Susilo, W.: On the CCA-1 security of somewhat homomorphic encryption over the integers. In: Ryan, M.D., Smyth, B., Wang, G. (eds.) ISPEC 2012. LNCS, vol. 7232, pp. 353–368. Springer, Heidelberg (2012)

Practical Attacks on AES-like Cryptographic Hash Functions

Stefan Kölbl$^{(\boxtimes)}$ and Christian Rechberger

Technical University of Denmark, Kongens Lyngby, Denmark
stek@dtu.dk

Abstract. Despite the great interest in rebound attacks on AES-like hash functions since 2009, we report on a rather generic, albeit keyschedule-dependent, algorithmic improvement: A new message modification technique to extend the inbound phase, which even for large internal states makes it possible to drastically reduce the complexity of attacks to very practical values for reduced-round versions. Furthermore, we describe new and practical attacks on Whirlpool and the recently proposed GOST R hash function with one or more of the following properties: more rounds, less time/memory complexity, and more relevant model. To allow for easy verification, we also provide a source-code for them.

Keywords: Hash functions · Cryptanalysis · Collisions · Whirlpool · GOST R · Streebog · Practical attacks

1 Introduction

Cryptographic hash functions are one of the most versatile primitives and have many practical applications like integrity checks, message authentication, digital signature or password protection. Often they are a critical part of more complex systems whose security might fall apart if hash a function does not provide the properties we expect it to have.

Cryptographic hash functions take as input a string of arbitrary finite length and produce a fixed-sized output of n bits called hash. As a consequence, the following main security requirements are defined for cryptographic hash functions:

- **Preimage Resistance:** For a given output y it should be computationally infeasible to find any input x' such that $y = h(x')$.
- **Second Preimage Resistance:** For given $x, y = h(x)$ it should be computationally infeasible to find any $x' \neq x$ such that $y = h(x')$.
- **Collision Resistance:** It should be computationally infeasible to find two distinct inputs x, x' such that $h(x) = h(x')$.

For any ideal hash function with n-bit output size, we can find preimages or second preimages with a complexity of 2^n, and collisions with a complexity of $2^{n/2}$ using generic attacks.

D.F. Aranha and A. Menezes (Eds.): LATINCRYPT 2014, LNCS 8895, pp. 259–273, 2015.
DOI: 10.1007/978-3-319-16295-9_14

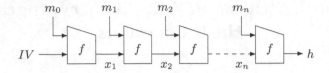

Fig. 1. Iterative construction for a cryptographic hash function.

Most cryptographic hash functions are constructed iteratively by splitting the message into evenly sized blocks m_i and using a compression function f to update the state. We call the intermediate results x_i chaining values and the final output h hash value (Fig. 1).

The security proofs for the hash function rely on the difficulty of finding a collision for this compression function, hence it is also of interest to consider the properties of the compression function and find properties which distinguish it from an ideal function.

- **semi-free start collision:** Find x, m, m' such that $f(x, m) = f(x, m')$.
- **free-start collision:** Find x, x', m, m' such that $f(x, m) = f(x', m')$.
- **near collision:** Find x, m, m' such that $f(x, m) \oplus f(x, m')$ has a low Hamming weight.

To sum up the various types with respect to their relevance: a semi-free-start collision is more interesting than a free-start collision, and a collision is more interesting than a near-collision.

1.1 Motivation

Cryptanalytic attacks are often hard to verify. Cryptanalysts often concentrate on the total running time of the attack, which is boiled down to a single number. While one can argue about the exact transition point between cryptanalytic attacks of practical and theoretical time complexity, it is often placed around an equivalent of 2^{64} calls to the primitive [1]. While this is a reasonable assumption for state-level adversaries, it is out of reach for academic research labs. However, the ability to fully implement and verify attacks is crucial, as this is often the only way to make sure that all details are modelled correctly in the theoretical analysis. In this paper we therefore aim at attacks that can actually be executed (and verified) with limited budget computing resources.

In this paper we show a new practical attack on a class of AES-like hash functions. We show attacks on reduced round versions of the ISO/IEC 10118-3 standard `Whirlpool` [2] and the new Russian federal standard `GOST R 34.11-2012` [3]. The model we consider is semi-free-start attacks on the compression function, which in contrast to the free-start attacks do not allow the attacker to choose different chaining values in a pair of inputs. This reduced degree of freedom makes the task of cryptanalysts harder, but is more relevant as it is closer to the actual use in the hash function.

1.2   Contribution

Despite a lot of attention on rebound-attacks of AES and AES-like primitives, we show that more improvements are possible in the inbound phase.

To the best of our knowledge, currently no practical attacks on reduced round GOST R have been published. However, there exists a practical 4-round free-start collision attack on the Whirlpool compression function [4]. It seems very hard to apply this specific attack directly to GOST R due to the extra round in the key schedule, which gives GOST R additional security against these free-start attacks.

In this paper we show a new method to carry out a 4-round practical attack on the Whirlpool and GOST R compression function. Additionally, and in contrast to many other attacks known on GOST R, we do not need the freedom to add half a round at the end to turn a near-collision into a collision. As the full hash function also does not end with a half round, we argue that a result on 4 rounds can actually be more informative than a result on 4.5 rounds.

New message modification technique. The attack is based on the rebound attack and start-in-the-middle techniques, and it carefully chooses the key input to significantly reduce the complexity resulting in a very low complexity[1]. We are also able to improve the results on 6.5 rounds by extending this attack. We give an actual example for such a collision, and have the source code of both the attack and the general framework publicly available to facilitate further research on practical attacks[2]. The method is not specific to a particular primitive, but is an algorithmic technique that however depends on two conditions in a primitive to hold (see also Sect. 4).

1.3   Related Work

In Table 1 we summarize the practical results on Whirlpool and GOST R. As the GOST R compression function uses a design similar to the Whirlpool hash function [2], many of the previous results on Whirlpool can be applied to GOST R. We would also like to note on adding half a round at the end for GOST R. This does not always make an attack more difficult, and in some cases it makes it easier, as it makes it possible to turn a near-collision into a collision, therefore we distinguish for our attacks if it applies for both cases.

There have also been practical attacks on other AES-based hash functions like Maelstroem (6 out of 10 rounds [7]), Grøstl (6 out of 10 rounds [8]) and Whirlwind (4.5 out of 12 rounds [9]).

1.4   Rebound Attacks

The rebound attack is a powerful tool in the cryptanalysis of hash functions, especially for finding collisions for AES-based designs [10,11]. The cipher is split

---

[1] Naturally, the improvement is not applicable for constructions or modes that do not allow modification of the key input.

[2] The source-code can be found at https://github.com/kste/aeshash.

Table 1. Summary of attacks with a complexity up to 2^{64} on AES-based hash functions. Time is given in compression function calls and memory in bytes.

Function	Rounds	Time	Memory	Type	Reference
GOST R	4.5	2^{64}	2^{16}	semi-free-start collision	[5]
	4.75	practical	2^8	semi-free-start near-collision	[6]
	4	$2^{19.8}$	2^{16}	semi-free-start collision	this work
	4.5	$2^{19.8}$	2^{16}	semi-free-start collision	this work
	5.5	2^{64}	2^{64}	semi-free-start collision	[5]
	6.5	2^{64}	2^{16}	semi-free-start collision	this work
Whirlpool	4	$2^{25.1}$	2^{16}	semi-free-start collision	this work
	6.5	$2^{25.1}$	2^{16}	semi-free-start near-collision	this work
	4	2^8	2^8	free-start collision	[5]
	7	2^{64}	2^8	free-start collision	[4]

into three sub-ciphers

$$E = E_{fw} \circ E_{in} \circ E_{bw}$$

and the attack proceeds in two steps. First, the inbound phase which is an efficient meet-in-middle in E_{in} using the available degree of freedom. This is followed by a probabilistic part, the outbound phase in E_{fw} and E_{bw} using the solutions from the inbound phase. The basic 4-round rebound attack uses a differential characteristic with $1 - 8 - 64 - 8 - 1$ active bytes per round and has a complexity of 2^{64}. There are many techniques extending and improving this attack. Some can even improve this very basic and simple setting of a square geometry, like start-from-the-middle [8], super S-Box [12,13] or solving three fully active states in the middle [14,15]. Other generic extensions exploit additional degrees of freedom or non-square geometries to improve results, like and using multiple inbounds [12,16]. In these settings, improved list-matching techniques [17,18] are also a generic improvement.

2   Description of GOST R

This section gives a short description of the GOST R compression function as we will use it for describing our attack in detail. As we are only looking at the compression function, we leave out some details not relevant for the upcoming attack in order to simplify the description. For a more detailed description of GOST R we refer to [3].

The compression function g uses two 512-bit inputs (the message block m and the chaining value h) to update the state in the following way (see Fig. 2)

$$g_N(h, m) = E(L \circ P \circ S(h), m) \oplus h \oplus m \tag{1}$$

where E is an AES-based block cipher using a state of 8×8 bytes and S, P, L are the same functions as used in this block cipher (see below).

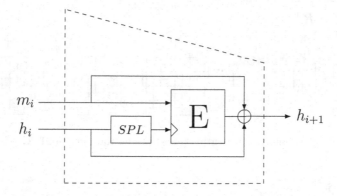

Fig. 2. An outline of the GOST R compression function. The chaining input is processed through an additional round before entering E

If we want to find a collision for the compression function, the following equation must hold

$$\Delta m_i \oplus \Delta h_i \oplus \Delta E(h_i, m_i) = 0 \tag{2}$$

2.1 Block Cipher E

The block cipher E takes two 512-bit inputs M and K^0 and produces a 512-bit output C. The state update consists of 12 rounds r and a final key addition.

$$L_1 = L \circ P \circ S \circ AK(M, K^0)$$
$$L_{i+1} = L \circ P \circ S \circ AK(L^i, K^i) \quad i = 1 \ldots 11$$
$$C = AK(L^{12}, K^{12})$$

The following four operations are used in one round (see Fig. 3):

- **AK** Adds the key byte-wise by XORing it to the state.
- **S** Substitutes each byte of the state independently using an 8-bit S-Box.
- **P** Transposes the state.
- **L** Multiplies each row by an 8×8 MDS matrix.

The 512-bit key input is expanded to 13 subkeys $K_0, \ldots, K_{12}$. This is done similar to the state update but AK is replaced with the addition of a round-dependent constant RC^r.

$$L_{i+1} = L \circ P \circ S \circ AK(K^0, RC^0) \quad i = 0 \ldots 11$$
$$K^{12} = AK(L^{12}, K^{12})$$

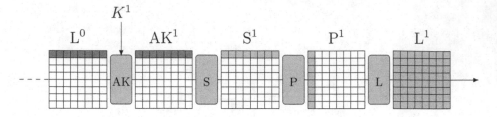

Fig. 3. The four operations used in one round of GOST R.

2.2 Notation

The notation we use for naming the states is:

- The state after applying the round function $\{AK, S, P, L\}$ in round r is named $\{AK^r, S^r, P^r, L^r\}$
- The byte at position x, y in state X^r is named $X^r_{x,y}$
- A row is denoted by $X^r_{*,y}$ and a column by $X^r_{x,*}$
- ■ and □ denote that there is a difference in a byte.
- ■ and □ are used for highlighting values of a byte.

2.3 Differential Properties

The attacks in this paper are based on differential cryptanalysis, and the resulting complexity correlates with the differential properties of the round functions. Therefore, to ease understanding, we give a short overview of the properties that are relevant for our attack.

The linear layer L has a strong influence on the number of active S-Boxes. There is no proof given that the linear layer L is MDS or has a branch number of 9 in the GOST R reference [3], but it was shown that this is the case in [19]. Hence, if we have one active byte at the input we will get 8 active bytes at the output with probability one. If we have a active bytes at the input the probability that there will be b active bytes at the output under the condition $a \neq 0, b \neq 0$ and $a + b \geq 9$ is $2^{(b-8)8}$.

The properties of the S-Box have a strong influence on the complexity of our attack, as will be seen later. Given a S-Box $S : \mathbb{F}_2^n \to \mathbb{F}_2^n$

$$\{x \mid S(x) \oplus S(x \oplus a) = b\} \tag{3}$$

is the number of solutions for an input a and output difference b. Table 2 gives the number of solutions for some S-Box designs used in practice.

To get a bound on the probability of the differential characteristics we are interested in the maximum value of Eq. 3 which we will refer to as the *maximum differential probability* (mdp) of an S-Box. A 4-round differential characteristic has at least 81 active bytes due to the properties of the linear layer, therefore any 4-round characteristic has a probability of $\leq \text{mdp}^{81}$.

Table 2. Comparison of different 8-bit S-Box designs used in AES-based hash functions.

Solutions	AES	Whirlpool	GOST R
0	33150	39655	38235
2	32130	20018	22454
4	255	5043	4377
6	-	740	444
8	-	79	25
256	1	1	1

For the rebound attack it is also important to know the average number of possible output differences, given a non-zero input difference. We will refer to this as the average number of solutions (ANS) for an S-Box which can be computed by constructing the *differential distribution table* (DDT). The ANS corresponds to the average number of non-zero entries in each row of the DDT.

This property influences the complexity for the matching step in the inbound phase and increases the costs of finding one solution. For the GOST R S-Box we get on average 107.05 solutions.

3   Attack on GOST R

In this section we describe our 4-round practical attack in detail and also show how it can be applied to more rounds. The description of the attack is split into two parts. First, we find a differential characteristic leading to a collision. Then we show how to construct a message pair following this characteristic in a very efficient way.

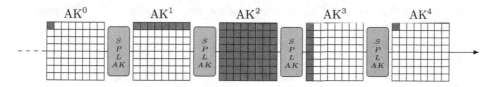

Fig. 4. The 4-round differential characteristic used in our attack.

3.1   Constructing the Differential Characteristic

For our 4-round attack we use a characteristic of the form $1 - 8 - 64 - 8 - 1$ (see Fig. 4). This truncated differential path has the minimal number of possible active S-Boxes for 4 rounds and is the starting point for many attacks. Next, we will determine the values of the differences before continuing with the construction of the message pair.

The approach we use for this is based on techniques from the rebound attack, like the start-in-the-middle technique used in [8]. This approach would also give us an efficient way to find both the characteristic and message pair for a characteristic of the form $1 - 8 - 64 - 8$. However this would still lead to a higher attack complexity if extended to 4 rounds. Hence, we only use ideas from this approach to determine the differential characteristic and do not assume the key input as constant.

Precomputation. First we pre-compute the differential distribution table (DDT) of the S-Box and we also construct a list M_{lin}. This list contains all possible 255 non-zero values of $P_{0,0}$ and the result after applying L (see Fig. 5).

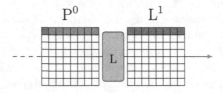

Fig. 5. Computing list M_{lin} for all 255 values of $P_{0,0}^0$ (blue) to find all possible transitions from 1 to 8 bytes. Gray bytes are set to zero (Color figure online).

Construction

1. Start with a random difference in $AK_{0,0}^4$ and propagate it back to S^2 through the inverse round functions. For the linear steps this is deterministic, and for propagating through the S-Box we choose a random possible input difference to the given output difference. After this step we will have a full active state in S^2.
2. For each difference in S^2 we look up the set of all possible input differences from the DDT for each byte of the state.
3. Check for each row of AK^2 whether there is a possible match with the rows stored in M_{lin} (see Fig. 6).
   - The probability that a single byte matches is $107.05/255 \approx 2^{-1.252}$ therefore a row matches with a probability of $2^{-10.018}$.
   - If we take into account that M_{lin} has 255 entries we expect to find a match with a probability of $1 - (1 - 2^{-10.018})^{255} \approx 2^{-2.2}$.
   - Therefore the probability for a match of all 8 rows is given by

$$(2^{-2.2})^8 = 2^{-17.6} \tag{4}$$

After this step we have found a characteristic spanning from S^1 to AK^4. Now we have to repeat the previous process for a single row to find the right differences in AK^1. This has a probability of $2^{-2.2}$ of succeeding. Hence we need to repeat the whole process $2^{19.8}$ times to obtain one solution.

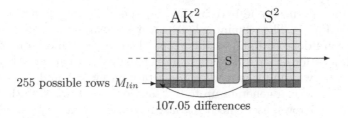

Fig. 6. The matching step in the middle is done on each row individually. There are 2^8 possible values for each row $AK^2_{*,j}$ for $j = 0, 1, \ldots, 7$.

Note that we can only choose 255 differences for $AK^4_{0,0}$, but we can also freely choose from the set of possible differences when propagating from S^3 to AK^3. This gives us an additional 107.05 choices for each row in S^2 leading to $\approx 2^{54}$ possible values for the state S^2. Hence, we have enough starting points for finding our differential characteristic.

3.2 Finding the Message Pair

Now we want to find a message pair which follows the previously constructed characteristic. At this point only the differences, but not the values of the state, are fixed. We start by fixing the values of AK^2 such that the 64 differential transitions $S^2 = S(AK^2)$ are fulfilled.

Next we use the key input to solve any further conditions on the active S-Boxes in order to lower the complexity of our attack. This step is split into solving the conditions on $S^1_{*,0} = S(AK^1_{*,0})$ and $S^3_{0,*} = S(AK^3_{0,*})$.

Solving Conditions at the Start. We have 8 conditions on $S^1_{*,0}$ which we need to solve. These conditions can be solved row-wise by choosing the corresponding values in K^2 such that $P^{-1}(L^{-1}(AK^2 \oplus K^2)) = S^1$. We can do this step row-wise by solving a linear equation. As there is only a single byte condition for

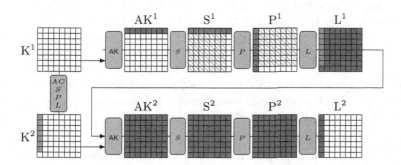

Fig. 7. The values of AK^2 are fixed. We solve 7 of the conditions on S^1 by using the freedom in K^2 (bytes marked orange), which allows us to influence the values on the bytes in S^1 (orange slash pattern).

each row, we only need one byte in the corresponding row of K^2 to solve the equation (see Fig. 7). The remaining bytes are fixed to arbitrary values as we have no further conditions to fulfill at this step. These bytes could be used to solve more conditions for other differential characteristics or to construct additional solutions, as we will do for extending the attack on more rounds.

In this step we can generate up to 2^{56} solutions per row. Note that we only do this step for 7 rows, as we need the last row in the next step.

Solving Conditions at the End. For solving the conditions $S^3 = S(AK^3)$, we can use the bytes in $K^2_{*,7}$. These bytes form a column in $\mathsf{KP}^3_{7,*}$ (see Fig. 8), which allows us to solve a single byte condition per row for AK^3.

1. Assume that $K^2_{*,0-6}$ are fixed and propagate them forward to KP^3.
2. We can now solve the conditions for each row individually. In each row there are 7 bytes fixed in KP^3 and a single byte in K^3 (from AK^3). This gives us a linear equation with one solution per row and allows us to solve all conditions on AK^3.

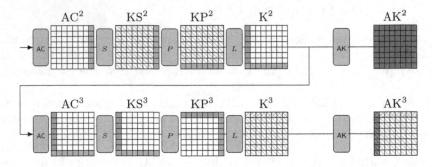

Fig. 8. Solving all the conditions on AK^3. The orange values are fixed from the previous step and the purple values are used to fulfill the conditions on AK^3.

Remaining Conditions. We still need to solve one byte condition on $S^1_{0,7}$, which can be done by repeating the previous procedure 2^8 times. The bytes which are used to solve the conditions on AK^3 form a row in K^2 and influence the values of L^1 resp. P^1 and S^1 (see Fig. 11 in Appendix A). This implies that we can change the value of $S^1_{0,7}$ by constructing different solutions for $K^2_{*,7}$.

The only remaining condition is $\Delta AK^0_{0,0} = \Delta AK^4_{0,0}$, which can again be solved by repeating the previous steps 2^8 times. It follows that we need to repeat the algorithm shown in Sect. 3.2 about 2^{16} times.

Complexity. We can construct the differential characteristic with a complexity of $2^{19.8}$. Finding a message pair following this characteristic requires 2^{16} steps using our message modification technique. Hence, the total complexity of the attack is $\approx 2^{19.9}$. We have implemented this attack and verified our results. The un-optimized proof-of-concept implementation in Python is publibly available [20]. An example for a 4-round collision can be found in Appendix B.

3.3   Extending the Attack

As we only need to control 15 bytes of the key, we can extend the attack on 6.5 rounds by using a characteristic of the form $8 - 1 - 8 - 64 - 8 - 1 - 8$. In this case we would use the same approach to find the differential characteristic for 4 rounds and in the message modification part we would construct more solutions by using the additional freedom in the key. This will influence the differences at the input/output of the 6.5 rounds. The complexity of this attack is $\approx 2^{64}$, as the 8-byte difference at the input/output needs to be equal (Fig. 9).

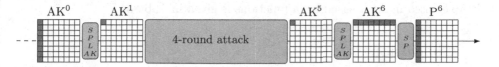

Fig. 9. The 4-round attack is extended by one round in the beginning and one round in the end to mount an attack on 6.5 rounds.

4   Application to Other AES-based Hash Functions

The message modification technique presented is not specific to GOST R, but requires a few criteria to be met. First the transposition layer has to have the property that every byte of a single row/column is moved to a different row/column (see Fig. 10). This is true for all AES-based hash functions we consider in this paper, as it is a desired property against other attacks.

The second criteria is that there is a key addition in every round, hence our attack is applicable to both Whirlpool and GOST R. Permutation-based designs like Grøstl do not have this property. The attacker has less control of the input for each round, which makes the hash function more resistant against these types of attacks.

The complexity of the attack depends on the choice of the S-Box, as this directly influences the costs of constructing the differential characteristic. Given the average number of solutions $\bar{s}$ for $\Delta out = S(\Delta in)$ with a fixed value Δin, this directly gives the complexity for the matching step of the attack

$$\left(1 - \left(1 - \frac{\bar{s}}{255}\right)^{255}\right)^{8} \tag{5}$$

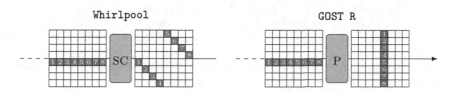

Fig. 10. The transposition layer used in Whirlpool and GOST R.

Table 3. Comparing the maximum differential probability (MDP) and average number of solutions (ANS) for different 8-bit S-Boxes in AES-based designs.

S-Box	MDP	ANS	Matching Costs	$\#S^2$
AES	2^{-6}	127	$2^{6.42}$	$2^{55.91}$
Whirlpool	2^{-5}	101.49	$2^{25.10}$	$2^{53.32}$
GOST-R	2^{-5}	107.05	$2^{19.77}$	$2^{53.94}$

and the number of possible states for S^2 is $\approx \overline{s}\ ^8$. A comparison of the different S-Boxes used in AES-based hash functions is given in Table 3.

5   Conclusion

In this paper, we have shown new practical attacks for both the `Whirlpool` and `GOST R` compression function. We presented a 4-round attack with very low complexity of $2^{25.10}$ resp. $2^{19.8}$. Importantly, the attack is fully verified and source-code for it is available. In the case of `GOST R` the attack can be extended to find collisions for 6.5 rounds with a complexity of 2^{64} and for `Whirlpool` we can extend it to construct a near-collision in 50 bytes with a complexity of $2^{25.10}$ for 6.5 rounds of the compression function. The difference in the results for `GOST R` and Whirlpool is due to the ShiftColumns operation which does not align the bytes to lead to a collision for the differential characteristic we use.

Our attack is applicable to all AES-based primitives where it is possible for the attacker to control the key input for a few rounds. This significantly reduces the complexity of previous attacks and might be useful to speed up other attacks on AES-based hash-function designs.

References

1. Biryukov, A., Dunkelman, O., Keller, N., Khovratovich, D., Shamir, A.: Key recovery attacks of practical complexity on AES-256 variants with up to 10 rounds. In: Gilbert, H. (ed.) EUROCRYPT 2010. LNCS, vol. 6110, pp. 299–319. Springer, Heidelberg (2010)
2. Barreto, P., Rijmen, V.: The Whirlpool hashing function. In: First open NESSIE Workshop, Leuven, Belgium, vol. 13, p. 14 (2000)
3. Dolmatov, V., Degtyarev, A.: GOST R 34.11-2012: Hash Function (2013). http://tools.ietf.org/html/rfc6986
4. Sasaki, Y., Wang, L., Wu, S., Wu, W.: Investigating fundamental security requirements on whirlpool: improved preimage and collision attacks. In: Wang, X., Sako, K. (eds.) ASIACRYPT 2012. LNCS, vol. 7658, pp. 562–579. Springer, Heidelberg (2012)
5. Wang, Z., Yu, H., Wang, X.: Cryptanalysis of GOST R Hash Function. Cryptology ePrint Archive, Report 2013/584 (2013). http://eprint.iacr.org/

6. AlTawy, R., Kircanski, A., Youssef, A.M.: Rebound Attacks on Stribog. Cryptology ePrint Archive, Report 2013/539 (2013). http://eprint.iacr.org/
7. Kölbl, S., Mendel, F.: Practical attacks on the maelstrom-0 compression function. In: Lopez, J., Tsudik, G. (eds.) ACNS 2011. LNCS, vol. 6715, pp. 449–461. Springer, Heidelberg (2011)
8. Mendel, F., Peyrin, T., Rechberger, C., Schläffer, M.: Improved cryptanalysis of the reduced Grøstl compression function, ECHO permutation and AES block cipher. In: Jacobson Jr., M.J., Rijmen, V., Safavi-Naini, R. (eds.) SAC 2009. LNCS, vol. 5867, pp. 16–35. Springer, Heidelberg (2009)
9. Barreto, P.S.L.M., Nikov, V., Nikova, S., Rijmen, V., Tischhauser, E.: Whirlwind: a new cryptoaphic hash function. Des. Codes Crypt. 56(2–3), 141–162 (2010)
10. Mendel, F., Rechberger, C., Schläffer, M., Thomsen, S.S.: The rebound attack: cryptanalysis of reduced whirlpool and Grøstl. In: Dunkelman, O. (ed.) FSE 2009. LNCS, vol. 5665, pp. 260–276. Springer, Heidelberg (2009)
11. Lamberger, M., Mendel, F., Schläffer, M., Rechberger, C., Rijmen, V.: The rebound attack and subspace distinguishers: application to whirlpool. J. Cryptol., 1–40 (2013)
12. Lamberger, M., Mendel, F., Rechberger, C., Rijmen, V., Schläffer, M.: Rebound Distinguishers: Results on the Full Whirlpool Compression Function. [21] 126–143
13. Gilbert, H., Peyrin, T.: Super-Sbox cryptanalysis: improved attacks for AES-like permutations. In: Hong, S., Iwata, T. (eds.) FSE 2010. LNCS, vol. 6147, pp. 365–383. Springer, Heidelberg (2010)
14. Jean, J., Naya-Plasencia, M., Peyrin, T.: Improved rebound attack on the finalist Grøstl. In: Canteaut, A. (ed.) FSE 2012. LNCS, vol. 7549, pp. 110–126. Springer, Heidelberg (2012)
15. Jean, J., Naya-Plasencia, M., Peyrin, T.: Improved cryptanalysis of AES-like permutations. J. Cryptology 27(4), 772–798 (2014)
16. Matusiewicz, K., Naya-Plasencia, M., Nikolic, I., Sasaki, Y., Schläffer, M.: Rebound Attack on the Full Lane Compression Function. [21] 106–125
17. Dinur, I., Dunkelman, O., Keller, N., Shamir, A.: Efficient dissection of composite problems, with applications to cryptanalysis, knapsacks, and combinatorial search problems. In: Safavi-Naini, R., Canetti, R. (eds.) CRYPTO 2012. LNCS, vol. 7417, pp. 719–740. Springer, Heidelberg (2012)
18. Naya-Plasencia, M.: How to improve rebound attacks. In: Rogaway, P. (ed.) CRYPTO 2011. LNCS, vol. 6841, pp. 188–205. Springer, Heidelberg (2011)
19. Kazymyrov, O., Kazymyrova, V.: Algebraic Aspects of the Russian Hash Standard GOST R 34.11-2012. Cryptology ePrint Archive, Report 2013/556 (2013). http://eprint.iacr.org/
20. https://github.com/kste/aeshash
21. Matsui, M. (ed.): ASIACRYPT 2009. LNCS, vol. 5912. Springer, Heidelberg (2009)

A   Solving Conditions

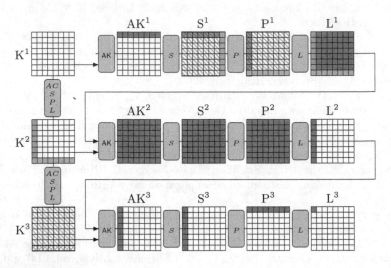

Fig. 11. Solving both conditions on S^1 and AK^3. The bytes marked purple solve the conditions on AK^3 and a single condition on S^1, whereas the orange bytes solve 7 conditions on S^1.

B   Colliding Message Pair

Here a colliding message pair (M, M') and the chaining value are given. The message pair has been found by using the 4-round characteristic and the difference in the messages is $\Delta AK^0_{0,0} = \Delta AK^4_{0,0} = fc$. All values are given in hexadecimal notation.

K^0

```
e80c313d6875c049
865604acdb024055
5cfd18c5d1a5a3f3
793359475ae90836
528e80e4aedfd674
20c777c0306ff6da
da4f9acc1e4fc9c4
04f486b91fa6bdeb
```

AK^0	AK'^0	ΔAK^0
81c2cf897a89e94f	7dc2cf897a89e94f	fc00000000000000
11e6c3ac6020a3c9	11e6c3ac6020a3c9	0000000000000000
ef44c4305cfade20	ef44c4305cfade20	0000000000000000
c04dc08f87db9f8f	c04dc08f87db9f8f	0000000000000000
b56983ec66229993	b56983ec66229993	0000000000000000
1bf0262cb3b92956	1bf0262cb3b92956	0000000000000000
dcb1e1511414ac83	dcb1e1511414ac83	0000000000000000
8efcd0ef76fa4671	8efcd0ef76fa4671	0000000000000000

AK^4	AK'^4	ΔAK^4
1444aab6eb146b46	e844aab6eb146b46	fc00000000000000
0692f61cf6b43022	0692f61cf6b43022	0000000000000000
935d2a8720ae7f0f	935d2a8720ae7f0f	0000000000000000
329542c23a812da1	329542c23a812da1	0000000000000000
eca3ba2984ab0219	eca3ba2984ab0219	0000000000000000
636b834737911049	636b834737911049	0000000000000000
0ce0d7edefe72c55	0ce0d7edefe72c55	0000000000000000
6dc8a467a81f8587	6dc8a467a81f8587	0000000000000000

Key Recovery Attacks on Recent Authenticated Ciphers

Andrey Bogdanov[1], Christoph Dobraunig[2], Maria Eichlseder[2],
Martin M. Lauridsen[1], Florian Mendel[2], Martin Schläffer[2],
and Elmar Tischhauser[1(✉)]

[1] DTU Compute, Technical University of Denmark, Kongens Lyngby, Denmark
ewti@dtu.dk
[2] IAIK, Graz University of Technology, Graz, Austria

Abstract. In this paper, we cryptanalyze three authenticated ciphers:
AVALANCHE, Calico, and RBS. While the former two are contestants
in the ongoing international CAESAR competition for authenticated
encryption schemes, the latter has recently been proposed for lightweight
applications such as RFID systems and wireless networks.

All these schemes use well-established and secure components such
as the AES, Grain-like NFSRs, ChaCha and SipHash as their building
blocks. However, we discover key recovery attacks for all three designs,
featuring square-root complexities. Using a key collision technique, we
can recover the secret key of AVALANCHE in $2^{n/2}$, where $n \in \{128, 192, 256\}$ is the key length. This technique also applies to the authentication
part of Calico whose 128-bit key can be recovered in 2^{64} time. For RBS,
we can recover its full 132-bit key in 2^{65} time with a guess-and-determine
attack. All attacks also allow the adversary to mount universal forgeries.

Keywords: Authenticated encryption · CAESAR · Key collision ·
Guess-and-determine · Universal forgery · AVALANCHE · Calico · RBS

1 Introduction

An authenticated cipher is a symmetric-key cryptographic algorithm that aims
to provide both the confidentiality and authenticity of data – the two most
fundamental cryptographic functionalities. Authenticated encryption has been
used extensively since decades by combining encryption algorithms with message
authentication algorithms. However, it was not until the 2000s that the separate
security goal of authenticated encryption (AE) has been formulated [6,13].

Authenticated encryption schemes can basically be constructed either as
a generic composition of an encryption algorithm (a block or stream cipher)
and a message authentication code (MAC), or as a dedicated AE design. Doing
both encryption and authentication in one cryptographic primitive also has the
advantage of attaining a potentially higher peformance. Indeed, such combined
schemes as OCB [14,19,20] and OTR [17] require only one block cipher call

© Springer International Publishing Switzerland 2015
D.F. Aranha and A. Menezes (Eds.): LATINCRYPT 2014, LNCS 8895, pp. 274–287, 2015.
DOI: 10.1007/978-3-319-16295-9_15

per block of data processed to produce both the ciphertext and the authentication tag.

Owing to its relatively recent origins, only very few AE designs have been included in international standards, the most prominent examples being CCM [9] and GCM [10,15] which have been included in ANSI/ISO, IEEE and IETF standards. Recently, the CAESAR competition [1] has been initiated in order to establish a portfolio of recommended AE algorithms during the next years, making authenticated encryption a major focus of the cryptographic community. A large number of diverse designs has been submitted to this competition, ranging from block cipher modes of operation to dedicated designs.

A substantial fraction of the CAESAR submissions build upon proven components, such as the AES, the SHA-2 family or the Keccak [8] permutation. An important reason for such a design decision is the assumption that the AE scheme will inherit the good security properties of the building blocks. For some designs, this is backed up by a formal security reduction. However this is not the case for all candidates, and even when security proofs for confidentiality and integrity are provided, they are often limited by the birthday bound, while at the same time any key recovery attack with complexity below 2^n would be highly undesirable.

We analyze three recent proposals for authenticated encryption: AVALANCHE [4] and Calico [21] are general-purpose AE schemes that have been submitted to CAESAR, and RBS [12] has recently been proposed for lightweight applications. All three algorithms are based on secure building blocks: AVALANCHE uses the AES, Calico builds upon ChaCha [7] and SipHash [5], and RBS a Grain-like [3] register-accumulator architecture [2].

Despite being based on these secure components, our analysis establishes that all three algorithms admit key recovery attacks with complexities significantly below exhaustive search. AVALANCHE and Calico lend themselves to attacks based on key collisions, leading to a key recovery with complexity $2^{n/2}$ for all versions of AVALANCHE ($n \in \{128, 192, 256\}$), and with complexity 2^{64} for the 128-bit authentication key of Calico. For RBS, recovering its full 132-bit key requires 2^{65} time with a guess-and-determine strategy. For all algorithms, the recovered key material enables the adversary to obtain universal forgeries. In the case of AVALANCHE and RBS, the adversary also obtains all necessary key material for decryption of arbitrary messages.

All these attacks are entirely structural and do not make use of any weaknesses of the building blocks themselves. We give an overview of the attacks and their complexities in Table 1.

The remainder of the paper is organized as follows. We introduce some common notation in Sect. 2. Section 3 describes the key recovery attack on AVALANCHE. Our attack on Calico is presented in Sect. 4. In Sect. 5, we describe our guess-and-determine attack on RBS. We conclude our findings in Sect. 6.

2   Notation

In the following, we use N, A, M and C to denote nonce, associated data (data which is authenticated but not encrypted), a message (to be encrypted and

Table 1. Overview of the attacks presented in this paper. For the attacks marked with (*), memoryless variants are possible [18], reducing the memory requirements to $\mathcal{O}(1)$, eliminating the offline computations, and increasing the time complexity by a factor of about 2.

Algorithm	Recovered key bits	Time		Memory	Data
		Offline	Online		
AVALANCHE-128	384	2^{64}	2^{64}	$2^{64}(*)$	2^{64}
AVALANCHE-192	448	2^{96}	2^{96}	$2^{96}(*)$	2^{96}
AVALANCHE-256	512	2^{128}	2^{128}	$2^{128}(*)$	2^{128}
Calico	128	2^{64}	2^{64}	$2^{64}(*)$	2^{64}
RBS	132	–	2^{65}	$\mathcal{O}(1)$	1

authenticated) and a ciphertext, respectively. For binary strings x and y, we let $|x|$ denote the bit-length of x and $x\|y$ is the concatenation of x and y. We use ϵ for the empty string, i.e. $|\epsilon| = 0$. Subscript usually denotes the bit index of a string, so x_i is the ith bit of x, with the convention that x_0 is the least significant, rightmost bit. We use $\oplus$ to denote the XOR operation and use $\text{hw}(x)$ to denote the Hamming weight of x. In the case where X is a binary string of blocks, and where the block size is understood to be b, we let $X[i]$ denote the ith block of X, i.e. $X[i] = x_{bi-1}\| \cdots \|x_{b(i-1)}$ for $i \geq 1$.

3   AVALANCHE

The AVALANCHE scheme [4] is a submission to the ongoing CAESAR competition for authenticated encryption ciphers. We note that the specification of AVALANCHE leaves some room for interpretation. In the aspects relevant to our attacks, we assume the following:

- The nonce has sizes $|N| \in \{80, 160, 128\}$ for key lengths $n \in \{128, 256, 192\}$, respectively.
- The nonce N is randomly generated.
- The counter c is initialized to $c = 0$.
- The tag length is $|T| = 128$ as well as the security parameters k and p are 128-bit.
- The $(n + 256)$-bit key K consists of three independent parts, $K = (K_P, k, p)$.

AVALANCHE uses the AES to process a message M of m blocks and associated data A of arbitrary length to produce a ciphertext C of $m + 1$ blocks and an authentication tag T. It does not support a public message number, instead a nonce N is generated by the encryption algorithm itself.

The input to AVALANCHE with a specified secret key $K = (K_P, k, p)$, is a 3-tuple (M, A, K) of message and associated data. The output is a 4-tuple (N, A, C, T) of nonce, associated data, ciphertext, and tag. The scheme uses two main algorithms described in the following; PCMAC for message processing and RMAC for processing associated data. The interfaces and outputs of the two algorithms are

$$(N, C, \tau_P) = PCMAC(M) \quad \text{and} \quad \tau_A = RMAC(A).$$

The final tag T is then computed as $T = \tau_P \oplus \tau_A$.

3.1   PCMAC

The encryption with PCMAC is illustrated in Fig. 1. The padded message is denoted $M[1] \cdots M[m]$ and the ciphertext $C[0] \cdots C[m]$. The number r is generated at random.

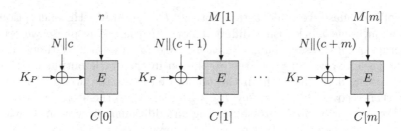

Fig. 1. Message processing with PCMAC

3.2   RMAC

The output of RMAC is an intermediate tag τ_A of 128 bits. RMAC uses the secrets k and p, p being a randomly chosen 128-bit prime and k chosen at random from $\{\lfloor p/2 \rfloor + 1, \ldots, p - 1\}$. The intermediate tag τ_A is determined as

$$\tau_A = (1\|A) \cdot k \mod p. \tag{1}$$

3.3   Recovering the PCMAC Key

The critical part of PCMAC is that the encryption key for E (see Fig. 1) depends on the nonce and counter. This facilitates key collision attacks, similar to the one on McOE-X [16]. Our attack works in an *offline phase* and an *online phase* (see Algorithms 1 and 2). Both are called with the same, arbitrary single-message block M. The offline phase outputs a list L which is used in the online phase. We also note that this technique allows a free trade-off between time and memory by choosing the list size accordingly.

Algorithm 1. OFFLINE(M)	**Algorithm 2.** ONLINE(M, L)
Data: Single-block M	**Data:** Single-block M, List L
1 $L \leftarrow \emptyset$	output from Algorithm 1
2 **for** $i = 1, \ldots, \ell$ **do**	1 **for** $i = 1, \ldots, 2^n/\ell$ **do**
3 Choose new	2 Obtain (N, ϵ, C, T) for (M, ϵ)
$K = (K_P, k, p) \in \{0,1\}^{n+256}$	from AE
4 $(N, \epsilon, C, T) \leftarrow$	3 **if** $\exists (x, y, z) \in L : x = C[1]$
AVALANCHE(M, ϵ, K)	**then**
5 $L \leftarrow L \cup \{(C[1], K_P, N)\}$	4 **return**
6 **end**	$y \oplus ((N \oplus z) \| 0^{n-|N|})$
7 **return** L	5 **end**
	6 **end**
	7 **return** Failure

In the offline phase we build a table of size ℓ of AVALANCHE encryptions of the same message block, using different keys. In the online phase we request the encryption of the same single-block message M in total $2^n/\ell$ times. By the birthday paradox, we expect to see a collision in the oracle output $C[1]$ in the online phase and the list L from the offline phase. As the nonce N is public, we can then recover the secret key K_P by adding it to the stored nonce z and key y. We can verify candidate keys using an additional encryption. Obviously, choosing $\ell = 2^{n/2}$ gives the best overall complexity, using just $2 \cdot 2^{n/2}$ time and memory in the order of $2^{n/2}$ to store L. Memoryless versions of the meet-in-the-middle technique can be used here as well [18].

3.4 Recovering the RMAC Secret Parameters

To recover (k, p), we use the attack described above to recover the secret K_P. We furthermore ask for encryption and tag of some arbitrary message block; once with empty associated data, i.e. $A = \epsilon$, and once with $A = 0$, i.e. a single zero bit. Let the corresponding outputs of AVALANCHE be (N, ϵ, C, T) and $(N', 0, C', T')$, where $T = \tau_P \oplus \tau_A$ and $T' = \tau'_P \oplus \tau'_A$.

With K_P in hand, we can ourselves compute τ_P and τ'_P using PCMAC. Consequently, we obtain τ_A and τ'_A. Using the definition of RMAC of Eq. (1), we observe that for the case where $A = \epsilon$ we directly obtain $\tau_A \equiv k \mod p$, but since $k \in \{\lfloor p/2 \rfloor + 1, \ldots, p - 1\}$ we have $k = \tau_A$. Now, for the case where $A = 0$, we find

$$\tau'_A \equiv (1\|0) \cdot k \mod p$$
$$\Leftrightarrow \tau'_A \equiv 2k \mod p$$
$$\Leftrightarrow p = 2k - \tau'_A.$$

We therefore obtain the secret parameters (k, p) of RMAC with a complexity of two one-block encryption queries.

In summary, we have recovered all $n + 256$ bits of secret key material in about $2^{n/2}$ time.

4 Calico

Calico [21] is an authenticated encryption design submitted to the CAESAR competition. For Calico in reference mode, ChaCha-14 [7] and SipHash-2-4 [5] work together in an Encrypt-then-MAC scheme [11]. The Calico design is depicted in Fig. 2.

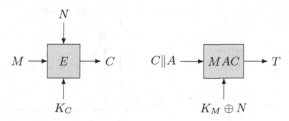

Fig. 2. The Calico scheme with encryption (left) and tag generation (right)

4.1 Specification

For the purpose of using ChaCha-14 and SipHash-2-4 in Calico, the 384-bit key K is split into two parts: a 256-bit encryption key K_C and a 128-bit authentication key K_M, s.t. $K = K_C \| K_M$. The plaintext is encrypted with ChaCha under K_C to obtain a ciphertext with the same length as the plaintext. Then, the tag is computed as the SipHash MAC of the concatenated ciphertext and associated data. The key used for SipHash is generated by XORing the nonce to the (lower, least significant part of the) MAC key K_M, so

$$(C, T) = \text{Enc}_{\text{Calico}}(K_C \| K_M, N, A, M),$$

where the ciphertext and tag, C and T respectively, are computed with

$$C = \text{Enc}_{\text{ChaCha-14}}(K_C, N, M)$$
$$T = \text{MAC}_{\text{SipHash-2-4}}(K_M \oplus N, C \| A).$$

The tag T and nonce N are both 64 bits long.

4.2 MAC Key Recovery

In Calico, SipHash is modified by XORing a nonce to the lower-significance bits of the key. This modification of SipHash facilitates an attack similar to the one described by Mendel et al. [16]. The attack targets the tag generation to recover the MAC key, which in turn allows to forge tags for arbitrary associated data and ciphertexts.

We can split the attack into an offline phase and an online phase (see Algorithms 3 and 4), where the online phase requires access to an encryption oracle. Algorithm 3 does 2^{64} online queries for an overall complexity of 2^{64}. Tradeoffs are possible to reduce the number of online queries at the cost of the overall complexity.

Algorithm 3. OFFLINE	**Algorithm 4.** ONLINE(L)

1 $L \leftarrow \emptyset$
 2 **for** $i = 0, \ldots, 2^{64} - 1$ **do**
 3   Compute tag T for $A, M = \epsilon$
      under MAC key $K_M = i\|0$
      and nonce $N = 0$
 4      $L \leftarrow L \cup \{(T, K_M)\}$
 5 **end**
	Data: List L output from
	         Algorithm 3
	 1 **for** $j = 0, \ldots, 2^{64} - 1$ **do**
	 2 Request tag T for $A, M = \epsilon$
	      under nonce $N = j$ from
	      encryption oracle
	 3 **if** $\exists (x, y) \in L : x = T, y = i\|0$
	      then
	 4 $i\|j$ is a candidate for K_M
	 5 **end**
	 6 **end**

This produces at least one MAC key candidate; if necessary, remaining candidates can be filtered with additional offline computations, though their expected number is very small.

Since Calico preserves the plaintext length for the ciphertext, an empty plaintext and associated data will produce an empty input for the MAC, independent of the cipher key or nonce. Thus, all offline computations and online queries give tags calculated from the same MAC input, only with varying keys fed to SipHash. The SipHash keys used in the offline phase all have the lower 64 bits set to 0 and the upper 64 bits iterating through all possible values. In the online phase, the SipHash keys have the upper 64 bits set to the original bits of the secret K_M, while the lower bits iterate through all possibilities. Thus, there is exactly one match between the two key lists, which will also produce a colliding tag (though other tag pairs may collide as well). The matching key stored in the offline list gives the upper 64 bits of the correct key, the colliding nonce from the online phase the lower 64 bits.

For tradeoffs with online complexity $2^N < 2^{64}$, replace 2^{64} by 2^N in the online phase and by $2^{(128-N)}$ in the offline phase; the success probability remains 1. We note that memoryless versions of the meet-in-the-middle technique apply also here [18].

5 RBS

RBS is an authenticated encryption scheme by Jeddi et al. [12] proposed for use in RFID tags. The idea of RBS is to insert the bits of a MAC on the message among the message bits, in key-dependent positions, to produce the authenticated ciphertext.

5.1 Specification

The RBS scheme is depicted in Fig. 3. It takes as input a 64-bit message M and a 132-bit key k to produce a 132-bit authenticated ciphertext C. Effectively, the key is split in two parts of sizes which we denote n and m respectively: the least n significant bits are used for clocking the MAC (which we described in detail later) while the most significant m bits are used for initializing the NFSR in the MAC. RBS uses $n = 64$ and $m = 68$, but we sometimes use n and m for generality in the following. Note that a requirement on the key k is that it has Hamming weight 68, and hence the size of the key space is $\binom{132}{68} \approx 2^{128.06}$.

The RBS MAC takes either a 64-bit or 68-bit input to be processed, along with the key k, and produces a 68-bit output. While RBS does not specify this, we assume (without influence on our attack) that the second MAC output is truncated by taking the least significant 64 bits to obtain the value S.

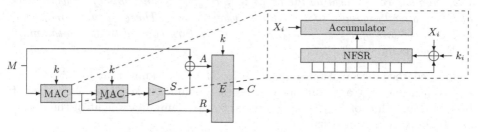

Fig. 3. The RBS scheme as an overview (left) and with the internals of the MAC (right)

Consider A and R of Fig. 3 as registers of 64 bits and 68 bits, respectively. For the function E, the ith ciphertext bit, denoted C_i, is obtained as

$$C_i = \begin{cases} \text{least significant bit of } A, & k_i = 0 \\ \text{least significant bit of } R, & k_i = 1 \end{cases}.$$

Each time a bit is taken from A or R, to produce a ciphertext bit, the corresponding register is right-rotated one position. As 132 bits are produced for the ciphertext, E effectively obtains C by inserting the bits of R (the MAC of the message), in order, into A at key-dependent positions.

The RBS MAC. The MAC used in RBS which we denote $MAC(X, k)$, (depicted in the right side of Fig. 3 where the input is denoted X) is a Grain-like design based on the MAC of Ågren et al. [2]. It is composed of a 68-bit NFSR and a 68-bit accumulator. In this work, we consider the NFSR with an arbitrary update function (and indeed the specification does not provide one). When a MAC is computed, the NFSR is loaded with the most significant 68 bits of the key, i.e. $k_{131}, \ldots, k_{64}$ and the accumulator is set to zero. To produce $MAC(X, k)$, the NFSR is clocked $|X|$ times, i.e. it is shifted left and the least significant bit is set to the feedback XORed with the input bit $X_i \oplus k_i$ where $i = 0, \ldots, |X| - 1$. If and only if $X_i = 1$, the accumulator is updated by XORing the current NFSR state to it (we assume this is done prior to clocking the NFSR). When $|X| > 64$, which is the case for the second MAC call, we assume that one re-uses $k_{63}, \ldots, k_0$ for clocking, until all of X is processed, although this makes no difference to our attack.

5.2 Cryptanalysis of RBS

The attack on the RBS scheme we present in the following uses a single chosen plaintext and has *expected worst case* time complexity 2^{65} and negligible memory complexity. The attack is based on the following observations:

Observation 1. *When computing $R = MAC(M, k)$, if $M = 1$, then it immediately follows from the definition of the MAC that $R = k_{131}\| \cdots \|k_{64}$, i.e. the 68 most significant bits of the key.*

Observation 2. *Assuming one knows $k_{a-1}\| \cdots \|k_0$ for some a with $1 \leq a \leq 132$, then one can determine the first $\ell := hw(k_{a-1}\| \cdots \|k_0)$ bits of R, as the bits of R are directly mapped to C by the k_i where $k_i = 1$. These in turn correspond to the first ℓ bits of $k_{131}\| \cdots \|k_{64}$. These can in turn be used to determine more of R, and so on.*

Combined, these observations imply that for $M = 1$, we know that $R = k_{131}\| \cdots \| k_{64}$. When guessing any number of the least significant key bits, a number of bits of R and thus of $k_{131}\| \cdots \|k_{64}$, equal to the Hamming weight of the guess, can be directly obtained from C.

Definition 1 (Free bit iteration). *The ith free bit iteration, with $i \geq 0$, refers to the number of bits obtained "for free" in one such iteration.*

Thus, the 0th free bit iteration refers to the analysis of how many free bits are obtained from the initially guessed key bits; the 1st free bit iteration refers to how many free bits are obtained from the ones obtained from the 0th free bit iteration, and so on.

For $i \geq 0$, in the ith free bit iteration, we let ℓ_i denote the *expected* number of free bits obtained and let δ_i denote the *expected* density of 1-bits in the remaining unknown bits, after obtaining the ℓ_i free bits.

Lemma 1. *Let $k_{a-1}\| \cdots \|k_0$ be the initially guessed key bits and let $\ell_0 = hw(k_{a-1}\| \cdots \|k_0)$. Then*

$$\delta_i = \frac{m - \sum_{j=0}^{i} \ell_j}{n + m - a - \sum_{j=0}^{i-1} \ell_j}, \qquad i \geq 0 \qquad \text{and}$$

$$\ell_i = \ell_{i-1}\delta_{i-1}, \qquad i \geq 1. \tag{2}$$

Proof. In the ith free bit iteration, $a + \sum_{j=0}^{i-1} \ell_j$ bits have already been guessed, so the denominator of δ_i is what remains unknown. The key bits guessed thus far have Hamming weight $\sum_{j=0}^{i} \ell_j$, so the 1-bits density among the last bits is δ_i.

The number of bits expected to obtained for free in iteration $i + 1$ is determined by the expected Hamming weight of the free bit portion just obtained in iteration i, which in turn is $\ell_i \delta_i$. ∎

We now derive a closed formula for the quantity ℓ_i by observing that the ratios ℓ_{i+1}/ℓ_i between consecutive elements of the sequence are actually constant, i.e. independent of i. We formally prove this in the following lemma.

Lemma 2. *Let a and ℓ_0 be such that $m - \ell_0 \neq n + m - a$ and $n + m - a \neq 0$. With the notations of Lemma 1, we have*

$$\ell_i = \left(\frac{m - \ell_0}{n + m - a} \right)^i \ell_0 \tag{3}$$

for $i \geq 1$.

Proof. We prove the claim by induction. For $i = 1$, Eq. (2) yields $\ell_1 = \frac{m - \ell_0}{n + m - a} \ell_0 = \left(\frac{m - \ell_0}{n + m - a} \right)^1 \ell_0$.

Assuming (3) holds for all $k \leq i$, we have

$$\ell_{i+1} = \frac{m - \sum_{j=0}^{i} \ell_j}{n + m - a - \sum_{j=0}^{i-1} \ell_j} \cdot \ell_i$$

$$= \frac{m - \ell_0 \sum_{j=0}^{i} \left(\frac{m - \ell_0}{n + m - a} \right)^j}{n + m - a - \ell_0 \sum_{j=0}^{i-1} \left(\frac{m - \ell_0}{n + m - a} \right)^j} \cdot \left(\frac{m - \ell_0}{n + m - a} \right)^i \ell_0. \tag{4}$$

For $r \neq 1$, the geometric series $\sum_{i=0}^{N} r^i$ is equal to $\frac{r^{N+1} - 1}{r - 1}$. Instantiating this with $r = \frac{a}{b}$ yields $\sum_{i=0}^{N} \left(\frac{a}{b} \right)^i = \frac{\left(\frac{a}{b} \right)^N a - b}{a - b}$ and $\sum_{i=0}^{N-1} \left(\frac{a}{b} \right)^i = \frac{\left(\frac{a}{b} \right)^N b - b}{a - b}$. Since $\left(\frac{m - \ell_0}{n + m - a} \right) \neq 1$, we can apply this to the two sums in Eq. (4), yielding

$$\ell_{i+1} = \frac{m - \ell_0 \left(\frac{\left(\frac{m - \ell_0}{n + m - a} \right)^i (m - l_0) - (n + m - a)}{-\ell_0 - n + a} \right)}{n + m - a - \ell_0 \left(\frac{\left(\frac{m - \ell_0}{n + m - a} \right)^i (n + m - a) - (n + m - a)}{-\ell_0 - n + a} \right)} \cdot \left(\frac{m - \ell_0}{n + m - a} \right)^i \ell_0,$$

which can be reformulated to

$$= \frac{\left(\frac{m(-\ell_0 - n + a) - \ell_0 \left(\frac{m - \ell_0}{n + m - a} \right)^i (m - \ell_0) + \ell_0 (n + m - a)}{-\ell_0 - n + a} \right)}{\left(\frac{(n + m - a)(-\ell_0 - n + a) - \ell_0 \left(\frac{m - \ell_0}{n + m - a} \right)^i (n + m - a) + \ell_0 (n + m - a)}{-\ell_0 - n + a} \right)} \cdot \left(\frac{m - \ell_0}{n + m - a} \right)^i \ell_0,$$

and collecting common terms gives

$$= \frac{(m - \ell_0) \left(\left(\frac{m - \ell_0}{n + m - a} \right)^i \ell_0 + n - a \right)}{\ell_0 + n - a} \cdot \frac{\ell_0 + n - a}{(n + m - a) \left(\left(\frac{m - \ell_0}{n + m - a} \right)^i \ell_0 + n - a \right)} \cdot \left(\frac{m - \ell_0}{n + m - a} \right)^i \ell_0$$

$$= \left(\frac{m - \ell_0}{n + m - a} \right) \cdot \left(\frac{m - \ell_0}{n + m - a} \right)^i \ell_0$$

$$= \left(\frac{m - \ell_0}{n + m - a} \right)^{i+1} \ell_0,$$

as claimed.   □

Note that the preconditions of the previous lemma are not imposing a limitation for the evaluation of the ℓ_i for relevant values of a. For instance, with $a = n$, the closed formula holds for any $1 \leq \ell_0 \leq m$, and the remaining case $\ell_0 = 0$ is trivial since all remaining unknown bits must be equal to one.

Optimal Choice of a. The closed formula of Lemma 3 also yields an estimate for the optimal number of key bits a that should be guessed initially. Specifically, we should choose $a < n + m$ such that $\ell_0 \sum_{i=0}^{\infty} \left(\frac{m - \ell_0}{n + m - a} \right)^i$ reaches $n + m - a$, the number of still unknown bits. Since

$$\ell_0 \sum_{i=0}^{\infty} \left(\frac{m - \ell_0}{n + m - a} \right)^i = \frac{1}{1 - \left(\frac{m - \ell_0}{n + m - a} \right)} \ell_0 = \left(\frac{n + m - a}{\ell_0 + n - a} \right) \ell_0,$$

this means that the optimal choice of a should be such that

$$\left(\frac{n + m - a}{\ell_0 + n - a} \right) \ell_0 = n + m - a$$

$$\Updownarrow$$

$$a = n \text{ or } a = n + m.$$

Note however that this only holds asymptotically, and it is expected that slightly more than n bits will need to be guessed to determine the remaining part of the.

For RBS this suggests that an initial guess of around $n = 64$ key bits should be sufficient to determine all remaining 68 key bits. In order to determine how many more bits than n we should guess, a more careful analysis of the progression of the ℓ_i's is needed. In the following, we develop a conservative estimate:

Lemma 3. *Let a and ℓ_i be as in Lemma 1. Let $L(a, \ell_0) = (\ell_0, \ldots, \ell_t)$ be the series of ℓ_i defined from a and ℓ_0 s.t. t is the largest integer s.t. $\ell_t \geq 1$. When guessing a initial key bits, the expected number of extra free bits obtained is determined as $\sum_{j=0}^{t-1} \ell_j$ and the expected Hamming weight of these bits is determined as $\sum_{j=0}^{t} \ell_j$.*

Proof. This follows directly from the definition of ℓ_i and $L(a, \ell_0)$. □

Theorem 1. *Let a, ℓ_i and $L(a, \ell_0)$ be as in Lemma 3. Let $w(a)$ denote the worst case expected complexity of key recovery when a is the number of key bits initially guessed. Then*

$$w(a) = \sum_{\ell_0 = \max\{0, a-64\}}^{\min\{68, a\}} \binom{a}{\ell_0} \binom{\max\{0, \lfloor 132 - a - \sum_{j=0}^{t-1} \ell_j \rfloor\}}{\max\{0, \lfloor 68 - \sum_{j=0}^{t} \ell_j \rfloor\}} \tag{5}$$

Proof. When initially guessing $k_{a-1} \| \cdots \| k_0$, the Hamming weight of this guess, ℓ_0, is bounded below by $\max\{0, a - 64\}$, because when $a > 0$, the Hamming weight must be positive by the pigeon-hole principle. The Hamming weight ℓ_0 is bounded above by either a or 68.

There are $\binom{a}{\ell_0}$ ways to distribute the ℓ_0 ones over $k_{a-1}\|\cdots\|k_0$. For each of these, the rightmost binomial coefficient of Eq. (5) gives the number of ways to place the remaining 1-bits among the unknown bits for this fixed combination of (a, ℓ_0). We take the sums of the ℓ_j as $\lfloor \sum_j \ell_j \rfloor$ for a conservative estimate of the complexity. Summing over all the possible ℓ_0 for a fixed a, the result follows. $\square$

The Key Recovery Attack. We summarize the resulting key recovery attack on RBS in Algorithm 5.

Algorithm 5. RBS-KEY-RECOVERY(a)

    Data: Number of initial key bits to guess, a

1 $C \leftarrow \mathsf{RBS}(1)$
2 **for** $\ell_0 = \max\{0, 64 - a\}, \ldots, \min\{68, a\}$ **do**
3   **forall the** guesses *of* $k'_{a-1}\|\cdots\|k'_0$ with Hamming weight ℓ_0 **do**
4   Let $L = (\ell_0, \ldots, \ell_t)$, where t is the largest integer s.t. $\ell_t \geq 1$
5         $\Xi \leftarrow \max\{0, 132 - a - \sum_{j=0}^{t-1} \ell_j\}$;   /* # of bits yet unknown */
6         $\Phi \leftarrow \max\{0, 68 - \sum_{j=0}^{t} \ell_j\}$;   /* # of 1-bits remaining */
7   **forall the** $\binom{\Xi}{\Phi}$ remaining candidates for $k'_{131}\|\cdots\|k'_{131-\Xi+1}$ **do**
8   **if** $C = \mathsf{RBS}(1)$ under the key $k'_{131}\|\cdots\|k'_0$ **then**
9   **return** $k'_{131}\|\cdots\|k'_0$ as the correct key k
10   **end**
11   **end**
12   **end**
13 **end**

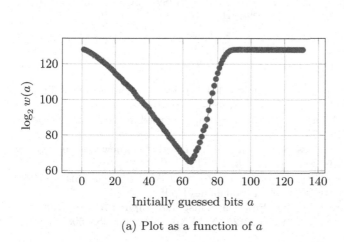

a	$\log_2 w(a)$
61	68.24
62	67.32
63	66.22
64	65.27
65	65.00
66	66.75
67	68.39
68	71.18
69	72.83
70	76.03

(a) Plot as a function of a

(b) Data points for the best values of a

Fig. 4. Expected worst time complexity for key recovery in RBS as a function of the number of bits initially guessed, denoted a

It remains to determine the number of key bits a that should be guessed initially. Figure 4 shows the base-2 logarithm of the expected worst case complexity $w(a)$. While Fig. 4a shows a plot of $w(a)$ with $a \in \{1, \ldots, 131\}$, Fig. 4b gives a numerical illustration of the best values for a giving the lowest complexity. From the data, we find that guessing $a = 65$ bits gives the lowest key recovery complexity of 2^{65}.

6   Conclusion

In this paper we presented key recovery attacks on three recent authenticated ciphers: AVALANCHE, Calico and RBS. The two former are submissions to the ongoing CAESAR competition for authenticated encryption schemes while the latter is a proposal for use in lightweight applications.

Common to all three designs is that they make use of solid primitives such as the AES, SipHash and ChaCha. We stress that the attacks presented here are purely structural, i.e. the weaknesses are present due to *the way the primitives are combined* and not the primitives themselves.

For AVALANCHE and Calico, the key recovery is possible due to the nonce being used as (part of) the key material, thus facilitating a key collision attack. For RBS, we used a guess-and-determine approach. In all cases, the key was recovered with a complexity of at most square root of the brute-force effort. Our attacks allows an adversary to perform universal forgeries in all three cases, and for AVALANCHE and RBS this extends to the ability to decrypt arbitrary ciphertexts.

Acknowledgments. The work has been supported in part by the Austrian government through the research program FIT-IT Trust in IT Systems (project 835919) and by the Austrian Science Fund (project P26494-N15).

References

1. CAESAR: Competition for Authenticated Encryption: Security, Applicability, and Robustness, March 2014. http://competitions.cr.yp.to/caesar.html
2. Ågren, M., Hell, M., Johansson, T.: On hardware-oriented message authentication. IET Inf. Secur. **6**, 329–336 (2012)
3. Ågren, M., Hell, M., Johansson, T., Meier, W.: Grain-128a: a new version of Grain-128 with optional authentication. IJWMC **5**, 48–59 (2011)
4. Alomair, B.: AVALANCHEv1. Submission to the CAESAR competition (2014). http://competitions.cr.yp.to/round1/avalanchev1.pdf
5. Aumasson, J.-P., Bernstein, D.J.: SipHash: a fast short-input PRF. In: Galbraith, S., Nandi, M. (eds.) INDOCRYPT 2012. LNCS, vol. 7668, pp. 489–508. Springer, Heidelberg (2012)
6. Bellare, M., Namprempre, C.: Authenticated encryption: relations among notions and analysis of the generic composition paradigm. In: Okamoto, T. (ed.) ASIACRYPT 2000. LNCS, vol. 1976, pp. 531–545. Springer, Heidelberg (2000)

7. Bernstein, D.J.: ChaCha, a variant of Salsa20. In: Workshop Record of SASC 2008: The State of the Art of Stream Ciphers (2008)
8. Bertoni, G., Daemen, J., Peeters, M., Van Assche, G.: The Keccak SHA-3 submission. Submission to NIST (2011)
9. Dworkin, M.J.: SP 800–38C. Recommendation for block cipher modes of operation: the CCM mode for authentication and confidentiality. Technical report, Gaithersburg, MD, United States (2004)
10. Dworkin, M.J.: SP 800–38D. Recommendation for block cipher modes of operation: galois/counter mode (GCM) and GMAC. Technical report, Gaithersburg, MD, United States (2007)
11. ISO 19772:2009. Information technology - Security techniques - Authenticated encryption (2009)
12. Jeddi, Z., Amini, E., Bayoumi, M.: A novel authenticated cipher for RFID systems. Int. J. Crypt. Inf. Secur. 4 (2014)
13. Katz, J., Yung, M.: Unforgeable encryption and chosen ciphertext secure modes of operation. In: Schneier, B. (ed.) FSE 2000. LNCS, vol. 1978, pp. 284–299. Springer, Heidelberg (2001)
14. Krovetz, T., Rogaway, P.: The software performance of authenticated-encryption modes. In: Joux, A. (ed.) FSE 2011. LNCS, vol. 6733, pp. 306–327. Springer, Heidelberg (2011)
15. McGrew, D.A., Viega, J.: The security and performance of the galois/counter mode (GCM) of operation. In: Canteaut, A., Viswanathan, K. (eds.) INDOCRYPT 2004. LNCS, vol. 3348, pp. 343–355. Springer, Heidelberg (2004)
16. Mendel, F., Mennink, B., Rijmen, V., Tischhauser, E.: A simple key-recovery attack on McOE-X. In: Pieprzyk, J., Sadeghi, A.-R., Manulis, M. (eds.) CANS 2012. LNCS, vol. 7712, pp. 23–31. Springer, Heidelberg (2012)
17. Minematsu, K.: Parallelizable rate-1 authenticated encryption from pseudorandom functions. In: Nguyen, P.Q., Oswald, E. (eds.) EUROCRYPT 2014. LNCS, vol. 8441, pp. 275–292. Springer, Heidelberg (2014)
18. Quisquater, J.-J., Delescaille, J.-P.: How easy is collision search. new results and applications to DES. In: Brassard, G. (ed.) CRYPTO 1989. LNCS, vol. 435, pp. 408–413. Springer, Heidelberg (1990)
19. Rogaway, P.: Efficient instantiations of tweakable blockciphers and refinements to modes OCB and PMAC. In: Lee, P.J. (ed.) ASIACRYPT 2004. LNCS, vol. 3329, pp. 16–31. Springer, Heidelberg (2004)
20. Rogaway, P., Bellare, M., Black, J., Krovetz, T.: OCB: a block-cipher mode of operation for efficient authenticated encryption. In: ACM Conference on Computer and Communications Security, pp. 196–205 (2001)
21. Taylor, C.: The Calico Family of Authenticated Ciphers Version 8. Submission to the CAESAR competition (2014). http://competitions.cr.yp.to/round1/calicov8.pdf

Tuning GaussSieve for Speed

Robert Fitzpatrick[1], Christian Bischof[2], Johannes Buchmann[3],
Özgür Dagdelen[3]([✉]), Florian Göpfert[3],
Artur Mariano[2], and Bo-Yin Yang[1]

[1] Institute of Information Science, Academia Sinica, Taipei, Taiwan
[2] Institute for Scientific Computing, TU Darmstadt, Darmstadt, Germany
[3] Cryptography and Computer Algebra, TU Darmstadt, Darmstadt, Germany
oezguer.dagdelen@cased.de

Abstract. The area of lattice-based cryptography is growing ever-more prominent as a paradigm for quantum-resistant cryptography. One of the most important hard problem underpinning the security of lattice-based cryptosystems is the *shortest vector problem* (SVP). At present, two approaches dominate methods for solving instances of this problem in practice: enumeration and sieving. In 2010, Micciancio and Voulgaris presented a heuristic member of the sieving family, known as GaussSieve, demonstrating it to be comparable to enumeration methods in practice. With contemporary lattice-based cryptographic proposals relying largely on the hardness of solving the shortest and closest vector problems in *ideal* lattices, examining possible improvements to sieving algorithms becomes highly pertinent since, at present, only sieving algorithms have been successfully adapted to solve such instances more efficiently than in the random lattice case. In this paper, we propose a number of heuristic improvements to GaussSieve, which can also be applied to other sieving algorithms for SVP.

1 Introduction

Lattice-based cryptography is gaining increasing traction and popularity as a basis for post-quantum cryptography, with the Shortest Vector Problem (SVP) being one of the most important computational problems on lattices. Its difficulty is closely related to the security of most lattice-based cryptographic constructions to date. The SVP consists in finding a shortest (with respect to a particular, usually Euclidean, norm) non-zero lattice point in a given lattice.

For solving SVP instances, we have a choice of algorithms available. In recent works, heuristic variants of Kannan's simple enumeration algorithm have dominated. The original algorithm [11] solves SVP (deterministically) with time complexity $n^{\frac{n}{2}+o(n)}$ (n being the lattice dimension). More recent works (such as [7]) allow probabilistic SVP solution, sacrificing guaranteed solution for run-time improvements.

A more recently-studied family of algorithms is known as lattice sieving algorithms, introduced in the 2001 work of Ajtai et al. [4]. In 2008, Nguyen and Vidick [21] presented a careful analysis of the algorithm of Ajtai et al., showing it to possess time complexity of $2^{5.90n+o(n)}$ and space complexity $2^{2.95n+o(n)}$.

© Springer International Publishing Switzerland 2015
D.F. Aranha and A. Menezes (Eds.): LATINCRYPT 2014, LNCS 8895, pp. 288–305, 2015.
DOI: 10.1007/978-3-319-16295-9_16

Heuristic variants of [21], which run significantly faster than proven lower bounds are presented in [21,27,28]. In 2010, Micciancio and Voulgaris [19] proposed two new algorithms: ListSieve and a heuristic derivation known as GaussSieve, with GaussSieve being the most practical sieving algorithm known at present. While no runtime bound is known for GaussSieve, the use of a simple heuristic stopping condition, in practice, appears effective with no cases being known (to the best of our knowledge) in which GaussSieve fails to return a shortest non-zero vector.

For purposes of enhanced communication, computation and memory complexity, many recent lattice-based cryptographic proposals employ *ideal* lattices rather than "random" lattices. Ideal lattices, in brief, possess significant additional structure which allows much more attractive implementation of said proposals. However, as with any introduction of structure, the question of any simultaneously-introduced weakening of the underlying problems arises. In 2011, Schneider [22] illustrated that (following a suggestion in the work of Micciancio and Voulgaris [19]) one can take advantage of the additional structure present in ideal lattices in a simple way to obtain substantial speedups for such cases. Interestingly, no such comparable techniques are known for other SVP algorithms, with only sieving algorithms appearing to be capable of exploiting the additional structure exposed in ideal lattices.

Another attractive feature of sieving algorithms is their relative amenability to parallelization. Also in 2011, Milde and Schneider [20] proposed a parallel implementation of GaussSieve, though the methodology used limited the number of threads to about ten, before no substantial further speedups could be obtained. In 2013, Ishiguro et al. [10] proposed a somewhat more natural parallelization of GaussSieve, allowing a much larger number of threads. Using such an approach, they report the solution of the 128-dimensional ideal lattice challenge [2] in 30,000 CPU hours. Currently, the most efficient GaussSieve implementation (of which details have been published) is due to Mariano et al. [16] who implemented GaussSieve with a particular effort to avoid resource contention. In this work, we exhibit several further speedups which can be obtained both in the random and ideal lattice cases.

While the security of most lattice-based cryptographic constructions relies on the difficulty of *approximate* versions of the related Closest Vector Problem (CVP) and SVP, the importance of improving *exact* SVP solvers stems from their use (following Schnorr's hierarchy [23]) in the construction of approximate CVP/SVP solvers. Thus, any improvements, both theoretically and experimentally, in exact SVP solvers can lead to a need for re-appraisal of proposed parameterizations.

Our Contribution. In this work, we highlight several practical improvements that are applicable to other sieving algorithms. In particular, we propose the following optimizations, which we incorporated into GaussSieve:

– We correct an error in the Gaussian sampler of the reference implementation of Voulgaris and propose an *optimized Gaussian sampler* in which we dynamically adapt the Gaussian parameter used during the execution of the algorithm. Our experiments show that GaussSieve with our optimized Gaussian sampler requires significantly fewer iterations to terminate and leads to a speedup of up to 3.0× over the corrected reference implementation in random lattices in dimension 60–70.

- The use of *multiple randomized bases* to seed the list before running the sieving process offers substantial efficiency gains. Indeed, the speedup appears to grow linearly in the dimension of the underlying lattice.
- We introduce a very efficient heuristic to compute *a first approximation to the angle between two vectors* in order to test cheaply whether there is the need to compute full inner products for the reduction process. This optimization is possibly of independent interest beyond sieving algorithms.

We note that our improvements can be integrated into parallel versions of GaussSieve without complication or restriction.

2   Background and Notation

A (full-rank) lattice Λ in $\mathbb{R}^n$ is a discrete additive subgroup. For a general introduction, the reader is referred to [18]. We view a lattice as being generated by a (non-unique) basis $\mathbf{B} = \{\mathbf{b}_0, \ldots, \mathbf{b}_{n-1}\} \subset \mathbb{R}^n$ of linearly-independent vectors. We assume that the vectors $\mathbf{b}_0, \ldots, \mathbf{b}_{n-1}$ form the rows of the $n \times n$ matrix $\mathbf{B}$. That is,

$$\Lambda = \mathcal{L}(\mathbf{B}) = \mathbb{Z}^n \cdot \mathbf{B} = \left\{ \sum_{i=0}^{n-1} x_i \cdot \mathbf{b}_i \mid x_0, \ldots, x_{n-1} \in \mathbb{Z} \right\}.$$

The rank of a lattice Λ is the dimension of the linear span span(Λ) of Λ. The basis $\mathbf{B}$ is not unique, and thus we call two bases $\mathbf{B}$ and $\mathbf{B}'$ *equivalent* if and only if $\mathbf{B}' = \mathbf{B}\mathbf{U}$ where $\mathbf{U}$ is a unimodular matrix, i.e., an integer matrix with $|\det(\mathbf{U})| = 1$. We note that such unimodular matrices form the general linear group $GL_n(\mathbb{Z})$. Being a discrete subgroup, in any lattice there exists a subset of vectors which possess minimal (non-zero) norm amongst all vectors. When asked to solve the shortest vector problem, we are given a lattice basis and asked to deliver a member of this subset. SVP is known to be NP-hard under randomized reductions [3].

Random Lattices. Throughout this work, we rely on experiments with "random" lattices. However, the question of what a "random" lattice is and how to generate a random basis of one are non-trivial. In a mathematical sense, an answer to the definition of a random lattice follows from a work in 1945 by Siegel [25], with efficient methods for sampling such random lattices being proposed, for instance, by Goldstein and Mayer [9]. In this work, all experiments were conducted with Goldstein-Mayer lattices, as provided by the TU Darmstadt Lattice Challenge project. For more details, the reader is directed to [8].

Definition 1. *Given two vectors* $\mathbf{v}, \mathbf{w}$ *in a lattice* Λ, *we say that* $\mathbf{v}, \mathbf{w}$ *are Gauss-reduced if*

$$\min(\|\mathbf{v} \pm \mathbf{w}\|) \geq \max(\|\mathbf{v}\|, \|\mathbf{w}\|) .$$

Lattice Basis Reduction. A given lattice has an infinite number of bases. The aim of lattice basis reduction is to transform a given lattice basis into one which contains vectors which are both relatively short and relatively orthogonal. Such bases, in some sense, allow easier and/or more accurate solutions of approximation variants of SVP or its related problem, the Closest Vector Problem (CVP). In practice, the most effective arbitrary-dimension lattice basis reduction algorithms are descendants of the LLL algorithm [14], with the Block-Korkine-Zolotarev (BKZ) family [5,23] (or framework) of algorithms being the most effective in practice. The LLL and BKZ algorithms rely on successive exact SVP solution in a number of projected lattices. These projected lattices are two-dimensional in the case of LLL and of arbitrary dimension in the case of BKZ – the (maximal) projected lattice dimension being termed the "blocksize" in BKZ. For more details, the reader is referred to [6].

Balls and Spheres. We define the Euclidean n-sphere $\mathcal{S}_n(\mathbf{x}, r)$ centered at $\mathbf{x} \in \mathbb{R}^{n+1}$ and of radius r by $\mathcal{S}_n(\mathbf{x}, r) := \{\mathbf{y} \in \mathbb{R}^{n+1}: \; \| \mathbf{x} - \mathbf{y} \| = r\}$. The (open) Euclidean n-ball $\mathcal{B}_n(\mathbf{x}, r)$ centered at $\mathbf{x} \in \mathbb{R}^n$ and of radius r is defined to be $\mathcal{B}_n(\mathbf{x}, r) := \{\mathbf{y} \in \mathbb{R}^n: \; \| \mathbf{x} - \mathbf{y} \| < r\}$.

Gaussians. The discrete Gaussian distribution with parameter s over a lattice Λ is defined to be the probability distribution with support Λ which, for each $\mathbf{x} \in \Lambda$, assigns probability proportional to $\exp(-\pi\|\mathbf{x}\|^2/s^2)$.

Miscellany. We use $\oplus$ to denote the bitwise XOR operation and use $\mathbf{a} \angle \mathbf{b}$ to denote the angle between vectors $\mathbf{a}$ and $\mathbf{b}$. Given a binary vector $\mathbf{a}$, we use $\mathrm{w}(\mathbf{a})$ to denote the Hamming weight of $\mathbf{a}$.

3   The GaussSieve Algorithm

In 2010, Micciancio and Voulgaris [19] introduced the GaussSieve algorithm. GaussSieve is a heuristic efficient variant of the ListSieve algorithm. In contrast to GaussSieve, for ListSieve there exist provable bounds on the running time and space requirements. An empirical comparison of ListSieve and GaussSieve is given in [17]. In this work, we focus on the most efficient variant GaussSieve. Algorithm 1 depicts the GaussSieve algorithm in more detail.

GaussSieve operates upon a supplied lattice basis **B**. It utilizes a dynamic list L of lattice points. At each iteration, GaussSieve samples a new lattice point – typically with Klein's algorithm [12] – and attempts to reduce that vector against vectors in the list L. By "reducing" we mean adding an integer multiple of a list vector such that the norm of the resulting vector is reduced. Once the vector cannot be reduced further by list members, the resulting vector is incorporated in the list. Afterwards, all the vectors in the list L are tested to determine if they can be reduced against this new vector. If so, those vectors are removed to a stack S, with the stack playing the role of Klein's algorithm in subsequent iterations till it is depleted. This ensures that all vectors in the list L remain pairwise

---

Algorithm 1: GaussSieve

1 **Input** : Basis **B**, collision limit c
  Output: $\mathbf{v} : \mathbf{v} \in \Lambda(\mathbf{B}) \wedge \|\mathbf{v}\| = \lambda_1(\mathbf{B})$
2 $\mathsf{L} \leftarrow \{\}, \mathsf{S} \leftarrow \{\}, \mathsf{col} \leftarrow 0$
3 **while** $\mathsf{col} < c$ **do**
4   **if** S is not empty **then**
5         $\mathbf{v} \leftarrow \mathsf{S}.\mathrm{pop}()$
6   **else**
7         $\mathbf{v} \leftarrow \mathrm{SampleKlein}(\mathbf{B})$
8      $j \leftarrow \mathrm{GaussReduce}(\mathbf{v}, \mathsf{L}, \mathsf{S})$
9   **if** $j = \mathrm{true}$ **then**
10         $\mathsf{col} \leftarrow \mathsf{col} + 1$
11 **return** $v \in \mathsf{L}$ s.t. $\|v\| = \min_{\mathbf{x} \in \mathsf{L}} \|\mathbf{x}\|$

function GaussReduce(p,L,S)
   was_reduced $\leftarrow$ true
   while was_reduced $= \mathit{true}$ **do**
     was_reduced $\leftarrow$ false
     for all $\mathbf{v}_i \in \mathsf{L}$ **do**
       if $\exists t \in \mathbb{Z} : \|\mathbf{p} + t\mathbf{v}_i\| < \|\mathbf{p}\|$ **then**
         $\mathbf{p} \leftarrow \mathbf{p} + t\mathbf{v}_i$
         was_reduced $\leftarrow$ true
   if $\|\mathbf{p}\| = 0$ **then**
     return true
   for all $\mathbf{v}_i \in \mathsf{L}$ **do**
     if $\exists u \in \mathbb{Z} : \|\mathbf{v}_i + u\mathbf{p}\| < \|\mathbf{v}_i\|$ **then**
       $\mathsf{L} \leftarrow \mathsf{L} \backslash \{\mathbf{v}_i\}$
       $\mathbf{v}_i \leftarrow \mathbf{v}_i + u\mathbf{p}$
       $\mathsf{S}.\mathrm{push}(\mathbf{v}_i)$
   $\mathsf{L} \leftarrow \mathsf{L} \cup \{\mathbf{p}\}$
   return false
end function

Gauss-reduced at any point during the execution of the algorithm. Eventually, by this iterative process, the shortest vector in the lattice is found (with high probability). In the following, we detail the GaussSieve algorithm in several aspects.

Sampling. In order to populate the list with reasonably short vectors, GaussSieve samples lattice points via Klein's randomized algorithm [12], following the suggestion in [21]. Klein's algorithm, upon input a lattice basis **B** outputs a lattice point distributed according to a zero-centered Gaussian of parameter s over the lattice $\Lambda(\mathbf{B})$. Since the vectors so derived are small integer combinations of the supplied basis vectors, the norms of these vectors are strongly dependent on the "quality" of the supplied basis. Hence, reducing the input basis with "stronger" lattice-reduction algorithms yields shorter vectors output and thus, intuitively and in practice, GaussSieve terminates earlier than when given a "less-reduced" basis from which to sample vectors. However, while the cost of enumeration algorithms is strongly affected by the strength of the lattice-reduction employed, such a strong correspondence does not appear to hold for the case of GaussSieve - such issues are discussed further in Sect. 5.

Reduction. We attempt to reduce the given vector **p** (obtained either from Klein's algorithm or from the stack) against all list vectors, i.e., we try to find a list vector **v** and integer t such that $\|\mathbf{p} + t\mathbf{v}\| < \|\mathbf{p}\|$, in which case we reduce **p** using **v**. Once no such **v** exists in the list, we attempt to reduce the extant list vectors against **p**. All list vectors which can be reduced using **p** are duly reduced, removed from the list and inserted to the stack S. As a result, GaussSieve maintains its list L in a pairwise reduced state at the close of every iteration. In the following iteration, if the stack contains at least one element, we pop a vector from the stack in lieu of employing Klein's algorithm.

Stopping criteria. Given that one cannot prove (at present) that GaussSieve terminates, stopping conditions for GaussSieve must be chosen in a heuristic way, chosen such that any further reduction in the norm of the shortest vector found is unlikely to occur. In [19], it is suggested to terminate the algorithm after a certain number of successively-sampled vectors are all reduced to zero using the extant list, with 500 such consecutive zero reductions being mentioned as a possible choice in practice. In Voulgaris' implementation, a stopping condition is employed which depends on the maximal list size encountered. In our experiments we follow the suggestions of [19] in this regard.

Complexity. As with all sieving algorithms, the complexity of GaussSieve is largely determined by arguments related to sphere packing and the *Kissing Number* - the maximum number of equivalent hyperspheres in n dimensions which are permitted to touch another equivalent hypersphere yet not intersect. With practical variants of GaussSieve, as dealt with here, no complexity bound is known due to the possibility of perpetual reductions of vectors to zero without a shortest vector being found. For further details, we direct the reader to [19].

4   Approximate Gauss Reduction

The motivation for our first contribution stems from the observation that, at least in moderate dimension, the overwhelming majority of vector pairs we consider are already Gauss-reduced, yet we expend the vast majority of effort in the algorithm in verifying that they are indeed Gauss-reduced. Thus, by "detecting" relatively cheaply whether such a pair is *almost-certainly* Gauss-reduced, we can

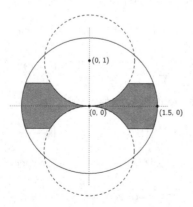

Fig. 1. Example Gauss-reduced region (shaded), dimension 2.

obtain substantial (polynomial) speedups at the cost of possibly erring (almost inconsequentially) with respect to a few pairs.

We now make an idealizing assumption, namely the random ball assumption (as appears in [26]) that we can gain insights into the behavior of lattice algorithms by assuming that lattice vectors are sampled uniformly at random from the surface of an Euclidean ball of a given radius. As in [26], we term this the "random ball model". For intuition, Fig. 1 shows the region (shaded) of vectors in the ball $\mathcal{B}_2(\mathbf{0}, 1.5)$ which are Gauss-reduced with respect to the vector $(0, 1)$.

Lemma 1. *Given a vector* $\mathbf{v} \in \mathbb{R}^n$ *of (Euclidean) norm* r *sampled from at random from* $\mathcal{S}_{n-1}(\mathbf{0}, r)$ *and a second vector* $\mathbf{w}$ *sampled independently at random from* $\mathcal{S}_{n-1}(\mathbf{0}, r')$ *(where* r' *is a second radius), the probability that* $\mathbf{w}$ *is Gauss-reduced with respect to* $\mathbf{v}$ *is*

$$1 - I_{1-(r/2r')^2}\left(\frac{n-1}{2}, \frac{1}{2}\right)$$

where $h_h := r' - r/2$ *and* $I_x(a, b)$ *denotes the regularized incomplete beta function.*

Proof. We assume, without loss of generality, that $r' \geq r$, otherwise, we swap $\mathbf{v}$ and $\mathbf{w}$. The surface area of the n-sphere of radius r' (denoted $\mathcal{S}_{n-1}(\mathbf{0}, r')$) is

$$\mathcal{S}_{n-1}(\mathbf{0}, r') = \frac{n\pi^{n/2}r'^{n-1}}{\Gamma(1 + \frac{n}{2})}.$$

Then, the points from this sphere which are Gauss-reduced with respect to $\mathbf{v}$ are determined by the relative complement of $\mathcal{S}_{n-1}(\mathbf{0}, r')$ with the hyper-cylinder of radius $\sqrt{r'^2 - r^2/4}$, of which both the origin and $\mathbf{v}$ lie on the center-line. We can calculate the surface area of this relative complement by subtracting the surface area of a certain hyperspherical cap from the surface area of a hemisphere of the hypersphere of radius r'. Specifically, let us consider only one hemisphere of $\mathcal{S}_{n-1}(\mathbf{0}, r')$. Considering the hyperspherical cap of height $h_h := r' - r/2$, this cap has surface area

$$\frac{1}{2}\mathcal{S}_{n-1}(\mathbf{0}, r')I_{1-(r/2r')^2}\left(\frac{n-1}{2}, \frac{1}{2}\right),$$

where $I_x(a, b)$ denotes the regularized incomplete beta function:

$$I_x(a, b) := \sum_{i=a}^{\infty} \binom{a+b-1}{i} x^i(1-x)^{a+b-1-i}.$$

Thus, the relative complement has surface area

$$\frac{1}{2}\mathcal{S}_{n-1}(\mathbf{0}, r')\left(1 - I_{1-(r/2r')^2}\left(\frac{n-1}{2}, \frac{1}{2}\right)\right)$$

and hence, the probability of obtaining a Gauss-reduced vector is

$$1 - I_{1-(r/2r')^2}\left(\frac{n-1}{2}, \frac{1}{2}\right).$$

□

For instance, Fig. 2 gives the probability of two vectors being *a priori* Gauss-reduced with increasing dimension in the case of $r = 1000$ and $r' = 1100$. By *a priori* Gauss-reduced, we mean that two vectors, sampled at random from zero-centered spheres of respective radii, are Gauss-reduced with respect to each other. These illustrative values are chosen to be representative of the similar-norm pairs of vectors which comprise the vast majority of attempted reductions in GaussSieve.

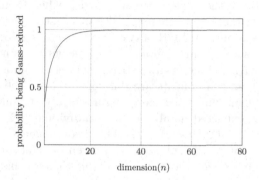

Fig. 2. Example probabilities of *a priori* Gauss-reduction, $r = 1000$, $r' = 1100$.

If we are given two such vectors, we can easily determine whether they are Gauss-reduced by considering the angle θ between them. It follows simply from elementary Euclidean geometry that if the following condition is satisfied, they are Gauss-reduced:

$$|\frac{\pi}{2} - \theta| \leq \arcsin(r/2r')$$

Thus, if we can "cheaply" determine an approximate angle, we can tell with good confidence whether they are indeed Gauss-reduced or not. We note that, while we do not believe one can prove similar arguments to the above in the context of lattices, the behavior appears indistinguishable for random lattices in practice. Indeed, we also experimented with vector pairs sampled from random lattice bases using Klein's algorithm and obtained identical behavior to that illustrated in Figs. 2 and 3. For determining such approximate angles, we investigated two approaches: (a) computing the angle between restrictions of vectors to subspaces and (b) exploiting correlations between the XOR + population count of the sign bits of a pair of vectors and the angle between them. We only report the latter approach, which appears to offer superior results in practice.

Using XOR and Population Count as a First Approximation to the Angle. Given a vector $\mathbf{a} \in \mathbb{Z}^n$ we define $\tilde{\mathbf{a}} \in \mathbb{Z}_2^n$ such that $\tilde{\mathbf{a}}_i = \mathrm{sgn}(\mathbf{a}_i)$. Here, we define

$$\text{sgn}(a) : \mathbb{R} \to \{0,1\} \quad \text{by sgn}(a) = \begin{cases} 0 & \text{if } a < 0 \\ 1 & \text{otherwise} \end{cases}$$

and define the normalized XOR followed by population count of $\mathbf{a}$ and $\mathbf{b}$ to be

$$\text{sip}(\mathbf{a}, \mathbf{b}) : \mathbb{R}^n \times \mathbb{R}^n \to \mathbb{R}^+ \quad \text{by sip}(\mathbf{a}, \mathbf{b}) = \text{w}(\tilde{\mathbf{a}} \oplus \tilde{\mathbf{b}})/n$$

Based on Assumption 1, we can use the XOR + population count of $\mathbf{a}$ and $\mathbf{b}$ as a first approximation to the angle between $\mathbf{a}$ and $\mathbf{b}$ when their norms are relatively similar. The attraction of using $\text{sip}(\mathbf{a}, \mathbf{b})$ as a first approximation to $\mathbf{a} \angle \mathbf{b}$ is the need to only compute an XOR of two binary vectors followed by a population count, operations which can be implemented efficiently. For intuition, consider the first components a_1, b_1 of vectors $\mathbf{a}$ and $\mathbf{b}$, respectively. If $\text{sgn}(a_1) \oplus \text{sgn}(b_1) = 1$ then the signs of these components are different and are the same otherwise. Clearly, in higher dimensions, when sampling uniformly at random from a zero-centered sphere, the expected number of such individual XORs would be $n/2$, hence $\text{E}[\text{sip}(\mathbf{a}, \mathbf{b})] = 1/2$. If $\text{sip}(\mathbf{a}, \mathbf{b}) = 1$, then all components of both vectors lie in the same intersection of the sphere with a given orthant and thus we might expect that the angle between these two vectors has a good chance of being relatively small. The analogous case of $\text{sip}(\mathbf{a}, \mathbf{b}) = 0$ corresponds to taking the negative of one of the vectors. Conversely, since the expected value of $\text{sip}(\mathbf{a}, \mathbf{b})$ is $1/2$, we expect this to coincide with the heuristic that, in higher dimensions, most vectors sampled uniformly at random from a zero-centered sphere are almost orthogonal. Again, we stress that these arguments are given purely for intuition and appear to work well in practice, as posited in Assumption 1:

Assumption 1. *[Informal] Let $n \gg 2$. Then, given a random (full-rank) lattice Λ of dimension n and two vectors $\mathbf{a}, \mathbf{b} \in \Lambda$ of "similar" norms sampled uniformly at random from the set of all such lattice vectors, the distribution of the normalized sign XOR + population count of these vectors $\text{sip}(\mathbf{a}, \mathbf{b})$ and the angle between them can be approximated by a bivariate Gaussian distribution.*

Note 1. We note that, in our experiments, we took "similar" norm to mean that $\max\{\|\mathbf{a}\|/\|\mathbf{b}\|, \|\mathbf{b}\|/\|\mathbf{a}\|\} \leq 1.2$, with a failure to satisfy this condition leading to full inner product calculation.

Application of Mardia's test [15] for multivariate normalcy yields confirmative results. As an example, the covariance matrix below provides a good approximation of this distribution, in dimension 96 as shown by our experiments.

$$\begin{bmatrix} 0.01200 & -0.00307 \\ -0.00307 & 0.00290 \end{bmatrix}$$

For example, Fig. 3 shows the result of 100,000 pairs of vectors sampled according to a discrete Gaussian from a 96-dimensional random lattice, with the region lying between the horizontal lines containing the cases in which we assume that the pair of vectors is Gauss-reduced and hence do not expend effort in computing the full inner-product to confirm this. More specifically, we choose an integer parameter k and, when we wish to compute the angle between vectors $\mathbf{a}$ and $\mathbf{b}$, we firstly compute $c = \mathrm{sip}(\mathbf{a}, \mathbf{b})$. If $(\lfloor n/2 \rfloor - k)/n \le c \le (\lceil n/2 \rceil + k)/n$ we assume that $\mathbf{a}$ and $\mathbf{b}$ are already Gauss-reduced. Otherwise, we compute $\langle \mathbf{a}, \mathbf{b} \rangle$.

Choosing k, i.e. determining the distance of the horizontal lines from $n/2$ to n was done heuristically, with values of 6 or 7 appearing to work best for the lattice dimensions with which we experimented. If k is too small, the heuristic loses value, while if it is too large we will commit too many false negatives (missed reductions) which will lead to a decreased speedup. In the case of Fig. 3, $k = 6$. The occurrence of a few false negatives arising from this approach appears to have little consequence for the algorithm - this assumption appears to be borne out by the experiments reported in Sect. 7. We also note that false positives cannot occur.

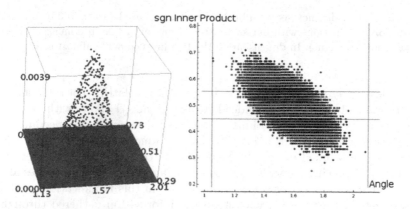

Fig. 3. Example distribution of $sip(\mathbf{a}, \mathbf{b})$ and angle between random unit vectors in dimension 96.

5   Using Multiple Randomized Bases

When examining the performance of enumeration-type algorithms in solving SVP instances, the level of preprocessing carried out on the basis is of prime importance, with the norms of the Gram-Schmidt vectors of the reduced basis being the main determinant of the running time. With sieving algorithms, however, this does not appear to hold - the level of preprocessing carried out on the basis has a far smaller impact on the running time of the sieving algorithms than might be expected at first.

We posit that a much more natural consideration is the number of randomized lattice bases which are reduced and used to "seed" the list. That is, instead of adding the input basis to the list before starting the sieving procedure, we randomize and reduce the given basis several times, appending all so-obtained lattice vectors to the list L by running GaussReduce($\mathbf{b}_i, \mathsf{L}, \mathsf{S}$) for all obtained vectors $\mathbf{b}_i$ (cf. Algorithm 1).

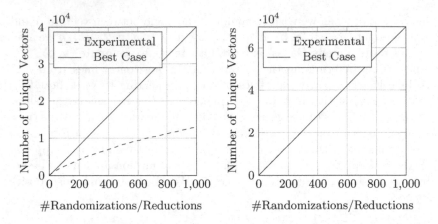

Fig. 4. Number of distinct vectors obtained under repeated (up to 1000) randomizations and BKZ reductions with blocksize 20 for random lattice in dimension 40 (left) and dimension 70 (right). In dimension 70 the two lines coincide almost exactly.

The idea of rerandomizing and reducing a given lattice basis for algorithmic improvements is not new. Indeed, Gama et al. [7] show, with respect to enumeration-based SVP algorithms, a theoretical exponential speedup if the input basis is rerandomized, reduced and the enumeration search tree for each reduced basis is pruned extremely. Experiments confirm this huge speedup in practice [13]. While in enumeration rerandomizing and reducing provides almost independent instances of (pruned) enumeration, in this modification to Gauss-Sieve we instead concurrently exploit all the information gathered through all generated bases in a single instance of GaussSieve rather than running multiple instances of GaussSieve.

However, a natural concern that arises in this setting is that of the number of unique lattice vectors we can hope to obtain by way of multiple randomization and reduction - we wish for this number to be as large as possible to maximize the size of our starting list. Our experiments indicate that, given a large enough lattice dimension, the number of duplicate vectors obtained by this approach is negligible even when performing a few thousand such randomizations and reductions. Figure 4 illustrates the number of distinct vectors obtained through this approach in dimensions 40 and 70, highlighting that, beyond toy dimensions, obtaining distinct vectors through this approach is not problematic. We also observe that such a "seeding" of the list is only slightly more costly in practice as this approach makes the first stage of the algorithm embarrassingly parallel, i.e. each thread can carry out an independent basis randomization and reduction, with a relatively fast merging of the resulting collection of vectors into a pairwise Gauss-reduced list.

After seeding the list using the vectors from the reduced bases, we additionally store these bases and, rather than sampling all vectors from a single basis, sample from our multiple bases in turn. We note that our optimizations have some similarities with the random sampling algorithm of Schnorr [24]. Here, short lattice vectors are sampled to update a given basis, thereby performing multiple lattice reductions. However, we add new vectors into the list while Schnorr's algorithm uses a fixed number of vectors throughout the execution.

In practice, this modification appears to give linear speedups based on our experimental timing results given in Sect. 7.

Given that parallel adaptations of GaussSieve are highly practical, especially for ideal lattices, we expect the approach of randomizing and reducing the basis to seed the list to be very effective in practice. For instance, the implementation of Ishiguro et al. employed more than 2,688 threads to solve the Ideal-SVP 128-dimensional challenge, with the number of thread-hours totaling 479,904. However, only a single basis was used, having been reduced with BKZ with a blocksize of 30. If each thread additionally performed, say three randomizations and reductions, over one million unique lattice vectors would be easily obtained. In comparison, in dimension 128, we would expect our final list to contain roughly 4.2 million vectors in the ideal lattice case.

6   Reducing the Gaussian Parameter

Recall that Klein's algorithm samples from a discrete Gaussian over the given lattice by taking (integer) linear combinations of the basis vectors, with each coefficient being sampled from a discrete Gaussian of parameter s over $\mathbb{Z}$. The parameter s is proportional to the norm of the Gram-Schmidt vector corresponding to that dimension. Unfortunately, in the implementation of Voulgaris [1], Klein's algorithm is implemented incorrectly, with the result that one either samples integers which are non-Gaussian, or one does sample from a Gaussian but very slowly.

In the original implementation of Voulgaris, an arbitrary Gaussian sampling parameter is chosen, while Ishiguro et al. choose an arbitrary though smaller parameter. We choose the Gaussian parameter dynamically in our experiments, i.e., by starting with an unfeasibly small (for example 500) parameter (which is guaranteed to return only the zero vector) and then incrementing this by one each time Klein's algorithm returns a zero vector. The intuition for this strategy is that, if the Gaussian parameter is too large, Klein's algorithm will generate unnecessarily long vectors, while if the Gaussian parameter is too small, the only vectors delivered by Klein's algorithm will be the zero vector and (occasionally) single vectors from the basis. Hence we need to choose a Gaussian parameter which is large enough that the number of lattice vectors obtainable is large enough to generate a list which satisfies the termination condition, but which is small enough that the vectors generated are not too long as to impose an additional unnecessary number of iterations on the algorithm. While it is probably possible to prove an optimal value for the Gaussian parameter (i.e. to provide a lower-bound on the Gaussian parameter which leads to enough entropy in Klein's algorithm to deliver a terminating list), we do not deal with this here as,

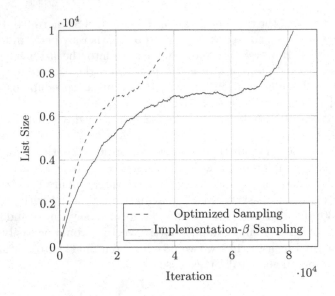

Fig. 5. Progressive list sizes in a 56-dimensional random lattice.

in practice, our approach of dynamically increasing the parameter upon seeing zero vectors appears to work very well.

Upon choosing the parameter in this way, a substantial change in behavior occurs, as illustrated in Fig. 5, with far fewer iterations (less than half as many in this case) being necessary to satisfy the termination condition. In contrast to the suggestions in Ishiguro et al. (in which it was suggested that a fixed Gaussian parameter should be used throughout the execution of the algorithm but that the optimal Gaussian parameter increases with lattice dimension), our experiments indicate that decreasing the Gaussian sampling parameter with increasing dimension delivers superior results[1].

7   Implementation and Experimental Results

To test the modifications outlined, we adapted the single-threaded implementation of Voulgaris [1], comparing minimally-modified versions to the reference implementation. While several obvious optimizations are possible, we did not implement these, for consistency. We stress, however, that the timings given here are purely for comparative purposes and in addition to our algorithmic optimizations further optimizations at the implementation level can significantly enhance the performance of the algorithm, for instance using 16-bit integers for vector entries rather than the 64-bit integers used in the reference implementation.

---

[1] We note that the speedups gained from dynamic choice of the Gaussian parameter are independent of the bug in the reference implementation, said bug leading to only a minor slowdown in most cases. See Table 1 further for details.

Table 1. Execution time (in seconds) of Voulgaris' implementation [1] and our optimized variants.

Dimension	60	62	64	66	68	70
Reference implementation [1]	464	1087	2526	5302	12052	23933
Reference implementation-β	455	1059	2497	5370	12047	24055
XOR + Pop. Count (Sect. 4)	203	459	1042	2004	4965	11161
Mult. Rand. Bases (Sect. 5)	210	555	1103	2023	3949	7917
Opt. Gaussian sampling (Sect. 6)	158	376	1023	2222	5389	10207
Combined (s)	79	146	397	868	2082	4500
Shortest norm $\approx$	1943	2092	2103	2099	2141	2143

All experiments were carried out using a single core of an AMD FX-8350 4.0 GHz CPU, 32 GB RAM, with all software (C++) compiled using the Gnu Compiler Collection, version 4.7.2-5. Throughout, we only experiment with the Goldstein-Mayer quasi-random lattices as provided by the TU Darmstadt SVP challenge [2].

7.1 Our Timings

In order to better assess the impact of our modifications to GaussSieve, we compare our implementations both to the original implementation of Voulgaris and a "corrected" version where we embed a correct implementation of a discrete Gaussian sampler[2]. We denote the original implementation by "Reference Implementation" and the original implementation + corrected Gaussian sampler by "Reference Implementation-β".

Table 1 shows timings for the original (unoptimized) implementation of Voulgaris [1], of Reference Implementation-β, and of our proposed optimizations explicitly. We also provide timings for an implementation which incorporates all the discussed optimizations for which the pseudocode can be found in Appendix A. For the multiple-bases optimization we display the timing with best efficiency, i.e., with the optimal (in terms of our limited experiments) number of bases. All timings exclude the cost of lattice reduction but we include the additional necessary lattice reduction via BKZ when considering multiple bases.

We observe that our optimized Gaussian sampler gives a speedup of up to 3.0x. However, with increasing dimension the speedup decreases slightly. Our integration of approximate inner-product computations increases performance by a factor of up to 2.7x, as compared to the original implementation of GaussSieve.

---

[2] In the implementation of Voulgaris, no lookup table is employed for Gaussian carrying out rejection sampling over a subset of the integers. Hence, the sampled integers are much closer to uniform than to the intended truncated Gaussian. In our corrected comparative implementation we employ the same Gaussian parameter from the Voulgaris implementation but ensure that the sampled vectors adhere to the prescribed Gaussian.

Table 2. Time for (sieving, initialization) in seconds.

Dim	Number of additional bases						
	0	10	20	40	80	160	320
60	(453, 0)	(274, 2)	(238, 4)	(210, 8)	(195, 15)	(185, 29)	(164, 59)
62	(1075, 0)	(810, 1)	(686, 3)	(612, 12)	(570, 12)	(533, 22)	(530, 43)
64	(2507, 0)	(1389, 17)	(1209, 36)	(1025, 75)	(877, 153)	(723, 322)	(461, 748)
66	(5302, 0)	(3193, 19)	(2716, 41)	(2328, 83)	(1961, 171)	(1659, 364)	(1233, 835)
68	(12052, 0)	(6842, 23)	(5852, 48)	(5071, 99)	(4360, 200)	(3652, 415)	(3015, 934)
70	(23933, 0)	(14933, 24)	(12641, 53)	(10933, 111)	(9561, 225)	(8139, 464)	(6871, 1046)

Similar speedups are obtained by considering multiple randomized bases; however, the speedup increases for larger dimensions. Indeed, if we ignore dimension 70, for which we did not consider an optimal number of bases due to time constraints, the speedup is approximated closely by the function $0.1838n - 9.471$. Figure 6 illustrates the speedups for several dimensions when increasing the number of bases considered.

When employing multiple randomized bases it is almost always the case that with increasing dimension employing more bases is preferable. Table 2 shows the runtime of our implementation when employing various numbers of randomized bases. It also depicts the amount of time necessary to reduce all the generated bases.

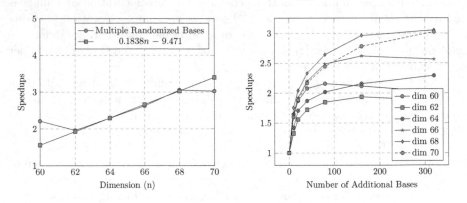

Fig. 6. Speedups with respect to the dimension (left) and the number of additional bases used to seed the list before sieving (right).

Acknowledgments. The authors would like to thank the anonymous reviewers of Latincrypt 2014 for their helpful comments and suggestions which substantially improved this paper. Özgür Dagdelen is supported by the German Federal Ministry of Education and Research (BMBF) within EC-SPRIDE.

References

1. Voulgaris, P.: GaussSieve Implementation. (http://cseweb.ucsd.edu/pvoulgar/impl.html)
2. TU Darmstadt Lattice Challenge. (http://www.latticechallenge.org)
3. Ajtai, M.: The shortest vector problem in L2 is NP-hard for randomized reductions (Extended Abstract). In: STOC 1998, pp. 10–19. ACM, New York (1998)
4. Ajtai, M., Kumar, R., Sivakumar, D.: A sieve algorithm for the shortest lattice vector problem. In: Vitter, J.S., Spirakis, P.G., Yannakakis, M. (eds.) STOC, pp. 601–610. ACM (2001)
5. Chen, Y., Nguyen, P.Q.: BKZ 2.0: Better lattice security estimates. In: ASIACRYPT, pp. 1–20 (2011)
6. Gama, N., Nguyen, P.Q.: Predicting lattice reduction. In: EUROCRYPT, pp. 31–51 (2008)
7. Gama, N., Nguyen, P.Q., Regev, O.: Lattice enumeration using extreme pruning. In: EUROCRYPT, pp. 257–278 (2010)
8. Gama, N., Schneider, M.: SVP Challenge (2010). (http://www.latticechallenge.org/svp-challenge)
9. Goldstein, D., Mayer, A.: On the equidistribution of hecke points. Forum Mathematicum 15, 165–190 (2003)
10. Ishiguro, T., Kiyomoto, S., Miyake, Y., Takagi, T.: Parallel gauss sieve algorithm: Solving the SVP challenge over a 128-Dimensional ideal lattice. In: Public Key Cryptography, pp. 411–428 (2014)
11. Kannan, R.: Improved algorithms for integer programming and related lattice problems. In: Johnson, D.S., Fagin, R., Fredman, M.L., Harel, D., Karp, R.M., Lynch, N.A., Papadimitriou, C.H., Rivest, R.L., Ruzzo, W.L., Seiferas, J.I. (eds.) STOC, pp. 193–206. ACM (1983)
12. Klein, P.N.: Finding the closest lattice vector when it's unusually close. In: SODA, pp. 937–941 (2000)
13. Kuo, P.-C., Schneider, M., Dagdelen, Ö., Reichelt, J., Buchmann, J., Cheng, C.-M., Yang, B.-Y.: Extreme enumeration on GPU and in clouds - How many dollars you need to break SVP challenges. In: CHES, pp. 176–191 (2011)
14. Lenstra, A.K., Lenstra, H.W., Lovász, L.: Factoring polynomials with rational coefficients. Mathematische Annalen 261(4), 515–534 (1982)
15. Mardia, K.V. (ed.): Tests of Univariate and Multivariate Normality. Handbook of Statistics. North-Holland, Amsterdam (1980)
16. Mariano, A., Timnat, S., Bischof, C.: Lock-free GaussSieve for linear speedups in parallel high performance SVP calculation. In: SBAC-PAD (2014)
17. Mariano, A., Dagdelen, O., Bischof, C.: A comprehensive empirical comparison of parallel ListSieve and GaussSieve. APCI&E (2014)
18. Micciancio, D., Goldwasser, S.: Complexity of Lattice Problems: A Cryptographic Perspective. Milken Institute Series on Financial Innovation and Economic Growth. Springer, US (2002)
19. Micciancio, D., Voulgaris, P.: Faster exponential time algorithms for the shortest vector problem. In: Proceedings of the Twenty-first Annual ACM-SIAM Symposium on Discrete Algorithms, SODA 2010, pp. 1468–1480. Society for Industrial and Applied Mathematics (2010)
20. Milde, B., Schneider, M.: A parallel implementation of gausssieve for the shortest vector problem in lattices. In: Malyshkin, V. (ed.) PaCT 2011. LNCS, vol. 6873, pp. 452–458. Springer, Heidelberg (2011)

21. Nguyen, P.Q., Vidick, T.: Sieve algorithms for the shortest vector problem are practical. J. Math. Cryptol. **2**(2), 181–207 (2008)
22. Schneider, M.: Sieving for shortest vectors in ideal lattices. IACR Cryptology ePrint Archive 2011, 458 (2011)
23. Schnorr, C.-P.: A hierarchy of polynomial time lattice basis reduction algorithms. Theor. Comput. Sci. **53**, 201–224 (1987)
24. Schnorr, C.-P.: Lattice reduction by random sampling and birthday methods. In: STACS, pp. 145–156 (2003)
25. Siegel, C.L.: A mean value theorem in geometry of numbers. Ann. Math. **46**(2), 340–347 (1945)
26. Vallée, B., Vera, A.: Probabilistic analyses of lattice reduction algorithms. In: Nguyen, P.Q., Vallée, B. (eds.) The LLL Algorithm, Information Security and Cryptography, pp. 71–143. Springer, Heidelberg (2010)
27. Wang, X., Liu, M., Tian, C., Bi, J.: Improved Nguyen-Vidick heuristic sieve algorithm for shortest vector problem. In: ASIACCS, pp. 1–9 (2011)
28. Zhang, F., Pan, Y., Hu, G.: A Three-level sieve algorithm for the shortest vector problem. In: Selected Areas in Cryptography, pp. 29–47 (2013)

A Pseudocode of our Optimized GaussSieve

Algorithm 2: Optimized GaussSieve

---

1 **Input**   : Basis $\mathbf{B}$, $k \in \mathbb{Z}^+$, $r \in \mathbb{Z}^+$, $a' \in \mathbb{R}^+$, $\delta \in \mathbb{R}^+$
   Output: $\mathbf{v} : \mathbf{v} \in \Lambda(\mathbf{B}) \wedge \|\mathbf{v}\| = \lambda_1(\mathbf{B})$

2   $\mathsf{L} \leftarrow \{\}, \mathsf{S} \leftarrow \{\}, \mathsf{col} \leftarrow 0, a \leftarrow a'$
  repeat

3     $\mathbf{B}' \leftarrow \text{RandomizeBasis}(\mathbf{B})$

4   **for** $\mathbf{v} \in \mathbf{B}'$ **do**

5       $\mathbf{v}' \leftarrow \text{GaussReduce}(\mathbf{v}, \mathsf{L}, \mathsf{S}, k)$

6   **if** $\|\mathbf{v}'\| \neq 0$ **then**

7         $\mathsf{L} \leftarrow \mathsf{L} \cup \{\mathbf{v}\}$

8 **until** r *times*

9 **while** $\mathsf{col} < c$ **do**

10   **if** S is not empty **then**

11     $\mathbf{v} \leftarrow \mathsf{S}.\text{pop}()$

12   **else**

13     $\mathbf{v} \leftarrow \text{SampleKlein}(\mathbf{B}', a)$

14   **while** $\mathbf{v} = 0$ **do**

15       $a \leftarrow a + \delta$

16       $\mathbf{v} \leftarrow \text{SampleKlein}(\mathbf{B}', a)$

17   $j \leftarrow \text{GaussReduce}(\mathbf{v}, \mathsf{L}, \mathsf{S}, k)$

18   **if** $j = \text{true}$ **then**

19     $\mathsf{col} \leftarrow \mathsf{col} + 1$

20 **return** $\mathbf{v} \in \mathsf{L}$ s.t. $\|v\| = \min_{\mathbf{x} \in \mathsf{L}} \|\mathbf{x}\|$

---

function GaussReduce($\mathbf{p}, \mathsf{L}, \mathsf{S}, k$)
  $\text{was\_reduced} \leftarrow \text{true}$
  while $\text{was\_reduced} = true$ **do**
    $\text{was\_reduced} \leftarrow \text{false}$
    for all $\mathbf{v}_i \in \mathsf{L}$ **do**
      if $\text{SIP}(\mathbf{v}_i, \mathbf{p}, k) = 1$ **then**
        if $\exists t \in \mathbb{Z} : \|\mathbf{p} + t\mathbf{v}_i\| < \|\mathbf{p}\|$ **then**
          $\mathbf{p} \leftarrow \mathbf{p} + t\mathbf{v}_i$
          $\text{was\_reduced} \leftarrow \text{true}$
  if $\|\mathbf{p}\| = 0$ **then**
    return true
  for all $\mathbf{v}_i \in \mathsf{L}$ **do**
    if $\text{SIP}(\mathbf{v}_i, \mathbf{p}, k) = 1$ **then**
      if $\exists u \in \mathbb{Z} : \|\mathbf{v}_i + u\mathbf{p}\| < \|\mathbf{v}_i\|$ **then**
        $\mathsf{L} \leftarrow \mathsf{L} \backslash \{\mathbf{v}_i\}$
        $\mathbf{v}_i \leftarrow \mathbf{v}_i + u\mathbf{p}$
        $\mathsf{S}.\text{push}(\mathbf{v}_i)$
  $\mathsf{L} \leftarrow \mathsf{L} \cup \{\mathbf{p}\}$
  return false
end function

The below pseudocode displays our proposed modifications to GaussSieve. In lines (3)–(9) we incorporate our multiple-randomized-bases optimization, and in the function $\mathsf{GaussReduce}(\mathbf{p},\mathsf{L},\mathsf{S},k)$ we embed the cheap test SIP implementing our XOR + population count computation for the approximation of the angle between two vectors. The optimized Gaussian sampler modifies the function $\mathsf{SampleKlein}$.

In the pseudocode, the parameter $k \in \mathbb{Z}^+$ defines the bounds on the XOR + population count, within which we assume that a pair of vectors is Gauss-reduced, i.e. if $n/2 - k \le \langle \tilde{\mathbf{a}}, \tilde{\mathbf{b}} \rangle \le n/2 + k$, we assume the pair $\mathbf{a}, \mathbf{b}$ are Gauss-reduced.

Analysis of NORX: Investigating Differential and Rotational Properties

Jean-Philippe Aumasson[1], Philipp Jovanovic[2]([⊠]), and Samuel Neves[3]

[1] Kudelski Security, Lausanne, Switzerland
jeanphilippe.aumasson@gmail.com
[2] University of Passau, Passau, Germany
jovanovic@fim.uni-passau.de
[3] University of Coimbra, Coimbra, Portugal
sneves@dei.uc.pt

Abstract. This paper presents a thorough analysis of the AEAD scheme NORX, focussing on differential and rotational properties. We first introduce mathematical models that describe differential propagation with respect to the non-linear operation of NORX. Afterwards, we adapt a framework previously proposed for ARX designs allowing us to automatise the search for differentials and characteristics. We give upper bounds on the differential probability for a small number of steps of the NORX core permutation. For example, in a scenario where an attacker can only modify the nonce during initialisation, we show that characteristics have probabilities of less than 2^{-60} (32-bit) and 2^{-53} (64-bit) after only one round. Furthermore, we describe how we found the best characteristics for four rounds, which have probabilities of 2^{-584} (32-bit) and 2^{-836} (64-bit), respectively. Finally, we discuss some rotational properties of the core permutation which yield some first, rough bounds and can be used as a basis for future studies.

Keywords: NORX · AEAD · LRX · Differential cryptanalysis · Rotational cryptanalysis

1   Introduction

NORX [4] is a new scheme for authenticated encryption with associated data (AEAD) and was recently submitted to CAESAR [1]. NORX is based on well-known building blocks but refines those components to provide certain desirable features. Its layout is a modified version of the monkeyDuplex construction [9], which allows to process data in parallel. The duplex construction is an alteration of sponge functions [10], which were introduced alongside KECCAK [12]. The core permutation F of NORX is derived from ChaCha [6] and BLAKE2 [5], which are parade examples for ARX primitives, i.e. cryptographic functions based solely on integer addition mod 2^n, bit rotations and XOR. However, the permutation F

© Springer International Publishing Switzerland 2015
D.F. Aranha and A. Menezes (Eds.): LATINCRYPT 2014, LNCS 8895, pp. 306–324, 2015.
DOI: 10.1007/978-3-319-16295-9_17

is a so-called LRX[1] construction, because integer addition, which can be written as $a + b = (a \oplus b) + ((a \wedge b) \ll 1)$ [21], is replaced by the approximation $(a \oplus b) \oplus ((a \wedge b) \ll 1)$, a purely logic-based operation. The aim is to increase hardware friendliness and simplify cryptanalysis. Despite its famous predecessors, that have already resisted extensive analysis [3, 19, 25] and are deemed secure, this new permutation F still lacks in-depth analysis and its security level is yet unclear.

Differential cryptanalysis [13] is one of the most powerful and versatile attack techniques usable against symmetric primitives and belongs to the standard repertoire of every cryptanalyst. Therefore, it is not surprising that every new symmetric primitive is examined upon its resistance against differential attacks. Usually, it is much easier to establish bounds for strongly aligned ciphers, like AES [16], than for weakly aligned ones [8]. NORX rather belongs to the latter category and, despite some successful inroads into deriving bounds for weakly aligned ciphers [15, 17], it is not obvious how to establish such bounds in the general case. Hence, in the first part of the paper, we investigate differential propagation in F and, based on that, introduce NODE [2], the NORX *Differential Search Engine*, a framework providing a way to search for differentials and characteristics in an automated way. Our approach is guided by the work of Mouha and Preneel [24], where a search framework was introduced for the ARX cipher Salsa20 [7]. Their framework constructs a description of the differential propagation behaviour of Salsa20, using well-known differential properties of integer addition [22]. The description is formulated in the CVC language, the standard input language of the constraint solver STP [18], which supports operations on bit vectors (like bitwise XOR, AND, modular addition, etc.) and therefore allows a straightforward modelling of the differential search problem. The resulting description has a simple shape, which facilitates cryptanalysis.

However, in order to use such a framework for NORX, some adjustments are necessary: The permutation F of NORX is not based on integer addition, and hence we can not rely upon already known results on the differential properties of the latter [22]. Therefore, we start with the mathematical modelling of differential propagation with respect to the non-linear operation $(a \oplus b) \oplus ((a \wedge b) \ll 1)$ of NORX. All of our claims are supported by rigorous proofs. Then, we use these results to show how to adapt the search framework to the NORX permutation, which requires some more modifications, since the original framework [24] was developed for Salsa20, whereas F is based on ChaCha [6]. Finally, we present the results from our extensive empirical analysis of F^R.

The second part of this paper is dedicated to the rotational cryptanalysis [20] of the core permutation F^R. Rotational cryptanalysis is another important aspect for the security evaluation of ARX/LRX-based primitives. We present some basic rotational properties of F and based on that derive bounds for a few simple rotational attacks.

---

[1] This is not an official term. We introduce it to easily distinguish between ARX- and purely logic-based primitives. Terminology-wise it is not entirely correct, though, as integer addition can be obviously modelled by bitwise logical operations as well.

Outline. The paper is structured as follows. Section 2 introduces notation and recalls the basic layout of NORX, with a focus on its core permutation F^R, as it is the main target of our cryptanalysis efforts. Sections 3 and 4 present differential and rotational cryptanalysis of NORX and Sect. 5 concludes the paper.

2 Preliminaries

2.1 Notation

Hexadecimal numbers are denoted in `typewriter`, e.g. `c9` = 201. A *word* is either a 32-bit or 64-bit string, depending on the context. Parsing of data streams (as byte arrays) to word arrays is done in little-endian order. The concatenation of strings x and y is denoted by $x \parallel y$. The length of a bit string x is written as $|x|$, and its Hamming weight as $\mathsf{hw}(x)$. We use the standard notation $\neg$, $\wedge$, $\vee$ and $\oplus$ for bitwise NOT, AND, OR and XOR, $x \ll n$ and $x \gg n$ for left- and right-shift, and $x \lll n$ and $x \ggg n$ for left- and right-rotation of x by n bits.

2.2 Core Components of NORX

The NORX family of AEAD schemes is based on the *monkeyDuplex construction* [9,11] and parametrised by a *word size* $W \in \{32, 64\}$, a *round number* $1 \leq R \leq 63$, a *parallelism degree* $0 \leq D \leq 255$ and a *tag size* $|A| \leq 10W$. The meaning of the parameters is basically self-explanatory, for more details see [4].

The state S of NORX consists of sixteen words $s_0, \ldots, s_{15}$ each of size W bits, which are arranged in a 4×4 matrix. Thus, the state has a size of 512 bits for $W = 32$ and a size of 1024 bits for $W = 64$. Due to the duplex construction, the words of the state are divided into two types: $s_0, \ldots, s_9$ are called the *rate words* and $s_{10}, \ldots, s_{15}$ are called the *capacity words*[2]. The rate words are used for data processing, whereas the capacity words remain untouched and ensure the security of the scheme. S is initialised by loading a *nonce* n_0, n_1, a *key* $k_0, \ldots, k_3$ and *constants* $u_0, \ldots, u_9$ in the following way

$$
\begin{pmatrix} s_0 & s_1 & s_2 & s_3 \\ s_4 & s_5 & s_6 & s_7 \\ s_8 & s_9 & s_{10} & s_{11} \\ s_{12} & s_{13} & s_{14} & s_{15} \end{pmatrix} \longleftarrow \begin{pmatrix} u_0 & n_0 & n_1 & u_1 \\ k_0 & k_1 & k_2 & k_3 \\ u_2 & u_3 & u_4 & u_5 \\ u_6 & u_7 & u_8 & u_9 \end{pmatrix}
$$

More information on the constants can be found in [4]. This initial state is transformed by F^{2R}, where F is the *round function*, interleaved with the injection of parameter and domain separation constants, before data processing starts, which uses F^R. Concrete instances of NORX, as given in [4], use $R \in \{4, 6\}$. The round function F of NORX is composed of a *column step*

$$
\mathsf{G}(s_0, s_4, s_8, s_{12})\ \mathsf{G}(s_1, s_5, s_9, s_{13})\ \mathsf{G}(s_2, s_6, s_{10}, s_{14})\ \mathsf{G}(s_3, s_7, s_{11}, s_{15})
$$

---

[2] These are also respectively known as the *outer* and *inner* part of the state [9,10].

followed by a *diagonal step*

$$G(s_0, s_5, s_{10}, s_{15})\ G(s_1, s_6, s_{11}, s_{12})\ G(s_2, s_7, s_8, s_{13})\ G(s_3, s_4, s_9, s_{14})$$

The function G transforms four words a, b, c, and d by doing

$$
\begin{array}{llll}
1: & a \longleftarrow (a \oplus b) \oplus \big((a \wedge b) \ll 1\big) & \quad 5: & a \longleftarrow (a \oplus b) \oplus \big((a \wedge b) \ll 1\big) \\
2: & d \longleftarrow (a \oplus d) \ggg r_0 & \quad 6: & d \longleftarrow (a \oplus d) \ggg r_2 \\
3: & c \longleftarrow (c \oplus d) \oplus \big((c \wedge d) \ll 1\big) & \quad 7: & c \longleftarrow (c \oplus d) \oplus \big((c \wedge d) \ll 1\big) \\
4: & b \longleftarrow (b \oplus c) \ggg r_1 & \quad 8: & b \longleftarrow (b \oplus c) \ggg r_3
\end{array}
$$

where rotation offsets (r_0, r_1, r_2, r_3) have the values $(8, 11, 16, 31)$ for 32-bit and $(8, 19, 40, 63)$ for 64-bit.

Since our analysis focusses on the core permutation F^R, we do not go into the details of NORX's mode of operation. For more information on these topics, we refer to the official specification [4].

2.3 Weak States

The NORX specification [4] contains a discussion about the all-zero state, which is mapped to itself by F^R for any $R > 0$, and why it is no problem for the security of the scheme. However, due to the layout of F, there is another class of weak states. These are of the form

$$
\begin{pmatrix}
w & w & w & w \\
x & x & x & x \\
y & y & y & y \\
z & z & z & z
\end{pmatrix}
$$

with w, x, y, and z being arbitrary W-bit sized words. The column-pattern is preserved by F^R for an arbitrary value of $R > 0$. The ability to hit such a state purposely, is equivalent to the ability of reconstructing the key and therefore breaking the entire scheme. While there are quite many of these states, namely 2^{4W}, their number is still negligible compared to the total number of 2^{16W} states. Thus, the probability to hit such a state is 2^{-12W}, which translates to probabilities of 2^{-384} $(W = 32)$ and 2^{-768} $(W = 64)$. Additionally, this attack does not take into account the extra protection provided through the duplex construction, the asymmetric constants used during initialisation, or the domain separation constants which are integrated into the state before each application of F^R. All of the above features should impede the exploitation of these states.

3   Differential Cryptanalysis

This section is dedicated to the differential cryptanalysis of NORX. First, we introduce the required mathematical models to describe differential propagation in F^R of NORX. Then we describe how to construct the search framework and finally apply it to NORX and present our results.

3.1 Mathematical Models

Let n denote the word size, let x and y denote bit strings of size n and let α, β and γ denote differences of size n. We identify by α_i, β_i, γ_i, x_i and y_i the individual bits of α, β, γ, x and y, with $0 \leq i \leq n - 1$.

Definition 1. *The non-linear operation* H *of* NORX *is the vector Boolean function defined by*

$$\mathsf{H} : \mathbb{F}_2^{2n} \longrightarrow \mathbb{F}_2^n, \ (x, y) \mapsto (x \oplus y) \oplus ((x \wedge y) \lll 1)$$

Definition 2. *Let* $f : \mathbb{F}_2^{2n} \longrightarrow \mathbb{F}_2^n$ *be a vector Boolean function and let* α, β *and* γ *be n-bit sized XOR-differences. We call* $(\alpha, \beta) \longrightarrow \gamma$ *a (XOR-)differential of* f *if there exist n-bit strings* x *and* y *such that the following equation holds:*

$$f(x \oplus \alpha, y \oplus \beta) = f(x, y) \oplus \gamma$$

Otherwise, if no such n-bit strings x *and* y *exist, we call* $(\alpha, \beta) \longrightarrow \gamma$ *an impossible (XOR-)differential of* f.

Plugging the non-linear operation H of NORX from Definition 1 into the formula of Definition 2, we see that an XOR-differential $(\alpha, \beta) \longrightarrow \gamma$ of H fulfils

$$\alpha \oplus \beta \oplus \gamma = ((x \wedge \beta) \oplus (y \wedge \alpha) \oplus (\alpha \wedge \beta)) \lll 1 \tag{1}$$

for n-bit strings x and y. Rewriting the above formula on bit level we get

$$0 = \alpha_0 \oplus \beta_0 \oplus \gamma_0$$
$$0 = (\alpha_i \oplus \beta_i \oplus \gamma_i) \oplus (\alpha_{i-1} \wedge \beta_{i-1}) \oplus (x_{i-1} \wedge \beta_{i-1}) \oplus (y_{i-1} \wedge \alpha_{i-1}), \quad i > 0$$

Lemma 3 is an important step towards expressing differential propagation in NORX and is the analogue to Theorem 1 for integer addition from [22]. The lemma eliminates the dependence of Eq. 1 on the bit strings x and y and therefore allows us to check in a constant amount of word operations if a given tuple (α, β, γ) of differences is an (impossible) XOR-differential of H.

Lemma 3. *For each XOR-differential* $(\alpha, \beta) \longrightarrow \gamma$ *of the non-linear operation* H *of* NORX *the following equation is satisfied:*

$$(\alpha \oplus \beta \oplus \gamma) \wedge (\neg((\alpha \vee \beta) \lll 1)) = 0 \tag{2}$$

Proof. See Appendix A.

Obviously, a tuple of differences (α, β, γ) not satisfying Lemma 3 is an impossible XOR-differential of H.

Definition 4. *Let* f *be a vector Boolean function and let* δ *be an XOR-differential in terms of Definition 2. The probability* xdp^f *of* δ *is defined as*

$$\mathsf{xdp}^f(\delta) = |\{x, y \in \mathbb{F}_2^n : f(x \oplus \alpha, y \oplus \beta) \oplus f(x, y) \oplus \gamma = 0\}| \cdot 2^{-2n}$$

The value $\mathsf{xdp}^f(\delta)$ *is also called the XOR-differential probability of* δ. *Moreover, for* $\mathsf{xdp}^f(\delta) = 2^{-w}$ *we call* w *the XOR-(differential) weight of* δ.

The differential probability of an impossible differential is always 0 by prerequisite, as $\{x, y \in \mathbb{F}_2^n : f(x \oplus \alpha, y \oplus \beta) \oplus f(x, y) \oplus \gamma = 0\}$ is then the empty set, see Definition 2. To compute the probability of a differential with respect to the non-linear operation H of NORX, we can use the following lemma.

Lemma 5. *Let δ be a XOR-differential with respect to the non-linear operation H of NORX. Its differential probability is then given by*

$$\mathsf{xdp}^{\mathsf{H}}(\delta) = 2^{-\mathsf{hw}((\alpha \vee \beta) \ll 1)}$$

Proof. See Appendix A.

Instead of looking at XOR-differences one could alternatively also analyse f-differentials, which is done in the following.

Definition 6. *Let $f : \mathbb{F}_2^{2n} \longrightarrow \mathbb{F}_2^n$ be a vector Boolean function and let α, β and γ be differences with respect to f. We call $(\alpha, \beta) \longrightarrow \gamma$ an f-differential of XOR if there exist n-bit strings x and y such that the following equation holds:*

$$f(x, \alpha) \oplus f(y, \beta) = f(x \oplus y, \gamma)$$

Otherwise, if no such n-bit strings x and y exist, we call $(\alpha, \beta) \longrightarrow \gamma$ an impossible f-differential of XOR.

Plugging the non-linear operation H of NORX into the formula of Definition 6 we obtain the following equation

$$\alpha \oplus \beta \oplus \gamma = ((x \wedge (\alpha \oplus \gamma)) \oplus (y \wedge (\beta \oplus \gamma))) \ll 1 \tag{3}$$

which can be expressed on bit level as

$$0 = \alpha_0 \oplus \beta_0 \oplus \gamma_0$$
$$0 = (\alpha_i \oplus \beta_i \oplus \gamma_i) \oplus (x_{i-1} \wedge (\alpha_{i-1} \oplus \gamma_{i-1})) \oplus (y_{i-1} \wedge (\beta_{i-1} \oplus \gamma_{i-1})), \quad i > 0$$

Lemma 7. *Let H denote the non-linear operation of NORX. For each H-differential in terms of Definition 6 the following equation is satisfied:*

$$(\alpha \oplus \beta \oplus \gamma) \wedge (\neg(\gamma \ll 1) \oplus (\alpha \ll 1)) \wedge (\neg(\beta \ll 1) \oplus (\gamma \ll 1)) = 0 \tag{4}$$

Proof. See Appendix A.

Definition 8. *Let f be a vector Boolean function and δ be an f-differential in terms of Definition 6. The probability $\mathsf{fdp}^{\oplus}$ of δ is defined as*

$$\mathsf{fdp}^{\oplus}(\delta) = |\{x, y \in \mathbb{F}_2^n : f(x, \alpha) \oplus f(y, \beta) \oplus f(x \oplus y, \gamma) = 0\}| \cdot 2^{-2n}$$

We call $\mathsf{fdp}^{\oplus}(\delta)$ the f-differential probability of δ. Moreover, for $\mathsf{fdp}^{\oplus}(\delta) = 2^{-w}$ we call w the f-(differential) weight of δ.

Lemma 9. *Let H denote the non-linear operation of NORX and let δ be an H-differential in terms of Definition 6. Its probability is then given by*

$$\mathsf{Hdp}^{\oplus}(\delta) = 2^{-\mathsf{hw}(((\alpha \oplus \gamma) \vee (\beta \oplus \gamma)) \ll 1)}$$

Proof. See Appendix A.

While we exclusively consider XOR-differentials and -characteristics in the rest of the paper, f-differentials might be of interest for future investigations.

3.2   NODE – The NORX Differential Search Engine

Now that we have introduced the mathematical model, we describe in this part the framework NODE for the search of differential characteristics of a predefined weight. Our tool is freely available at [2] under a public domain-like license. We focus here on XOR-differentials, as introduced in Definition 2, i.e. differences are computed with respect to XOR and for the vector Boolean function we use the non-linear operation H of NORX. If we speak in the following of differentials we always refer to the above type. Below we show the general approach, and refer to Appendix B for the CVC code.

For modelling the differential propagation through a sequence of operations, we use a technique well known from algebraic cryptanalysis: For every output of an operation a new set of variables is introduced. These output variables are then modelled as a function of its input variables. Moreover, the former are used as input to the next operation. This is repeated until all required operations have been integrated into the problem description. Before we show how the differential propagation in F^R is modelled concretely, we introduce the required variables.

Let s denote the number of (column and diagonal) steps to be analysed and let $0 \leq i \leq 15$ and $0 \leq j \leq 2(s-1)$. For example, if we analyse F^2, we have $s = 4$. Let x_i, $y_{i,j}$ and z_i be W-bit sized variables, which model the input, internal and output XOR differences of a differential characteristic. Recall that $W \in \{32, 64\}$ denotes the word size of NORX. Moreover, let $w_{i,k}$, with $0 \leq k \leq s - 1$, be W-bit sized helper variables which are used for differential weight computations or equivalently to determine the probability of a differential characteristic. We assume that the probability of a differential characteristic is the sum of weights of each non-linear operation H. Furthermore, let d denote a W-bit sized variable, which fixes the total weight of the characteristic we plan to search for. The description of the search problem is generated through the following steps:

1. Every time the function G applies the non-linear operation H we add two expressions to our description:
   (a) Append the equation $0 = (\alpha \oplus \beta \oplus \gamma) \wedge (\neg((\alpha \vee \beta) \ll 1))$ from Lemma 3, with α, β and γ each substituted by one of the variables x_i, $y_{i,j}$ or z_i. This ensures that only non-impossible characteristics are considered.
   (b) Add the expression $w_{i,k} = (\alpha \vee \beta) \ll 1$ from Lemma 5, with α and β substituted by the same variables x_i, $y_{i,j}$ or z_i as in step (a). This expression keeps track of the weight of the characteristic.
2. Every time the function G applies a rotation we apply the same rotation to the corresponding XOR difference, i.e. we add $\gamma = (\alpha \oplus \beta) \ggg r$ to the problem description, with α, β and γ substituted appropriately. Note that the rotation is a linear operation and thus does not change the differential probability.
3. Add an expression corresponding to the following equation:

$$d = \sum_{k=0}^{s-1} \sum_{i=0}^{15} \mathsf{hw}(w_{i,k}) \tag{5}$$

This equation ensures that indeed a characteristic of weight d is found. Depending on the technique how Hamming weights are computed, additional variables might be necessary. Refer to Appendix B for one possible implementation to compute Hamming weights in the CVC language.

4. Set the variable d to the target differential weight and append it to the problem description.

5. Exclude the trivial characteristic mapping an all-zero input difference to an all-zero output difference. To do so, it is sufficient to exclude the all-zero input difference. Therefore, append an expression equivalent to $\neg((x_0 = 0) \wedge ... \wedge (x_{15} = 0))$ to the CVC description.

After the generation of the problem description is finished, it can be used to search for differential characteristics using STP. Alternatively, STP allows to convert the representation of the problem to SMT-LIB2 or CNF, enabling searches with other SMT or SAT solvers, like Boolector [14] or CryptoMiniSat [23].

3.3 Applications of NODE

In this part we describe the application of the search framework to the permutation F^R of NORX. Depending on the concrete attack model, there are different ways an attacker could inject differences into the NORX state. During initialisation an adversary is allowed to modify either the nonce words s_1 and s_2 (init_N) or nonce and key words $s_1, s_2, s_4, \dots, s_7$ ($\mathrm{init}_{N,K}$). During data processing an attacker can inject differences into the words of the rate $s_0, \dots, s_9$ (rate). Last but not least, we also investigate the case where an attacker can manipulate the whole state $s_0, \dots, s_{15}$ (full). While an attacker is not able to influence the entire state at any point directly due to the duplex construction, the full scenario is nevertheless useful to estimate the general strength of F^R, because all of the other settings described above are special cases of the latter. Additionally, it could be useful for the chaining of characteristics: For example, an attacker could start with a search in the data processing part (i.e. under the rate setting) over a couple of steps, say F^{R_1}, and continue afterwards with a second search, starting from the full state for another couple of steps, say F^{R_2}, so that differentials from the second search connect to those from the first, resulting in differentials for $\mathsf{F}^{R_1+R_2}$. We will explore this Divide&Conquer strategy in more detail below.

For the rest of the paper, we denote a differential characteristic as a tuple of differences $(\delta_0, \dots, \delta_n)$, where δ_0 is the *input difference* and δ_n is the *output difference*. The values δ_i for $0 < i < n$ are called *internal differences*. The weight of the probability that difference δ_i is transformed into difference δ_{i+1} by the r_i-fold iteration of F is denoted by w_i for $0 \leq i \leq n - 1$. Recall, that we assume that the probability of the entire characteristic is equal to the multiplication of probabilities of the partial characteristics, and thus we have $w = \sum_{i=0}^{n-1} w_i$ for the total weight of the characteristic. The notation $\mathsf{F}^{R+0.5}$ describes that we do R full rounds followed by one more column step, e.g. $\mathsf{F}^{1.5}$ corresponds to one full round plus one additional column step.

Experimental Verification of the Search Framework. The goal of the experimental verification is to show that the framework indeed does what it is supposed to do, namely find differentials of a predetermined weight w in F^R. Therefore, we generated differentials for $\mathsf{F}^{1.5}$ (full) and verified them against a C reference implementation of $\mathsf{F}^{1.5}$. Under these prerequisites our framework found the first differentials at a weight of 12, for both $W = 32$ and $W = 64$, which thus should have a probability of about 2^{-12}. To get a better coverage of our verification test, we did not use only differentials of that particular weight, but generated random differentials of weights $w \in \{12, \ldots, 18\}$, which are listed in Appendix C.1 for both 32- and 64-bit. Then we applied them to the C implementation of $\mathsf{F}^{1.5}$ for 2^{w+16} pairs of randomly chosen input states having the input difference of the characteristic. In each case, we checked if the output difference had the predicted pattern. The number of pairs adhering the characteristic should be around 2^{16}. The results are illustrated in the first table of Appendix C.2 and show that the search framework indeed finds characteristics with the expected properties.

Lower Bounds for Differential Weights of F^R. We made an extensive analysis on the weight bounds of differential paths in F^R, where we investigated $1 \leq s \leq 4$ steps for our four different scenarios init_N, $\text{init}_{N,K}$, rate and full. We tried to find the lowest weights where differentials appear for the first time. These cases are listed in Table 1 as entries without brackets. For example, in case of NORX32 under the setting full, there are no differentials in $\mathsf{F}^{1.5}$ with a weight smaller than 12. Entries in brackets are the maximal weights we were capable of examining without finding any differentials. Due to memory constraints, our methods failed for differential weights higher than those presented in Table 1. For example, our search routine did not find any characteristics of weight smaller than 40 (i.e. of probability higher than 2^{-40}) for the scenario $\mathsf{F}^{1.5}$, $\text{init}_{N,K}$ and $W = 32$. The required amount of RAM, to execute this check, was approximately 49 GiB (using CryptoMiniSat with 16 threads) with a running time of 8 h.

The security of NORX depends heavily on the security of the initialisation, which transforms the initial state by F^{2R}. As init_N is the most realistic attack scenario, we conducted a search over all possible 1- and 2-bit differences in the nonce words. Our search revealed that the best characteristics have weights of

Table 1. Lower bounds for differential trail weights

	NORX32				NORX64			
	init_N	$\text{init}_{N,K}$	rate	full	init_N	$\text{init}_{N,K}$	rate	full
$\mathsf{F}^{0.5}$	6	2	2	0	6	2	2	0
$\mathsf{F}^{1.0}$	(60)	22	10	2	(53)	22	12	2
$\mathsf{F}^{1.5}$	(60)	(40)	(31)	12	(53)	(35)	(27)	12
$\mathsf{F}^{2.0}$	(61)	(45)	(34)	(27)	(51)	(37)	(30)	(23)

67 (32-bit) and 76 (64-bit) under those prerequisites. Obviously, these weights are not too far away from the computationally verified values of 60 (32-bit) and 53 (64-bit) from Table 1, showing that the bounds for F (init$_N$) are quite tight.

Extrapolating the above results to F^8 (i.e. $R = 4$), we get lower weights of $61 + 3 \cdot 27 = 142$ (init$_N$) or $45 + 3 \cdot 27 = 126$ (init$_{N,K}$) for NORX32 and $51 + 3 \cdot 23 = 132$ (init$_N$) or $37 + 3 \cdot 23 = 106$ (init$_{N,K}$) for NORX64. However, these are only loose bounds and we expect the real ones to be considerably higher.

Search for Differential Characteristics in F^4. This part shows how we constructed differential characteristics in F^4 under the setting full for both versions of the permutation, i.e. 32- and 64-bit. Unsurprisingly, a direct approach to find such characteristics turned out to be infeasible, hence we decomposed the search into multiple parts and constructed the entire path step by step.

At first we made searches that only stretched over $R \leq 2$ rounds. After tens of thousands of iterations using many different search parameter combinations we found differentials having internal differences of Hamming weight 1 and 2 after one application of F. We also used a probability-1 differential in G, which is listed as the first entry in the table of Appendix A, as a starting place. We expanded all those characteristics for both word sizes, in forward and backward direction one column or diagonal step at a time, until their paths stretched the entire 4 rounds. The best differential paths we found this way have weights of 584 (32-bit) and 836 (64-bit), respectively. Both are depicted in Appendix C.3.

Iterative Differentials. We also performed extensive searches for iterative differentials in F for the setting full. Using our framework, we could show that there are no such differentials up to a weight of 29 (32-bit) and 27 (64-bit), before our methods failed due to computational constraints. Extrapolating these results to F^8 and F^{12}, i.e. the number of initialisation rounds for $R = 4$ and $R = 6$, we get lower weight bounds of 232 and 348, for 32-bit, or of 216 and 324 for 64-bit. The best iterative differentials we could find for F, have weights of 512 (32-bit) and 843 (64-bit) and are depicted in Appendix C.4. These weights are obviously much higher than our guaranteed lower bounds, and hence we expect that the latter are much better compared to the values we were able to verify computationally.

Differentials with Equal Columns. The class of weak states from Sect. 2.3 can be obviously transformed into XOR-differentials having four equal columns. The best differentials we could find for one round F have weight 44 for both 32-bit and 64-bit. They exploit an already well known probability-1 differential in G, see Appendix C.2. The 64-bit variant was also used in the construction of the characteristics with weight 836 in F^4 above. Concrete representations of these differentials can be found in Appendix C.5.

3.4   Further Applications

The techniques presented in this section are obviously not restricted to NORX only. In principle, every function based on integer addition, as shown for Salsa20 in [24], and/or bitwise logical operations, like OR, NAND, NOR and so on, can be analysed just as easily. For LRX ciphers, all one has to do is rewrite their non-linear operations in terms of bitwise logical AND, which then allows to reuse the results from above.

4   Rotational Cryptanalysis

Definition 10. *Let f be a vector Boolean function $f : \mathbb{F}_2^{2n} \longrightarrow \mathbb{F}_2^n$ and let x, y be n-bit strings. We call (x, y) a rotational pair with respect to f if the following equation holds:*

$$f(x, y) \ggg r = f(x \ggg r, y \ggg r)$$

Lemma 11. *Let H be the non-linear function of NORX, and let x, y be n-bit strings. The probability of (x, y) being a rotational pair is:*

$$\Pr(\mathsf{H}(x, y) \ggg r = \mathsf{H}(x \ggg r, y \ggg r)) = \frac{9}{16} \ (\approx 2^{-0.83})$$

Proof. See Appendix D.

Now we can use Lemma 11 and Theorem 1 from [20] (under the assumption that the latter holds for H, too) to compute the probability of $\Pr(\mathsf{F}^R(S) \ggg r = \mathsf{F}^R(S \ggg r))$ for a state S and a number of rounds R. It is given by:

$$\Pr(\mathsf{F}^R(S) \ggg r = \mathsf{F}^R(S \ggg r) = (9/16)^{4 \cdot 4 \cdot 2 \cdot R}$$

Table 2 summarizes the (rounded) weights (i.e. the negative logarithms of the probabilities) for different values of R, which are relevant for NORX.

Table 2. Weights for rotational distinguishers of F^R

R	4	6	8	12
w	106	159	212	318

As a consequence, the permutation F^R on a $16W$ state is indistinguishable from a random permutation for $R \geq 20$ if $W = 32$ and for $R \geq 39$ if $W = 64$ with probabilities of $\Pr \leq 2^{-531}$ and $\Pr \leq 2^{-1035}$ respectively.

Definition 12. *Let f be a vector Boolean function $f : \mathbb{F}_2^{2n} \longrightarrow \mathbb{F}_2^n$ and let x, y be n-bit strings. We call (x, y) a rotational fixed point with respect to f if the following equation holds:*

$$f(x, y) \ggg r = f(x, y)$$

Lemma 13. *Let f be a vector Boolean function $f : \mathbb{F}_2^{2n} \longrightarrow \mathbb{F}_2^n$, $(x, y) \mapsto f(x, y)$, which is a permutation on $\mathbb{F}_2^n$, if either x or y is fixed. The probability that (x, y) is a rotational fixed point is:*

$$\Pr(f(x, y) \ggg r = f(x, y)) = 2^{-(n - \gcd(r, n))}$$

Proof. See Appendix D.

A direct consequence of Lemma 13 is that for n even and $r = n/2$ the probability that (x, y) is a rotational fixed point is $2^{-n/2}$. The rotation $r = n/2$, which swaps the two halves of a bit string, is especially interesting for cryptanalysis as it results in the highest probability among all $0 < r < n$.

The non-linear function H of NORX obviously satisfies the requirement of being a permutation on $\mathbb{F}_2^n$, when one of its inputs is fixed. Therefore we get probabilities of 2^{-16} (32-bit, $r = 16$) and 2^{-32} (64-bit, $r = 32$), that (x, y) is a rotational fixed point of H.

5 Conclusion

In this paper, we provide an extensive analysis of the differential and rotational properties of NORX's core permutation F^R and derive some first bounds for attacks on the complete scheme. We introduce the mathematical models required to describe XOR- and H-differentials with respect to F^R. All mathematical claims are verified by rigorous proofs. Moreover, we present NODE, a framework, which allows to automatise the search for XOR-differentials and -characteristics. We show the results of our extensive experiments and can conclude that there is a large gap between those differential bounds that are computationally verifiable and the weights of the best differentials that we were able to find. In particular, when considering initialisation with F^8, the verifiable but extrapolated weight bounds have values of 126 (NORX32) and 106 (NORX64) for an attacker in the related key model. On the other hand, the best differentials for F^4 have weights of 584 (32-bit) and 836 (64-bit). Thus, initialisation with F^8 ($R = 4$) and F^{12} ($R = 6$) seems to have a high security margin against differential attacks.

For rotational cryptanalysis, we are able to derive lower weight bounds of 212 and 318 for distinguishers on F^8 and F^{12} using a mix of new and already known results. We stress that these distinguishers only hold for the bare permutation. They do not take into account the additional protection provided by the duplex construction of NORX or the asymmetric constants used during initialisation.

Acknowledgements. The authors would like to thank the anonymous reviewers for their comprehensive commentaries which helped to improve the quality of this paper.

References

1. CAESAR – Competition for Authenticated Encryption: Security, Applicability, and Robustness (2014). http://competitions.cr.yp.to/caesar.html

2. NODE – The NORX Differential Search Engine (2014). https://github.com/norx/NODE
3. Aumasson, J.-P., Fischer, S., Khazaei, S., Meier, W., Rechberger, C.: New Features of Latin Dances: Analysis of Salsa, ChaCha, and Rumba. In: Nyberg, K. (ed.) FSE 2008. LNCS, vol. 5086, pp. 470–488. Springer, Heidelberg (2008)
4. Aumasson, J.-P., Jovanovic, P., Neves, S.: NORX: Parallel and Scalable AEAD. In: Kutyłowski, M., Vaidya, J. (eds.) ESORICS 2014, Part II. LNCS, vol. 8713, pp. 19–36. Springer, Heidelberg (2014)
5. Aumasson, J.-P., Neves, S., Wilcox-O'Hearn, Z., Winnerlein, C.: BLAKE2: Simpler, Smaller, Fast as MD5. In: Jacobson, M., Locasto, M., Mohassel, P., Safavi-Naini, R. (eds.) ACNS 2013. LNCS, vol. 7954, pp. 119–135. Springer, Heidelberg (2013)
6. Bernstein, D.J.: ChaCha, a Variant of Salsa20. In: Workshop Record of SASC 2008: The State of the Art of Stream Ciphers (2008). http://cr.yp.to/chacha.html
7. Bernstein, D.J.: The Salsa20 Family of Stream Ciphers. In: Robshaw, M., Billet, O. (eds.) New Stream Cipher Designs. LNCS, vol. 4986, pp. 84–97. Springer, Heidelberg (2008)
8. Bertoni, G., Daemen, J., Peeters, M., Assche, G.V.: On Alignment in Keccak. In: ECRYPT II Hash Workshop, May 2011
9. Bertoni, G., Daemen, J., Peeters, M.,Assche, G.V.: Permutation-based Encryption, Authentication and Authenticated Encryption, presented at DIAC, Stockholm, Sweden, 05–06 July 2012
10. Bertoni, G., Daemen, J., Peeters, M., Assche, G.V.: Cryptographic Sponge Functions, January 2011. http://sponge.noekeon.org/CSF-0.1.pdf
11. Bertoni, G., Daemen, J., Peeters, M., Van Assche, G.: Duplexing the Sponge: Single-pass Authenticated Encryption and Other Applications. In: Miri, A., Vaudenay, S. (eds.) SAC 2011. LNCS, vol. 7118, pp. 320–337. Springer, Heidelberg (2012)
12. Bertoni, G., Daemen, J., Peeters, M., Assche, G.V.: The Keccak Reference, January 2011. http://keccak.noekeon.org/
13. Biham, E., Shamir, A.: Differential Cryptanalysis of DES-like Cryptosystems. J. Cryptol. 4(1), 3–72 (1991)
14. Brummayer, R., Biere, A.: Boolector: An Efficient SMT Solver for Bit-vectors and Arrays. In: Kowalewski, S., Philippou, A. (eds.) TACAS 2009. LNCS, vol. 5505, pp. 174–177. Springer, Heidelberg (2009)
15. Daemen, J., Peeters, M., Assche, G.V., Rijmen, V.: Nessie Proposal: the Block Cipher Noekeon. Nessie submission (2000). http://gro.noekeon.org/
16. Daemen, J., Rijmen, V.: The Wide Trail Design Strategy. In: Honary, B. (ed.) Cryptography and Coding 2001. LNCS, vol. 2260, pp. 222–238. Springer, Heidelberg (2001)
17. Daemen, J., Van Assche, G.: Differential Propagation Analysis of Keccak. In: Canteaut, A. (ed.) FSE 2012. LNCS, vol. 7549, pp. 422–441. Springer, Heidelberg (2012)
18. Ganesh, V., Govostes, R., Phang, K.Y., Soos, M., Schwartz, E.: STP – A Simple Theorem Prover (2006–2013). http://stp.github.io/stp
19. Guo, J., Karpman, P., Nikolić, I., Wang, L., Wu, S.: Analysis of BLAKE2. In: Benaloh, J. (ed.) CT-RSA 2014. LNCS, vol. 8366, pp. 402–423. Springer, Heidelberg (2014)
20. Khovratovich, D., Nikolić, I.: Rotational Cryptanalysis of ARX. In: Hong, S., Iwata, T. (eds.) FSE 2010. LNCS, vol. 6147, pp. 333–346. Springer, Heidelberg (2010)

21. Knuth, D.E.: The Art of Computer Programming, Volume 4A: Combinatorial Algorithms, Part 1, vol. 4A. Addison-Wesley, Upper Saddle River (2011). http://www-cs-faculty.stanford.edu/~uno/taocp.html

22. Lipmaa, H., Moriai, S.: Efficient Algorithms for Computing Differential Properties of Addition. In: Matsui, M. (ed.) FSE 2001. LNCS, vol. 2355, pp. 336–350. Springer, Heidelberg (2002)

23. Mate Soos: CryptoMinisat (2009–2014). http://www.msoos.org/cryptominisat2

24. Mouha, N., Preneel, B.: Towards Finding Optimal Differential Characteristics for ARX: Application to Salsa20. Cryptology ePrint Archive, Report 2013/328 (2013)

25. Shi, Z., Zhang, B., Feng, D., Wu, W.: Improved Key Recovery Attacks on Reduced-round Salsa20 and ChaCha. In: Kwon, T., Lee, M.-K., Kwon, D. (eds.) ICISC 2012. LNCS, vol. 7839, pp. 337–351. Springer, Heidelberg (2013)

26. Shoup, V.: Computational Introduction to Number Theory and Algebra, 2nd edn. Cambridge University Press, Cambridge (2009). http://shoup.net/ntb

A Addenda to Differential Cryptanalysis

Proof of Lemma 3. On bit level Eq. 2 has the form

$$0 = \alpha_0 \oplus \beta_0 \oplus \gamma_0$$
$$0 = (\alpha_i \oplus \beta_i \oplus \gamma_i) \wedge (\alpha_{i-1} \oplus 1) \wedge (\beta_{i-1} \oplus 1), \quad i > 0$$

Obviously, the least significant bits (i.e. $i = 0$) are identical for Eqs. 1 and 2. For $i > 0$ let $t = (\alpha_i \oplus \beta_i \oplus \gamma_i) \oplus (\alpha_{i-1} \wedge \beta_{i-1})$. If $t = 0$ then Eq. 1 has always the solution $x_{i-1} = y_{i-1} = 0$. Otherwise, if $t = 1$, Eq. 1 is only solvable if $\alpha_{i-1} = 1$ or $\beta_{i-1} = 1$, and these are exactly the cases captured in Eq. 2.

Proof of Lemma 5. Without loss of generality we assume that $\alpha \neq 0$ or $\beta \neq 0$. Looking at Eq. 1, we see that the term $(\alpha \oplus \beta \oplus \gamma)$ has no effect on the probability of the differential δ, since it does not depend on either x or y. It has therefore probability 1.

Analysing the bit level representation of Eq. 1, we observe that the term $(x_{i-1} \wedge \alpha_{i-1}) \oplus (y_{i-1} \wedge \beta_{i-1}) \oplus (\alpha_{i-1} \wedge \beta_{i-1})$ is balanced (i.e., is 1 with probability $1/2$) if $\alpha_{i-1} = 1$ or $\beta_{i-1} = 1$. Therefore, under the assumption of independence of α_i and β_i, the overall probability of δ can be computed by counting the number of 1s in the first $n - 1$ bits of $\alpha \vee \beta$ or, equivalently, of $(\alpha \vee \beta) \ll 1$, which proves the lemma.

Proof of Lemma 7. It is easy to see that the least significant bits (i.e. $i = 0$) of Eqs. 3 and 4 are the same. Therefore, we will consider them no longer. Looking at the bit level representation of Eq. 3 (for $i > 0$) we consider two cases:

- $\alpha_i \oplus \beta_i \oplus \gamma_i = 0$: Here, Eq. 3 has always the solution $x_{i-1} = y_{i-1} = 0$.
- $\alpha_i \oplus \beta_i \oplus \gamma_i = 1$: In this case, the bit level representation of Eq. 3 is only solvable if either $\alpha_{i-1} \neq \gamma_{i-1}$ or $\beta_{i-1} \neq \gamma_{i-1}$. Furthermore, the bit level representation of Eq. 4 is given by

$$(\alpha_i \oplus \beta_i \oplus \gamma_i) \wedge (\alpha_{i-1} \oplus \gamma_{i-1} \oplus 1) \wedge (\beta_{i-1} \oplus \gamma_{i-1} \oplus 1) = 0, \qquad i > 0$$

It is evident that the latter equation only holds if $(\alpha_i \oplus \beta_i \oplus \gamma_i) = 0$, $\alpha_{i-1} \neq \gamma_{i-1}$, or $\beta_{i-1} \neq \gamma_{i-1}$. As seen above, these are the very same conditions that define a H-differential.

Proof of Lemma 9. The claim can be proven analogously to Lemma 5. It follows from the fact that in the bit level representation of Eq. 3 the expression

$$(x_{i-1} \wedge (\alpha_{i-1} \oplus \gamma_{i-1})) \oplus (y_{i-1} \wedge (\beta_{i-1} \oplus \gamma_{i-1}))$$

is balanced if $\alpha_{i-1} \oplus \gamma_{i-1} = 1$ or $\beta_{i-1} \oplus \gamma_{i-1} = 1$.

B   CVC Code

Below we show exemplarily for NORX64 how to translate the differential search operations to the CVC language. Variables have the datatype `BITVECTOR(W)`, where $W = 64$ is the wordsize.

$0 = (\alpha \oplus \beta \oplus \gamma) \wedge (\neg((\alpha \vee \beta) \ll 1))$	`ASSERT(0 = BVXOR(BVXOR(α,β),γ) & (-(((α | β)≪1)[63:0])));`
$w = (\alpha \vee \beta) \ll 1$	`ASSERT(w = (((α | β)≪1)[63:0]));`
$\gamma = (\alpha \oplus \beta) \ggg 8$	`ASSERT(γ = (BVXOR(α,β)≫8)|((BVXOR(α,β) ≪56)[63:0]));`

Computation of hw(w) using helper variables $h_0, \ldots, h_5$, where hw$(w) = h_5$:

`ASSERT(m1 = 0x5555555555555555);`	`ASSERT(h0 = BVPLUS(64,(w & m1), (((w≫1)[63:0]) & m1)));`
`ASSERT(m2 = 0x3333333333333333);`	`ASSERT(h1 = BVPLUS(64,(h0 & m2), (((h0≫2)[63:0]) & m2)));`
`ASSERT(m4 = 0x0f0f0f0f0f0f0f0f);`	`ASSERT(h2 = BVPLUS(64,(h1 & m4), (((h1≫4)[63:0]) & m4)));`
`ASSERT(m8 = 0x00ff00ff00ff00ff);`	`ASSERT(h3 = BVPLUS(64,(h2 & m8), (((h2≫8)[63:0]) & m8)));`
`ASSERT(m16 = 0x0000ffff0000ffff);`	`ASSERT(h4 = BVPLUS(64,(h3 & m16), (((h3≫16)[63:0]) & m16)));`
`ASSERT(m32 = 0x00000000ffffffff);`	`ASSERT(h5 = BVPLUS(64,(h4 & m32), (((h4≫32)[63:0]) & m32)));`

C   Selected Differentials

C.1   Experimental Verification of NODE

The first table shows the results from our verification of NODE, see Sect. 3.3. Notation is used as follows. w_e: expected weight, $\#S$: number of samples, v_e: expected value of input/output pairs adhering the differential, v_m: measured value of input/-output pairs adhering the differential, w_m: measured weight. After that we list the differentials in 32- and 64-bit $\mathsf{F}^{1.5}$ that we used to perform the verification.

			NORX32			NORX64		
w_e	$\#S$	v_e	v_m	$v_m - v_e$	w_m	v_m	$v_m - v_e$	w_m
12	2^{28}	65536	65652	+116	11.997	65627	+91	11.997
13	2^{29}	65536	65788	+252	12.994	65584	+48	12.998
14	2^{30}	65536	65170	−366	14.008	65476	−60	14.001
15	2^{31}	65536	65441	−95	15.002	65515	−21	15.000
16	2^{32}	65536	65683	+147	15.996	65563	+27	15.999
17	2^{33}	65536	65296	−240	17.005	65608	+72	16.998
18	2^{34}	65536	65389	−147	18.003	65565	+29	17.999

δ_0				δ_1				w
00000000	00000400	80000080	80000000	00000000	00000000	00000000	80001000	
00000000	80000400	80000080	00000000	00000000	00000000	00000000	21012100	12
00000000	80000000	80808080	80000000	00000000	00000000	00000000	10808080	
00000000	80000000	80800000	80000080	00000000	00000000	00000000	10008080	
80000000	00000000	00000400	80000180	80001000	00000000	00000000	00000000	
00000000	00000000	80000400	80000080	21012100	00000000	00000000	00000000	13
80000000	00000000	00000000	80808080	10808080	00000000	00000000	00000000	
80000080	00000000	80000000	80800000	10008080	00000000	00000000	00000000	
80000080	80000000	00000000	00000400	00000000	80001000	00000000	00000000	
80000180	00000000	00000000	80000400	00000000	21012100	00000000	00000000	14
80808080	80000000	00000000	80000000	00000000	10808080	00000000	00000000	
80800000	80000080	00000000	80000000	00000000	10008080	00000000	00000000	
00000400	80000000	00000400	40100000	00100000	00000000	00000000	00000000	
80000400	80000000	00000000	00100200	00200021	00000000	00000000	00000000	15
80000000	80018000	00000400	00000000	80000010	00000000	00000000	00000000	
80000000	00800000	00040400	40000600	00000010	00000000	00000000	00000000	
00000400	80000080	80000000	00000000	00000000	00000000	80003000	00000000	
80000400	80000080	00000000	00000000	00000000	00000000	63016100	00000000	16
80000000	81808080	80000000	00000000	00000000	00000000	31808080	00000000	
80000000	80800000	80000080	00000000	00000000	00000000	30008080	00000000	
00000000	00000400	80000080	80000000	00000000	00000000	00000000	80001000	
00000000	80000400	80000080	00000000	00000000	00000000	00000000	21012100	17
00000000	80000000	80838780	80000000	00000000	00000000	00000000	10808080	
00000000	80000000	80800000	80000080	00000000	00000000	00000000	10008080	
00000400	00000000	80000000	C0000200	00100000	00000000	00000000	00606001	
80000400	00000000	00000000	00000200	00200021	00000000	00000000	C24242C0	18
80000000	00000000	80000000	00000000	80000010	00000000	00000000	61010160	
80000000	00000000	80000080	C0000000	00000010	00000000	00000000	60010160	

δ_0				δ_1				w
8000000000000000	0000000000000000	0000000000040000	8000000000000080	8000001000000000	0000000000000000	0000000000000000	0000000000000000	
0000000000000000	0000000000000000	8000000000040000	8000000000000080	2100002001010000	0000000000000000	0000000000000000	0000000000000000	12
8000000000000000	0000000000000000	8000000000000000	80008080000000B0	1080000000808000	0000000000000000	0000000000000000	0000000000000000	
8000000000000080	0000000000000000	8000000000000000	0080800000000080	1000000000808000	0000000000000000	0000000000000000	0000000000000000	
4000001000000000	0000000000040000	8000000000000000	0000000000040000	0000000000000000	0001000000000000	0000000000000000	0000000000000000	
0000001000020000	8000000000040000	8000000000000000	0000000000000000	0000000000000000	0002000000000021	0000000000000000	0000000000000000	13
0000000000000000	8000000000000000	8000800000000000	0000000000040000	0000000000000000	8000000000000010	0000000000000000	0000000000000000	
4000000000020000	8000000000000000	0000800000000000	0000000004040000	0000000000000000	0000000000000010	0000000000000000	0000000000000000	
0000000000040000	8000000000000080	8000000000000000	0000000000000000	0000000000000000	0000000000000000	8000001000000000	0000000000000000	
8000000000040000	0000000000000000	0000000000000000	0000000000000000	0000000000000000	0000000000000000	2100002001010000	0000000000000000	14
8000000000000000	8003808000000080	8000000000000000	0000000000000000	0000000000000000	0000000000000000	1080000000808000	0000000000000000	
8000000000000000	0080800000000000	8000000000000080	0000000000000000	0000000000000000	0000000000000000	1000000000808000	0000000000000000	
0000000000000000	000000000000C000	8000000000000080	8000000000000000	0000000000000000	0000000000000000	0000000000000000	8000001000000000	
0000000000000000	8000000000040000	8000000000000080	0000000000000000	0000000000000000	0000000000000000	0000000000000000	2300006001010000	15
0000000000000000	8000000000000000	8000800000000080	8000000000000000	0000000000000000	0000000000000000	0000000000000000	1180000000808000	
0000000000000000	8000000000000000	0080800000000000	8000000000000080	0000000000000000	0000000000000000	0000000000000000	1000000000808000	
0000000000040000	4000001000080000	0000000000040000	8000000000000000	0000000000000000	0000000000000000	0001000000000000	0000000000000000	
0000000000000000	0000001000020000	8000000000040000	8000000000000000	0000000000000000	0000000000000000	0002000000000021	0000000000000000	16
0000000000040000	0000000000000000	8000000000000000	8000000800000000	0000000000000000	0000000000000000	8000000000000010	0000000000000000	
0000000004040000	C0000000000E0000	8000000000000000	0000800000000000	0000000000000000	0000000000000000	0000000000000010	0000000000000000	
8000000000000080	0000000000000000	0000000000000000	0000000000040000	0000000000000000	8000007000000000	0000000000000000	0000000000000000	
8000000000000080	0000000000000000	0000000000000000	8000000000040000	0000000000000000	E300006001010000	0000000000000000	0000000000000000	17
8000808000000180	0000000000000000	0000000000000000	8000000000000000	0000000000000000	7180000000808000	0000000000000000	0000000000000000	
0080800000000000	8000000000000080	0000000000000000	8000000000000000	0000000000000000	7000000000808000	0000000000000000	0000000000000000	
0000000000040000	8000000000000000	0000000000040000	400000F000000000	0001000000000000	0000000000000000	0000000000000000	0000000000000000	
8000000000040000	8000000000000000	0000000000000000	0000001000020000	0002000000000021	0000000000000000	0000000000000000	0000000000000000	18
8000000000000000	8000080000000000	0000000000040000	0000000000000000	8000000000000010	0000000000000000	0000000000000000	0000000000000000	
8000000000000000	0000800000000000	000000000C040000	4000000000020000	0000000000000010	0000000000000000	0000000000000000	0000000000000000	

C.2   Probability-1 Differentials in G

Using NODE we could show that there are exactly 3 probability-1 differentials in both versions (32- and 64-bit) of G.

Differences			
δ_0 80000000	80000000	80000000	00000000
δ_1 00000000	00000001	80000000	00000000
δ_0 80000000	00000000	80000000	80000080
δ_1 80000000	00000000	00000000	00000000
δ_0 00000000	80000000	00000000	80000080
δ_1 80000000	00000001	80000000	00000000

Differences			
δ_0 8000000000000000	8000000000000000	8000000000000000	0000000000000000
δ_1 0000000000000000	0000000000000001	8000000000000000	0000000000000000
δ_0 8000000000000000	0000000000000000	8000000000000000	8000000000000080
δ_1 8000000000000000	0000000000000000	0000000000000000	0000000000000000
δ_0 0000000000000000	8000000000000000	0000000000000000	8000000000000080
δ_1 8000000000000000	0000000000000001	8000000000000000	0000000000000000

C.3   Best Differential Characteristics for F^4

The following two tables show the best differential characteristics in F^4 that we were capable to find with NODE. The values δ_0 and δ_4 are in- and output difference, respectively, and δ_1, δ_2, and δ_3 are internal differences. The differences are listed after a single application of F, respectively, and the values w_i, with $i \in \{0, \dots, 3\}$, are the corresponding differential weights.

δ_0				w_0	δ_1				w_1
80140100	90024294	84246020	92800154		40100000	00000400	80000000	00000400	
e4548300	52240214	e0202424	d0004054	172	00100200	80000000	80000000	00000000	11
c4464046	00a08480	c1008108	90d43134		00000000	80000000	80008000	00000400	
e200c684	e2eac480	a4848881	06915342		40000200	80000000	00800000	00040400	

δ_2				w_2	δ_3				w_3
00000000	00000000	00000000	00000000		04042425	00100002	00020000	02100000	
00000000	00000000	00000000	00000000	44	04200401	42024200	20042024	20042004	357
00000000	80000000	00000000	00000000		10001002	80000200	25250504	10021010	
00000000	00000000	00000000	00000000		10020010	00001002	00000210	04252504	

δ_4				
c4001963	804da817	0c05b60e	12220503	
9072b909	185b792a	cc0d56cd	7e0ac646	total weight: 584
80116300	100c2800	8f003320	3b270222	
01056104	88000041	92002824	04210001	

δ_0	w_0	δ_1	w_1
00900824010288c5 4000443880011086 224012044220ac43 e004044484049520		8000000800050000 8000000000000000 4000000000000000 0000001000020080	
4080882001010885 4600841880821086 a3c0721444632c43 c224440007849504	349	8000000800040000 8000000000000000 c000000000040000 8000001000020080	27
81600850830b0484 840080c080868000 8004449040c14400 810210184090 8a80		0000000000000000 8000008000000000 c0000400040000 400080 8000020080	
6191548c08000581 0200004006038044 8104f01c8702c0e0 60605084938886e3		0000000000010080 0000080000000000 8000400004040000 8080800002000 0c0	

δ_2	w_2	δ_3	w_3
8000000000000000 0000000000000000 0000000000000000 0000000000000000		0000000000000000 0000000000000000 0000100000000000 0000202000000001	
8000000000000000 0000000000000000 0000000000000000 0000000000000000	12	4200404002020040 0000000000000000 0000000000000000 0000200000000021	448
8000000000000000 0000000000000000 0000000000000000 0000000000000000		8000000000000010 2100000001010020 0000000000000000 0000000000000000	
0000000000000000 0000000000000000 0000000000000000 0000000000000000		0000000000000000 0000000000000010 2000000001010020 0000000000000000	

δ_4	
321a4500060e4e2e 27404405026e500e 3806422387200a08 8c40f4a0884c0820	
71540fb858cb9902 ee018cc282747980 c714164174ce3eb9 1a49a091101191e1	total weight: 836
786680d0e46406cb 14440844013274e6 03a843203f071b7c 09a840c00c0ccc78	
4000404a22120005 07220c4202016240 2aa4200a0a041a62 84a468682000601c	

C.4   Best Iterative Differentials for **F**

	Differences	w	Differences	w
δ_1 $\parallel$ δ_0	818c959b 00186049 eb5b7984 791c6da1 677b513d 80000400 00000227 5293655f 00809a2b bfa98bff c08b8e89 0000711c 800027c3 f984eb5b 6d81f915 b5aaa99d	512	0000000100000000 0000000000000000 f77c78b200000d04 0000000000000000 be7fffeffe0f349f 0000000000000000 6c07fbd200000001 ff1ab5be4e7500be 0060c54927018000 0000000000000000 0000000000000000 b603fde900000000 b6035caf00000000 0000000000000000 0000000000000000 0000000000000000	843

C.5   Best Differentials Having Equal Columns of Weight 44 in **F**

	Differences		Differences
δ_0	80000000 80000000 80000000 80000000 80000000 80000000 80000000 80000000 80000000 80000000 80000000 80000000 00000000 00000000 00000000 00000000	δ_0	8000000000000000 8000000000000000 8000000000000000 8000000000000000 8000000000000000 8000000000000000 8000000000000000 8000000000000000 8000000000000000 8000000000000000 8000000000000000 8000000000000000 0000000000000000 0000000000000000 0000000000000000 0000000000000000
δ_1	00102001 00102001 00102001 00102001 42624221 42624221 42624221 42624221 a1010110 a1010110 a1010110 a1010110 20010110 20010110 20010110 20010110	δ_1	0000102000000001 0000102000000001 0000102000000001 0000102000000001 4200604002020021 4200604002020021 4200604002020021 4200604002020021 a100000001010010 a100000001010010 a100000001010010 a100000001010010 2000000001010010 2000000001010010 2000000001010010 2000000001010010

D   Addenda to Rotational Cryptanalysis

Proof of Lemma 11. After evaluating and simplifying the equation $\mathsf{H}(x, y) \ggg r = \mathsf{H}(x \ggg r, y \ggg r)$ we get $((x \wedge y) \lll 1) \ggg r = ((x \ggg r) \wedge (y \ggg r)) \lll 1$. Translating this equation to bit vectors results in

$$(x_{r-1} \wedge y_{r-1}, \ldots, x_0 \wedge y_0, 0, x_{n-2} \wedge y_{n-2}, \ldots, x_r \wedge y_r)$$
$$= (x_{r-1} \wedge y_{r-1}, \ldots, x_0 \wedge y_0, x_{n-1} \wedge y_{n-1}, x_{n-2} \wedge y_{n-2}, \ldots, 0)$$

The probability that those two vectors match is $(3/4)^2 = 9/16$, as $a \wedge b = 0$ with probability $3/4$ for bits a and b chosen uniformly at random.

Proof of Lemma 13. The first important observation is that the statement of this lemma is independent of the function f, as it only makes a claim on the image of f. Thus it is sufficient to prove the lemma for $z \ggg r = z$, where $z = f(x, y)$ and x or y was fixed.

We identify the indices of an n-bit string by the elements in $G := \mathbb{Z}/n\mathbb{Z}$. Let $\tau : G \longrightarrow G$, $i \bmod n \mapsto (i + 1) \bmod n$. Then τ obviously generates the cyclic group G, i.e. $\mathrm{ord}(\tau) = n$. Moreover, for an arbitrary $r \in \mathbb{Z}$ we have $\mathrm{ord}(\tau^r) = n/\gcd(r, n)$, see [26, §§6.2]. In other words, the subgroup $H := \langle \tau^r \rangle$ of G has order $n/\gcd(r, n)$. By Lagrange's theorem we have $\mathrm{ord}(G) = [G : H] \cdot \mathrm{ord}(H)$ and it follows for the group index $[G : H] = \gcd(r, n)$, which corresponds to the number of (left) cosets of H in G. These cosets contain the indices of a bit string which are mapped onto each other by a rotation $\ggg r$. This means that there are $2^{\gcd(r,n)}$ n-bit strings z which satisfy $z \ggg r = z$. Thus the probability, that an n-bit string z, chosen uniformly at random among all n-bit strings, satisfies $z \ggg r = z$ is $2^{-(n-\gcd(r,n))}$. This proves the lemma.

Cryptographic Protocols

Efficient Distributed Tag-Based Encryption and Its Application to Group Signatures with Efficient Distributed Traceability

Essam Ghadafi[✉]

Computer Science Department, University of Bristol, Bristol, UK
essam_gha@yahoo.com

Abstract. In this work, we first formalize the notion of dynamic group signatures with distributed traceability, where the capability to trace signatures is distributed among n managers without requiring any interaction. This ensures that only the participation of all tracing managers permits tracing a signature, which reduces the trust placed in a single tracing manager. The threshold variant follows easily from our definitions and constructions. Our model offers strong security requirements. Our second contribution is a generic construction for the notion which has a concurrent join protocol, meets strong security requirements, and offers efficient traceability, i.e. without requiring tracing managers to produce expensive zero-knowledge proofs for tracing correctness. To dispense with the expensive zero-knowledge proofs required in the tracing, we deploy a distributed tag-based encryption with public verifiability. Finally, we provide some concrete instantiations, which, to the best of our knowledge, are the first efficient provably secure realizations in the standard model simultaneously offering all the aforementioned properties. To realize our constructions efficiently, we construct an efficient distributed (and threshold) tag-based encryption scheme that works in the efficient Type-III asymmetric bilinear groups. Our distributed tag-based encryption scheme yields short ciphertexts (only 1280 bits at 128-bit security), and is secure under an existing variant of the standard decisional linear assumption. Our tag-based encryption scheme is of independent interest and is useful for many applications beyond the scope of this paper. As a special case of our distributed tag-based encryption scheme, we get an efficient tag-based encryption scheme in Type-III asymmetric bilinear groups that is secure in the standard model.

Keywords: Group signatures · Distributed traceability · Distributed public-key encryption · Standard model

1 Introduction

Group signatures, introduced by Chaum and van Heyst [25], are a fundamental cryptographic primitive allowing a member of a group (administered by a designated manager) to anonymously sign messages on behalf of the group.

© Springer International Publishing Switzerland 2015
D.F. Aranha and A. Menezes (Eds.): LATINCRYPT 2014, LNCS 8895, pp. 327–347, 2015.
DOI: 10.1007/978-3-319-16295-9_18

In the case of a dispute, a designated tracing manager can revoke anonymity by revealing the signer. The downside of granting a single entity the capability to trace signatures is the high trust placed in such an entity. As a result, anonymity in group signatures relying on a single tracing authority only holds if the tracing authority is fully honest. More precisely, a misbehaving tracing authority could abuse the power granted to it and open signatures need not be opened. Therefore, reducing the trust placed in the tracing manager by distributing the tracing capability among different parties is desirable. While some of the existing schemes can be translated into the distributed traceability setting by utilizing standard secret-sharing techniques, e.g. [10,50], unfortunately, most of those secure in the strong Bellare et al. model [13], would become impractical due to the expensive zero-knowledge proofs required in the tracing.

Related Work. After their introduction, a long line of research on group signatures has emerged. Bellare, Micciancio and Warinschi [11] formalized the security definitions for group signatures supporting static groups. In such a notion, the group population is fixed at the setup phase. Moreover, the group manager (which also provides the traceability feature) needs to be fully trusted. Later, Bellare, Shi and Zhang [13] provided formal security definitions for the more practical dynamic case where members can enroll at any time. Also, [13] separated the tracing role from the group management.

Besides correctness, the model of [13] defines three other requirements: anonymity, traceability and non-frameability. Informally, anonymity requires that signatures do not reveal the identity of the signer; traceability requires that the tracing manager is always able to identify the signer and prove such a claim; non-frameability ensures that even if the group and tracing managers collude with the rest of the group, they cannot frame an honest member. More recently, Sakai et al. [49] strengthened the security definitions of group signatures by adding the *opening soundness* requirement.The stronger variant of opening soundness ensures that even if all entities are corrupt, it is infeasible to produce a signature that traces to two different members.

Constructions of group signatures in the random oracle model [12] include [8,15,18,21–24,26,29,41,46]. Constructions not relying on random oracles include [5,7,19,20,35,36,43,44]. Other measures in which the above mentioned constructions differ are: the security they offer, the size of the signatures they yield and the round complexity of the join protocol.

Different approaches have been proposed to minimize the trust placed in the tracing manager. Sakai et al. [48] recently proposed the notion of group signatures with message-dependent opening. In such a notion, an admitter specifies what messages signatures upon which can be traced. This prevents the tracing manager from opening signatures on messages not admitted by the admitter. In [38], the authors informally highlighted how to extend their linkable group signature scheme (secure in the random oracle model) to provide distributed traceability. In [30], the authors presented a scheme where the roles of the managers can be distributed. Their scheme is only non-frameable against honest tracing managers, and requires both random oracles and the generic group model.

Benjumea et al. [14] introduced the notion of fair traceable multi-group signatures which combines the features of group signatures and traceable signatures [40] and in which traceability requires the co-operation of a judge with designated parties known as fairness authorities. The authors also provided a construction of their primitive in the random oracle model.

Zheng et al. [51] extended Manulis's notion of democratic group signatures [45] to add threshold traceability where group members must collude to trace signatures. Democratic group signatures differ from group signatures in many aspects. In the former, the roles of the group and tracing managers are eliminated and the group is managed by the members themselves. In addition, signatures are only anonymous to non-members.

Our Contribution. We offer the following contributions:

1. A formal security model for group signatures with distributed traceability without requiring any interaction. Only the participation of all n tracing managers makes it possible to trace a signature. The more general k out of n threshold case follows easily from our definitions. Our model offers strong security including the notion of tracing soundness [49].
2. A generic framework for constructing group signatures with efficient (i.e. without requiring expensive zero-knowledge proofs in the tracing) distributed traceability that supports dynamic groups with a concurrent join protocol and which is provably secure w.r.t. our strong security model.
3. Instantiations of the generic framework in the standard model. To the best of our knowledge, they are the first provably secure realizations not relying on idealized assumptions offering all the aforementioned properties.
4. An efficient distributed/threshold selective-tag weakly IND-CCA tag-based encryption scheme that is based on an existing variant of the standard decisional linear assumption. Our scheme is non-interactive (i.e. requires no interaction between the decryption servers) and is robust, i.e. the validity of the decryption shares as well as the ciphertext is publicly verifiable. The scheme works in the efficient Type-III bilinear groups setting and yields short ciphertexts which are much shorter than those of the original Kiltz's tag-based encryption scheme [42] and its threshold variant of [6]. By combining our scheme with a strongly unforgeable one-time signature scheme as per the transformation in [42], we obtain an efficient fully secure IND-CCA distributed/threshold encryption scheme, which is useful for many applications beyond the scope of this paper.

Paper Organization. In Sect. 2, we give some preliminary definitions. We present our model for group signatures with distributed traceability in Sect. 3. We present the building blocks we use in Sect. 4. In Sect. 5, we present our generic construction and provide a proof of its security. In Sect. 6, we present instantiations in the standard model.

Notation. A function $\nu(.) : \mathbb{N} \to \mathbb{R}^+$ is negligible in c if for every polynomial $p(.)$ and all sufficiently large values of c, it holds that $\nu(c) < \frac{1}{p(c)}$. Given a probability distribution Y, we denote by $x \leftarrow Y$ the operation of selecting an element according to Y. If M is a probabilistic machine, we denote by $M(x_1, \ldots, x_n)$ the output distribution of M on inputs $(x_1, \ldots, x_n)$. By $[n]$ we denote the set $\{1, \ldots, n\}$. By PPT we mean running in probabilistic polynomial time in the relevant security parameter.

2 Preliminaries

In this section we provide some preliminary definitions.

Bilinear Groups. A bilinear group is a tuple $\mathcal{P} := (\mathbb{G}, \tilde{\mathbb{G}}, \mathbb{T}, p, G, \tilde{G}, e)$ where $\mathbb{G}$, $\tilde{\mathbb{G}}$ and $\mathbb{T}$ are groups of a prime order p, and G and $\tilde{G}$ generate $\mathbb{G}$ and $\tilde{\mathbb{G}}$, respectively. The function e is a non-degenerate bilinear map $e : \mathbb{G} \times \tilde{\mathbb{G}} \longrightarrow \mathbb{T}$. We use multiplicative notation for all the groups. We let $\mathbb{G}^\times := \mathbb{G} \setminus \{1_\mathbb{G}\}$ and $\tilde{\mathbb{G}}^\times := \tilde{\mathbb{G}} \setminus \{1_{\tilde{\mathbb{G}}}\}$. In this paper, we focus on the efficient Type-III setting [32], where $\mathbb{G} \neq \tilde{\mathbb{G}}$ and there is no isomorphism between the groups in either direction. We assume there is an algorithm BGrpSetup taking as input a security parameter λ and outputting a description of bilinear groups.

Complexity Assumptions. We use the following existing assumptions:

Symmetric External Decisional Diffie-Hellman (SXDH). The Decisional Diffie-Hellman (DDH) assumption holds in both groups $\mathbb{G}$ and $\tilde{\mathbb{G}}$.

Decisional Linear in $\mathbb{G}$ ($DLIN_\mathbb{G}$) Assumption [1,33]. Given $\mathcal{P}$ and a tuple $(G^h, G^v, G^u, G^{rh}, G^{sv}, G^{ut}) \in \mathbb{G}^6$ for unknown $h, r, s, t, u, v \in \mathbb{Z}_p$, it is hard to determine whether or not $t = r + s$.

External Decisional Linear in $\mathbb{G}$ ($XDLIN_\mathbb{G}$) Assumption [1][1]. Given $\mathcal{P}$ and a tuple $(G^h, G^v, G^u, G^{rh}, G^{sv}, G^{ut}, \tilde{G}^h, \tilde{G}^v, \tilde{G}^u, \tilde{G}^{rh}, \tilde{G}^{sv}) \in \mathbb{G}^6 \times \tilde{\mathbb{G}}^5$ for unknown $h, r, s, t, u, v \in \mathbb{Z}_p$, it is hard to determine whether or not $t = r + s$.

q-Strong Diffie-Hellman (q-SDH) Assumption in $\mathbb{G}$ [17]. Given the tuple $(G, G^x, \ldots, G^{x^q}) \in \mathbb{G}^{q+1}$ for $x \leftarrow \mathbb{Z}_p$, it is hard to output a pair $(c, G^{\frac{1}{x+c}}) \in \mathbb{Z}_p \times \mathbb{G}$ for an arbitrary $c \in \mathbb{Z}_p \setminus \{-x\}$.

q-AGHO [3]. Given a random tuple $(G, \tilde{G}, \tilde{W}, \tilde{X}, \tilde{Y}) \in \mathbb{G} \times \tilde{\mathbb{G}}^4$, and q uniformly random tuples $(A_i, B_i, R_i, \tilde{D}_i) \in \mathbb{G}^3 \times \tilde{\mathbb{G}}$, each satisfying $e(A_i, \tilde{D}_i) = e(G, \tilde{G})$ and $e(G, \tilde{X}) = e(A_i, \tilde{W})e(B_i, \tilde{G})e(R_i, \tilde{Y})$, it is hard to output a new tuple $(A^*, B^*, R^*, \tilde{D}^*)$ satisfying the above equations.

Group Signatures. Here we briefly review the model of Bellare et al. [13] for dynamic group signatures with a single tracing authority. A dynamic group signature scheme consists of the following algorithms:

- $\mathsf{GKg}(1^\lambda)$ outputs a group public key gpk, a group manager's secret key msk and a tracing key tsk.
- $\mathsf{UKg}(1^\lambda)$ outputs a secret/public key pair $(\mathbf{usk}[\mathsf{uid}], \mathbf{upk}[\mathsf{uid}])$ for user uid.
- $\langle \mathsf{Join}(\mathsf{gpk}, \mathsf{uid}, \mathbf{usk}[\mathsf{uid}]), \mathsf{Issue}(\mathsf{msk}, \mathsf{uid}, \mathbf{upk}[\mathsf{uid}]) \rangle$ is an interactive protocol between a user uid and the group manager GM via which the user joins the group. If successful, the final state of the Issue algorithm is stored in the registration table at index uid (i.e. $\mathbf{reg}[\mathsf{uid}]$), whereas that of the Join algorithm is stored in $\mathbf{gsk}[\mathsf{uid}]$ and is used as the user's group signing key.

---

[1] $XDLIN_{\tilde{\mathbb{G}}}$ can be defined analogously by giving $\tilde{G}^{ut} \in \tilde{\mathbb{G}}$ instead of $G^{ut} \in \mathbb{G}$.

- Sign(gpk, **gsk**[uid], m) outputs a group signature Σ on the message m by member uid.
- Verify(gpk, m, Σ) verifies whether or not Σ is a valid group signature on m outputting a bit.
- Trace(gpk, tsk, m, Σ, **reg**) is the tracing algorithm in which the tracing manager uses its tracing key tsk to identify the group member uid who produced the signature Σ plus a proof π_{Trace} for such a claim.
- TraceVerify(gpk, uid, π_{Trace}, **upk**[uid], m, Σ) verifies the tracing proof π_{Trace} outputting a bit accordingly.

Besides correctness, the security requirements defined by [13] are:

- *Anonymity:* A group signature does not reveal the identity of the member who produced it even when the keys of the group manager and all group members are all revealed. This requirement relies on the tracing manager being fully honest.
- *Non-Frameability:* Even if the group and tracing managers collude with the rest of the group, they cannot frame an honest group member.
- *Traceability:* Even if the tracing manager and all group members are corrupt, they cannot produce a signature that does not trace to a member of the group.

3 Syntax and Security of Dynamic Group Signatures with Distributed Traceability

The parties involved in a Dynamic Group Signature with Distributed Traceability ($\mathcal{DGSDT}$) are: a group manager GM who authorizes who can join the group; κ tracing managers $\mathsf{TM}_1, \ldots, \mathsf{TM}_\kappa$ which only the participation of all of which makes it possible to identify who produced a signature; a set of users who can join group at any time by contacting the group manager. A $\mathcal{DGSDT}$ scheme consists of the following polynomial-time algorithms:

- GKg(1^λ, κ) is run by a trusted third party. On input a security parameter λ and the number of tracing managers κ, it outputs a group public key gpk, a group manager's secret key msk and secret tracing keys $\{\mathsf{tsk}_i\}_{i=1}^\kappa$.
- UKg(1^λ) outputs a secret/public key pair (**usk**[uid], **upk**[uid]) for user uid. We assume that the public key table **upk** is publicly available (possibly via some PKI) so that anyone can obtain authentic copies of the public keys.
- $\langle$Join(gpk, uid, **usk**[uid]), Issue(msk, uid, **upk**[uid])$\rangle$ is an interactive protocol between a user uid and the group manager GM. Upon successful completion, uid becomes a member of the group. The final state of the Issue algorithm is stored in the registration table at index uid (i.e. **reg**[uid]), whereas that of the Join algorithm is stored in **gsk**[uid]. We assume that the communication in this interactive protocol takes place over a secure (i.e. private and authentic) channel. The protocol is initiated by a call to Join.
- Sign(gpk, **gsk**[uid], m) on input the group public key gpk, a user's group signing key **gsk**[uid] and a message m, outputs a group signature Σ on m by the group member uid.

- Verify(gpk, m, Σ) is a deterministic algorithm which checks whether or not Σ is a valid group signature on m outputting a bit.
- TraceShare(gpk, tsk_i, m, Σ) on input the group public key gpk, a tracing key tsk_i belonging to tracing manager TM_i, a message m and a signature Σ, it outputs (ν, π_{Trace}) where ν is the tracing share of TM_i of Σ and π_{Trace} is a proof for the correctness of the tracing share. If TM_i is unable to compute her share, she outputs $(\perp, \perp)$. If the validity of the shares are publicly verifiable, we just omit π_{Trace}.
- ShareVerify(gpk, tid, ν, π_{Trace}, m, Σ) verifies whether the share ν is a valid tracing share of Σ by tracing manager TM_{tid} outputting a bit accordingly.
- TraceCombine(gpk, $\{(\nu, \pi_{Trace})_i\}_{i=1}^{\kappa}$, m, Σ, $\mathbf{reg}$) on input the group public key gpk, κ tracing shares and their proofs, a message m, a signature Σ, and the users' registration table, it outputs an identity uid > 0 of the user who produced Σ plus a proof θ_{Trace} attesting to this claim. If the algorithm is unable to trace the signature to a user, it returns $(0, \theta_{Trace})$. This algorithm does not require any secret information and hence could be run by any party.
- TraceVerify(gpk, uid, θ_{Trace}, $\mathbf{upk}[uid]$, m, Σ) on input the group public key gpk, a user identity uid, a tracing proof θ_{Trace}, the user's public key $\mathbf{upk}[uid]$, a message m, and a signature Σ, outputs 1 if θ_{Trace} is a valid proof that uid has produced Σ or 0 otherwise.

Security of Dynamic Group Signatures with Distributed Traceability.
Our model extends Bellare's et al. model [13] to provide distributed traceability and additionally captures tracing soundness as recently defined by [49] in the context of group signatures with a single tracing manager, which is vital for many applications as we explain later. Moreover, our non-frameability definition is slightly stronger than that of [13].

The security requirements of a dynamic group signature with distributed traceability are: *correctness*, *anonymity*, *non-frameability*, *traceability* and *tracing soundness*. To define those requirements, we use a set of games in which the adversary has access to a set of oracles. The following global lists are maintained: HUL is a list of honest users; CUL is a list of corrupt users whose personal secret keys have been chosen by the adversary; BUL is a list of bad users whose personal and group signing keys have been revealed to the adversary; SL is a list of signatures obtained from the Sign oracle; CL is a list of challenge signatures obtained from the challenge oracle.

The details of the following oracles are given in Fig. 1.

AddU(uid) adds an honest user uid to the group.
CrptU(uid, pk) adds a new corrupt user whose public key $\mathbf{upk}[uid]$ is chosen by the adversary. This is called in preparation for calling the SndM oracle.
SndM(uid, M_{in}) used to engage in the Join-Issue protocol with the honest, Issue-executing group manager.
SndU(uid, M_{in}) used to engage in the Join-Issue protocol with an honest, Join-executing user uid on behalf of the corrupt group manager.
RReg(uid) returns the registration information $\mathbf{reg}[uid]$ of user uid.

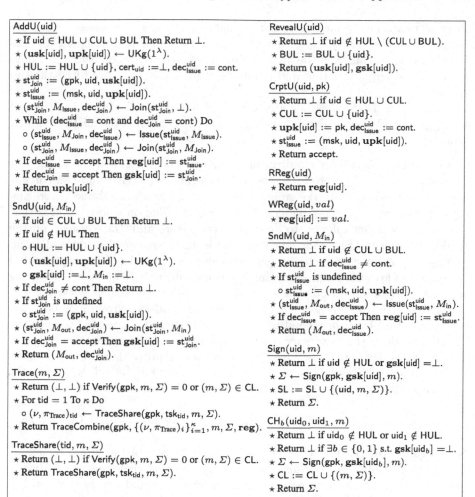

Fig. 1. Details of the oracles used in the security games

WReg(uid, val) modifies the entry $\mathbf{reg}[uid]$ by setting $\mathbf{reg}[uid] := val$.

RevealU(uid) returns the personal secret key $\mathbf{usk}[uid]$ and the group signing key $\mathbf{gsk}[uid]$ of group member uid.

Sign(uid, m) returns a signature on the message m by the group member uid.

$CH_b(uid_0, uid_1, m)$ is a left-right oracle for defining anonymity. The adversary sends a couple of identities (uid_0, uid_1) and a message m and receives a group signature by member uid_b for $b \leftarrow \{0, 1\}$.

TraceShare(tid, m, Σ) returns the tracing share of signature Σ of tracing manager TM_{tid}.

Trace(m, Σ) returns the identity of the signer of the signature Σ, i.e. first obtains the different tracing shares and then combines them.

The following security requirements are defined by the games in Fig. 2.

Correctness. This guarantees that: signatures produced by honest users are accepted by the Verify algorithm, the tracing shares are accepted by the ShareVerify algorithm, and the final tracing outcome of TraceCombine is accepted by the TraceVerify algorithm and points out to the user who produced the signature.

Formally, a $\mathcal{DGSDT}$ scheme is *correct* if for all $\lambda, \kappa \in \mathbb{N}$, the advantage $\mathsf{Adv}^{\mathrm{Corr}}_{\mathcal{DGSDT}, \mathcal{A}, \kappa}(\lambda) := \Pr[\mathsf{Exp}^{\mathrm{Corr}}_{\mathcal{DGSDT}, \mathcal{A}, \kappa}(\lambda) = 1]$ is negligible for all PPT adversaries $\mathcal{A}$.

(Full) Anonymity. This requires that signatures do not reveal the identity of the group member who produced them. In the game, the adversary can corrupt any user and fully corrupt the group manager. It can also learn the secret tracing keys of up to $\kappa - 1$ tracing managers of its choice. The only restriction is that the adversary is not allowed to query the TraceShare and Trace oracles on the challenge signature. Since the adversary can learn the personal secret and group signing keys of any user, including the challenge users, our definition captures full key exposure attacks. Also, since the adversary can corrupt up to $\kappa - 1$ tracing managers, it can obtain up to $\kappa - 1$ tracing shares of the challenge signature.

In the game, the adversary chooses a message and two group members and gets a signature by either member and wins if it correctly guesses the member. WLOG we allow the adversary a single call to the challenge oracle. A hybrid argument (similar to that used in [13]) can be used to prove that this is sufficient.

Formally, a $\mathcal{DGSDT}$ scheme is *(fully) anonymous* if for all $\lambda, \kappa \in \mathbb{N}$, the advantage $\mathsf{Adv}^{\mathrm{Anon}}_{\mathcal{DGSDT}, \mathcal{A}, \kappa}(\lambda)$ is negligible for all PPT adversaries $\mathcal{A}$, where

$$\mathsf{Adv}^{\mathrm{Anon}}_{\mathcal{DGSDT}, \mathcal{A}, \kappa}(\lambda) := \left| \Pr[\mathsf{Exp}^{\mathrm{Anon\text{-}0}}_{\mathcal{DGSDT}, \mathcal{A}, \kappa}(\lambda) = 1] - \Pr[\mathsf{Exp}^{\mathrm{Anon\text{-}1}}_{\mathcal{DGSDT}, \mathcal{A}, \kappa}(\lambda) = 1] \right|.$$

Non-Frameability. This ensures that even if the rest of the group as well as the group and all tracing managers are fully corrupt, they cannot produce a signature that traces to an honest group member who did not produce such a signature. Our definition is stronger than that used in other group signature models, e.g. [13], in the sense that the adversary in our game wins even if it produces a new signature on a message that was queried to the signing oracle, i.e. analogous to strong unforgeability in traditional signatures. The definition can in a straightforward manner be adapted to the weaker variant by requiring that the adversary's signature is on a new message that the framed user did not sign.

Formally, a $\mathcal{DGSDT}$ scheme is *non-frameable* if for all $\lambda, \kappa \in \mathbb{N}$, the advantage $\mathsf{Adv}^{\mathrm{Non\text{-}Frame}}_{\mathcal{DGSDT}, \mathcal{A}, \kappa}(\lambda) := \Pr[\mathsf{Exp}^{\mathrm{Non\text{-}Frame}}_{\mathcal{DGSDT}, \mathcal{A}, \kappa}(\lambda) = 1]$ is negligible for all PPT adversaries $\mathcal{A}$.

Traceability. This ensures that the adversary cannot produce a signature that cannot be traced to a member in the group. In the game, the adversary can corrupt any user and learn the tracing keys of all tracing managers. The only restriction is that the adversary is not given the group manager's secret key as this would allow it to create dummy users which are thus untraceable.

Experiment: $\text{Exp}_{\mathcal{DGSDT},\mathcal{A},\kappa}^{\text{Corr}}(\lambda)$

- $(\text{gpk}, \text{msk}, \{\text{tsk}_i\}_{i=1}^{\kappa}) \leftarrow \text{GKg}(1^{\lambda}, \kappa); \text{HUL} := \emptyset.$
- $(\text{uid}, m) \leftarrow \mathcal{A}(\text{gpk} : \text{AddU}(\cdot), \text{RReg}(\cdot)).$
- If $\text{uid} \notin \text{HUL}$ or $\text{gsk}[\text{uid}] = \perp$ Then Return 0.
- $\Sigma \leftarrow \text{Sign}(\text{gpk}, \text{gsk}[\text{uid}], m).$
- If $\text{Verify}(\text{gpk}, m, \Sigma) = 0$ Then Return 1.
- For $\text{tid} = 1$ To κ Do
  - $\circ$ $(\nu, \pi_{\text{Trace}})_{\text{tid}} \leftarrow \text{TraceShare}(\text{gpk}, \text{tsk}_{\text{tid}}, m, \Sigma); \text{If } \nu_{\text{tid}} = \perp \text{ or } \pi_{\text{Trace}_{\text{tid}}} = \perp \text{ Then Return 1.}$
  - $\circ$ If $\text{ShareVerify}(\text{gpk}, \text{tid}, \nu_{\text{tid}}, \pi_{\text{Trace}_{\text{tid}}}, m, \Sigma) = 0$ Then Return 1.
- $(\text{uid}^*, \theta_{\text{Trace}}) \leftarrow \text{TraceCombine}(\text{gpk}, \{(\nu, \pi_{\text{Trace}})_{\text{tid}}\}_{\text{tid}=1}^{\kappa}, m, \Sigma, \text{reg}); \text{If uid} \neq \text{uid}^* \text{ Then Return 1.}$
- If $\text{TraceVerify}(\text{gpk}, \text{uid}, \theta_{\text{Trace}}, \text{upk}[\text{uid}], m, \Sigma) \neq 1$ Then Return 1 Else Return 0.

Experiment: $\text{Exp}_{\mathcal{DGSDT},\mathcal{A},\kappa}^{\text{Anon-}b}(\lambda)$

- $(\text{st}_{\text{init}}, \text{BTL} \subset [\kappa]) \leftarrow \mathcal{A}_{\text{init}}(\lambda, \kappa).$
- $(\text{gpk}, \text{msk}, \{\text{tsk}_i\}_{i=1}^{\kappa}) \leftarrow \text{GKg}(1^{\lambda}, \kappa).$
- $\text{HUL}, \text{CUL}, \text{BUL}, \text{SL}, \text{CL} := \emptyset.$
- $b^* \leftarrow \mathcal{A}_{\text{guess}}(\text{st}_{\text{init}}, \text{gpk}, \text{msk}, \{\text{tsk}_i\}_{i\in\text{BTL}} : \text{AddU}(\cdot), \text{CrptU}(\cdot, \cdot), \text{SndU}(\cdot, \cdot), \text{WReg}(\cdot, \cdot,), \text{RevealU}(\cdot),$
$\qquad\qquad\qquad\qquad\qquad\qquad\quad \text{TraceShare}(\cdot, \cdot, \cdot), \text{Trace}(\cdot, \cdot), \text{CH}_b(\cdot, \cdot, \cdot)).$
- Return $b^*.$

Experiment: $\text{Exp}_{\mathcal{DGSDT},\mathcal{A},\kappa}^{\text{Non-Frame}}(\lambda)$

- $(\text{gpk}, \text{msk}, \{\text{tsk}_i\}_{i=1}^{\kappa}) \leftarrow \text{GKg}(1^{\lambda}, \kappa); \text{HUL}, \text{CUL}, \text{BUL}, \text{SL} := \emptyset.$
- $(m^*, \Sigma^*, \text{uid}^*, \theta_{\text{Trace}}^*) \leftarrow \mathcal{A}(\text{gpk}, \text{msk}, \{\text{tsk}_i\}_{i=1}^{\kappa} : \text{CrptU}(\cdot, \cdot), \text{SndU}(\cdot, \cdot), \text{WReg}(\cdot, \cdot), \text{RevealU}(\cdot),$
$\qquad\qquad\qquad\qquad\qquad\qquad\qquad\qquad\quad \text{Sign}(\cdot, \cdot)).$
- If $\text{Verify}(\text{gpk}, m^*, \Sigma^*) = 0$ Then Return 0.
- If $\text{TraceVerify}(\text{gpk}, \text{uid}^*, \theta_{\text{Trace}}^*, \text{upk}[\text{uid}^*], m^*, \Sigma^*) \neq 1$ Then Return 0.
- If $\text{uid}^* \notin \text{HUL} \setminus \text{BUL}$ or $(\text{uid}^*, m^*, \Sigma^*) \in \text{SL}$ Then Return 0 Else Return 1.

Experiment: $\text{Exp}_{\mathcal{DGSDT},\mathcal{A},\kappa}^{\text{Trace}}(\lambda)$

- $(\text{gpk}, \text{msk}, \{\text{tsk}_i\}_{i=1}^{\kappa}) \leftarrow \text{GKg}(1^{\lambda}, \kappa); \text{HUL}, \text{CUL}, \text{BUL}, \text{SL} := \emptyset.$
- $(m^*, \Sigma^*) \leftarrow \mathcal{A}(\text{gpk}, \{\text{tsk}_i\}_{i=1}^{\kappa} : \text{AddU}(\cdot), \text{CrptU}(\cdot, \cdot), \text{SndM}(\cdot, \cdot), \text{RevealU}(\cdot), \text{Sign}(\cdot, \cdot), \text{RReg}(\cdot)).$
- If $\text{Verify}(\text{gpk}, m^*, \Sigma^*) = 0$ Then Return 0.
- For $\text{tid} = 1$ To κ Do
  - $\circ$ $(\nu, \pi_{\text{Trace}})_{\text{tid}} \leftarrow \text{TraceShare}(\text{gpk}, \text{tsk}_{\text{tid}}, m^*, \Sigma^*); \text{If } \nu_{\text{tid}} = \perp \text{ or } \pi_{\text{Trace}_{\text{tid}}} = \perp \text{ Then Return 1.}$
  - $\circ$ If $\text{ShareVerify}(\text{gpk}, \text{tid}, \nu_{\text{tid}}, \pi_{\text{Trace}_{\text{tid}}}, m^*, \Sigma^*) = 0$ Then Return 1.
- $(\text{uid}^*, \theta_{\text{Trace}}) \leftarrow \text{TraceCombine}(\text{gpk}, \{(\nu, \pi_{\text{Trace}})_i\}_{i=1}^{\kappa}, m^*, \Sigma^*, \text{reg}).$
- If $\text{uid}^* = 0$ or $\text{TraceVerify}(\text{gpk}, \text{uid}^*, \theta_{\text{Trace}}, \text{upk}[\text{uid}^*], m^*, \Sigma^*) = 0$ Then Return 1 Else Return 0.

Experiment: $\text{Exp}_{\mathcal{DGSDT},\mathcal{A},\kappa}^{\text{Trace-Sound}}(\lambda)$

- $(\text{gpk}, \text{msk}, \{\text{tsk}_i\}_{i=1}^{\kappa}) \leftarrow \text{GKg}(1^{\lambda}, \kappa); \text{HUL}, \text{CUL}, \text{BUL}, \text{SL} := \emptyset.$
- $(m^*, \Sigma^*, \text{uid}_1^*, \theta_{\text{Trace}_1}^*, \text{uid}_2^*, \theta_{\text{Trace}_2}^*) \leftarrow \mathcal{A}(\text{gpk}, \text{msk}, \{\text{tsk}_i\}_{i=1}^{\kappa} : \text{CrptU}(\cdot, \cdot), \text{WReg}(\cdot, \cdot)).$
- If $\text{Verify}(\text{gpk}, m^*, \Sigma^*) = 0$ or $\text{uid}_1^* = \text{uid}_2^*$ or $\text{uid}_1^* = \perp$ or $\text{uid}_2^* = \perp$ Then Return 0.
- If $\exists i \in \{1, 2\}$ s.t. $\text{TraceVerify}(\text{gpk}, \text{uid}_i^*, \theta_{\text{Trace}_i}^*, \text{upk}[\text{uid}_i^*], m^*, \Sigma^*) = 0$ Then Return 0 Else Return 1.

Fig. 2. Security games for dynamic group signatures with distributed traceability

Formally, a $\mathcal{DGSDT}$ scheme is *traceable* if for all $\lambda, \kappa \in \mathbb{N}$, the advantage $\text{Adv}_{\mathcal{DGSDT},\mathcal{A},\kappa}^{\text{Trace}}(\lambda) := \Pr[\text{Exp}_{\mathcal{DGSDT},\mathcal{A},\kappa}^{\text{Trace}}(\lambda) = 1]$ is negligible for all PPT adversaries $\mathcal{A}$.

Tracing Soundness. Tracing soundness, as recently defined by [49] in the context of group signatures with a single tracing manager, requires that even if the group and all tracing managers as well as all members of the group collude, they cannot produce a valid signature that traces to two different members. As shown (in the single tracing authority setting) in [49], such a property is important for many applications. For example, applications where signers might get rewarded or where abusers might be prosecuted. In such applications, it is important that signatures can only trace to one user. Refer to [49] for more details.

Formally, a $\mathcal{DGSDT}$ scheme has *tracing soundness* if for all $\lambda, \kappa \in \mathbb{N}$, the advantage $\mathsf{Adv}_{\mathcal{DGSDT},\mathcal{A},\kappa}^{\text{Trace-Sound}}(\lambda) := \Pr[\mathsf{Exp}_{\mathcal{DGSDT},\mathcal{A},\kappa}^{\text{Trace-Sound}}(\lambda) = 1]$ is negligible for all PPT adversaries $\mathcal{A}$.

4 Building Blocks

In this section we present the building blocks that we use in our constructions.

Distributed Tag-Based Encryption with Public Verification. A Distributed Tag-Based Encryption scheme DTBE is a special case of threshold tag-based encryption [6] where n out of n decryption servers must compute their decryption shares honestly for the decryption to succeed. *Public verification* requires that checking the well-formedness of the ciphertext only require public information. We say the scheme is *non-interactive* if decrypting a ciphertext requires no interaction among the decryption servers. Also, the scheme is *robust* if invalid decryption shares can be identified by the combiner.

Formally, a DTBE scheme for a message space $\mathcal{M}_{\mathcal{DTBE}}$ and a tag space $\mathcal{T}_{\mathcal{DTBE}}$ is a tuple of polynomial-time algorithms (Setup, Enc, IsValid, ShareDec, ShareVerify, Combine), where $\mathsf{Setup}(1^\lambda, n)$ outputs a public key pk and vectors $\mathbf{svk} = (\mathsf{svk}_1, \ldots, \mathsf{svk}_n)$ and $\mathbf{sk} = (\mathsf{sk}_1, \ldots, \mathsf{sk}_n)$ of verification/secret keys for the decryption servers; $\mathsf{Enc}(\mathsf{pk}, t, m)$ outputs a ciphertext C_{dtbe} on the message m under the tag t; $\mathsf{IsValid}(\mathsf{pk}, t, C_{\text{dtbe}})$ outputs 1 if the ciphertext is valid under the tag t w.r.t. the pubic key pk or 0 otherwise; $\mathsf{ShareDec}(\mathsf{pk}, \mathsf{sk}_i, t, C_{\text{dtbe}})$ takes as input the public key pk, the i-th server secret key sk_i, a tag t, and the ciphertext C_{dtbe}, and outputs the i-th server decryption share ν_i of C_{dtbe} or the reject symbol $\perp$; $\mathsf{ShareVerify}(\mathsf{pk}, \mathsf{svk}_i, t, C_{\text{dtbe}}, \nu_i)$ takes as input the public key pk, the i-th server verification key svk_i, a tag t, the ciphertext C_{dtbe}, and the i-th server decryption share ν_i and outputs 1 if the decryption share ν_i is valid or 0 otherwise. $\mathsf{Combine}(\mathsf{pk}, \{\mathsf{svk}_i\}_{i=1}^n, \{\nu_i\}_{i=1}^n, C_{\text{dtbe}}, t)$ outputs either the message m or the reject symbol $\perp$.

We say the scheme is *correct* if for every message $m \in \mathcal{M}_{\mathcal{DTBE}}$, every tag $t \in \mathcal{T}_{\mathcal{DTBE}}$ and every $(\mathsf{pk}, \{\mathsf{svk}\}_{i=1}^n, \{\mathsf{sk}\}_{i=1}^n)$ output by Setup, if $C_{\text{dtbe}} \leftarrow \mathsf{Enc}(\mathsf{pk}, t, m)$ then we have that:

1. $\forall i \in [n]$, if $\nu_i \leftarrow \mathsf{ShareDec}(\mathsf{pk}, \mathsf{sk}_i, t, C_{\text{dtbe}})$ then $\mathsf{ShareVerify}(\mathsf{pk}, \mathsf{svk}_i, t, C_{\text{dtbe}}, \nu_i) = 1$.
2. $m \leftarrow \mathsf{Combine}(\mathsf{pk}, \{\mathsf{svk}_i\}_{i=1}^n, \{\nu_i\}_{i=1}^n, C_{\text{dtbe}}, t)$.

Besides correctness, we require two security properties: *Selective-Tag weak Indistinguishability against Adaptive Chosen Ciphertext Attacks (ST-wIND-CCA)* [42] and *Decryption Consistency (DEC-CON)*. Informally, the former

---

$\mathcal{DTBE}.\mathsf{Setup}(1^\lambda, n).$

o $\mathcal{P} \leftarrow \mathsf{BGrpSetup}(1^\lambda).$

o $h, w, z, \{u_i\}_{i=1}^n, \{v_i\}_{i=1}^n \leftarrow \mathbb{Z}_p.$

   $u := \sum_{i=1}^n u_i, v := \sum_{i=1}^n v_i, (H, \tilde{H}) := (G^h, \tilde{G}^h),$

   $(U, \tilde{U}) := (H^u, \tilde{H}^u), (V, \tilde{V}) := (U^{\frac{1}{v}}, \tilde{U}^{\frac{1}{v}}),$

   $(W, \tilde{W}) := (H^w, \tilde{H}^w), (Z, \tilde{Z}) := (V^z, \tilde{V}^z).$

o $\mathsf{sk}_i := (u_i, v_i), \mathsf{svk}_i := (\tilde{U}_i := \tilde{H}^{u_i}, \tilde{V}_i := \tilde{V}^{v_i}).$

o $\mathsf{pk} := (\mathcal{P}, H, \tilde{H}, U, \tilde{U}, V, \tilde{V}, W, \tilde{W}, Z, \tilde{Z}).$

$\mathcal{DTBE}.\mathsf{Enc}(\mathsf{pk}, t, M)$

o $r_1, r_2 \leftarrow \mathbb{Z}_p; C_1 := H^{r_1}, C_2 := V^{r_2},$

   $C_3 := MU^{r_1+r_2}, C_4 := (U^t W)^{r_1}, C_5 := (U^t Z)^{r_2}.$

o $C_{\mathrm{dtbe}} := \Big(C_1, C_2, C_3, C_4, C_5 \Big).$

$\mathcal{DTBE}.\mathsf{Combine}(\mathsf{pk}, \{\mathsf{svk}_i\}_{i=1}^n, \{\nu_i\}_{i=1}^n, C_{\mathrm{dtbe}}, t)$

o If $\mathcal{DTBE}.\mathsf{IsValid}(\mathsf{pk}, t, C_{\mathrm{dtbe}}) = 0$ Then Return $\perp$.

o Parse C_{dtbe} as $(C_1, C_2, C_3, C_4, C_5).$

o Parse ν_i as $(C_{i,1}, C_{i,2})$ and svk_i as $(\tilde{U}_i, \tilde{V}_i).$

o Return $\perp$ if $\exists i$ s.t. $\mathcal{DTBE}.\mathsf{ShareVerify}(\mathsf{pk}, \mathsf{svk}_i, t, C_{\mathrm{dtbe}}, \nu_i) = 0.$

o Return $M := \frac{C_3}{\prod_{i=1}^n C_{i,1} C_{i,2}}.$

$\mathcal{DTBE}.\mathsf{ShareDec}(\mathsf{pk}, \mathsf{sk}_i, t, C_{\mathrm{dtbe}})$

o Return $\perp$ if $\mathcal{DTBE}.\mathsf{IsValid}(\mathsf{pk}, t, C_{\mathrm{dtbe}}) = 0.$

o Parse C_{dtbe} as $(C_i)_{i=1}^5$ and sk_i as $(u_i, v_i).$

o Return $\nu_i := (C_{i,1} := C_1^{u_i}, C_{i,2} := C_2^{v_i}).$

$\mathcal{DTBE}.\mathsf{ShareVerify}(\mathsf{pk}, \mathsf{svk}_i, t, C_{\mathrm{dtbe}}, \nu_i)$

o Parse svk_i as $(\tilde{U}_i, \tilde{V}_i)$ and ν_i as $(C_{i,1}, C_{i,2}).$

o Parse C_{dtbe} as $(C_1, C_2, C_3, C_4, C_5).$

o Return 0 If $\mathcal{DTBE}.\mathsf{IsValid}(\mathsf{pk}, t, C_{\mathrm{dtbe}}) = 0.$

o If $e(C_{i,1}, \tilde{H}) \neq e(C_1, \tilde{U}_i)$ Or

   $e(C_{i,2}, \tilde{V}) \neq e(C_2, \tilde{V}_i)$ Then Return 0.

o Else Return 1.

$\mathcal{DTBE}.\mathsf{IsValid}(\mathsf{pk}, t, C_{\mathrm{tbe}})$

o Parse C_{dtbe} as $(C_1, C_2, C_3, C_4, C_5).$

o If $e(C_1, \tilde{U}^t \tilde{W}) \neq e(C_4, \tilde{H})$ Or

   $e(C_2, \tilde{U}^t \tilde{Z}) \neq e(C_5, \tilde{V})$ Then Return 0.

o Else Return 1.

Fig. 3. Our distributed tag-based encryption scheme

requires that an adversary who gets a decryption oracle for any ciphertext under a tag different from the target tag (which is chosen beforehand), cannot distinguish which challenge message was encrypted. The latter requires that an adversary cannot output two different sets of decryption shares of a ciphertext which open differently. The formal definitions of those can be found in the full version.

We provide in Fig. 3 a new efficient construction of a distributed tag-based encryption scheme with public verification that works in the efficient Type-III bilinear group setting. Our scheme which is secure in the standard model under a variant of the DLIN assumption, namely, the $\mathsf{XDLIN_G}$ assumption is based on Kiltz's tag-based encryption scheme [42] and its Type-I threshold variant in [6]. Our scheme is efficient and yields ciphertexts of size $\mathbb{G}^5$. Note that in Type-III bilinear groups, elements of $\mathbb{G}$ are much smaller than their Type-I counterparts, especially now that small-characteristic symmetric bilinear groups are rendered insecure [9,34]. To give a sense of comparison, we outline that at 128-bit security, the size of elements of $\mathbb{G}$ is 256 bits whereas that of their large-characteristic symmetric groups counterparts is 1536 bits. Therefore, our construction yields much shorter ciphertexts than the variants of Kiltz's scheme in symmetric bilinear groups. Our scheme is of independent interest and has other applications beyond the scope of this paper. For instance, combining it with a strongly unforgeable one-time signature scheme (e.g. the full Boneh-Boyen scheme) as per the transformation in [42], we get an efficient distributed (or threshold) IND-CCA secure encryption scheme [27,28] in Type-III groups which is secure in the standard model under the $\mathsf{XDLIN_G}$ and q-SDH assumptions. In addition, when $n = 1$, we obtain a tag-based encryption scheme in the efficient Type-III setting with 29 % shorter ciphertexts than the Type-III variant of Kiltz's scheme in [39] (which yields ciphertexts of size $\mathbb{G}^3 \times \tilde{\mathbb{G}}^2$). Unlike the scheme in [39], which only works

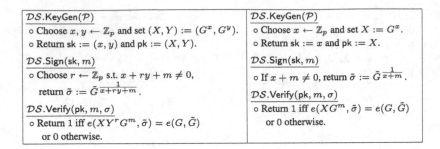

$\mathcal{DS}$.KeyGen($\mathcal{P}$)	$\mathcal{DS}$.KeyGen($\mathcal{P}$)
○ Choose $x, y \leftarrow \mathbb{Z}_p$ and set $(X, Y) := (G^x, G^y)$. ○ Return sk $:= (x, y)$ and pk $:= (X, Y)$.	○ Choose $x \leftarrow \mathbb{Z}_p$ and set $X := G^x$. ○ Return sk $:= x$ and pk $:= X$.
$\mathcal{DS}$.Sign(sk, m) ○ Choose $r \leftarrow \mathbb{Z}_p$ s.t. $x + ry + m \neq 0$, return $\tilde{\sigma} := \tilde{G}^{\frac{1}{x+ry+m}}$.	$\mathcal{DS}$.Sign(sk, m) ○ If $x + m \neq 0$, return $\tilde{\sigma} := \tilde{G}^{\frac{1}{x+m}}$.

$\mathcal{DS}$.Verify(pk, m, σ)
 ○ Return 1 iff $e(XY^rG^m, \tilde{\sigma}) = e(G, \tilde{G})$
   or 0 otherwise.
	$\mathcal{DS}$.Verify(pk, m, σ)
	 ○ Return 1 iff $e(XG^m, \tilde{\sigma}) = e(G, \tilde{G})$
	   or 0 otherwise.

Fig. 4. The Full Boneh-Boyen (Left) and the Weak Boneh-Boyen (Right) signatures

for a polynomial (in the security parameter) message space, our scheme has no restriction on the message space.

For simplicity we consider the n-out-of-n case. However, our scheme can, in a straightforward manner, be adapted to the k-out-of-n case by deploying any k-out-of-n secret sharing scheme to compute the servers' secret keys.

We prove the following Theorem in the full version of the paper.

Theorem 1. *The construction in Fig. 3 is a secure distributed tag-based encryption scheme if the $XDLIN_{\mathbb{G}}$ assumption holds.*

Digital Signatures. A *digital signature* for a message space $\mathcal{M_{DS}}$ is a tuple of polynomial-time algorithms $\mathcal{DS} := (\mathsf{KeyGen}, \mathsf{Sign}, \mathsf{Verify})$ where KeyGen outputs a pair of secret/public keys $(\mathsf{sk}, \mathsf{pk})$; $\mathsf{Sign}(\mathsf{sk}, m)$ outputs a signature σ on the message m; $\mathsf{Verify}(\mathsf{pk}, m, \sigma)$ outputs 1 if σ is a valid signature on m.

Besides correctness, we require existential unforgeability under adaptive chosen-message attack which demands that all PPT adversaries getting the public key and access to a sign oracle, have a negligible advantage in outputting a valid signature on a message that was not queried to the sign oracle. A weaker variant of existential unforgeability (i.e. existential unforgeability under a weak chosen-message attack) requires that the adversary sends all its queries before seeing the public key.

In this paper, we use two digital signatures by Boneh and Boyen [17] which we refer to as the Full Boneh-Boyen signature (Fig. 4 (Left)) and the Weak Boneh-Boyen signature (Fig. 4 (Right)), respectively. Both schemes are secure under the q-SDH assumption. The weaker scheme is secure under a weak chosen-message attack.

Structure-Preserving Signatures. Structure-preserving signatures [2] are signature schemes where the message, the public key and the signature are all group elements, and signatures are verified by evaluating pairing product equations.

In this paper, we use two structure-preserving signatures from the literature. The first scheme is by Abe et al. [4] which offers controlled re-randomizability where a signature can only be re-randomized if the user has a special randomization token. The scheme in the asymmetric setting is illustrated in Fig. 5 (Left). The unforgeability of the scheme relies on an interactive assumption. Refer to [4]

$\mathcal{SPDSS}$.Setup(1^λ)	$\mathcal{SPDSS}$.Setup(1^λ)
$\circ$ $\mathcal{P} \leftarrow$ BGrpSetup(1^λ).	$\circ$ $\mathcal{P} \leftarrow$ BGrpSetup(1^λ).
$\circ$ $F \leftarrow \mathbb{G}$. Return param $:= (\mathcal{P}, F)$.	$\circ$ Return param $:= \mathcal{P}$.
$\mathcal{SPDSS}$.KeyGen(param)	$\mathcal{SPDSS}$.KeyGen(param)
$\circ$ Choose $x \leftarrow \mathbb{Z}_p$ and set $\tilde{X} := \tilde{G}^x$.	$\circ$ $w, x, y_1, y_2 \leftarrow \mathbb{Z}_p, \tilde{W} := \tilde{G}^w, \tilde{X} := \tilde{G}^x,$
$\circ$ Set sk $:= x$ and pk $:= \tilde{X}$. Return (sk, pk).	$\qquad \tilde{Y}_1 := \tilde{G}^{y_1}, \tilde{Y}_2 := \tilde{G}^{y_2}.$
$\mathcal{SPDSS}$.Sign(sk, M)	$\circ$ sk $:= (w, x, y_1, y_2)$, pk $:= (\tilde{W}, \tilde{X}, \tilde{Y}_1, \tilde{Y}_2)$.
$\circ$ $r \leftarrow \mathbb{Z}_p, \tilde{\Omega}_1 := \tilde{G}^r, \Omega_2 := M^{\frac{x}{r}} F^{\frac{1}{r}},$	$\circ$ Return (sk, pk).
$\quad \Omega_3 := \Omega_2^{\frac{x}{r}} G^{\frac{1}{r}}$, the token is $\Omega_4 := G^{\frac{1}{r}}$.	$\mathcal{SPDSS}$.Sign(sk, M)
$\circ$ Return $\sigma := \left(\tilde{\Omega}_1, \Omega_2, \Omega_3 \right)$.	$\circ$ $\Omega_1 \leftarrow \mathbb{G}, a \leftarrow \mathbb{Z}_p, \Omega_2 := G^a, \tilde{\Omega}_3 := \tilde{G}^{\frac{1}{a}},$
$\mathcal{SPDSS}$.Randomize(M, σ, Ω_4)	$\quad \Omega_4 := G^{x-aw} \Omega_1^{-y_1} M^{-y_2}.$
$\circ$ $r' \leftarrow \mathbb{Z}_p, \tilde{\Omega}_1' := \Omega_1^{\frac{1}{r'}}, \Omega_2' := \Omega_2^{r'},$	$\circ$ Return $\sigma := \left(\Omega_1, \Omega_2, \tilde{\Omega}_3, \Omega_4 \right)$.
$\quad \Omega_3' := \Omega_3^{r'^2} \Omega_4^{r'(1-r')}.$	$\mathcal{SPDSS}$.Verify(pk, M, σ)
$\mathcal{SPDSS}$.Verify(pk, M, σ)	$\circ$ Return 1 if $e(\Omega_2, \tilde{\Omega}_3) = e(G, \tilde{G})$ and
$\circ$ Return 1 if $e(\Omega_2, \tilde{\Omega}_1) = e(M, \tilde{X})e(F, \tilde{G})$	$\quad e(G, \tilde{X}) = e(\Omega_2, \tilde{W})e(\Omega_4, \tilde{G})e(\Omega_1, \tilde{Y}_1)e(M, \tilde{Y}_2).$
$\quad$ and $e(\Omega_3, \tilde{\Omega}_1) = e(\Omega_2, \tilde{X})e(G, \tilde{G}).$	

Fig. 5. The structure-preserving signature of [4] (Left) and that of [3] (Right)

for details. The second scheme we use is that of Abe et al. [3]. The scheme in the asymmetric setting is given in Fig. 5 (Right). The strong unforgeability of the scheme relies on the non-interactive q-AGHO assumption.

Strongly Unforgeable One-Time Signatures. A one-time signature scheme is a signature scheme that is unforgeable against an adversary who is only allowed a single signing query. *Strong Unforgeability* requires that the adversary cannot even forge a new signature on a message that she queried the sign oracle on. In this paper, we will instantiate the one-time signature using the Full Boneh-Boyen signature scheme from Fig. 4.

Non-Interactive Zero-Knowledge Proofs. Let $\mathcal{R}$ be an efficiently computable relation. For pairs $(x, w) \in \mathcal{R}$, we call x the statement and w the witness. We define the language $\mathcal{L}$ as all the statements x in $\mathcal{R}$. A Non-Interactive Zero-Knowledge (NIZK) proof system [16] for $\mathcal{R}$ is defined by a tuple of algorithms $\mathcal{NIZK} := (\mathsf{Setup}, \mathsf{Prove}, \mathsf{Verify}, \mathsf{Extract}, \mathsf{SimSetup}, \mathsf{SimProve})$.

$\mathsf{Setup}(1^\lambda)$ outputs a common reference string crs and an extraction key xk which allows for witness extraction. $\mathsf{Prove}(\mathsf{crs}, x, w)$ outputs a proof π that $(x, w) \in \mathcal{R}$. $\mathsf{Verify}(\mathsf{crs}, x, \pi)$ outputs 1 if the proof is valid, or 0 otherwise. $\mathsf{Extract}(\mathsf{crs}, \mathsf{xk}, x, \pi)$ outputs a witness. $\mathsf{SimSetup}(1^\lambda)$ outputs a simulated reference string $\mathsf{crs}_{\mathsf{Sim}}$ and a trapdoor key tr that allows for proof simulation. $\mathsf{SimProve}(\mathsf{crs}_{\mathsf{Sim}}, \mathsf{tr}, x)$ outputs a simulated proof π_{Sim} without a witness.

We require: completeness, soundness and zero-knowledge. Completeness requires that honestly generated proofs are accepted; Soundness requires that it is infeasible (but for a small probability) to produce a valid proof for a false statement; Zero-knowledge requires that a proof reveals no information about the witness used. For formal definitions refer to [16].

- Pairing Product Equation (PPE): $\prod_{i=1}^{n} e(A_i, \underline{\tilde{Y}_i}) \prod_{i=1}^{m} e(\underline{X_i}, \tilde{B}_i) \prod_{i=1}^{m} \prod_{j=1}^{n} e(\underline{X_i}, \underline{\tilde{Y}_j})^{k_{i,j}} = t_T.$

- Multi-Scalar Multiplication Equation (MSME) in $\mathbb{G}$: $\prod_{i=1}^{n'} A_i^{\underline{y_i}} \prod_{i=1}^{m} \underline{X_i}^{b_i} \prod_{i=1}^{m} \prod_{j=1}^{n'} \underline{X_i}^{k_{i,j}\underline{y_j}} = T.$

- Multi-Scalar Multiplication Equation (MSME) in $\tilde{\mathbb{G}}$: $\prod_{i=1}^{m'} \tilde{B}_i^{\underline{x_i}} \prod_{i=1}^{n} \underline{\tilde{Y}_i}^{a_i} \prod_{i=1}^{m'} \prod_{j=1}^{n} \underline{\tilde{Y}_i}^{k_{i,j}\underline{x_j}} = \tilde{T}.$

- Quadratic Equation (QE) in $\mathbb{Z}_p$: $\sum_{i=1}^{n'} a_i\underline{y_i} + \sum_{i=1}^{m'} \underline{x_i}b_i + \sum_{i=1}^{m'} \sum_{j=1}^{n'} \underline{x_i}\underline{y_j} = t.$

Fig. 6. Types of equations one can use Groth-Sahai proofs for

GROTH-SAHAI PROOFS. Groth-Sahai (GS) proofs [37] are efficient non-interactive proofs in the Common Reference String (CRS) model. The language for the system has the form

$$\mathcal{L} := \{\text{statement} \mid \exists \, \text{witness} : E_i(\text{statement}, \text{witness})_{i=1}^{n} \text{ hold }\},$$

where $E_i(\text{statement}, \cdot)$ is one of the types of equation summarized in Fig. 6, where $X_1, \ldots, X_m \in \mathbb{G}, \tilde{Y}_1, \ldots, \tilde{Y}_n \in \tilde{\mathbb{G}}, x_1, \ldots, x_{m'}, y_1, \ldots, y_{n'} \in \mathbb{Z}_p$ are secret variables (hence underlined), whereas $A_i, T \in \mathbb{G}, \tilde{B}_i, \tilde{T} \in \tilde{\mathbb{G}}, a_i, b_i, k_{i,j}, t \in \mathbb{Z}_p,$ $t_T \in \mathbb{G}_T$ are public constants.

The proof system has perfect completeness, (perfect) soundness, composable witness-indistinguishability/zero-knowledge. We use the SXDH-based instantiation of the proofs. Refer to [37] for details.

5   Our Generic Construction

In this section, we present our generic construction for dynamic group signatures with distributed traceability.

Overview of the construction. The idea behind our generic construction has some in common with Groth's scheme [36] in that we combine a standard NIZK proof system with a weakly secure tag-based encryption scheme and a strong one-time signature scheme to eliminate the need for the more expensive simulation-sound NIZK systems [47] and IND-CCA public-key encryption schemes which were required by the construction of Bellare et al. [13]. However, unlike [36], our framework provides distributed traceability, has a concurrent join protocol and achieves tracing soundness.

Our generic construction requires three digital signatures $\mathcal{DS}_1, \mathcal{DS}_2$ and $\mathcal{WDS}$ where the first two have to be unforgeable against a standard adaptive chosen-message attack, whereas it suffices for the third scheme to be unforgeable against a weak chosen-message attack. We also require a strongly unforgeable one-time signature scheme $\mathcal{OTS}$ that is secure against an adaptive chosen-message attack. Additionally, we require a NIZK proof of knowledge system $\mathcal{NIZK}$ and a ST-wIND-CCA distributed tag-based encryption scheme $\mathcal{DTBE}$. In order to have efficient tracing, we ask that $\mathcal{DTBE}$ is non-interactive and robust. As was noted

Join(gpk, uid, **usk**[uid])		Issue(msk, uid, **upk**[uid])
$(\mathsf{pk}_{\mathsf{uid}}, \mathsf{sk}_{\mathsf{uid}}) \leftarrow \mathcal{WDS}.\mathsf{KeyGen}(1^\lambda)$		
$\mathsf{sig}_{\mathsf{uid}} \leftarrow \mathcal{DS}_2.\mathsf{Sign}(\mathbf{usk}[\mathsf{uid}], \mathsf{pk}_{\mathsf{uid}})$	$\xrightarrow{\mathsf{sig}_{\mathsf{uid}}, \, \mathsf{pk}_{\mathsf{uid}}}$	
		Abort if $\mathcal{DS}_2.\mathsf{Verify}(\mathbf{upk}[\mathsf{uid}], \mathsf{pk}_{\mathsf{uid}}, \mathsf{sig}_{\mathsf{uid}}) = 0$
	$\xleftarrow{\mathsf{cert}_{\mathsf{uid}}}$	$\mathsf{cert}_{\mathsf{uid}} \leftarrow \mathcal{DS}_1.\mathsf{Sign}(\mathsf{msk}, \mathsf{pk}_{\mathsf{uid}})$
Abort if $\mathcal{DS}_1.\mathsf{Verify}(\mathsf{pk}_{\mathsf{GM}}, \mathsf{pk}_{\mathsf{uid}}, \mathsf{cert}_{\mathsf{uid}}) = 0$		
$\mathbf{gsk}[\mathsf{uid}] := (\mathsf{sk}_{\mathsf{uid}}, \mathsf{pk}_{\mathsf{uid}}, \mathsf{cert}_{\mathsf{uid}})$		$\mathbf{reg}[\mathsf{uid}] := (\mathsf{pk}_{\mathsf{uid}}, \mathsf{sig}_{\mathsf{uid}})$

Fig. 7. The Join/Issue protocol for our construction

by [31], such a property simplifies tracing even in traditional group signatures with a single tracing manager. Finally, we require a collision-resistant hash function $\mathcal{H}$: $\{0,1\}^* \rightarrow \mathcal{T}_{\mathcal{DTBE}}$. For simplicity and WLOG we assume that $\mathcal{T}_{\mathcal{DTBE}} = \mathcal{M}_{\mathcal{WDS}}$. Otherwise, one could use a second hash function. Note that one can use the same signature scheme for both $\mathcal{DS}_1$ and $\mathcal{DS}_2$ but using different key pairs.

The GKg algorithm runs $\mathcal{NIZK}.$Setup to generate a common reference string crs for $\mathcal{NIZK}$. It also runs $\mathcal{DTBE}.$Setup to generate $(\mathsf{pk}_{\mathcal{DTBE}}, \{\mathsf{svk}_i\}_{i=1}^\kappa, \{\mathsf{sk}_i\}_{i=1}^\kappa)$, and $\mathcal{DS}_1.$KeyGen to generate $(\mathsf{pk}_{\mathsf{GM}}, \mathsf{sk}_{\mathsf{GM}})$. The group public key is $\mathsf{gpk} := (1^\lambda, \mathsf{crs}, \mathsf{pk}_{\mathsf{GM}}, \mathsf{pk}_{\mathcal{DTBE}}, \{\mathsf{svk}_i\}_{i=1}^\kappa, \mathcal{H})$. The group managers' secret key is $\mathsf{msk} := \mathsf{sk}_{\mathsf{GM}}$, whereas the tracing key of tracing manager TM_i is $\mathsf{tsk}_i := \mathsf{sk}_i$.

A new user creates her personal key pair by running $\mathcal{DS}_2.$KeyGen to generate $(\mathbf{upk}[\mathsf{uid}], \mathbf{usk}[\mathsf{uid}])$. When the user wishes to join the group, she generates a key pair $(\mathsf{pk}_{\mathsf{uid}}, \mathsf{sk}_{\mathsf{uid}})$ for the signature scheme $\mathcal{WDS}$ and then signs $\mathsf{pk}_{\mathsf{uid}}$ using $\mathcal{DS}_2$ and her personal secret key $\mathbf{usk}[\mathsf{uid}]$ to obtain a signature $\mathsf{sig}_{\mathsf{uid}}$. We use $\mathsf{sig}_{\mathsf{uid}}$ as a proof when proving that the user has produced a group signature.

To join the group, the user sends $\mathsf{pk}_{\mathsf{uid}}$ and $\mathsf{sig}_{\mathsf{uid}}$ to the group manager. If $\mathsf{sig}_{\mathsf{uid}}$ is valid, the group manager issues a membership certificate $\mathsf{cert}_{\mathsf{uid}}$ (which is a $\mathcal{DS}_1$ signature on $\mathsf{pk}_{\mathsf{uid}}$ that verifies w.r.t. $\mathsf{pk}_{\mathsf{GM}}$).

To sign a message m, the member chooses a fresh key pair $(\mathsf{otsvk}, \mathsf{otssk})$ for the one-time signature $\mathcal{OTS}$ and encrypts her public key $\mathsf{pk}_{\mathsf{uid}}$ with $\mathcal{DTBE}$ using $\mathcal{H}(\mathsf{otsvk})$ as tag (and possibly some randomness τ) to obtain a ciphertext C_{dtbe}. She then signs $\mathcal{H}(\mathsf{otsvk})$ using the digital signature scheme $\mathcal{WDS}$ and her secret key $\mathsf{sk}_{\mathsf{uid}}$ to obtain a signature σ. She then uses $\mathcal{NIZK}$ to produce a proof π proving that: she did the encryption correctly, she has a signature σ on $\mathcal{H}(\mathsf{otsvk})$ that verifies w.r.t. her public key $\mathsf{pk}_{\mathsf{uid}}$ and she has a certificate $\mathsf{cert}_{\mathsf{uid}}$ on $\mathsf{pk}_{\mathsf{uid}}$ from the group manager. Finally, she signs $(m, C_{\mathrm{dtbe}}, \mathsf{otsvk}, \pi)$ using the one-time signature $\mathcal{OTS}$ to obtain σ_{ots}. The group signature is $\Sigma := (\sigma_{\mathrm{ots}}, \pi, C_{\mathrm{dtbe}}, \mathsf{otsvk})$. To verify the signature, one verifies the proof π and the one-time signature σ_{ots}, and ensures that the ciphertext C_{dtbe} is well-formed.

We remark here if $\mathcal{DS}_1$ and/or $\mathcal{WDS}$ schemes are re-randomizable, one can reveal the signature components which are independent of their respective messages after re-randomization. This simplifies the proof π and subsequently improves the efficiency. The revealed parts of those signatures can then be included as the part of the message to be signed by $\mathcal{OTS}$ to ensure that one achieves the stronger notion of non-frameability.

$\mathsf{GKg}(1^\lambda, \kappa)$

○ $(\mathsf{pk}_{\mathsf{GM}}, \mathsf{sk}_{\mathsf{GM}}) \leftarrow \mathcal{DS}_1.\mathsf{KeyGen}(1^\lambda)$; $(\mathsf{pk}_{\mathcal{DTBE}}, \{\mathsf{svk}_i\}_{i=1}^\kappa, \{\mathsf{sk}_i\}_{i=1}^\kappa) \leftarrow \mathcal{DTBE}.\mathsf{Setup}(1^\lambda, \kappa)$.

○ $(\mathsf{crs}, \mathsf{xk}) \leftarrow \mathcal{NIZK}.\mathsf{Setup}(1^\lambda)$; Choose a collision-resistant hash function $\mathcal{H} : \{0,1\}^* \to T_{\mathcal{DTBE}}$.

○ Let $\mathsf{tsk}_i := \mathsf{sk}_i$, $\mathsf{gpk} := (1^\lambda, \mathsf{crs}, \mathsf{pk}_{\mathsf{GM}}, \mathsf{pk}_{\mathcal{DTBE}}, \{\mathsf{svk}_i\}_{i=1}^\kappa, \mathcal{H})$ and $\mathsf{msk} := \mathsf{sk}_{\mathsf{GM}}$.

○ Return $(\mathsf{gpk}, \{\mathsf{tsk}_i\}_{i=1}^\kappa, \mathsf{msk})$.

$\mathsf{UKg}(1^\lambda)$

○ $(\mathbf{upk}[i], \mathbf{usk}[i]) \leftarrow \mathcal{DS}_2.\mathsf{KeyGen}(1^\lambda)$. Return $(\mathbf{upk}[i], \mathbf{usk}[i])$.

$\mathsf{Sign}(\mathsf{gpk}, \mathbf{gsk}[\mathsf{uid}], m)$

○ $(\mathsf{otsvk}, \mathsf{otssk}) \leftarrow \mathcal{OTS}.\mathsf{KeyGen}(1^\lambda)$; $C_{\mathsf{dtbe}} \leftarrow \mathcal{DTBE}.\mathsf{Enc}(\mathsf{pk}_{\mathcal{DTBE}}, \mathcal{H}(\mathsf{otsvk}), \mathsf{pk}_{\mathsf{uid}}; \tau)$.

○ $\sigma \leftarrow \mathcal{WDS}.\mathsf{Sign}(\mathsf{sk}_{\mathsf{uid}}, \mathcal{H}(\mathsf{otsvk}))$.

○ $\pi \leftarrow \mathcal{NIZK}.\mathsf{Prove}(\mathsf{crs}, \{\mathsf{pk}_{\mathsf{uid}}, \tau, \sigma, \mathsf{cert}_{\mathsf{uid}}\} : (C_{\mathsf{dtbe}}, \mathcal{H}(\mathsf{otsvk}), \mathsf{pk}_{\mathsf{GM}}, \mathsf{pk}_{\mathcal{DTBE}}) \in \mathcal{L})$.

○ $\sigma_{\mathsf{ots}} \leftarrow \mathcal{OTS}.\mathsf{Sign}(\mathsf{otssk}, (m, C_{\mathsf{dtbe}}, \mathsf{otsvk}, \pi))$.

○ Return $\Sigma := (\sigma_{\mathsf{ots}}, \pi, C_{\mathsf{dtbe}}, \mathsf{otsvk})$.

$\mathsf{Verify}(\mathsf{gpk}, m, \Sigma)$

○ Parse Σ as $(\sigma_{\mathsf{ots}}, \pi, C_{\mathsf{dtbe}}, \mathsf{otsvk})$ and gpk as $(1^\lambda, \mathsf{crs}, \mathsf{pk}_{\mathsf{GM}}, \mathsf{pk}_{\mathcal{DTBE}}, \{\mathsf{svk}_i\}_{i=1}^\kappa, \mathcal{H})$.

○ Return 1 if all the following verify; otherwise, return 0:

  ⋆ $\mathcal{OTS}.\mathsf{Verify}(\mathsf{otsvk}, (m, C_{\mathsf{dtbe}}, \mathsf{otsvk}, \pi), \sigma_{\mathsf{ots}}) = 1 \wedge \mathcal{NIZK}.\mathsf{Verify}(\mathsf{crs}, \pi) = 1$.

  ⋆ $\mathcal{DTBE}.\mathsf{IsValid}(\mathsf{pk}_{\mathcal{DTBE}}, \mathcal{H}(\mathsf{otsvk}), C_{\mathsf{dtbe}}) = 1$.

$\mathsf{TraceShare}(\mathsf{gpk}, \mathsf{tsk}_i, m, \Sigma)$

○ Parse Σ as $(\sigma_{\mathsf{ots}}, \pi, C_{\mathsf{dtbe}}, \mathsf{otsvk})$ and gpk as $(1^\lambda, \mathsf{crs}, \mathsf{pk}_{\mathsf{GM}}, \mathsf{pk}_{\mathcal{DTBE}}, \{\mathsf{svk}_i\}_{i=1}^\kappa, \mathcal{H})$.

○ Return $\perp$ if $\mathsf{Verify}(\mathsf{gpk}, m, \Sigma) = 0$ Else Return $\mathcal{DTBE}.\mathsf{ShareDec}(\mathsf{pk}_{\mathcal{DTBE}}, \mathsf{tsk}_i, \mathcal{H}(\mathsf{otsvk}), C_{\mathsf{dtbe}})$.

$\mathsf{ShareVerify}(\mathsf{gpk}, \mathsf{tid}, \nu, m, \Sigma)$

○ Parse Σ as $(\sigma_{\mathsf{ots}}, \pi, C_{\mathsf{dtbe}}, \mathsf{otsvk})$ and gpk as $(1^\lambda, \mathsf{crs}, \mathsf{pk}_{\mathsf{GM}}, \mathsf{pk}_{\mathcal{DTBE}}, \{\mathsf{svk}_i\}_{i=1}^\kappa, \mathcal{H})$.

○ Return $\mathcal{DTBE}.\mathsf{ShareVerify}(\mathsf{pk}_{\mathcal{DTBE}}, \mathsf{svk}_{\mathsf{tid}}, \mathcal{H}(\mathsf{otsvk}), C_{\mathsf{dtbe}}, \nu)$.

$\mathsf{TraceCombine}(\mathsf{gpk}, \{\nu_i\}_{i=1}^\kappa, m, \Sigma, \mathbf{reg})$

○ Parse Σ as $(\sigma_{\mathsf{ots}}, \pi, C_{\mathsf{dtbe}}, \mathsf{otsvk})$ and gpk as $(1^\lambda, \mathsf{crs}, \mathsf{pk}_{\mathsf{GM}}, \mathsf{pk}_{\mathcal{DTBE}}, \{\mathsf{svk}_i\}_{i=1}^\kappa, \mathcal{H})$.

○ Return $(0, (\{\nu\}_{i=1}^\kappa, \perp, \perp))$ if for any $i \in [\kappa]$ $\mathcal{DTBE}.\mathsf{ShareVerify}(\mathsf{pk}_{\mathcal{DTBE}}, \mathsf{svk}_i, \mathcal{H}(\mathsf{otsvk}), C_{\mathsf{dtbe}}, \nu_i) = 0$.

○ $\mathsf{pk}_{\mathsf{uid}} \leftarrow \mathcal{DTBE}.\mathsf{Combine}(\mathsf{pk}_{\mathcal{DTBE}}, \{\mathsf{svk}_i\}_{i=1}^\kappa, \{\nu_i\}_{i=1}^\kappa, C_{\mathsf{dtbe}}, \mathcal{H}(\mathsf{otsvk}))$.

○ Return $(0, (\{\nu\}_{i=1}^\kappa, \perp, \perp))$ if $\mathsf{pk}_{\mathsf{uid}} = \perp$.

○ If $\exists j$ s.t. $\mathbf{reg}[j].\mathsf{pk} = \mathsf{pk}_{\mathsf{uid}}$ Then Return $(j, (\{\nu\}_{i=1}^\kappa, \mathsf{pk}_{\mathsf{uid}}, \mathbf{reg}[j].\mathsf{sig}))$.

○ Return $(0, (\{\nu\}_{i=1}^\kappa, \mathsf{pk}_{\mathsf{uid}}, \perp))$.

$\mathsf{TraceVerify}(\mathsf{gpk}, \mathsf{uid}, \theta_{\mathsf{Trace}}, \mathbf{upk}[\mathsf{uid}], m, \Sigma)$

○ Parse θ_{Trace} as $(\{\nu\}_{i=1}^\kappa, \mathsf{pk}_{\mathsf{uid}}, \mathsf{sig}_{\mathsf{uid}})$ and Σ as $(\sigma_{\mathsf{ots}}, \pi, C_{\mathsf{dtbe}}, \mathsf{otsvk})$.

○ Parse gpk as $(1^\lambda, \mathsf{crs}, \mathsf{pk}_{\mathsf{GM}}, \mathsf{pk}_{\mathcal{DTBE}}, \{\mathsf{svk}_i\}_{i=1}^\kappa, \mathcal{H})$.

○ $\mathsf{pk}'_{\mathsf{uid}} \leftarrow \mathcal{DTBE}.\mathsf{Combine}(\mathsf{pk}_{\mathcal{DTBE}}, \{\mathsf{svk}_i\}_{i=1}^\kappa, \{\nu_i\}_{i=1}^\kappa, C_{\mathsf{dtbe}}, \mathcal{H}(\mathsf{otsvk}))$.

○ If $\mathsf{uid} = 0$ Then Return $(\mathsf{pk}'_{\mathsf{uid}} = \perp \ \vee \ \mathsf{Verify}(\mathsf{gpk}, m, \Sigma) = 0)$.

○ If $\mathsf{pk}'_{\mathsf{uid}} \neq \mathsf{pk}_{\mathsf{uid}}$ Then Return 0 Else Return $\mathcal{DS}_2.\mathsf{Verify}(\mathbf{upk}[\mathsf{uid}], \mathsf{pk}_{\mathsf{uid}}, \mathsf{sig}_{\mathsf{uid}})$.

Fig. 8. Our generic construction

To trace a signature, the decryption shares ν_i of the ciphertext C_{dtbe} are obtained from the respective tracing managers and then combined together in order to recover the plaintext $\mathsf{pk}_{\mathsf{uid}}$. Then one just needs to search in the registration table $\mathbf{reg}$ to see if any entry $\mathbf{reg}[j].\mathsf{pk}$ matches $\mathsf{pk}_{\mathsf{uid}}$. If this is the case, $(j, (\{\nu\}_{i=1}^\kappa, \mathsf{pk}_{\mathsf{uid}}, \mathbf{reg}[j].\mathsf{sig}))$ is returned. Otherwise, $(0, (\{\nu\}_{i=1}^\kappa, \mathsf{pk}_{\mathsf{uid}}, \perp))$ is returned. Note that combining the different tracing shares does not require the knowledge of any secret key and hence this could be performed by any party. To verify the correctness of the tracing, one just needs to ensure that all decryption shares verify correctly and in the case that $j > 0$, one needs to verify that the signature sig verifies w.r.t. $\mathbf{upk}[j]$.

The construction is detailed in Fig. 8, whereas the Join/Issue protocol is given in Fig. 7. The language associated with the NIZK proof is as follows, where for clarity we underline the elements of the witness:

$$\mathcal{L} : \Big\{ \big((C_{\text{dtbe}}, \mathcal{H}(\text{otsvk}), \text{pk}_{\text{GM}}, \text{pk}_{\mathcal{DTBE}}), (\underline{\text{pk}_{\text{uid}}}, \underline{\tau}, \underline{\sigma}, \underline{\text{cert}_{\text{uid}}}) \big) :$$

$$\mathcal{DS}_1.\text{Verify}(\text{pk}_{\text{GM}}, \underline{\text{pk}_{\text{uid}}}, \underline{\text{cert}_{\text{uid}}}) = 1 \ \wedge \ \mathcal{WDS}.\text{Verify}(\underline{\text{pk}_{\text{uid}}}, \mathcal{H}(\text{otsvk}), \underline{\sigma}) = 1$$

$$\wedge \ \mathcal{DTBE}\text{-Enc}(\text{pk}_{\mathcal{DTBE}}, \mathcal{H}(\text{otsvk}), \underline{\text{pk}_{\text{uid}}}; \underline{\tau}) = C_{\text{dtbe}} \Big\}.$$

Theorem 2. *The construction in Figs. 7 and 8 is a secure dynamic group signature with distributed traceability providing that the building blocks are secure w.r.t. their security requirements.*

The full proof of this Theorem can be found in the full paper.

Next, we present two example instantiations of the generic construction in the standard model.

6 Instantiations in the Standard Model

Here we provide two example instantiations of the generic framework in the standard model.

Instantiation I. Here we instantiate the signature schemes $\mathcal{DS}_1$ and $\mathcal{DS}_2$ using the recent structure-preserving signature scheme by Abe et al. [4] (shown in Fig. 5 (Left)), and instantiate $\mathcal{WDS}$ and $\mathcal{OTS}$ with the weak and full Boneh-Boyen signature schemes, respectively. We also instantiate $\mathcal{NIZK}$ using the Groth-Sahai proof system. For the distributed tag-based encryption scheme, we use our new scheme from Fig. 3. The size of the signature of this instantiation is $\mathbb{G}^{24} \times \tilde{\mathbb{G}}^{21} \times \mathbb{Z}_p^5$. The details are in the full paper. The proof for the following Theorem follows from that of Theorem 2.

Theorem 3. *Instantiation I is secure if the Abe et al. signature scheme [4] is unforgeable and the SXDH, XDLIN$_\mathbb{G}$ and q-SDH assumptions hold.*

Instantiation II. To eliminate the need for interactive intractability assumptions, we instead use the strongly unforgeable signature scheme by Abe et al. [3] (Fig. 5 (Right)) to instantiate $\mathcal{DS}_1$ and $\mathcal{DS}_2$ signature schemes. The rest of the tools remain the same as in Instantiation I. The size of the group signature of this instantiation is $\mathbb{G}^{28} \times \tilde{\mathbb{G}}^{24} \times \mathbb{Z}_p^3$. The details are in the full paper. The proof for the following Theorem follows from that of Theorem 2.

Theorem 4. *Instantiation II is secure if the SXDH, q-AGHO, XDLIN$_\mathbb{G}$ and q-SDH assumptions hold.*

Efficiency Comparison. Since there are no existing constructions which simultaneously offer all the properties as our constructions, we compare the size of the signature of our instantiations with that of Groth's scheme [35] for the single tracing manager setting which is considered the-state-of-the-art. Groth's scheme yields signatures of size $\mathbb{G}^{46} \times \mathbb{Z}_p$ in symmetric groups. Besides the extra distributed traceability feature, our instantiations involve fewer rounds in the join protocol than [35] and, in addition, satisfy tracing soundness.

Acknowledgments. We thank anonymous reviewers for their comments. The author was supported by ERC Advanced Grant ERC-2010-AdG-267188-CRIPTO and EPSRC via grant EP/H043454/1.

References

1. Abe, M., Chase, M., David, B., Kohlweiss, M., Nishimaki, R., Ohkubo, M.: Constant-size structure-preserving signatures: generic constructions and simple assumptions. In: Wang, X., Sako, K. (eds.) ASIACRYPT 2012. LNCS, vol. 7658, pp. 4–24. Springer, Heidelberg (2012)
2. Abe, M., Fuchsbauer, G., Groth, J., Haralambiev, K., Ohkubo, M.: Structure-preserving signatures and commitments to group elements. In: Rabin, T. (ed.) CRYPTO 2010. LNCS, vol. 6223, pp. 209–236. Springer, Heidelberg (2010)
3. Abe, M., Groth, J., Haralambiev, K., Ohkubo, M.: Optimal structure-preserving signatures in asymmetric bilinear groups. In: Rogaway, P. (ed.) CRYPTO 2011. LNCS, vol. 6841, pp. 649–666. Springer, Heidelberg (2011)
4. Abe, M., Groth, J., Ohkubo, M., Tibouchi, M.: Unified, minimal and selectively randomizable structure-preserving signatures. In: Lindell, Y. (ed.) TCC 2014. LNCS, vol. 8349, pp. 688–712. Springer, Heidelberg (2014)
5. Abe, M., Haralambiev, K., Ohkubo, M.: Signing on Elements in Bilinear Groups for Modular Protocol Design. Cryptology ePrint Archive, Report 2010/133. http://eprint.iacr.org/2010/133
6. Arita, S., Tsurudome, K.: Construction of threshold public-key encryptions through tag-based encryptions. In: Abdalla, M., Pointcheval, D., Fouque, P.-A., Vergnaud, D. (eds.) ACNS 2009. LNCS, vol. 5536, pp. 186–200. Springer, Heidelberg (2009)
7. Ateniese, G., Camenisch, J., Hohenberger, S., de Medeiros, B.: Practical group signatures without random oracles. Cryptology ePrint Archive. Report 2005/385. http://eprint.iacr.org/2005/385
8. Ateniese, G., Camenisch, J.L., Joye, M., Tsudik, G.: A practical and provably secure coalition-resistant group signature scheme. In: Bellare, M. (ed.) CRYPTO 2000. LNCS, vol. 1880, pp. 255–270. Springer, Heidelberg (2000)
9. Barbulescu, R., Gaudry, P., Joux, A., Thomé, E.: A heuristic quasi-polynomial algorithm for discrete logarithm in finite fields of small characteristic. In: Nguyen, P.Q., Oswald, E. (eds.) EUROCRYPT 2014. LNCS, vol. 8441, pp. 1–16. Springer, Heidelberg (2014)
10. Blakley, G.R.: Safeguarding cryptographic keys. In: AFIPS National Computer Conference, vol. 48, pp. 313–317. AFIPS Press (1979)

11. Bellare, M., Micciancio, D., Warinschi, B.: Foundations of group signatures: formal definitions, simplified requirements, and a construction based on general assumptions. In: Biham, E. (ed.) EUROCRYPT 2003. LNCS, pp. 614–629. Springer, Heidelberg (2003)

12. Bellare, M., Rogaway, P.: Random oracles are practical: a paradigm for designing efficient protocols. In: ACM-CCS 1993, pp. 62–73. ACM (1993)

13. Bellare, M., Shi, H., Zhang, C.: Foundations of group signatures: the case of dynamic groups. In: Menezes, A. (ed.) CT-RSA 2005. LNCS, vol. 3376, pp. 136–153. Springer, Heidelberg (2005)

14. Benjumea, V., Choi, S.G., Lopez, J., Yung, M.: Fair traceable multi-group signatures. In: Tsudik, G. (ed.) FC 2008. LNCS, vol. 5143, pp. 231–246. Springer, Heidelberg (2008)

15. Bichsel, P., Camenisch, J., Neven, G., Smart, N.P., Warinschi, B.: Get shorty via group signatures without encryption. In: Garay, J.A., De Prisco, R. (eds.) SCN 2010. LNCS, vol. 6280, pp. 381–398. Springer, Heidelberg (2010)

16. Blum, M., Feldman, P., Micali, S.: Non-interactive zero-knowledge and its applications. In: STOC 1988, pp. 103–112 (1988)

17. Boneh, D., Boyen, X.: Short signatures without random oracles and the SDH assumption in bilinear groups. J. Cryptology 21(2), 149–177 (2008)

18. Boneh, D., Boyen, X., Shacham, H.: Short group signatures. In: Franklin, M. (ed.) CRYPTO 2004. LNCS, vol. 3152, pp. 41–55. Springer, Heidelberg (2004)

19. Boyen, X., Waters, B.: Compact group signatures without random oracles. In: Vaudenay, S. (ed.) EUROCRYPT 2006. LNCS, vol. 4004, pp. 427–444. Springer, Heidelberg (2006)

20. Boyen, X., Waters, B.: Full-domain subgroup hiding and constant-size group signatures. In: Okamoto, T., Wang, X. (eds.) PKC 2007. LNCS, vol. 4450, pp. 1–15. Springer, Heidelberg (2007)

21. Camenisch, J.L., Groth, J.: Group signatures: better efficiency and new theoretical aspects. In: Blundo, C., Cimato, S. (eds.) SCN 2004. LNCS, vol. 3352, pp. 120–133. Springer, Heidelberg (2005)

22. Camenisch, J.L., Lysyanskaya, A.: Signature schemes and anonymous credentials from bilinear maps. In: Franklin, M. (ed.) CRYPTO 2004. LNCS, vol. 3152, pp. 56–72. Springer, Heidelberg (2004)

23. Camenisch, J.L., Michels, M.: A group signature scheme with improved efficiency. In: Ohta, K., Pei, D. (eds.) ASIACRYPT 1998. LNCS, vol. 1514, pp. 160–174. Springer, Heidelberg (1998)

24. Camenisch, J.L., Stadler, M.A.: Efficient group signature schemes for large groups. In: Kaliski Jr., B.S. (ed.) CRYPTO 1997. LNCS, vol. 1294, pp. 410–424. Springer, Heidelberg (1997)

25. Chaum, D., van Heyst, E.: Group signatures. In: Davies, D.W. (ed.) EUROCRYPT 1991. LNCS, vol. 547, pp. 257–265. Springer, Heidelberg (1991)

26. Delerablée, C., Pointcheval, D.: Dynamic fully anonymous short group signatures. In: Nguyên, P.Q. (ed.) VIETCRYPT 2006. LNCS, vol. 4341, pp. 193–210. Springer, Heidelberg (2006)

27. Desmedt, Y.G., Frankel, Y.: Threshold cryptosystems. In: Brassard, G. (ed.) CRYPTO 1989. LNCS, vol. 435, pp. 307–315. Springer, Heidelberg (1990)

28. Frankel, Y.: A practical protocol for large group oriented networks. In: Quisquater, J.-J., Vandewalle, J. (eds.) EUROCRYPT 1989. LNCS, vol. 434, pp. 56–61. Springer, Heidelberg (1990)

29. Furukawa, J., Imai, H.: An efficient group signature scheme from bilinear maps. In: Boyd, C., González Nieto, J.M. (eds.) ACISP 2005. LNCS, vol. 3574, pp. 455–467. Springer, Heidelberg (2005)

30. Furukawa, J., Yonezawa, S.: Group signatures with separate and distributed authorities. In: Blundo, C., Cimato, S. (eds.) SCN 2004. LNCS, vol. 3352, pp. 77–90. Springer, Heidelberg (2005)

31. Galindo, D., Libert, B., Fischlin, M., Fuchsbauer, G., Lehmann, A., Manulis, M., Schröder, D.: Public-key encryption with non-interactive opening: new constructions and stronger definitions. In: Bernstein, D.J., Lange, T. (eds.) AFRICACRYPT 2010. LNCS, vol. 6055, pp. 333–350. Springer, Heidelberg (2010)

32. Galbraith, S., Paterson, K., Smart, N.P.: Pairings for cryptographers. Discrete Appl. Math. **156**, 3113–3121 (2008)

33. Ghadafi, E., Smart, N.P., Warinschi, B.: Groth–Sahai proofs revisited. In: Nguyen, P.Q., Pointcheval, D. (eds.) PKC 2010. LNCS, vol. 6056, pp. 177–192. Springer, Heidelberg (2010)

34. Granger, R., Kleinjung, T., Zumbrägel, J.: Breaking '128-bit Secure' supersingular binary curves (or how to solve discrete logarithms in $\mathbb{F}_{2^{4 \cdot 1223}}$ and $\mathbb{F}_{2^{12 \cdot 367}}$). In: Cryptology ePrint Archive, Report 2014/119. http://eprint.iacr.org/2014/119.pdf

35. Groth, J.: Simulation-sound NIZK proofs for a practical language and constant size group signatures. In: Lai, X., Chen, K. (eds.) ASIACRYPT 2006. LNCS, vol. 4284, pp. 444–459. Springer, Heidelberg (2006)

36. Groth, J.: Fully anonymous group signatures without random oracles. In: Kurosawa, K. (ed.) ASIACRYPT 2007. LNCS, vol. 4833, pp. 164–180. Springer, Heidelberg (2007)

37. Groth, J., Sahai, A.: Efficient non-interactive proof systems for bilinear groups. SIAM J. Comput. **41**(5), 1193–1232 (2012)

38. Hlauschek, C., Black, J., Vigna, G., Kruegel, C.: Limited-linkable Group Signatures with Distributed-Trust Traceability. Technical Report, Vienna University of Technology (2012) http://www.iseclab.org/people/haku/tr-llgroupsig12.pdf

39. Kakvi, S.A.: Efficient fully anonymous group signatures based on the Groth group signature scheme. Masters thesis, University College London (2010). http://www5.rz.rub.de:8032/mam/foc/content/publ/thesis_kakvi10.pdf

40. Kiayias, A., Tsiounis, Y., Yung, M.: Traceable signatures. In: Cachin, C., Camenisch, J.L. (eds.) EUROCRYPT 2004. LNCS, vol. 3027, pp. 571–589. Springer, Heidelberg (2004)

41. Kiayias, A., Yung, M.: Group signatures with efficient concurrent join. In: Cramer, R. (ed.) EUROCRYPT 2005. LNCS, vol. 3494, pp. 198–214. Springer, Heidelberg (2005)

42. Kiltz, E.: Chosen-Ciphertext security from tag-based encryption. In: Halevi, S., Rabin, T. (eds.) TCC 2006. LNCS, vol. 3876, pp. 581–600. Springer, Heidelberg (2006)

43. Libert, B., Peters, T., Yung, M.: Scalable group signatures with revocation. In: Pointcheval, D., Johansson, T. (eds.) EUROCRYPT 2012. LNCS, vol. 7237, pp. 609–627. Springer, Heidelberg (2012)

44. Libert, B., Peters, T., Yung, M.: Group signatures with almost-for-free revocation. In: Safavi-Naini, R., Canetti, R. (eds.) CRYPTO 2012. LNCS, vol. 7417, pp. 571–589. Springer, Heidelberg (2012)

45. Manulis, M.: Democratic group signatures: on an example of joint ventures. In: ASIACCS 2006, pp. 365–365. ACM (2006)

46. Nguyen, L., Safavi-Naini, R.: Efficient and provably secure trapdoor-free group signature schemes from bilinear pairings. In: Lee, P.J. (ed.) ASIACRYPT 2004. LNCS, vol. 3329, pp. 372–386. Springer, Heidelberg (2004)
47. Sahai, A.: Non-malleable non-interactive zero knowledge and adaptive chosen-ciphertext security. In: FOCS 1999, pp. 543–553 (1999)
48. Sakai, Y., Emura, K., Hanaoka, G., Kawai, Y., Matsuda, T., Omote, K.: Group signatures with message-dependent opening. In: Abdalla, M., Lange, T. (eds.) Pairing 2012. LNCS, vol. 7708, pp. 270–294. Springer, Heidelberg (2013)
49. Sakai, Y., Schuldt, J.C.N., Emura, K., Hanaoka, G., Ohta, K.: On the security of dynamic group signatures: preventing signature hijacking. In: Fischlin, M., Buchmann, J., Manulis, M. (eds.) PKC 2012. LNCS, vol. 7293, pp. 715–732. Springer, Heidelberg (2012)
50. Shamir, A.: How to share a secret. Commun. ACM **22**(11), 612–613 (1979)
51. Zheng, D., Li, X., Ma, C., Chen, K., Li, J.: Democratic group signatures with threshold traceability. In: Cryptology ePrint Archive, Report 2008/112. http://eprint.iacr.org/2008/112.pdf

How to Leak a Secret
and Reap the Rewards Too

Vishal Saraswat$^{(\boxtimes)}$ and Sumit Kumar Pandey

C.R.Rao Advanced Institute of Mathematics Statistics and Computer Science,
Hyderabad, India
{vishal.saraswat,emailpandey}@gmail.com

Abstract. We introduce the notion of the designated identity verifier
ring signature (DIVRS) and give a generic construction from any given
ordinary ring signature scheme. In a DIVRS scheme, the signer S of a
message has the additional capability to prove, at time of his choice,
to a *designated identity verifier* V that S is the actual signer without
revealing his identity to anyone else. Our definition of a DIVRS retains
applicability for all previous applications of a ring signature with an
additional capability which can be seen as mix of a designated verifier
signature [7] and an anonymous signature [14,18]. Our generic transformation preserves all the properties of the original ring signature without
significant overhead.

Keywords: Designated identity verifier · Ring signature · Signer anonymity · Signing proof · Unpretendability

1 Introduction

The notion of *ring signature* was introduced by Rivest, Shamir and Tauman [13]
in 2001 to enable a user to leak a secret without revealing his identity. It allows a
signer to form a ring of members (including himself) arbitrarily without collabo-
ration of any of those ring members and sign a message so that the message can
be authenticated to have been signed by a member of the ring without revealing
exactly which member of the ring is the actual signer. Unlike a *group signature* [4], a ring signature does not need any centralized authority or co-ordination
among users and there is no revocation of anonymity by anyone. The users in
the ring are not fixed and it can be generated on the fly at the time of signature
generation and it is possible to achieve *unconditional anonymity* [3,5,13], so that
even an adversary with an unlimited computational power cannot find out who
among the ring members is the actual signer. Thus ring signatures and its variants are a useful tool for whistle blowing [13], anonymous authentication in ad
hoc groups [5], and many other applications which require signer anonymity but
a group formation is not possible. In the next subsection, we present yet another
application of ring signatures but which cannot be solved completely with the
existing primitives and motivates the need of our defined notion of *designated
identity verifier ring signature*.

© Springer International Publishing Switzerland 2015
D.F. Aranha and A. Menezes (Eds.): LATINCRYPT 2014, LNCS 8895, pp. 348–367, 2015.
DOI: 10.1007/978-3-319-16295-9_19

1.1 Motivation

Many applications of ring signatures like whistle blowing or anonymity-preserving auctions come with a 'reward' for the signer (whistle blower or auction winner). To claim the reward the signer would need to reveal their identity to a designated authority (government official or auctioneer) and prove that they indeed produced the signature. Consider the following two examples:

Motivation I: Suppose the government announces a reward for information leading directly to the apprehension or conviction of the boss B of a criminal gang. B is a powerful person feared by all and has many officials in the government secretly working for him. A member S of the gang is tempted by the award and wants to reveal some information about B.

– S needs to make the information public and not tip off a select few officials since S does not know if they might be working for B and might hand him over to B.
– S cannot reveal his identity to anyone for same reasons.
– S wants/needs to reveal his identity to claim the reward
    - only after B has been convicted; and
    - only to a designated official V.
– Finally, V should not be able to prove that S had leaked the information.

Motivation II: Consider an auction scenario.

– A bidder A wants to anonymously participate in an auction without revealing his identity even to the auctioneer.
– The auctioneer B is not willing to accept 'completely anonymous' bids. For example, bids from millionaires or people from certain organization, may be accepted.
– When A wins the auction, A needs to prove his identity to claim the win while preserving his anonymity.

To achieve anonymity, A could possibly use a ring signature (with a sufficiently anonymizing ring formed of potential millionaire bidders) but a standard ring signature does not satisfy the last property and after winning the bid, A will not be able to prove that he indeed placed the winning bid. A solution is to use our proposed DIVRS with B as the designated verifier. If A loses the auction, there is no loss of anonymity. If A wins the auction, only B gets to identify A. Note that B cannot prove to others that A bid for the item. The winning bid though can be publicly verified.

1.2 Our Contribution

Designated Identity Verifier Ring Signature. We propose the notion of a *designated identity verifier ring signature* (DIVRS) to provide a solution to above and related problems. In a DIVRS scheme, the signer u_s of a (public) message has the capability to prove, at a time of his choice, to a *designated identity verifier* (DIV) u_d that u_s is the actual signer without revealing their identity to anyone else. Even after u_s has proved to u_d that he is indeed the signer, u_d cannot prove to anyone else that u_s is the actual signer. So, u_s is able to claim the reward while preserving his anonymity. Our definition of a DIVRS retains applicability for all previous applications of a ring signature with an additional capability which can be seen as mix of a designated verifier signature [7] and an anonymous signature [14,18].

We also give a construction of DIVRS scheme building on the sign and commit approach used in [14]. Our construction is a generic transformation from any given ordinary ring signature scheme(s), any encryption scheme and any commitment scheme. Our construction not only preserves all the properties of the original ring signature but also provides many additional desirable properties without significant overhead.

The universality and conceptual simplicity of our scheme and the flexibility in the choice of the component primitives lead to efficient and application relevant instantiations and makes our construction very attractive. One may construct a DIVRS based on the factoring problem or the discrete log problem by choosing the component primitives based on respective problems. One can have an identity-based DIVRS or even a post-quantum DIVRS by choosing the component primitives which are identity-based or post-quantum respectively. The generic and modular nature of our scheme allows many extensions such as a threshold scheme with multiple signatories and/or multiple designated ring signature verifiers and/or multiple designated identity verifiers.

Security Properties of DIVRS

1. **Unforgeability:** No one should be able to produce a valid message-signature pair which he has not signed (for a ring of which he is not a member).
2. **Public message authentication:** Message authentication provides credibility to the leaked information compared to "blind" tip-offs. Also, it prevents a corrupt official from rewarding someone unfairly on the conviction of a criminal without a tip-off.
3. **Signer Anonymity:**
   i. No one, including the DIV until the publication of the identity verification token, should be able to compute who among the ring members is the actual signer of the message from the message-signature pair. This allows the user to leak any information anonymously.
   ii. Even after the publication of the identity verification token no one excluding the DIV should be able to know anything about the signer.
   iii. Even the DIV should not be able to prove the identity of the signer even after the publication of the identity verification token.
4. **Designated Identity Verifier Anonymity:**
   i. No one other than the original signer, including the DIV, should be able to reveal the identity of the DIV.

   ii. Even after the publication of the identity verification token, no one other than the original signer and the DIV should be able to reveal the identity of the DIV.

5. **Signing Provability:** The actual signer should be able to prove to the DIV about the authorship of the signature and thus be able to claim the rewards.
6. **Unpretendability:** No one other than the actual signer should be able to convince anyone about the authorship of the signature. This ensures that only the actual whistle-blower/bidder/signer can claim the rewards.

1.3   Related Work

A lot of research, including group signatures [4] and ring signatures [13], has been done to achieve signer anonymity. Rivest et al. [13] noted this application and provided a mechanism for their specific scheme to achieve this but did not formalize the notion or the relevant security properties. Lv et al. [11] also provided a mechanism for the specific scheme in [13] but they too did not formalize the notion or the relevant security properties. We mention some of the related work in the areas.

– *(Universal) Designated verifier (ring) signature* [7, 15] allows the signer to *designate* a verifier at the time of signature generation so that the designated verifier, but no one else, can verify the validity of the message-signature pair and may know the identity of the signer. In the strong versions of these schemes the designated verifier cannot prove to a third party about the actual signer of the message.

   - In all the variations of designated verifier signatures, the message cannot be publicly authenticated so the Property (2) is not satisfied. In case of ring signature variants, the signer cannot really prove that he actually is the signer, thus Property (5) is missing.

– *Anonymous signatures* [14, 18] are signature schemes where the signature of a message does not reveal the identity of the signer. A signer signs a message m and publishes message-signature pair (m, σ) but the signature cannot be verified until and unless the signer reveals a *verification token* τ and thus the anonymity of the signer is preserved. It also guarantees *unpretendability*, that is, infeasibility for someone other than the correct signer to prove authorship of the message and signature.

   - Since the message m cannot be authenticated until the token τ is revealed the Property (2) of DIVRS is not satisfied and once the τ is revealed everyone can identify the signer so Property (3) is lost. (The scheme can achieve the designated identity verifier property by encrypting the signature with public-key of the DIV or by using a designated verifier signature as the underlying signature scheme but in that case the message cannot be publicly authenticated even after the publication of the verification token).

– *Deniable ring authentication/signature* [12, 16] enables the signer to allow any verifier (the verifier need not be designated at the time of signature generation) to authenticate a message without revealing which member has issued the signature and the verifier cannot convince any third party that message was indeed authenticated.

    - In such schemes, the message cannot be publicly authenticated so the Property (2) is not satisfied. Also, the opposite of Property (5) is guaranteed instead.

- *Step-out ring signatures* [8] are ring signatures with two additional procedures: *confession* and *step-out*. The confession procedure allows the signer to prove his authorship whenever he wants and the step-out procedure allows non-signer member of the ring to prove that he is not the signer. Similar property is also achieved by *deniable ring signatures* [9] and *conditionally anonymous ring signatures* [19].

    - In these schemes, the anonymity of the signer can be compromised by non-signer members of the ring which is not desirable in our scenario.

- *Convertible ring signatures* [10] allow the signer to prove their authorship of the ring signature by converting the ring signature into an ordinary signature.

    - In these schemes, the anonymity of the signer can be compromised by non-signer members of the ring which is not desirable in our scenario.

- *Accountable ring signatures* [17] have a designated trusted entity, a system-wide participant independent of any possible ring of users, which can identify the actual signer.

    - In these schemes, the designated trusted entity can find out the identity of the signer as soon as the message is leaked so the Property (3.i) of DIVRS is not satisfied. Further, it is not in the hands of the signer to designate the identity verifier.

All the above mentioned schemes and related schemes which allow for identity verifiability either lack immediate public verifiability of the message or do not provide (unconditional) anonymity or the signer cannot delay the identity verification. In the next subsection, we present a brief outline of this paper.

1.4 Outline of the Paper

Section 2 gives some basic definitions used in this paper. In Sect. 3, formal definitions of DIVRS and its security notions have been presented. In Sect. 4, we present our generic construction of a DIVRS scheme. Finally, the correctness and security proof of our proposed scheme are presented in Sect. 5.

2 Preliminaries

We denote by $v \leftarrow A(x, y, z, \ldots)$ the operation of running a randomized or deterministic algorithm $A(x, y, z, \ldots)$ and storing the output to the variable v. If X is a set, then $v \xleftarrow{\$} X$ denotes the operation of choosing an element v of X according to the uniform random distribution on X. We say that a given function $f : N \rightarrow [0, 1]$ is *negligible in* n if $f(n) < 1/p(n)$ for any polynomial p for sufficiently large n. $A^{\mathsf{ORCL}}(inp)$ denotes an adversary A which has an access to the oracle ORCL and takes input inp.

Unless stated otherwise, all algorithms are probabilistic and polynomial-time. Further, all adversaries are polynomial-time and are allowed to make at most polynomial number of queries to the oracle(s) they have access to. For the sake of brevity, we will assume the "setup" algorithms of the schemes mentioned and will not mention those explicitly. Same about the "spaces" — *message-space, key-space*, and any other domains of the respective algorithms. Same again, about the public parameters of the scheme — in all the algorithms, input of the respective public parameters will be implicit.

In the security notions and other properties mentioned in this work we encompass general definitions — depending on the requirements of the DIVRS, the exact security notions and any other property of the basic components may be chosen. We now present the standard definitions related to the components used in our scheme.

2.1 Public-Key Encryption Scheme

A *public-key encryption scheme* PKE is a tuple of algorithms (PK-Gen, PK-Enc, PK-Dec) where

1. the key generation algorithm PK-Gen takes as input the security parameter k and outputs a key pair $(pk_{\mathsf{PKE}}, sk_{\mathsf{PKE}}) \leftarrow \mathsf{PK\text{-}Gen}(1^k)$;
2. the encryption algorithm PK-Enc takes as input the recipient's public-key pk_{PKE} and a message m and outputs its encryption enc $\leftarrow \mathsf{PK\text{-}Enc}(pk_{\mathsf{PKE}}, m)$; and
3. the deterministic decryption algorithm PK-Dec takes as input the recipient's secret-key sk_{PKE} and the ciphertext enc and outputs the plaintext/decryption $\tilde{m} \leftarrow \mathsf{PK\text{-}Dec}\,(sk_{\mathsf{PKE}}, \mathsf{enc})$ or a symbol $\perp$ if enc was not a valid encryption.

We require the following three properties from an encryption scheme:

Correctness: For any valid encryptable message m,

$$\mathbf{Pr}\left[(pk_{\mathsf{PKE}}, sk_{\mathsf{PKE}}) \leftarrow \mathsf{PK\text{-}Gen}(1^k) : \mathsf{PK\text{-}Dec}(sk_{\mathsf{PKE}}, \mathsf{PK\text{-}Enc}(pk_{\mathsf{PKE}}, m)) = m\right] = 1\,.$$

Semantic Security: (aka, ciphertext indishtinguishability [6]) For any adversary $\mathcal{A} = (\mathcal{A}_1, \mathcal{A}_2)$ the advantage

$$\mathbf{Adv}_{\mathsf{PKE},\mathcal{A}}^{\mathrm{SEC}}(k) \stackrel{\mathrm{def}}{=} \left| \mathbf{Pr}[\mathbf{Game}_{\mathsf{PKE},\mathcal{A}}^{\mathrm{SEC\text{-}1}}(k) = 1] - \mathbf{Pr}[\mathbf{Game}_{\mathsf{PKE},\mathcal{A}}^{\mathrm{SEC\text{-}0}}(k) = 1] \right|$$

is negligible in the security games $\mathbf{Game}_{\mathsf{PKE},\mathcal{A}}^{\mathrm{SEC\text{-}}b}$, $b = 0, 1$, defined in Game 1.

Recipient Anonymity: (aka, Key Privacy [1]) For any adversary $\mathcal{A} = (\mathcal{A}_1, \mathcal{A}_2)$ the advantage

$$\mathbf{Adv}_{\mathsf{PKE},\mathcal{A}}^{\mathrm{ANON}}(k) \stackrel{\mathrm{def}}{=} \left| \mathbf{Pr}[\mathbf{Game}_{\mathsf{PKE},\mathcal{A}}^{\mathrm{ANON\text{-}1}}(k) = 1] - \mathbf{Pr}[\mathbf{Game}_{\mathsf{PKE},\mathcal{A}}^{\mathrm{ANON\text{-}0}}(k) = 1] \right|$$

is negligible in the anonymity games $\mathbf{Game}_{\mathsf{PKE},\mathcal{A}}^{\mathrm{ANON\text{-}}b}$, $b = 0, 1$, defined in Game 2.

$$\mathbf{Game}_{\mathsf{PKE},\mathcal{A}}^{\mathrm{SEC}\text{-}b}(k)$$
$$(pk_{\mathsf{PKE}}, sk_{\mathsf{PKE}}) \leftarrow \mathsf{PK\text{-}Gen}(1^k)$$
$$(m_0, m_1, st) \leftarrow \mathcal{A}_1^{\mathsf{PK\text{-}Dec}(sk_{\mathsf{PKE}},\cdot)}(pk_{\mathsf{PKE}})$$
$$\mathsf{enc} \leftarrow \mathsf{PK\text{-}Enc}(pk_{\mathsf{PKE}}, m_b)$$
$$b' \leftarrow \mathcal{A}_2^{\mathsf{PK\text{-}Dec}(sk_{\mathsf{PKE}},\cdot)}(\mathsf{enc}, st)$$
$$\textbf{return } b'$$

where $\mathcal{A}_2$ is not allowed to query enc from the decryption oracle $\mathsf{PK\text{-}Dec}(sk_{\mathsf{PKE}}, \cdot)$.

Game 1: PKE Security Game

$$\mathbf{Game}_{\mathsf{PKE},\mathcal{A}}^{\mathrm{ANON}\text{-}b}(k)$$
$$(pk_{\mathsf{PKE},0}, sk_{\mathsf{PKE},0}) \leftarrow \mathsf{PK\text{-}Gen}(1^k), \ (pk_{\mathsf{PKE},1}, sk_{\mathsf{PKE},1}) \leftarrow \mathsf{PK\text{-}Gen}(1^k)$$
$$(m, st) \leftarrow \mathcal{A}_1^{\mathsf{PK\text{-}Dec}(sk_{\mathsf{PKE},0},\cdot),\mathsf{PK\text{-}Dec}(sk_{\mathsf{PKE},1},\cdot)}(pk_{\mathsf{PKE},0}, pk_{\mathsf{PKE},1})$$
$$\mathsf{enc} \leftarrow \mathsf{PK\text{-}Enc}(pk_{\mathsf{PKE},b}, m)$$
$$b' \leftarrow \mathcal{A}_2^{\mathsf{PK\text{-}Dec}(sk_{\mathsf{PKE},0},\cdot),\mathsf{PK\text{-}Dec}(sk_{\mathsf{PKE},1},\cdot)}(\mathsf{enc}, st)$$
$$\textbf{return } b'$$

where $\mathcal{A}_2$ is not allowed to query enc from the decryption oracles $\mathsf{PK\text{-}Dec}(sk_{\mathsf{PKE},i}, \cdot)$.

Game 2: PKE Anonymity Game

2.2   Commitment Scheme

A *commitment scheme* CS consists of a pair of algorithms (CS-Com,CS-Ver) where

1. the commitment algorithm CS-Com takes as input a bit-string s and outputs its committal-decommittal pair

$$(\mathsf{com}, \mathsf{dec}) \leftarrow \mathsf{CS\text{-}Com}(s);$$

   and
2. the deterministic commitment verification algorithm CS-Ver takes as input the committal com, decommittal dec and a bit-string s and outputs a bit

$$b \leftarrow \mathsf{CS\text{-}Ver}(\mathsf{com}, \mathsf{dec}, s).$$

We require the following three properties from a commitment scheme:

Correctness: For any bit-string s,

$$\mathbf{Pr}[(\mathsf{com}, \mathsf{dec}) \leftarrow \mathsf{CS\text{-}Com}(s) : \mathsf{CS\text{-}Ver}(\mathsf{com}, \mathsf{dec}, s) = 1] = 1.$$

Hiding: For any adversary $\mathcal{A} = (\mathcal{A}_1, \mathcal{A}_2)$ the advantage

$$\mathbf{Adv}_{\mathsf{CS},\mathcal{A}}^{\mathrm{HIDE}}(k) \stackrel{\mathrm{def}}{=} \left| \mathbf{Pr}[\mathbf{Game}_{\mathsf{CS},\mathcal{A}}^{\mathrm{HIDE}\text{-}1}(k) = 1] - \mathbf{Pr}[\mathbf{Game}_{\mathsf{CS},\mathcal{A}}^{\mathrm{HIDE}\text{-}0}(k) = 1] \right|$$

is negligible in games $\mathbf{Game}_{\mathsf{CS},\mathcal{A}}^{\mathrm{HIDE}\text{-}b}$, $b = 0, 1$, defined in Game 3.

Binding: For any adversary $\mathcal{A}$ the advantage

$$\mathbf{Adv}_{\mathsf{CS},\mathcal{A}}^{\mathrm{BIND}}(k) \stackrel{\mathrm{def}}{=} \mathbf{Pr}\left[\mathbf{Game}_{\mathsf{CS},\mathcal{A}}^{\mathrm{BIND}}(k) = 1\right]$$

is negligible in the game $\mathbf{Game}_{\mathsf{CS},\mathcal{A}}^{\mathrm{BIND}}$ defined in Game 3.

$\mathbf{Game}_{CS,\mathcal{A}}^{HIDE}\text{-B}(k)$	$\mathbf{Game}_{CS,\mathcal{A}}^{BIND}(k)$
$(s_0, s_1, st) \leftarrow \mathcal{A}_1(1^k)$	$(\text{com}, \text{dec}, s, \text{dec}', s') \leftarrow \mathcal{A}(1^k)$
$(\text{com}, \text{dec}) \leftarrow \text{CS-Com}(s_b)$	$p \leftarrow \text{CS-Ver}(\text{com}, \text{dec}, s)$
$b' \leftarrow \mathcal{A}_2(\text{com}, st)$	$p' \leftarrow \text{CS-Ver}(\text{com}, \text{dec}', s')$
return b'	**return** $p \wedge p' \wedge (s \neq s')$

Game 3: Commitment Scheme Hiding and Binding Games

2.3   Ring Signature

A *ring* $R = (R[1], \ldots, R[n])$ is an ordered list of *ring members* $R[i]$. To each ring member $R[i]$, there is an associated public-key & secret-key pair $(pk[i], sk[i])$ and we can identify the user $R[i]$ with its public-key $pk[i]$. We refer to the tuple $pk := (pk[1], \ldots, pk[n])$ as the public-key of the ring and identify the ring with it.

A *ring signature scheme* RS consists of a tuple of algorithms (RS-Gen, RS-Sig, RS-Ver) where

1. the ring generation algorithm RS-Gen takes as input the security parameter k and forms a ring $R = (R[1], \ldots, R[n])$, with the signer $R[s]$ as its member, where $s \overset{\$}{\leftarrow} \{1, \ldots, n\}$, and outputs

$$pk = (pk[1], \ldots, pk[n]) \text{ and } sk = sk[s]$$

where $sk[s]$ is the secret-key of the signer and $pk[i]$ is the public-key of the ring member $R[i]$ for $i \in \{1, \ldots, n\}$.

Remark 1. In the ring generation algorithm RS-Gen, if the signer $R[s]$ already has an existing key-pair $(pk[s], sk[s])$, it is used. Otherwise, RS-Gen generates a valid key pair for the signer by invoking the key generation algorithm PK-Gen of a suitable public-key encryption scheme. Other ring members, including the DIV, must already have a valid public key with respect to some suitable public-key encryption scheme. Note that, in case of identity based ring signature schemes, the last condition is automatically fulfilled since the public-key can be derived for any identity.

2. the signature generation algorithm RS-Sig takes as input the ring's public-key pk, the signer's secret-key $sk[s]$ and a message m and outputs its signature $\sigma \leftarrow \text{RS-Sig}(pk, sk[s], m)$ with respect to the ring R; and
3. the deterministic ring signature verification algorithm RS-Ver takes as input the public-key pk of the ring R, a message m and a signature σ and outputs a bit $b \leftarrow \text{RS-Ver}(pk, m, \sigma)$.

We require the following three properties from a ring signature scheme RS:

Correctness: For any valid signable message m and a ring $R = \{R[1], \ldots, R[n]\}$,

$$\mathbf{Pr}\big[(pk, sk) \leftarrow \text{RS-Gen}(1^k), R[i] \leftarrow R : \text{RS-Ver}(pk, \text{RS-Sig}(pk, sk[i], m)) = 1\big] = 1.$$

Unforgeability: For any adversary $\mathcal{A}$ the advantage

$$\mathbf{Adv}_{\mathsf{RS},\mathcal{A}}^{\text{UF-CMA}}(k) \stackrel{\text{def}}{=} \mathbf{Pr}[\mathbf{Game}_{\mathsf{RS},\mathcal{A}}^{\text{UF-CMA}}(k) = 1]$$

is negligible in the game $\mathbf{Game}_{\mathsf{RS},\mathcal{A}}^{\text{UF-CMA}}$ defined in Game 4 where the attacker $\mathcal{A}$ cannot have queried the signing oracle $\mathsf{RS\text{-}Sig}(pk, sk, \cdot)$ with m. Relaxing this restriction to: "the attacker $\mathcal{A}$ may query the signing oracle $\mathsf{RS\text{-}Sig}(pk, sk, \cdot)$ with even m but must not have received σ as an answer" *strong unforgeability* is defined. In this paper we work with fixed rings but *unforgeability against chosen subring attacks* and *unforgeability w.r.t. insider corruption* [2] may be similarly defined in the obvious fashion.

Signer Anonymity: For any adversary $\mathcal{A} = (\mathcal{A}_1, \mathcal{A}_2)$ the advantage

$$\mathbf{Adv}_{\mathsf{RS},\mathcal{A}}^{\text{ANON}}(k) \stackrel{\text{def}}{=} \left| \mathbf{Pr}[\mathbf{Game}_{\mathsf{RS},\mathcal{A}}^{\text{ANON-1}}(k) = 1] - \mathbf{Pr}[\mathbf{Game}_{\mathsf{RS},\mathcal{A}}^{\text{ANON-0}}(k) = 1] \right|$$

is negligible in the games $\mathbf{Game}_{\mathsf{RS},\mathcal{A}}^{\text{ANON-}b}$, $b = 0, 1$, defined in Game 4. Allowing the attacker to have all the secret-keys sk as additional input, *anonymity against full key exposure* is defined.

	$\mathbf{Game}_{\mathsf{RS},\mathcal{A}}^{\text{ANON-}b}(k)$
$\mathbf{Game}_{\mathsf{RS},\mathcal{A}}^{\text{UF-CMA}}(k)$	$(pk, sk) \leftarrow \mathsf{RS\text{-}Gen}(1^k)$
$\quad (pk, sk) \leftarrow \mathsf{RS\text{-}Gen}(1^k)$	$\quad (m, pk_0, pk_1, st) \leftarrow \mathcal{A}_1^{\mathsf{RS\text{-}Sig}(pk, sk, \cdot)}(pk)$
$\quad (m, \sigma) \leftarrow \mathcal{A}^{\mathsf{RS\text{-}Sig}(pk, sk, \cdot)}(pk)$	$\quad \sigma \leftarrow \mathsf{RS\text{-}Sig}(pk, sk_b, m)$
$\quad \mathbf{return} \ \mathsf{RS\text{-}Ver}(pk, m, \sigma)$	$\quad b' \leftarrow \mathcal{A}_2^{\mathsf{RS\text{-}Sig}(pk, sk, \cdot)}(\sigma, st)$
	$\quad \mathbf{return} \ b'$

Game 4: Ring Signature Unforgeability and Anonymity Games

---

3   Formal Model and Security Notions

In this section, we define the formal model of a designated identity verifier ring signature (DIVRS) and the related security notions, namely,

1. correctness of a DIVRS –
   i. message authentication,
   ii. signing provability,
2. unforgeability,
3. signer anonymity,
4. designated identity verifier anonymity, and
5. unpretendability.

We point out that in all the security games defined below, even the designated identity verifier (DIV) is a potential adversary except only in the anonymity games played after the identity verification token has been released. Further, all adversaries are polynomial-time.

3.1 Designated Identity Verifier Ring Signature

Let $R = (R[1], \ldots, R[n])$ be a ring with the signer $u_s = R[s]$ and the designated identity verifier (DIV) $u_d = R[d]$ as its members where the indices s and d are chosen randomly from $\{1, \ldots, n\}$. Without loss of generality we assume that $u_s \neq u_d$, that is, a signer does not designate himself/herself for identity verification.

We define a *designated identity verifier ring signature* as a tuple of algorithms

$$\mathsf{DIVRS} = (\mathsf{DIV\text{-}Gen}, \mathsf{DIV\text{-}Sig}, \mathsf{DIV\text{-}SVf}, \mathsf{DIV\text{-}IVf})$$

where

1. DIV-Gen is the ring generation algorithm which takes as input the security parameter k and forms a ring $R = (R[1], \ldots, R[n])$ with the signer $R[s]$ and the designated identity verifier $R[d]$ as its members, $s, d \xleftarrow{\$} \{1, \ldots, n\}$, and outputs

$$pk = (pk[1], \ldots, pk[n]) \text{ and } sk = sk[s]$$

   where $sk[s]$ is the secret-key of the signer and $pk[i]$ is the public-key of the ring member $R[i]$ for $i \in \{1, \ldots, n\}$;

Remark 2. The observations made in Remark 1 apply here too.

2. DIV-Sig is the signature generation algorithm which takes as input the ring's public-key pk, the signer's secret-key $sk[s]$, the DIV's public-key $pk[d] = pk_{\mathsf{PKE}}$ and a message m and outputs a pair $(\sigma, \tau) \leftarrow \mathsf{DIV\text{-}Sig}(pk, sk[s], pk[d], m)$ where σ is the signature with respect to the ring R and τ is the *identity verification token*;
3. DIV-SVf is the deterministic ring signature verification algorithm which takes as input the public-key pk of the ring R, a message m and a signature σ and outputs a bit $b \leftarrow \mathsf{DIV\text{-}SVf}(pk, m, \sigma)$.
4. DIV-IVf is the deterministic identity verification algorithm which takes as input a message m, a signature σ, the "claimed" signer's public-key $pk[s]$, the DIV's secret-key $sk[d] = sk_{\mathsf{PKE}}$ (corresponding to $pk[d] = pk_{\mathsf{PKE}}$) and the identity verification token τ revealed by the signer and outputs a bit $b \leftarrow \mathsf{DIV\text{-}IVf}(pk[s], sk[d], m, \sigma, \tau)$.

3.2 Correctness

A DIVRS is said to be *correct* if for any valid signable message m and any two ring members $R[i], R[j] \leftarrow R$, the following two conditions hold:

$$\mathbf{Pr}\left[\begin{array}{l} (pk, sk) \leftarrow \mathsf{DIV\text{-}Gen}(1^k), (\sigma, \tau) \leftarrow \mathsf{DIV\text{-}Sig}(pk, sk[i], pk[j], m) : \\ \mathsf{DIV\text{-}SVf}(pk, m, \sigma) = 1 \end{array}\right] = 1. \quad (1)$$

$$\mathbf{Pr}\left[\begin{array}{l} (pk, sk) \leftarrow \mathsf{DIV\text{-}Gen}(1^k), (\sigma, \tau) \leftarrow \mathsf{DIV\text{-}Sig}(pk, sk[i], pk[j], m) : \\ \mathsf{DIV\text{-}IVf}(pk[i], sk[j], m, \sigma, \tau) = 1 \end{array}\right] = 1. \quad (2)$$

The Property (1) is referred to as the *correctness of the signature verification* and in combination with unforgeability property (Sect. 3.3), it allows reliable message authentication. The Property (2) is referred to as the *correctness of the identity verification* and in combination with unpretendability property (Sect. 3.6), it allows reliable signing provability.

3.3 Unforgeability

A DIVRS is said to be *unforgeable* if for any adversary $\mathcal{A}$ the advantage

$$\mathbf{Adv}_{\mathrm{DIVRS},\mathcal{A}}^{\mathrm{UF\text{-}CMA}}(k) \stackrel{\mathrm{def}}{=} \mathbf{Pr}[\mathbf{Game}_{\mathrm{DIVRS},\mathcal{A}}^{\mathrm{UF\text{-}CMA}}(k) = 1]$$

is negligible in the game $\mathbf{Game}_{\mathrm{DIVRS},\mathcal{A}}^{\mathrm{UF\text{-}CMA}}$ defined in Game 5 where the attacker $\mathcal{A}$ cannot have queried the signing oracle DIV-Sig$(pk, sk, \cdot, \cdot)$ with m.

A DIVRS is said to be *strongly unforgeable* if for any adversary $\mathcal{A}$ the advantage

$$\mathbf{Adv}_{\mathrm{DIVRS},\mathcal{A}}^{\mathrm{SUF\text{-}CMA}}(k) \stackrel{\mathrm{def}}{=} \mathbf{Pr}[\mathbf{Game}_{\mathrm{DIVRS},\mathcal{A}}^{\mathrm{SUF\text{-}CMA}}(k) = 1]$$

is negligible in the game $\mathbf{Game}_{\mathrm{DIVRS},\mathcal{A}}^{\mathrm{SUF\text{-}CMA}}$ defined in Game 5 where the attacker $\mathcal{A}$ may query the signing oracle DIV-Sig$(pk, sk, \cdot, \cdot)$ with even m but must not have received σ as an answer.

$\mathbf{Game}_{\mathrm{DIVRS},\mathcal{A}}^{\mathrm{UF\text{-}CMA}}(k)$	$\mathbf{Game}_{\mathrm{DIVRS},\mathcal{A}}^{\mathrm{SUF\text{-}CMA}}(k)$
$(pk, sk) \leftarrow$ DIV-Gen(1^k)	$(pk, sk) \leftarrow$ DIV-Gen(1^k)
$(m, \sigma, \tau) \leftarrow \mathcal{A}^{\mathrm{DIV\text{-}Sig}(pk,sk,\cdot,\cdot)}(pk)$	$(m, \sigma, \tau) \leftarrow \mathcal{A}^{\mathrm{DIV\text{-}Sig}(pk,sk,\cdot,\cdot)}(pk)$
return DIV-SVf(pk, m, σ)	**return** DIV-SVf(pk, m, σ)

Game 5: DIVRS (Strong) Unforgeability Games

3.4 Signer Anonymity

A DIVRS is said to be *signer anonymous* if for any adversary $\mathcal{A} = (\mathcal{A}_1, \mathcal{A}_2)$ the advantage

$$\mathbf{Adv}_{\mathrm{DIVRS},\mathcal{A}}^{\mathrm{ANON}}(k) \stackrel{\mathrm{def}}{=} \left| \mathbf{Pr}[\mathbf{Game}_{\mathrm{DIVRS},\mathcal{A}}^{\mathrm{ANON\text{-}1}}(k) = 1] - \mathbf{Pr}[\mathbf{Game}_{\mathrm{DIVRS},\mathcal{A}}^{\mathrm{ANON\text{-}0}}(k) = 1] \right|$$

is negligible in the games $\mathbf{Game}_{\mathrm{DIVRS},\mathcal{A}}^{\mathrm{ANON\text{-}}b}$, $b = 0, 1$, defined in Game 6 with $d_0 = d_1$.

A DIVRS is said to be *signer anonymous against full key exposure* if for any adversary $\mathcal{A} = (\mathcal{A}_1, \mathcal{A}_2)$ the advantage

$$\mathbf{Adv}_{\mathrm{DIVRS},\mathcal{A}}^{\mathrm{AFKE}}(k) \stackrel{\mathrm{def}}{=} \left| \mathbf{Pr}[\mathbf{Game}_{\mathrm{DIVRS},\mathcal{A}}^{\mathrm{AFKE\text{-}1}}(k) = 1] - \mathbf{Pr}[\mathbf{Game}_{\mathrm{DIVRS},\mathcal{A}}^{\mathrm{AFKE\text{-}0}}(k) = 1] \right|$$

is negligible in the games $\mathbf{Game}_{\mathrm{DIVRS},\mathcal{A}}^{\mathrm{AFKE\text{-}}b}$, $b = 0, 1$, defined in Game 6 with $d_0 = d_1$.

Remark 3. (a) It is required that $s_0, s_1, d_0, d_1 \in R$, else the games in 6 abort.
(b) Until the identification verification token has been released, τ is assumed to be null in these security games and even the DIV is a potential adversary.

Signer Anonymity Against Rogue DIV. The DIV should not be able to prove the identity of the actual signer even by releasing his secret-key even after the release of the identification verification token. This property is similar to the property required of a strong designated verifier signature [7].

$\textbf{Game}_{\text{DIVRS},\mathcal{A}}^{\text{ANON}-b}(k)$	$\textbf{Game}_{\text{DIVRS},\mathcal{A}}^{\text{AFKE}-b}(k)$
$(pk, sk) \leftarrow \text{DIV-Gen}(1^k)$	$(pk, sk) \leftarrow \text{DIV-Gen}(1^k)$
$(m,(s_0,d_0),(s_1,d_1), st) \leftarrow \mathcal{A}_1^{\text{DIV-Sig}(pk,sk,\cdot,\cdot)}(pk)$	$(m,(s_0,d_0),(s_1,d_1), st) \leftarrow \mathcal{A}_1^{\text{DIV-Sig}(pk,sk,\cdot,\cdot)}(pk, sk)$
$(\sigma,\tau) \leftarrow \text{DIV-Sig}(pk, sk[s_b], pk[d_b], m)$	$(\sigma,\tau) \leftarrow \text{DIV-Sig}(pk, sk[s_b], pk[d_b], m)$
$b' \leftarrow \mathcal{A}_2^{\text{DIV-Sig}(pk,sk,\cdot,\cdot)}(\sigma, \tau, st)$	$b' \leftarrow \mathcal{A}_2^{\text{DIV-Sig}(pk,sk,\cdot,\cdot)}(\sigma, \tau, st)$
return $b' \wedge ((s_0, d_0) \neq (s_1, d_1))$	**return** $b' \wedge ((s_0, d_0) \neq (s_1, d_1))$

Game 6: DIVRS (Full Key Exposure) Anonymity Games

3.5 Designated Identity Verifier Anonymity

A DIVRS is said to be *designated identity verifier anonymous* if for any adversary $\mathcal{A} = (\mathcal{A}_1, \mathcal{A}_2)$ the advantage

$$\textbf{Adv}_{\text{DIVRS},\mathcal{A}}^{\text{DIV-ANON}}(k) \stackrel{\text{def}}{=} \left| \Pr[\textbf{Game}_{\text{DIVRS},\mathcal{A}}^{\text{ANON}-1}(k) = 1] - \Pr[\textbf{Game}_{\text{DIVRS},\mathcal{A}}^{\text{ANON}-0}(k) = 1] \right|$$

is negligible in the games $\textbf{Game}_{\text{DIVRS},\mathcal{A}}^{\text{ANON}-b}$, $b = 0, 1$, defined in Game 6 with $s_0 = s_1$.

A DIVRS is said to be *designated identity verifier anonymous against full key exposure* if for any adversary $\mathcal{A} = (\mathcal{A}_1, \mathcal{A}_2)$ the advantage

$$\textbf{Adv}_{\text{DIVRS},\mathcal{A}}^{\text{DIV-AFKE}}(k) \stackrel{\text{def}}{=} \left| \Pr[\textbf{Game}_{\text{DIVRS},\mathcal{A}}^{\text{AFKE}-1}(k) = 1] - \Pr[\textbf{Game}_{\text{DIVRS},\mathcal{A}}^{\text{AFKE}-0}(k) = 1] \right|$$

is negligible in the games $\textbf{Game}_{\text{DIVRS},\mathcal{A}}^{\text{AFKE}-b}$, $b = 0, 1$, defined in Game 6 with $s_0 = s_1$.

Remark 4. The observations made in Remark 3 apply here too.

3.6 Unpretendability

A DIVRS is said to be *unpretendable* if for any adversary $\mathcal{A} = (\mathcal{A}_1, \mathcal{A}_2)$ the advantage

$$\textbf{Adv}_{\text{DIVRS},\mathcal{A}}^{\text{up}}(k) \stackrel{\text{def}}{=} \Pr[\textbf{Game}_{\text{DIVRS},\mathcal{A}}^{\text{UP}}(k) = 1]$$

is negligible in the game $\textbf{Game}_{\text{DIVRS},\mathcal{A}}^{\text{UP}}$ defined in Game 7.

As in the case of anonymity, we say that DIVRS is *unpretendable against full key exposure* if the advantage $\textbf{Adv}_{\text{DIVRS},\mathcal{A}}^{\text{UP-FKE}}(k)$ of any adversary is negligible in the game $\textbf{Game}_{\text{DIVRS},\mathcal{A}}^{\text{UP-FKE}}$ which is similar to the game $\textbf{Game}_{\text{DIVRS},\mathcal{A}}^{\text{UP}}(k)$ except that now the adversary also gets the target secret-key $sk[s_0]$ as an additional input.

$$\begin{array}{|l|}
\hline
\mathbf{Game}_{\mathrm{DIVRS},\mathcal{A}}^{\mathrm{UP}}(k) \\
\quad s_0 \leftarrow \{1,\ldots,n\},\; d_0 \leftarrow \{1,\ldots,n\} \\
\quad (pk, sk) \leftarrow \mathsf{DIV\text{-}Gen}(1^k) \\
\quad (m, st) \leftarrow \mathcal{A}_1^{\mathsf{DIV\text{-}Sig}(pk,sk,\cdot,\cdot)}(pk) \\
\quad (\sigma, \tau) \leftarrow \mathsf{DIV\text{-}Sig}(pk, sk[s_0], pk[d_0], m) \\
\quad (s_1, d_1, \tau_1) \leftarrow \mathcal{A}_2^{\mathsf{DIV\text{-}Sig}(pk,sk,\cdot,\cdot)}(\sigma, \tau, st) \\
\quad \mathbf{return}\ \mathsf{DIV\text{-}IVf}(pk[s_1], sk[d_1], m, \sigma, \tau_1) \wedge (s_0 \neq s_1) \\
\hline
\end{array}$$

Game 7: DIVRS (Strong) Unpretendability Games

Intuitively, the adversary aims to claim the authorship of the message m signed by the target secret-key $sk[s_0]$ with the ring signature σ and the identity-verification token τ. The adversary tries to produce an appropriate τ_1 satisfying

$$\mathsf{DIV\text{-}IVf}(pk[s_1], sk[d_1], m, \sigma, \tau_1) = 1,$$

so that his chosen identity $pk[s_1]$ is verified by his chosen DIV $pk[d_1]$, both of which could be chosen after τ has been released. Our definition of unpretendability guarantees that the probability of success for this attempt is negligible.

We can define various weaker versions of unpretendability:

1. The adversary does not get τ as input, that is, he tries to claim the reward before the original signer.
2. The DIV's identity is fixed *a priori* and the adversary is unable to choose it adaptively as in $\mathbf{Game}_{\mathrm{DIVRS},\mathcal{A}}^{\mathrm{UP}}$ or $\mathbf{Game}_{\mathrm{DIVRS},\mathcal{A}}^{\mathrm{UP\text{-}FKE}}$.

We can also define stronger notions of unpretendability as in the case of unforgeability. For example *unpretendability against chosen subring attacks* might be defined but that would be inapplicable to the objective of this paper.

4   Generic Construction of a DIVRS

We now present a generic construction of a designated identity verifier ring signature

$$\mathsf{DIVRS} = (\mathsf{DIV\text{-}Gen}, \mathsf{DIV\text{-}Sig}, \mathsf{DIV\text{-}SVf}, \mathsf{DIV\text{-}IVf})$$

using an ordinary ring signature scheme $\mathsf{RS} = (\mathsf{RS\text{-}Gen}, \mathsf{RS\text{-}Sig}, \mathsf{RS\text{-}Ver})$. Our construction uses an ordinary ring signature scheme $\mathsf{RD} = (\mathsf{RD\text{-}Gen}, \mathsf{RD\text{-}Sig}, \mathsf{RD\text{-}Ver})$ with possibly $\mathsf{RD} = \mathsf{RS}$, an encryption scheme $\mathsf{PKE} = (\mathsf{PK\text{-}Gen}, \mathsf{PK\text{-}Enc}, \mathsf{PK\text{-}Dec})$, and a commitment scheme $\mathsf{CS} = (\mathsf{CS\text{-}Com}, \mathsf{CS\text{-}Ver})$. We assume these schemes satisfy the properties defined in Sect. 2.

To leak an information m, the signer u_s chooses a set L of potential "leakers" $\{u_{l_1}, \ldots, u_{l_n}\}$, $n \geq 1$, and a *designated identity verifier* (DIV) u_d, sets $D := \{u_d\}$, forms two rings $R_L = (R_L[0], R_L[1], \ldots, R_L[n])$ and $R_D = (R_D[0], R_D[1])$ from the sets $\{u_s\} \cup D \cup L$ and $\{u_s\} \cup D$ respectively (with random orderings). He then chooses ring signature schemes RS and RD (with possibly $\mathsf{RD} = \mathsf{RS}$) for signing messages w.r.t. the rings R_L and R_D respectively and computes respective ring signatures σ_L on m and σ_D on $m||\sigma_L$. The signer u_s then obtains an encryption enc of the string $m_0 = m||\sigma_L||\sigma_D||u_s||u_d$ with the public-key pk_{PKE} of the DIV u_d and a commitment com of enc with respect to a decommittal dec. The signer finally leaks the information m and its DIVRS, $\sigma := \sigma_L||\mathsf{com}$.

Given σ, anyone can parse σ_L from it and verify that the message m was indeed signed by a member of the ring R_L.

When u_s wants to prove that he was the actual signer to u_d, he reveals the *identity verification token* $\tau = \mathsf{dec}||\mathsf{enc}$. Given, τ anyone can obtain enc and confirm its authenticity by parsing com from σ but only u_d is able to decrypt enc to find out the identity u_s. u_d can then verify it was indeed u_s who signed m by verifying the ring signature σ_D with respect to the ring R_D.

---

function $\text{DIV-GEN}(1^k)$
  $(pk_{\mathsf{RS}}, sk_{\mathsf{RS}}) \leftarrow \mathsf{RS\text{-}Gen}(1^k)$
  $pk \leftarrow pk_{\mathsf{RS}}$
  $sk \leftarrow sk_{\mathsf{RS}}||pk_{\mathsf{RS}}$
  return (pk, sk)
function $\text{DIV-SIG}(pk, sk[s], pk[d], m)$
  Parse sk as $sk_{\mathsf{RS}}||pk_{\mathsf{RS}}$
  $\sigma_L \leftarrow \mathsf{RS\text{-}Sig}(pk_{\mathsf{RS}}, sk_{\mathsf{RS}}[s], m)$
  $(pk_{\mathsf{RD}}, sk_{\mathsf{RD}}) \leftarrow \mathsf{RD\text{-}Gen}(1^k)$
  $\sigma_D \leftarrow \mathsf{RD\text{-}Sig}(pk_{\mathsf{RD}}, sk_{\mathsf{RD}}[s], m||\sigma_L)$
  $\mathsf{enc} \leftarrow \mathsf{PK\text{-}Enc}(pk_{\mathsf{PKE}}, m||\sigma_L||\sigma_D||u_s||u_d)$
  $(\mathsf{com}, \mathsf{dec}) \leftarrow \mathsf{CS\text{-}Com}(\mathsf{enc})$
  $\sigma \leftarrow \sigma_L||\mathsf{com};\ \tau \leftarrow \mathsf{dec}||\mathsf{enc}$
  return (σ, τ)

function $\text{DIV-SVF}(pk, m, \sigma)$
  Parse σ as $\sigma_L'||\mathsf{com}'$
  return $\mathsf{RS\text{-}Ver}(pk, m, \sigma_L')$
function $\text{DIV-IVF}(pk[s], sk[d], m, \sigma, \tau)$
  Parse σ as $\sigma_L'||\mathsf{com}'$
  If $\mathsf{RS\text{-}Ver}(pk, m, \sigma_L') = 0$ **return** 0
  Parse τ as $\mathsf{dec}'||\mathsf{enc}'$
  If $\mathsf{CS\text{-}Ver}(\mathsf{com}', \mathsf{dec}', \mathsf{enc}') = 0$ **return** 0
  $m_0 \leftarrow \mathsf{PK\text{-}Dec}(sk_{\mathsf{PKE}}, \mathsf{enc})$
  If $m_0 = \perp$ **return** 0
  Parse m_0 as $m'||\sigma_L''||\sigma_D'||u_s'||u_d'$
  If $m' \neq m$ **return** 0
  If $\sigma_L'' \neq \sigma_L'$ **return** 0
  If $u_d' \neq u_d$ **return** 0
  return $\mathsf{RD\text{-}Ver}(pk_{\mathsf{RD}}, m||\sigma_L', \sigma_D')$

Description 1: DIVRS description

---

5   Security Proof

In this section we present the security proof of our scheme. We show that the security of the scheme is tied with security of the component schemes. That is, if a certain property of our scheme is violated then some security property of one of the component schemes is also violated. In particular, in the following subsections, we prove the following theorem:

Theorem 1. *Let*

$$RS = (RS\text{-}Gen, RS\text{-}Sig, RS\text{-}Ver) \tag{3}$$

and

$$RD = (RD\text{-}Gen, RD\text{-}Sig, RD\text{-}Ver) \tag{4}$$

be correct, unforgeable and anonymous ring signature schemes (with possibly $RD = RS$),

$$PKE = (PK\text{-}Gen, PK\text{-}Enc, PK\text{-}Dec) \tag{5}$$

be a correct, semantically secure anonymous (key-private) encryption scheme, and

$$CS = (CS\text{-}Com, CS\text{-}Ver) \tag{6}$$

be a correct hiding and binding commitment scheme. Then the

$$DIVRS = (DIV\text{-}Gen, DIV\text{-}Sig, DIV\text{-}SVf, DIV\text{-}IVf) \tag{7}$$

constructed using these as components is correct, unforgeable, signer anonymous, designated identity verifier anonymous and unpretendable.

5.1 Correctness of the Proposed Scheme

Correctness of the Signature Verification. This is immediate from the correctness of the ring signature RS since for any $\sigma = \sigma'_L \| \text{com}'$ where σ'_L, is the RS-signature, $\text{DIV-SVf}(pk, m, \sigma) = 1$ if and only if $\text{RS-Ver}(pk, m, \sigma'_L) = 1$.

Correctness of the Identity Verification. This is immediate from the correctness of the ring signature(s) RD (and RS), the encryption scheme PKE and the commitment scheme CS. Since $\text{DIV-IVf}(pk[s], sk[d], m, \sigma, \tau) = 0$ if and only if at least one of the following holds:

- $\text{RS-Ver}(pk, m, \sigma'_L) = 0$ where $\sigma = \sigma'_L \| \text{com}'$
- $\text{CS-Ver}(\text{com}', \text{dec}', \text{enc}') = 0$ where $\tau = \text{dec}' \| \text{enc}'$
- $m_0 \leftarrow \text{PK-Dec}(sk_{\text{PKE}}, \text{enc})$ and $m_0 \neq m \| \sigma'_L \| \sigma'_D \| u'_s \| u_d$
- $\text{RD-Ver}(pk_{\text{RD}}, m \| \sigma'_L, \sigma'_D) = 0$ where $m_0 = m \| \sigma'_L \| \sigma'_D \| u'_s \| u_d$.

Thus if the DIVRS was correctly computed and the identity verification returns 0, the correctness of the component schemes is violated.

5.2 Unforgeability

The unforgeability of DIVRS is tied with the unforgeability of the ring signature RS. Suppose that $\mathcal{A}$ is an adversary attacking the unforgeability of DIVRS. Then using $\mathcal{A}$, we construct an adversary $\mathcal{B}$, with essentially the same time complexity as that of $\mathcal{A}$, which attacks the unforgeability of RS, and satisfying

$$\mathbf{Adv}_{\text{DIVRS}, \mathcal{A}}^{\text{UF-CMA}}(k) \leq \mathbf{Adv}_{\text{RS}, \mathcal{B}}^{\text{UF-CMA}}(k).$$

The adversary $\mathcal{B}$ is given a public-key pk_{RS} of RS, and the corresponding signing oracle $\text{RS-Sig}(sk_{\text{RS}}, \cdot)$. $\mathcal{B}$ sets $pk = pk_{\text{RS}}$, and gives it to $\mathcal{A}$ and answers the signing query of $\mathcal{A}$ as follows: for signing query of m, $\mathcal{B}$ calls its own signing oracle with query m to obtain σ_L. $\mathcal{B}$ then computes

$$\sigma_D \leftarrow \text{RD-Sig}(')(pk_{\text{RD}}, sk_{\text{RD}}[s], m \| \sigma_L),$$
$$\text{enc} \leftarrow \text{PK-Enc}(pk_{\text{PKE}}, m \| \sigma_L \| \sigma_D \| u_s \| u_d) \text{ and}$$
$$(\text{com}, \text{dec}) \leftarrow \text{CS-Com}(\text{enc})$$

and returns $(\sigma = \sigma_L \| \mathsf{com}, \tau = \mathsf{dec} \| \mathsf{enc})$ to $\mathcal{A}$.

Note that this simulation of the unforgeability game for $\mathcal{A}$ by $\mathcal{B}$ is perfectly done according to the description of DIVRS.

Suppose that $\mathcal{A}$ halts with output (m^*, σ^*). Then $\mathcal{B}$ parses σ^* as $\sigma_1 \| \sigma_2$, and halts with output (m^*, σ_1). Whenever the output (m^*, σ^*) of $\mathcal{A}$ is a successful forgery for DIVRS, then $\mathcal{B}$ outputs a successful forgery (m^*, σ_1) for RS since from the definition of DIV-SVf, DIV-SVf$(pk, m^*, \sigma^*) = 1$ holds only if RS-Ver$(pk, m^*, \sigma_1) = 1$ holds. This proves the claimed inequality.

5.3   Unpretendability

Finally, we show that DIVRS satisfies unpretendability with respect to full key exposure. Suppose that $\mathcal{A} = (\mathcal{A}_1, \mathcal{A}_2)$ is an adversary attacking unpretendability of DIVRS. Using $\mathcal{A}$, we construct an adversary $\mathcal{B}$, with essentially the same time complexity as that of $\mathcal{A}$, attacking the binding property of the commitment scheme CS, satisfying

$$\mathbf{Adv}^{\mathrm{UP\text{-}FKE}}_{\mathsf{DIVRS},\mathcal{A}}(k) \leq \mathbf{Adv}^{\mathrm{BIND}}_{\mathsf{CS},\mathcal{B}}(k).$$

Given the security parameter k, $\mathcal{B}$ obtains $(pk, sk) \leftarrow$ DIV-Gen(1^k) and gives it to $\mathcal{A}$. $\mathcal{B}$ answers $\mathcal{A}$'s queries as in DIVRS except that for the commitment computation, it calls the CS-Com oracle in its own challenge. Finally $\mathcal{B}$ obtains $(m, st) \leftarrow \mathcal{A}_1(pk, sk)$. $\mathcal{B}$ then computes $(\sigma, \tau) \leftarrow$ DIV-Sig$(pk, sk[s_0], pk[d_0], m)$ and obtains $(s_1, d_1, \tau_1) \leftarrow \mathcal{A}_2^{\mathrm{DIV\text{-}Sig}(sk, \cdot, \cdot, \cdot)}(\sigma, \tau, st)$.

Then $\mathcal{B}$ parses σ as $\sigma_L \| \mathsf{com}$, τ as $\mathsf{dec} \| \mathsf{enc}$ and τ_1 as $\mathsf{dec}_1 \| \mathsf{enc}_1$ and halts with output $(\mathsf{com}, \mathsf{dec}, \mathsf{enc}, \mathsf{dec}_1, \mathsf{enc}_1)$.

Note that this simulation of the full-key exposure unpretendability game for $\mathcal{A}$ by $\mathcal{B}$ is perfect and whenever $\mathcal{A}$ succeeds at breaking the unpretendability of DIVRS, that is,

$$\mathrm{DIV\text{-}SVf}(pk[s_1], sk[d_1], m, \sigma, \tau_1) = 1$$

and $s_0 \neq s_1$, then $\mathcal{B}$ also succeeds in breaking the binding property of CS. From the definition of DIV-SVf,

$$\mathrm{DIV\text{-}SVf}(pk[s_1], sk[d_1], m, \sigma, \tau_1) = 1$$

holds implies that CS-Ver$(\mathsf{com}, \mathsf{dec}_1, \mathsf{enc}_1)$ holds. Moreover, since

$$(\sigma, \tau) = \mathrm{DIV\text{-}Sig}(pk, sk[s_0], pk[d_0], m),$$

by correctness of the DIVRS,

$$\mathrm{DIV\text{-}SVf}(pk[s_0], sk[d_0], m, \sigma, \tau) = 1$$

holds and hence CS-Ver$(\mathsf{com}, \mathsf{dec}_1, \mathsf{enc}_1)$ must hold too. Now, $s_0 \neq s_1$ so that $\mathsf{enc} \neq \mathsf{enc}_1$ and hence $\mathcal{B}$ has successfully violated the binding property of CS.

5.4   Signer Anonymity

The signer anonymity of DIVRS is tied to the signer anonymity of the ring signature RS, the semantic security of the encryption scheme PKE and the hiding property of the commitment scheme CS.

The reduction of the signer anonymity to the signer anonymity of the ring signature RS is straight forward and we assume that the adversary gets a non-negligible advantage only due to com component of $\sigma = \sigma_L \| \text{com}$. That is, without com, σ_L is just a random string to $\mathcal{A}$.

We first prove the signer anonymity before the release of the identity verification token τ, when even the DIV is a potential adversary.

Suppose the advantage $\mathbf{Adv}_{\text{RS},\mathcal{C}}^{\text{ANON-FKE}}(k)$ of an adversary $\mathcal{C}$ attacking the signer anonymity of RS is zero. Suppose that $\mathcal{A} = (\mathcal{A}_1, \mathcal{A}_2)$ is an adversary attacking the signer anonymity of DIVRS. Using $\mathcal{A}$, we construct $\mathcal{B}$, with essentially the same time complexity as that of $\mathcal{A}$, attacking the hiding property of the commitment scheme CS, satisfying

$$\mathbf{Adv}_{\text{DIVRS},\mathcal{A}}^{\text{ANON-FKE}}(k) \leq \mathbf{Adv}_{\text{CS},\mathcal{B}}^{\text{HIDE}}(k).$$

Consider the game $\mathbf{Game}_{\text{CS},\mathcal{B}}^{\text{HIDE-}b}$ with respect to this adversary $\mathcal{B}$. Given the security parameter k, $\mathcal{B}$ obtains $(pk, sk) \leftarrow \text{DIV-Gen}(1^k)$ and gives it to $\mathcal{A}$. $\mathcal{B}$ answers $\mathcal{A}$'s queries as in DIVRS except that for the commitment computation, it calls the CS-Com oracle in its own challenge. Finally $\mathcal{B}$ obtains

$$(m, (s_0, d_0), (s_1, d_1), st) \leftarrow \mathcal{A}_1^{\text{DIV-Sig}(pk, sk, \cdot, \cdot)}(pk)$$

with $d_0 = d_1 = d$. $\mathcal{B}$ then computes for $b = 0, 1$,

$$\sigma_{L,b} \leftarrow \text{RS-Sig}(pk_{\text{RS}}, sk_{\text{RS}}[s_b], m),$$

$$\sigma_{D,b} \leftarrow \text{RD-Sig}^b(pk_{\text{RD}_b}, sk_{\text{RD}_b}[s_b], m \| \sigma_{L,b}) \text{ and}$$

$$\text{enc}_b \leftarrow \text{PK-Enc}(pk_{\text{PKE}}, m \| \sigma_{L,b} \| \sigma_{D,b} \| u_{s_b} \| u_d)$$

and gives enc_0 and enc_1 as challenge strings to the challenger of the $\mathbf{Game}_{\text{CS},\mathcal{B}}^{\text{HIDE-}b}$ and obtains $(\text{com}, \text{dec}) \leftarrow \text{CS-Com}(s_b)$ from the challenger. $\mathcal{B}$ then picks a random bit b_1 and gives $\sigma = \sigma_{L,b_1} \| com$ as the challenge signature to the attacker $\mathcal{A}$ and obtains its guess b_2 and returns b_2 as its guess to the challenger in its own challenge.

Note that this simulation of the anonymity game for $\mathcal{A}$ by $\mathcal{B}$ is perfect, and the output of $\mathcal{B}$ is the same as the output of $\mathcal{A}$. Hence,

$$\Pr[\mathbf{Game}_{\text{CS},\mathcal{B}}^{\text{HIDE-}b}(k)] = \Pr[\mathbf{Game}_{\text{DIVRS},\mathcal{A}}^{\text{ANON-FKE}}\text{-B}(k)],$$

for $b = 0, 1$. Therefore,

$$\mathbf{Adv}_{\text{DIVRS},\mathcal{A}}^{\text{ANON-FKE}}(k) = \left| \Pr[\mathbf{Game}_{\text{DIVRS},\mathcal{A}}^{\text{ANON-FKE-}1}(k) = 1] - \Pr[\mathbf{Game}_{\text{DIVRS},\mathcal{A}}^{\text{ANON-FKE-}0}(k) = 1] \right|$$

$$= \left| \Pr[\mathbf{Game}_{\text{CS},\mathcal{B}}^{\text{HIDE-}1}(k) = 1] - \Pr[\mathbf{Game}_{\text{CS},\mathcal{B}}^{\text{HIDE-}0}(k) = 1] \right|$$

$$= \mathbf{Adv}_{\text{CS},\mathcal{B}}^{\text{HIDE}}(k).$$

Intuitively, if com is the commitment with respect to enc_{b_1} then $\mathcal{A}$ returns $b_2 = b_1$ with a non-negligible advantage. But if com is not the commitment with respect to enc_{b_1} then $\mathcal{A}$ does not have the right σ and could possibly only return a random b_2. Thus when $\mathcal{B}$ returns b_2 as its guess, it achieves an equal advantage as that of $\mathcal{A}$. So, before the release of the identity verification token τ,

$$\mathbf{Adv}_{\mathsf{DIVRS},\mathcal{A}}^{\mathrm{ANON\text{-}FKE}}(k) \leq \mathbf{Adv}_{\mathsf{RS},\mathcal{C}}^{\mathrm{ANON\text{-}FKE}}(k) + \mathbf{Adv}_{\mathsf{CS},\mathcal{B}}^{\mathrm{HIDE}}(k).$$

Now we prove the signer anonymity after the release of the identity verification token τ, when anyone other than the DIV can be an adversary. The signer anonymity is now straight forward from semantic security of the encryption scheme PKE.

Given, τ (and σ) anyone can obtain enc by parsing com from σ. So now the signer anonymity can be attacked by using enc (only, assuming the RS anonymity). That is, one can try to obtain the u_s component of $m||\sigma_L||\sigma_D||u_s||u_d$ from its encryption enc $=$ PK-Enc($pk_{\mathsf{PKE}}, m||\sigma_L||\sigma_D||u_s||u_d$). But the semantic security of the encryption scheme PKE guarantees a negligible success probability of any such attack. So, after the release of the identity verification token τ,

$$\mathbf{Adv}_{\mathsf{DIVRS},\mathcal{A}}^{\mathrm{ANON\text{-}FKE}}(k) \leq \mathbf{Adv}_{\mathsf{RS},\mathcal{C}}^{\mathrm{ANON\text{-}FKE}}(k) + \mathbf{Adv}_{\mathsf{PKE},\mathcal{B}}^{\mathrm{SEC}}(k).$$

Note that DIVRS satisfies signer anonymity with respect to full key exposure as long as the underlying ring signature RS also satisfies the same.

Signer Anonymity Against Rogue DIV. Note that even if the DIV releases the decryption $m||\sigma_L||\sigma_D||u_s||u_d$ of enc (which is public after release of τ), one can be convinced that (m, σ, τ) was produced by a member of the ring $R_D \subset R_L$ but as long as the underlying ring signature RS and the ring signature RD are signer anonymous, they cannot distinguish which member of R_D was the actual signer and the DIV itself could equally likely have been the signer. We refer the reader to [13] where this property of a ring signature is discussed in detail.

5.5   Designated Identity Verifier Anonymity

The designated identity verifier anonymity of DIVRS is tied to the semantic security and the recipient anonymity of the encryption scheme PKE and the hiding property of the commitment scheme CS.

Until the release of the identity verification token τ, when even the DIV is a potential adversary, the designated identity verifier anonymity follows from an similar proof as in above proof of the signer anonymity, with the only difference that now $s_0 = s_1$ and $d_0 \neq d_1$. So, before the release of the identity verification token τ,

$$\mathbf{Adv}_{\mathsf{DIVRS},\mathcal{A}}^{\mathrm{AFKE}}(k) \leq \mathbf{Adv}_{\mathsf{CS},\mathcal{B}}^{\mathrm{HIDE}}(k).$$

Let us now consider the case after the release of the identity verification token τ, when anyone other than the DIV can be an adversary.

Given, τ (and σ) anyone can obtain enc by parsing com from σ. So now the DIV anonymity can be attacked by using enc. The recipient anonymity of the encryption scheme guarantees that the anonymity of the DIV is not revealed from enc directly. An adversary $\mathcal{C}$ can also try to obtain the u_d component of $m||\sigma_L||\sigma_D||u_s||u_d$ from its encryption enc = PK-Enc($pk_{\mathsf{PKE}}, m||\sigma_L||\sigma_D||u_s||u_d$). But the semantic security of the encryption scheme PKE guarantees a negligible success probability of any such attack. So, after the release of the identity verification token τ,

$$\mathbf{Adv}_{\mathsf{DIVRS},\mathcal{A}}^{\mathrm{DIV\text{-}AFKE}}(k) \leq \mathbf{Adv}_{\mathsf{PKE},\mathcal{C}}^{\mathrm{ANON}}(k) + \mathbf{Adv}_{\mathsf{PKE},\mathcal{B}}^{\mathrm{SEC}}(k).$$

Note that DIVRS satisfies designated identity verifier anonymity with respect to full key exposure as long as the underlying ring signature RS also satisfies the same.

References

1. Bellare, M., Boldyreva, A., Desai, A., Pointcheval, D.: Key-privacy in public-key encryption. In: Boyd, C. (ed.) ASIACRYPT 2001. LNCS, vol. 2248, pp. 566–582. Springer, Heidelberg (2001). 353
2. Bender, A., Katz, J., Morselli, R.: Ring signatures: stronger definitions, and constructions without random oracles. In: Halevi, S., Rabin, T. (eds.) TCC 2006. LNCS, vol. 3876, pp. 60–79. Springer, Heidelberg (2006). 356
3. Bresson, E., Stern, J., Szydlo, M.: Threshold ring signatures and applications to ad-hoc groups. In: Yung, M. (ed.) CRYPTO 2002. LNCS, vol. 2442, pp. 465–480. Springer, Heidelberg (2002). 348
4. Chaum, D., van Heyst, E.: Group signatures. In: Davies, D.W. (ed.) EUROCRYPT 1991. LNCS, vol. 547, pp. 257–265. Springer, Heidelberg (1991). 348, 351
5. Dodis, Y., Kiayias, A., Nicolosi, A., Shoup, V.: Anonymous identification in *ad hoc* groups. In: Cachin, C., Camenisch, J.L. (eds.) EUROCRYPT 2004. LNCS, vol. 3027, pp. 609–626. Springer, Heidelberg (2004). 348
6. Goldwasser, S., Micali, S.: Probabilistic encryption. J. Comput. Syst. Sci. **28**(2), 270–299 (1984). 353
7. Jakobsson, M., Sako, K., Impagliazzo, R.: Designated verifier proofs and their applications. In: Maurer, U.M. (ed.) EUROCRYPT 1996. LNCS, vol. 1070, pp. 143–154. Springer, Heidelberg (1996). 348, 350, 351, 359
8. Klonowski, M., Krzywiecki, Ł., Kutyłowski, M., Lauks, A.: Step-out ring signatures. In: Ochmański, E., Tyszkiewicz, J. (eds.) MFCS 2008. LNCS, vol. 5162, pp. 431–442. Springer, Heidelberg (2008). 352
9. Komano, Y., Ohta, K., Shimbo, A., Kawamura, S.: Toward the fair anonymous signatures: deniable ring signatures. IEICE Trans. **90–A**(1), 54–64 (2007). 352
10. Lee, K.-C., Wen, H.-A., Hwang, T.: Convertible ring signature. IEE Proc. Commun. **152**(4), 411–414 (2005). 352
11. Lv, J., Wang, X.: Verifiable ring signature. In: DMS 2003 - The 9th International Conference on Distribted Multimedia Systems, pp. 663–667 (2003) 351
12. Naor, M.: Deniable ring authentication. In: Yung, M. (ed.) CRYPTO 2002. LNCS, vol. 2442, pp. 481–498. Springer, Heidelberg (2002). 351

13. Rivest, R.L., Shamir, A., Tauman, Y.: How to leak a secret. In: Boyd, C. (ed.) ASIACRYPT 2001. LNCS, vol. 2248, pp. 552–565. Springer, Heidelberg (2001). 348, 351, 365

14. Saraswat, V., Yun, A.: Anonymous signatures revisited. In: Pieprzyk, J., Zhang, F. (eds.) ProvSec 2009. LNCS, vol. 5848, pp. 140–153. Springer, Heidelberg (2009). 348, 350, 351

15. Steinfeld, R., Bull, L., Wang, H., Pieprzyk, J.: Universal designated-verifier signatures. In: Laih, C.-S. (ed.) ASIACRYPT 2003. LNCS, vol. 2894, pp. 523–542. Springer, Heidelberg (2003). 351

16. Susilo, W., Mu, Y.: Deniable ring authentication revisited. In: Jakobsson, M., Yung, M., Zhou, J. (eds.) ACNS 2004. LNCS, vol. 3089, pp. 149–163. Springer, Heidelberg (2004). 351

17. Xu, S., Yung, M.: Accountable ring signatures: a smart card approach. In: Quisquater, J.-J., Paradinas, P., Deswarte, Y., Kalam, A. (eds.) Smart Card Research and Advanced Applications VI. IFIP, vol. 153, pp. 271–286. Springer, New York (2004). 352

18. Yang, G., Wong, D.S., Deng, X., Wang, H.: Anonymous signature schemes. In: Yung, M., Dodis, Y., Kiayias, A., Malkin, T. (eds.) PKC 2006. LNCS, vol. 3958, pp. 347–363. Springer, Heidelberg (2006). 348, 350, 351

19. Zeng, S., Jiang, S., Qin, Z.: A new conditionally anonymous ring signature. In: Fu, B., Du, D.-Z. (eds.) COCOON 2011. LNCS, vol. 6842, pp. 479–491. Springer, Heidelberg (2011). 352

Extending Oblivious Transfer Efficiently
or - How to Get Active Security with Constant Cryptographic Overhead

Enrique Larraia[(✉)]

Department of Computer Science, University of Bristol, Bristol, UK
cseldv@bristol.ac.uk

Abstract. On top of the passively secure extension protocol of [IKNP03] we build a new construction secure against active adversaries. We can replace the invocation of the hash function that is used to check the receiver is well-behaved with the XOR of bit strings. This is possible by applying a cut-and-choose technique on the length of the bit strings that the receiver sends in the reversed OT. We also improve on the number of seeds required for the extension, both asymptotically and practically. Moreover, the protocol used to test receiver's behaviour enjoys unconditional security.

1 Introduction

Oblivious Transfer (OT), concurrently introduced by Rabin [Rab81] and Wiesner [Wie83] (the latter under the name of multiplexing) is a two-party protocol between a sender Alice and a receiver Bob. In its most useful version the sender has two secret bit strings, and the receiver wants to obtain one of the secrets at his choosing. After the interaction the receiver has not learnt anything about the secret string he has not chosen, and the sender has not learnt anything about the receiver's choice. Several flavours have been considered and they turn out to be equivalent [EGL85, BCR86a, BCR86b, Cré87].

In the *Universally Composable Framework* [Can01], OT has been rigorously formalized and proved secure [CLOS02] under the assumption of trapdoor permutations (static adversaries) and non-committing encryption (adaptive adversaries). It was further realized [PVW08] under several hard assumptions (DDH, QR or worst-case lattice problems).

OT is a powerful cryptographic primitive that may be used to implement a wide range of other cryptographic primitives [Kil88, IPS08, Yao82, GMW86, GV87, EGL85]. Unfortunately, the results of Impagliazzo and Rudich [IR89] make it very unlikely that one can base OT on one-way functions (as a black-box).

As a second best solution, Beaver showed in its seminal paper [Bea96] that one can implement a large number of oblivious transfers assuming that only a small number of OTs are available. This problem is known as *Extended Oblivious Transfer*. The OTs that one starts with are sometimes called the *seeds* of

© Springer International Publishing Switzerland 2015
D.F. Aranha and A. Menezes (Eds.): LATINCRYPT 2014, LNCS 8895, pp. 368–386, 2015.
DOI: 10.1007/978-3-319-16295-9_20

the extension. Beaver showed that if one starts with say n seeds, it is possible to obtain any polynomial number (in n) of extended OTs. His solution is very elegant and concerns feasibility, but it is inherently non-efficient. Later, Ishai et al. [IKNP03] showed a very efficient reduction for semi-honest adversaries. Since then other works have focused on extensions with active adversaries [IKNP03,HIKN08,IPS08,NNOB12]. This paper continues this line of research.

State of the Art. The approach initiated in [IKNP03] runs at his core a reversed OT to implement the extension. As already noted in [IKNP03], proving security against a cheating receiver Bob* is not trivial, as nothing refrains him from inputting whatever he likes in the reversed OT, allowing him to recover both secrets on Alice's side.

In terms of efficiency, the passive version of [IKNP03] needs $O(s)$ OT seeds, where s is a security parameter, with cut-and-choose techniques and the combiner of [CK88] active security comes at the cost of using $\Omega(s)$ seed OTs[1]. In [HIKN08] active security is achieved at no extra cost in terms of seed expansion (and communication), they apply OT-combiners worsening the computational cost. In [NNOB12] the expansion factor is $\frac{8}{3} \approx 2.66$, which is already quite good. Recently, it has been shown [LZ13] that meaningful extensions only exist if one starts with $\omega(\log s)$ seeds, (for $\log s$ seeds one would have to construct an OT protocol from the scratch). The constructions of [Bea96,IKNP03] can be instantiated with superlogarithmic seeds, so are optimal in this respect.

The communication cost is not really an issue, due to known almost-free reductions of $\mathcal{OT}^n_{poly(n)}$ to $\mathcal{OT}^n_n$, using a *pseudo random generator*, and running the small OT on the seeds. The computational cost of [IKNP03] is extremely efficient (passive version), it needs $O(s)$ work, i.e. constant work per extended OT (precisely it needs three invocations of the cryptographic primitive). All active extensions need at least *amortized* $\Omega(s)$ work.

Our Contributions. A technique that has proven to be quite useful [Nie07] is to split the extension protocol in two: an outer protocol ρ, and an inner protocol π. The former implements the actual extended transfers, whereas the latter wraps the reversed OT, ensuring at the same time that the receiver Bob is well-behaved in some sense. We follow the same idea, the novelty of our construction being in how the inner protocol π is realized. More concretely, for a fixed security level s we give a family of protocols $\pi_{m,n,t}$, where n is the number of seeds, m is the number of extended transfers, and $t \in [\frac{1}{n}, 1)$. Values of t close to $\frac{1}{n}$ render less OT seeds, and values close to 1 less computational and communication cost. We obtain

- The overall construction has *amortized constant cost in terms of cryptographic computation*. Active security is obtained at the cost of XORing $O((1-t)n^2)$ bits. The construction has similar communication complexity. The previous best [NNOB12] need to hash $O(n)$ bits per extended transfer.

---

[1] The hidden constant is quite big.

- The seed expansion factor of the reduction, with respect to the passive version of [IKNP03] is asymptotically close to 2, and this convergence is quite fast, for example for security level $s = 128$ one needs about $n = 323$ seeds to produce about $1, 00, 000$ extended OTs. This means that our construction essentially suffers an overhead factor of 2 in the security parameter, with respect to the passive protocol of [IKNP03].
- The reduction of π to the inner OT is *information-theoretic*. Other constructions either required computational assumptions e.g. [IKNP03, HIKN08, IPS08], or were in the random oracle [Nie07, NNOB12]. The outer protocol ρ is the standard protocol of [IKNP03], thus it uses a *correlation robust function*.

Our proof technique is, to some extent, similar to those of [Nie07, NNOB12] in the sense that it is combinatorial. Instead of working with permutations, we are able to connect security with set partitions. In [NNOB12] adversarial behaviour was quantified through what the authors called *leakage functions*. We take a different approach, and measure adversarial behaviour with the *thickness* of a partition. Details are in Sect. 4.3.

Paper Organization. Notation and basic background is introduced in Sect. 2. Section 3 discusses the approach of [IKNP03] and fits it in our context. In Sect. 4 we present the inner protocol π and prove it secure. In Sect. 5 the final construction is concluded, we discuss complexity and further directions.

2 Preliminaries

2.1 Notation

We denote with $[n]$ the set of natural number less or equal than n. Let $\mathbb{F}_2$ be the field of telements, binary vectors $\mathbf{x}$ are written in bold lowercase and binary matrices $\mathbf{M}$ in bold uppercase. When $\mathbf{M}$ is understood from the context, its rows will be denoted with subindices $\mathbf{m}_i$, and its columns with superindices $\mathbf{m}^j$. The entry at position (i, j) is denoted with m_i^j. Accordingly, the jth bit of a row vector $\mathbf{r} \in \mathbb{F}_2^n$ will be denoted with r^j, and the ith bit of a column vector $\mathbf{c} \in \mathbb{F}_2^m$ with c_i. For any two matrices $\mathbf{M}, \mathbf{N}$, of dimension $m \times n$, we let $[\mathbf{M}, \mathbf{N}]$ be the $m \times 2n$ matrix whose first n columns are $\mathbf{m}^j$ and last n columns are $\mathbf{n}^j$. The symbol $\mathbf{a}_{|J}$ stands for the vector obtained by restricting $\mathbf{a}$ at positions indexed by J.

2.2 Set Partitions

Given a finite set X of n objects, for any $p \leq n$, a *partition* $\mathcal{P}$ of X is a collection of p pairwise disjoint subsets $\{P_k\}_{k=1}^p$ of X whose union is X. Each P_k is a *part* of X. We say that part P_k is maximal if its size is the largest one. Let $\mathcal{ER}(X)$ denote the set of all possible *equivalence relations* in X. There is a one-to-one correspondence between partitions of X and equivalence relations in X, given by the mapping $\mathcal{P} \mapsto \mathcal{R}$, where $x\mathcal{R}y$ iff $x \in P_k$ and $y \in P_k$. We write $\mathcal{P}^X$ to denote the set of all partitions of X. In this work we will be concerned with partitions of the set $[n]$, where n is the number of OT seeds.

2.3   Universally Composable Framework

Due to lack of space we assume the reader is familiar with the UC Framework [Can01], especially with the notions of environment, ideal and real adversaries, indistinguishability, protocol emulation, and the composition theorem. Functionalities will be denoted with calligraphic $\mathcal{F}$. As an example $\mathcal{OT}_n^m$ denotes the OT functionality, in which the sender inputs m pairs of secret strings $(\mathbf{l}_i, \mathbf{r}_i)_{i \in [m]}$, each string of length n. The receiver inputs vector $\boldsymbol{\sigma} \in \mathbb{F}_2^m$, and as a result obtains the ith left secret $\mathbf{l}_i$ if $\sigma_i = 0$, or the ith right secret $\mathbf{r}_i$ if $\sigma_i = 1$. We will also make use of a *correlation robust function*. We name the output of the CRF as the *hash* of the input. Some times we will write H instead of CRF. The definition can be found in [IKNP03].

3   The IKNP Approach

In 2003, in their breakthrough, Ishai, Kilian, Nissim and Petrank [IKNP03] opened the door for practical OT extensions. They provided two protocols for this task. Throughout this paper we will sometimes refer to the passive version as the IKNP extension. We consider the standard OT functionality [CLOS02] in its multi session version, the only difference is that the adversary is allowed to abort the execution. This is necessary because of how we deal with a cheating sender (see Fig. 3).

3.1   IKNP in a Nutshell

For any $m = \text{poly}(n)$, the ideal functionality $\mathcal{OT}_n^m$ is realized making a single call to $\mathcal{OT}_m^n$, where the security parameter of the reduction depends on n. This in turn implies a reduction to $\mathcal{OT}_n^n$ using a pseudorandom generator. It works as follows: Let $\boldsymbol{\sigma} \in \mathbb{F}_2^m$ be the input of Bob to $\mathcal{OT}_n^m$, he chooses a $m \times 2n$ binary matrix $[\mathbf{L}, \mathbf{R}]$ for which it holds $\mathbf{l}^j \oplus \mathbf{r}^j = \boldsymbol{\sigma}$, $j \in [n]$, but is otherwise random, and inputs it to an inner $\mathcal{OT}_m^n$ primitive. Alice inputs a random vector $\mathbf{a} \in \mathbb{F}_2^n$. As a result of the call Alice obtains (row) vectors $\{\mathbf{q}_i\}_{i \in [m]}$, for which hold $\mathbf{q}_i = \mathbf{l}_i \oplus \sigma_i \cdot \mathbf{a}$. Now, if Alice wants to obliviously transfer one of her two ith secrets $(\mathbf{x}_i^{(0)}, \mathbf{x}_i^{(1)})$, she XORs them with $\mathbf{p}_i^{(0)} = \mathbf{q}_i$ and $\mathbf{p}_i^{(1)} = \mathbf{q}_i \oplus \mathbf{a}$ respectively, and sends masks $\mathbf{y}_i^{(0)}$, $\mathbf{y}_i^{(1)}$ to Bob, who can obtain $\mathbf{x}_i^{(b_i)}$ from $\mathbf{y}_i^{(b_i)}$ and $\mathbf{l}_i$. This can be used to implement one transfer out of the m that Bob wishes to receive, but can not cope with more: the OTP used for the ith transfer, with pads $(\mathbf{p}_i^{(0)}, \mathbf{p}_i^{(1)})$, prohibits to use $(\mathbf{p}_j^{(0)}, \mathbf{p}_j^{(1)})$ in the jth transfer, because they are correlated (the *same* $\mathbf{a}$ is implicit in both pairs[2]). To move from a situation with correlated pads to a situation with uncorrelated ones, IKNP uses a CRF; i.e. Alice masks $\mathbf{x}_i^{(c)}$ with the hash of $\mathbf{p}_i^{(c)}$. The construction is perfectly secure

---

[2] Bob would learn e.g. the distance of two non-transmitted secrets. It is trivial to check that if two correlated pairs are used by Alice, then $\mathbf{x}_i^{(1+b_i)} \oplus \mathbf{x}_j^{(1+b_j)} = \mathbf{y}_i^{(1+b_i)} \oplus \mathbf{y}_j^{(1+b_j)} \oplus \mathbf{l}_i \oplus \mathbf{l}_j$.

against a malicious sender Alice*, and statistically secure against a *semi-honest* receiver Bob*.

Intuitively, each input bit, σ_i, of Bob is protected by using n independent additive sharings as inputs to the inner $\mathcal{OT}_m^n$. As for Alice's privacy, the crucial point being that as long as $\mathbf{a}$ is not known to Bob, then $\mathbf{x}^{(1+b_i)}$ remains hidden from him; in that situation, one of the pads in each pair is independent of Bob's view. Unfortunately, the above crucially relies on Bob following the protocol specifications. In fact, it is shown in [IKNP03] how Bob* can break privacy if he chooses carefully what he gives to the inner $\mathcal{OT}_m^n$.

3.2   Modularizing the Extension

We define an ideal functionality that acts as a wrapper of the inner call to the $\mathcal{OT}$ primitive.[3] It behaves as follows: (1) On an honest input $\mathbf{B} = [\mathbf{L}, \mathbf{R}]$ from Bob (i.e. $\mathbf{B}$ defines n sharings of some vector $\boldsymbol{\sigma}$), the functionality gives to Alice a pair $(\mathbf{a}, \mathbf{Q})$ that she will use to implement the extended transfers. The secret $\mathbf{a}$ is randomly distributed in Bob's view. (2) An ideal adversary $\mathcal{S}$ can guess d bits of $\mathbf{a}$, in this case the functionality takes the guesses with probability 2^{-d}. The secret $\mathbf{a}$ has $n - d$ bits randomly distributed in Bob's view.

The functionality is denoted with $c\mathcal{PAD}_{m,n}$ to emphasize that it gives m correlated pairs of pads, under the same $\mathbf{a}$ to Alice (of length n). See Fig. 1 for a formal description. We emphasize that $c\mathcal{PAD}$ without the malicious behaviour was implicit in [IKNP03], and with the malicious behaviour in [Nie07]. We have just made the probability of aborting more explicit. The novelty of our approaches lies in how is realized.

For completeness, we have included the IKNP extension protocol, see Fig. 2 for details. The only difference is that the pads $(\mathbf{p}_i^{(0)}, \mathbf{p}_i^{(1)})_{i \in [m]}$ that Alice uses to generate uncorrelated ones via the CRF are assumed to be given by $c\mathcal{PAD}_{m,n}$.

3.3   The Reduction

The proof is on the same lines of the reduction of [IKNP03]. For the case the receiver is actively corrupted, with $c\mathcal{PAD}_{m.n}$ at play, Bob* is forced to take a guess *before* the actual extended transfers are executed. He is not caught with probability 2^{-d}, in which case $n - d$ bits of $\mathbf{a}$ are completely unknown to him. This correspondence between adversarial advantage and uncertainty (observed in [Nie07]) is the key to argue security in the active case. What we observe is that within the set F that indexes the $n - d$ unknown bits, either $\mathbf{a}$ or the flipped vector $\mathbf{a} \oplus \mathbf{1}$ has at least $(n-d)/2$ bits set to one. Consequently, the same number of bits of one of the pads that Alice uses remains unknown to Bob*. Using bounding techniques borrowed from [Nie07] it is not difficult to simulate ρ with security *essentially half* the security of the IKNP extension.

---

[3] The purpose of the otherwise seemingly artificial functionality is to give a neat security analysis, both inwardly and outwardly.

Functionality $c\mathcal{PAD}_{m,n}$

$c\mathcal{PAD}$ runs with a pad's receiver Alice, a pad's creator Bob, and an adversary $\mathcal{S}$. It is parametrized with the numbers of transfers m, and the length of the bit strings n.

- Upon receiving input (receiver, sid) from Alice and (creator, sid, $[\mathbf{L}, \mathbf{R}]$) from Bob, where $[\mathbf{L}, \mathbf{R}] \in \mathcal{M}_{m \times 2n}$ defines n sharings of the same vector $\boldsymbol{\sigma}$, sample at random $\mathbf{a} \in \mathbb{F}_2^n$. Then record the tuple $(\mathbf{a}, [\mathbf{L}, \mathbf{R}])$, send (sid) to $\mathcal{S}$ and halt.
- Upon receiving message (deliver, sid) from $\mathcal{S}$, compute matrix $\mathbf{Q} \in \mathcal{M}_{m \times n}$ as

$$\mathbf{q}_i = \mathbf{l}_i \oplus (l_i^1 \oplus r_i^1) \cdot \mathbf{a}.$$

   Output (delivered, sid, $\mathbf{a}$, $\mathbf{Q}$) to Alice, (delivered, sid) to Bob and $\mathcal{S}$, and halt.
- Upon receiving (corruptAlice, sid, $\tilde{\mathbf{a}}$) from $\mathcal{S}$, where $\tilde{\mathbf{a}} \in \mathbb{F}_2^n$, give $\tilde{\mathbf{Q}} = [\mathbf{l}^j \oplus \tilde{a}^j (\mathbf{l}^j \oplus \mathbf{r}^j]_{j \in [n]}$ to $\mathcal{S}$. If additionally $\mathcal{S}$ sends (corruptAlice, sid, $\perp$), output (abort, sid) to Alice and Bob and halt.
- Upon receiving message (corruptBob, sid, $[\tilde{\mathbf{L}}, \tilde{\mathbf{R}}]$, $\tilde{\mathbf{a}}$, G) from $\mathcal{S}$, where $G \subseteq [n]$ is of size d, and $\tilde{\mathbf{a}} \in \mathbb{F}_2^d$, do:
    - with probability $p = 1 - 2^{-d}$, output (corruptBob, sid) to Alice and $\mathcal{S}$ and halt. Else,
    - replace $\mathbf{a}_{|G}$ with $\tilde{\mathbf{a}}$, and compute matrix $\mathbf{Q}$ subject to

$$\mathbf{q}_i = \tilde{\mathbf{l}}_i \oplus (\tilde{\mathbf{l}}_i \oplus \tilde{\mathbf{r}}_i) * \mathbf{a}.$$

   Output (delivered, sid, $\mathbf{a}$, $\mathbf{Q}$) to Alice, (delivered, sid) to Bob and $\mathcal{S}$, and halt.

Fig. 1. Modeling creation of correlated pads

Protocol ρ

The protocol is parametrized with the number of extended transfers m, and the length of the transmitted vectors n.

Primitive: A $c\mathcal{PAD}_{m,n}$ functionality.
Inputs: Alice inputs (sender, $(\mathbf{x}_{i,0}, \mathbf{x}_{i,1})_{i \in [m]}$, sid), where $\mathbf{x}_{i,c} \in \mathbb{F}_2^n$, and Bob inputs (receiver, $\boldsymbol{\sigma}$, sid) with $\boldsymbol{\sigma} \in \mathbb{F}_2^m$.
Protocol:
1. Bob samples n independent sharings of $\boldsymbol{\sigma}$. Denote this sharings as $[\mathbf{L}, \mathbf{R}]$ (i.e. $\mathbf{l}^j \oplus \mathbf{r}^j = \boldsymbol{\sigma}$).
2. The parties call $c\mathcal{PAD}$. Bob inputs (creator, sid, $[\mathbf{L}, \mathbf{R}]$), and Alice inputs (receiver, sid), as a result Alice gets $(\mathbf{a}, \mathbf{Q})$ where $\mathbf{Q}$ is a $m \times n$ binary matrix, and $\mathbf{a} \in \mathbb{F}_2^n$.
3. Let $\mathbf{p}_i^{(0)} = \mathbf{q}_i$ and $\mathbf{p}_i^{(1)} = \mathbf{q}_i \oplus \mathbf{a}$. Alice computes $\mathbf{y}_i^{(c)} = \mathbf{x}_i^{(c)} \oplus H_i^{(c)}(\mathbf{p}_i^{(c)})$ for $c = 0, 1$, and sends pairs $(\mathbf{y}_i^{(0)}, \mathbf{y}_i^{(1)})_{i \in [m]}$ to Bob.
Outputs: Bob computes $\mathbf{h}_i = H_i^{(\sigma_i)}(\mathbf{l}_i)$ and outputs $\mathbf{x}_i' = \mathbf{y}_i^{(\sigma_i)} \oplus \mathbf{h}_i$. Alice outputs nothing.

Fig. 2. IKNP extension

Claim (Restatement of [IKNP03, Lemma1] for Active Adversaries). In the $c\mathcal{PAD}_{m,n}$-hybrid model, in the presence of static active adversaries, with access to at most $2^{o(n)}$ queries of a CRF, the output of protocol ρ, and the output of the ideal process involving $\mathcal{OT}_n^m$, are $2^{-n/2+o(n)+2}$-close.

For completeness it follows a proof sketch that combines the proofs of [IKNP03, Nie07]. Later, in Sect. 4.4 we will elaborate on an alternative idea for the simulation.

We focus on the case Bob* is corrupted, simulating a malicious Alice* is easy, and we refer the reader to [IKNP03] for details. To simulate a real execution of ρ,

---

Functionality $\mathcal{OT}_n^m$

The functionality is parametrized by the number of transfers m, and the length of bit strings n. It runs between a sender Alice, a receiver Bob and an adversary $\mathcal{S}$.

1. Upon receiving (sender, sid, $(\mathbf{x}_i^{(0)}, \mathbf{x}_i^{(1)})_{i \in [m]}$) from Alice, where $\mathbf{x}_i^{(c)} \in \mathbb{F}_2^n$, record tuple $(\mathbf{x}_i^{(0)}, \mathbf{x}_i^{(1)})_{i \in [m]}$. (The length n and number of transfers t is fixed and known to all parties)
2. Upon receiving (receiver, sid, σ) from Bob, where $\sigma \in \mathbb{F}_2^m$, send ($sid$) to $\mathcal{S}$, record σ and halt.
3. Upon receiving (deliver, sid) from $\mathcal{S}$, send (delivered, sid, $(\mathbf{x}_i^{(\sigma_i)})_{i \in [m]}$) to Bob and (delivered, sid) to Alice and halt.
4. Upon receiving (abort, sid) from $\mathcal{S}$, and only if (deliver, sid) was not previously received, send (fail, sid) to Alice and Bob and halt.

Fig. 3. The functionality of [CLOS02] augmented with aborts

- $\mathcal{S}$ internally runs steps 1 and 2 of ρ. If $\mathcal{A}$ sends (deliver, sid) to $c\mathcal{PAD}$, then $\mathcal{S}$ sets $\mathbf{r} \stackrel{\text{def}}{=} \sigma^*$, where σ^* is what $\mathcal{A}$ specified as input to $c\mathcal{PAD}$.
  Otherwise, $\mathcal{S}$ internally gets message (corruptBob, sid, $[\tilde{\mathbf{L}}, \tilde{\mathbf{R}}], \tilde{\mathbf{a}}, G$) from $\mathcal{A}$, then $c\mathcal{PAD}$ either rejects, in which case $\mathcal{S}$ externally sends (abort, sid) to $\mathcal{OT}_n^m$, outputs what ρ_B outputs and halts.
  If $c\mathcal{PAD}$, does not abort, let $F = [n] \backslash G$, then (for each $i \in [m]$) split it in two disjoint subsets, F_1, F_0 such that the bits of $\tilde{\mathbf{l}}_i \oplus \tilde{\mathbf{r}}_i$ indexed with F_{c_i} are equal to bit c_i. Say F_{r_i} is the largest set. $\mathcal{S}$ sets $\mathbf{r} \stackrel{\text{def}}{=} (r_1, \dots, r_m)$.
- Next, $\mathcal{S}$ externally calls $\mathcal{OT}_n^m$ on input $\mathbf{r}$ getting output $(\mathbf{z}_i)_{i \in [m]}$. It then fills the input tape of ρ_A with $\mathbf{x}_i^{(r_i)} = \mathbf{z}_s$ and $\mathbf{x}_i^{(r_i+1)} = \mathbf{0}^n$, executes step 3 of ρ, outputs what ρ_B outputs and halts.

Fig. 4. The ideal adversary for actively corrupted receivers

an ideal adversary $\mathcal{S}$ starts setting an internal copy of the real adversary $\mathcal{A}$, and runs the protocol between $\mathcal{A}$ and dummy parties ρ_A and ρ_B. The communication with the environment $\mathcal{E}$ is delegated to $\mathcal{A}$. Recall t hat $\mathcal{S}$ is also interacting with the (augmented with aborts) ideal functionality $\mathcal{OT}_n^m$ (see Fig. 3). A description of $\mathcal{S}$ for a malicious Bob* is in Fig. 4.

Let Dist be the event that $\mathcal{E}$ distinguishes between the ideal process and the real process, we examine the simulation conditioned on three disjoint events: SH is the event defined as "$\mathcal{A}$ *sends* (deliver, sid) to $c\mathcal{PAD}$", Active is the event "$\mathcal{A}$ *sends* (corruptBob, sid) and $c\mathcal{PAD}$ does not abort", and Abort is the event "$\mathcal{A}$ *sends* (corruptBob, sid) and $c\mathcal{PAD}$ aborts". It is clear that conditioned on Abort the simulation is perfect (with unbounded environments), because no transfers are actually done. Now, say that $|G| = d$, then $c\mathcal{PAD}_{m,n}$ does not abort with probability 2^{-d}, so we write

$$Pr[\text{Dist}] \leq Pr[\text{Dist}|\text{SH}] + Pr[\text{Dist}|\text{Active}] \cdot 2^{-d} \tag{1}$$

Conditioning on Active. In this case, the only difference between the ideal and the real process is that $\mathcal{S}$ fills with garbage the secret $\mathbf{x}_i^{(r_i+1)}$ of ρ_A, thus, the transcripts are indistinguishable *provided* $\mathcal{E}$ (or $\mathcal{A}$) does not submit $Q = \mathbf{p}_i^{(r_i+1)}$ to the CRF (in that case, $\mathcal{E}$ sees the zero vector in the ideal process, and the actual input of Alice in the real process). It is enough to see that this happens with negligible probability: First, pad $\mathbf{p}_i^{(r_i+1)}$ restricted at positions indexed with F_{r_i} can be expressed as

$$\mathbf{p}_{i|F_{r_i}}^{(r_i+1)} = \mathbf{q}_{i|F_{r_i}} \oplus (r_i \oplus 1) \cdot \mathbf{a}_{|F_{r_i}} = (\tilde{\mathbf{l}}_i \oplus (\tilde{\mathbf{l}}_i \oplus \tilde{\mathbf{r}}_i) * \mathbf{a})_{|F_{r_i}}) \oplus (r_i \oplus 1) \cdot \mathbf{a}_{|F_{r_i}}$$

$$= \tilde{\mathbf{l}}_{i|F_{r_i}} \oplus r_i \cdot \mathbf{a}_{|F_{r_i}} \oplus (r_i \oplus 1) \cdot \mathbf{a}_{|F_{r_i}}$$

$$= \tilde{\mathbf{l}}_{i|F_{r_i}} \oplus \mathbf{a}_{|F_{r_i}}.$$

Second, the size of F_{r_i} is at least $(n-d)/2$, because $F = F_0 \vee F_1$ and F_{r_i} is maximal. Third, $c\mathcal{PAD}_{m,n}$ generates $\mathbf{a}_{|F_{r_i}}$ using his own random bits. It follows that $\mathbf{p}_i^{(r_i+1)}$ has $(n-d)/2$ bits randomly distributed in $\mathcal{E}$'s view.

He may still guess such bits searching through the query space and using the CRF to compare. We next bound the probability of this happening. If $\mathcal{E}$ (or $\mathcal{A}$) guess correctly such bits, they would have handed to the CRF query $Q = \mathbf{p}_i^{(r_i+1)}$. As $(n-d)/2$ bits are unknown, the CRF returns random answers on $\mathcal{E}$'s view, the probability of hitting all the bits in $\mathbf{p}_i^{(r_i+1)}$ is bounded by $p_i \leq h_{r_i+1} 2^{(d-n)/2}$ where h_{r_i+1} is the number of queries made to $H_i^{(r_i+1)}$. By the union bound, given h denoting the total number of queries, $\mathcal{E}$ and $\mathcal{A}$ jointly hit query $Q = \mathbf{p}_i^{(r_i+1)}$ for some $i \in [m]$, with probability

$$Pr[\mathsf{Dist}|\mathsf{Active}] \leq 2\Big(\sum_{i \in [m]} h_{r_i+1} 2^{(d-n)/2} \Big) \leq h 2^{d/2+1-n/2}. \qquad (2)$$

Conditioning on SH. This case corresponds to semi-honest adversaries. We refer the reader to the proof of [IKNP03] for details. The only difference is that now *also* $\mathcal{A}$ can submit arbitrary queries to the CRF, hitting the offending one with the same probability than the environment would, thus

$$Pr[\mathsf{Dist}|\mathsf{SH}] \leq h 2^{-n+1}. \qquad (3)$$

Plugging inequalities 2 and 3 into 1, we obtain that the simulation fails with probability

$$Pr[\mathsf{Dist}] \leq h 2^{-n+1} + h 2^{d/2+1-n/2} \cdot 2^{-d} \leq h 2^{-n/2+2}.$$

The Claim follows setting $h = 2^{o(n)}$.   □

4   Generating Correlated Pads

The result of Sect. 3.3 (and previous works) shows that the IKNP extension can be upgraded to active security assuming that any adversarial strategy, on the receiver's side, amounts to guessing some of the bits of the sender's secret $\mathbf{a}$ *before* the extended transfers are executed. In this section we realize the $c\mathcal{PAD}$ functionality in a way where the only computational cost involved, beyond the underlying OT primitive on which it builds, is XORing bit strings.

Protocol $\pi_{m,n,t}$

The protocol is parametrized with the length of the input m, the number of OT seeds n, and a parameter $t \in [\frac{1}{n}, 1)$.

Primitive: An $\mathcal{OT}^n_{m(r+1)}$ functionality with $r = \lceil \frac{1-t}{2} n \rceil$.

Inputs: Bob inputs $[\mathbf{L}_0, \mathbf{R}_0] \in \mathcal{M}_{m \times 2n}$ defining n sharings of some vector $\boldsymbol{\sigma}_0 \in \mathbb{F}_2^m$ (i.e. $\mathbf{l}_0^j \oplus \mathbf{r}_0^j = \boldsymbol{\sigma}_0$ for $j \in [n]$). Alice inputs nothing.

Commit Phase:

1. Alice samples $\mathbf{a} \in \mathbb{F}_2^n$ at random, and Bob randomly samples r matrices $[\mathbf{L}_i, \mathbf{R}_i]$ in $\mathcal{M}_{m \times 2n}$ (i.e. $i \in [r]$). Each defining n sharings of (say) vectors $\boldsymbol{\sigma}_1, \ldots, \boldsymbol{\sigma}_r$.

2. The parties call the $\mathcal{OT}^n_{r(m+1)}$ functionality. Alice inputs $\mathbf{a}$, and Bob offers matrix $[[\mathbf{L}_0, \ldots, \mathbf{L}_r], [\mathbf{R}_0, \ldots, \mathbf{R}_r]]$ as his matrix of n (left,right) secrets of length $r(m+1)$ (towering up the $\mathbf{L}_i$'s together, idem with the $\mathbf{R}_i$'s). As a result Alice obtains output matrix $[\mathbf{Q}_0, \ldots, \mathbf{Q}_r] \in \mathcal{M}_{r(m+1) \times n}$.

Prove Phase: Alice challenges Bob to make sure he used a good enough input matrix.

3. Alice sends to Bob a random challenge vector $\mathbf{e} \in \mathbb{F}_2^r$.

4. For $i \in [r]$, Bob computes $\tilde{\boldsymbol{\sigma}}_i = \boldsymbol{\sigma}_i \oplus e_i \cdot \boldsymbol{\sigma}_0$, $\tilde{\mathbf{L}}_i = \mathbf{L}_i \oplus e_i \cdot \mathbf{L}_0$, and $\tilde{\mathbf{R}}_i = \mathbf{R}_i \oplus e_i \cdot \mathbf{R}_0$. It sends the r proofs $(\tilde{\boldsymbol{\sigma}}_i, [\tilde{\mathbf{L}}_i, \tilde{\mathbf{R}}_i])_{i \in [r]}$ to Alice.

5. For each $i \in [r]$, and $j \in [n]$ Alice prepares witnesses $\tilde{W}_i = (\mathbf{a}, \tilde{\mathbf{Q}}_i = \mathbf{Q}_i \oplus e_i \cdot \mathbf{Q}_0)$, and checks whether $\tilde{\mathbf{q}}_i^j \stackrel{?}{=} \tilde{\mathbf{l}}_i^j + a^j \cdot (\tilde{\mathbf{l}}_i^j \oplus \tilde{\mathbf{r}}_i^j)$, and $\tilde{\boldsymbol{\sigma}}_i \stackrel{?}{=} \tilde{\mathbf{l}}_i^j \oplus \tilde{\mathbf{r}}_i^j$ If not, she outputs corruptBob and halts.

Outputs: If Alice did not abort, she outputs $(\mathbf{a}, \mathbf{Q}_0)$ and Bob outputs nothing.

Fig. 5. Realizing $c\mathcal{PAD}_{m,n}$

4.1   Warming Up: Committing Bob to His Input

The inner $\mathcal{OT}^n_m$ of the IKNP extension can be seen, in a way, as a commitment for Bob's input $\boldsymbol{\sigma}$ to the outer $\mathcal{OT}^n_n$. The idea resembles the commitment scheme of [Cré89] generalized to m-bit strings. We split the protocol in two phases: A "commit" phase and a "prove" phase. To commit to $\boldsymbol{\sigma}$, Bob chooses n independent sharings $\mathbf{B} = [\mathbf{L}, \mathbf{R}]$ (i.e. $\mathbf{l}^j \oplus \mathbf{r}^j = \boldsymbol{\sigma}$ for $j \in [n]$) and offers them to an $\mathcal{OT}^n_m$ primitive. For the jth sharing, Alice obliviously retrieves one of the shares using her secret bit a^j. She obtains a "witness" matrix $\mathbf{Q} = [\mathbf{q}^j]_{j \in [n]}$. To prove his input Bob reveals $(\boldsymbol{\sigma}, \tilde{\mathbf{B}})$, and Alice checks she got the right share in the first place, (i.e. she checks $\mathbf{q}^j \stackrel{?}{=} \tilde{\mathbf{l}}^j \oplus a^j \cdot (\tilde{\mathbf{l}}^j \oplus \tilde{\mathbf{r}}^j)$), and that $\tilde{\mathbf{B}}$ is consistent with $\boldsymbol{\sigma}$ (i.e. $\tilde{\mathbf{l}}^j \oplus \tilde{\mathbf{r}}^j \stackrel{?}{=} \boldsymbol{\sigma}$).

Witnessing. The above protocol is of no use in our context, as for Bob to show he behaved correctly, he would have to reveal his input $\boldsymbol{\sigma}$ to the outer $\mathcal{OT}^m_n$. Nevertheless, we retain the concept of Alice obtaining a "witness" of what Bob* gave to the inner $\mathcal{OT}$. Such object is a pair $W = (\mathbf{a}, \mathbf{Q})$ obtained as the output of an $\mathcal{OT}^n_m$ primitive. Two witnesses W, W' are *consistent* if $\mathbf{a} = \mathbf{a}'$. Similarly, a "proof for witness W" is a pair $(\boldsymbol{\sigma}, \tilde{\mathbf{B}})$ such that $\tilde{\mathbf{B}}$ defines n sharings of $\boldsymbol{\sigma}$. We say the proof is *valid* if it is consistent with W, in the sense Alice would accept in the above protocol, when she is presented the proof and uses W to check it.

We emphasize that with this terminology, the output of $c\mathcal{PAD}_{m,n}$ is precisely a witness (see Fig. 1).

4.2   The Protocol

Suppose Alice has obtained a witness W_0 and she wants to use it to implement the extended transfers (as in protocol ρ). She is not sure if in order to give her the witness Bob used a good matrix $\mathbf{B}_0 = [\mathbf{L}_0, \mathbf{R}_0]$ or a bad one (loosely speaking a good matrix defines almost n sharings of a fixed vector σ, whereas a bad matrix has many (left,right) pairs adding up to distinct vectors.). Now, say that Alice has not one but two witnesses W_0, W_1. If they are consistent it is not difficult to see that she also knows a witness for $\mathbf{B}_+ = \mathbf{B}_0 \oplus \mathbf{B}_1$. So what Alice can do is to ask Bob to "decommit" to $\mathbf{B}_+$ as explained in Sect. 4.1. Intuitively Bob^* is able to "decommit" if $\mathbf{B}_+$ is not a bad matrix. It is also intuitive that $\mathbf{B}_+$ is not a bad matrix provided $\mathbf{B}_0$ and $\mathbf{B}_1$ are both good, or both bad. To rule out the latter possibility, Alice flips a coin and asks Bob to either "decommit" to $\mathbf{B}_1$ or to $\mathbf{B}_+$ accordingly. The process is repeated r times to achieve real soundness. Observe that a malicious Alice^* can not tell anything from σ, as an honest Bob *always* sends either σ_1 or masked $\sigma_0 \oplus \sigma_1$ when he is "decommitting".

Generating r consistent witnesses with W_0 can be done very efficiently[4] using an $\mathcal{OT}^n_{r(m+1)}$ primitive. The details of the protocol are in Fig. 5.

Correctness. If the parties follow the protocol it is not difficult to see that $\pi_{m,n,t}$ outputs exactly the same as $c\mathcal{PAD}_{m,n}$. By the homomorphic property, Alice does not reject on honest inputs. Output correctness is due to the fundamental relation exploited in the IKNP extension.

4.3   Security Analysis

The rest of the section is dedicated to prove that the output of $\pi_{m,n,t}$ and the output of $c\mathcal{PAD}_{m,n}$ are statistically close. The road-map is as follows: we first explain why we can work with partitions of $[n]$, then we state some useful results, and lastly we use them to show indistinguishability.

Taxonomy of Receiver's Input. Here we are after a classification of Bob's matrix $\mathbf{B} = [\mathbf{L}, \mathbf{R}] \in \mathcal{M}_{m \times 2n}$. As an illustration consider an honest situation where Bob gives matrix $\mathbf{B}$ such that it defines n additive sharings of some vector σ of his choosing. This means that $\mathbf{l}^j \oplus \mathbf{r}^j = \sigma$ for all indices in $[n]$. Clearly, the relation $j_1 \mathcal{R} j_2$ iff $\mathbf{l}^{j_1} \oplus \mathbf{r}^{j_2} = \sigma$ is the trivial equivalence relation in $[n]$ where all indices are related to each other. In other words, the matrix $[\mathbf{L}, \mathbf{R}]$ defines the trivial partition of $[n]$, i.e. $\mathcal{P} = \{[n]\}$.

Underlying Partition. For any binary matrix $\boldsymbol{\Delta}$ in $\mathcal{M}_{m \times n}$, its underlying relation is the subset $\mathcal{R}_{\boldsymbol{\Delta}} \in [n] \times [n]$ defined as

$$\mathcal{R}_{\boldsymbol{\Delta}} = \{(i,j) \in [n] \times [n] \mid \boldsymbol{\delta}^i = \boldsymbol{\delta}^j\}.$$

As usual, we write $i \mathcal{R}_{\boldsymbol{\Delta}} j$ to mean $(i,j) \in \mathcal{R}_{\boldsymbol{\Delta}}$. It is not difficult to see that $\mathcal{R}_{\boldsymbol{\Delta}}$ is an equivalence relation[5], in particular each $\boldsymbol{\Delta}$ defines a unique partition $\mathcal{P}_{\boldsymbol{\Delta}}$

---

[4] The cost to pay is increasing the length of the input bit strings to the $\mathcal{OT}$, using a PRG one would only need to obliviously transfer the PRG seed.

[5] The reader can check the relation is reflexive, symmetric and transitive.

of $[n]$. Also, for any partition of $[n]$, we say is ℓ-thick if the size of its maximal parts are ℓ. Now it becomes clear that any (possibly malicious) receiver's input $\mathbf{B} = [\mathbf{L}, \mathbf{R}]$ implicitly defines a partition of $[n]$, given by matrix $\boldsymbol{\Delta} = [\mathbf{l}^1 \oplus \mathbf{r}^1, \ldots, \mathbf{l}^n \oplus \mathbf{r}^n]$. The input is ℓ-thick if its partition is ℓ-thick.

Parametrizing the Thickness. One can take a parametric definition, saying that $\mathcal{P}$ is ℓ-thick if $\ell = \frac{M}{n}$, where M is the size of a maximal part[6]. In the security analysis this notion will prove to be useful. For example, honest inputs have (high) thickness level $\ell = 1$. We will always adopt the parametric perspective.

Witnessing and Thickness. Let $W = (\mathbf{a}, \mathbf{Q})$ be a witness that Alice has. If Bob* used an ℓ-thick $\mathbf{B}$ to give W to Alice, then W is said to be ℓ-thick.

Rejecting Thin Inputs. Now we formalize the intuition that Bob* is caught with high probability if he inputs a matrix with their columns adding up to many distinct vectors.

The first lemma deals with rejections on a particular "proof" handed by Bob. The second lemma upper bounds the thickness of a witness derived from the XOR operation. The proof of the rejection lemma exploits the correctness of the $\mathcal{OT}$ primitive. Both proofs make heavy use of the underlying partition defined in Sect. 4.3. The reader might want to skip them in the first lecture, and go directly to Proposition 1.

Lemma 1. *Let* $W = (\mathbf{a}, \mathbf{Q})$ *be a witness that is known to* Alice. *Then,* Bob *knows a valid proof for* W *only if he knows at least* $n(1 - \ell)$ *bits of* $\mathbf{a}$, *where* ℓ *is the thickness of* W. *In particular, if* Alice *is honest this happens with probability* $p \leq 2^{-n(1-\ell)}$.

Proof. Let $\mathbf{B} = [\mathbf{L}, \mathbf{R}]$ be the input of Bob to the $\mathcal{OT}$ from which Alice obtained witness $(\mathbf{a}, \mathbf{Q})$, and let $(\boldsymbol{\sigma}, \tilde{\mathbf{B}})$ be the proof held by Bob. Also, let $\boldsymbol{\Delta} = [\mathbf{l}^1 \oplus \mathbf{r}^1, \ldots, \mathbf{l}^n \oplus \mathbf{r}^n]$, and say that $\boldsymbol{\Delta}$ defines partition $\mathcal{P} = \{P_1, \ldots, P_p\}$ of $[n]$.

If the proof $(\boldsymbol{\sigma}, \tilde{\mathbf{B}})$ is *valid*, then for all $j \in [n]$ we can derive the equations

$$(1) \ \mathbf{q}^j = \mathbf{l}^j \oplus a^j \cdot (\mathbf{l}^j \oplus \mathbf{r}^j), \quad (2) \ \boldsymbol{\delta}^j = \mathbf{l}^j \oplus \mathbf{r}^j,$$
$$(3) \ \mathbf{q}^j = \tilde{\mathbf{l}}^j \oplus a^j \cdot (\tilde{\mathbf{l}}^j \oplus \tilde{\mathbf{r}}^j), \quad (4) \ \boldsymbol{\sigma} = \tilde{\mathbf{l}}^j \oplus \tilde{\mathbf{r}}^j.$$

where (1) and (2) are given by the correctness of the $\mathcal{OT}_m^n$ executed on Bob's input $\mathbf{B} = [\mathbf{L}, \mathbf{R}]$, and (3) and (4) follow from assuming $(\boldsymbol{\sigma}, \tilde{\mathbf{B}})$ is valid. Adding (1) and (3), and plugging (2) and (4) in the result, we write $\mathbf{l}^j \oplus \tilde{\mathbf{l}}^j = a^j \cdot (\boldsymbol{\delta}^j \oplus \boldsymbol{\sigma})$. Assume first there exist $j_0 \in [n]$, such that $\boldsymbol{\sigma} = \boldsymbol{\delta}^{j_0}$. Say wlog. that $j_0 \in P_1$. Now, by definition of $\mathcal{P}$, we have $\boldsymbol{\sigma} = \boldsymbol{\delta}^j$ iff $j\mathcal{R}_{\boldsymbol{\Delta}}j_0$. In other words, for $2 \leq k \leq p$ and $j \in P_k$ we have $\boldsymbol{\sigma} \neq \boldsymbol{\delta}^j$. It follows that there exists $i \in [m]$ such that $\delta_i^j \neq \sigma_i$, and therefore $a^j = l_i^j \oplus \tilde{l}_i^j$. The RHS of the last equation is known to Bob, so is a^j. This is true for all $j \in P_k$, and all $k \geq 2$, therefore Bob knows $|P_2 \vee \ldots \vee P_p| = n - |P_1| \geq n(1 - \ell)$ bits of $\mathbf{a}$, where the last inequality follows

---

[6] Parameter ℓ lies in $[\frac{1}{n}, 1]$.

because $\mathcal{P}$ is ℓ-thick. On the other hand, if $\boldsymbol{\sigma} \neq \boldsymbol{\delta}^j$ for all $j \in [n]$, then Bob knows the entire vector $\mathbf{a}$. Adding up, Bob* knows at least $n(1 - \ell)$ bits of $\mathbf{a}$.

Since $\mathbf{a}$ is secured via the $\mathcal{OT}_m^n$, Bob knows such bits by guessing them at random. We conclude that Alice accepts any $\boldsymbol{\sigma}$ with probability $p < 2^{n(1-\ell)}$, provided Alice samples $\mathbf{a}$ at random, which is indeed the case.

Lemma 2. *If $W = (\mathbf{a}, \mathbf{Q})$ is ℓ-thick and $\tilde{W} = (\mathbf{a}, \tilde{\mathbf{Q}})$ is $\tilde{\ell}$-thick, then $W_+ = (\mathbf{a}, \mathbf{Q} \oplus \tilde{\mathbf{Q}})$ is ℓ_+-thick with $\ell_+ \leq 1 - |\ell - \tilde{\ell}|$.*

Proof. Say that $\epsilon = |\ell - \tilde{\ell}|$. and let $[\mathbf{L}, \mathbf{R}]$, $[\tilde{\mathbf{L}}, \tilde{\mathbf{R}}]$ be the Bob's inputs from which Alice obtained witnesses W and $\tilde{W}$. Say that they define partitions $\mathcal{P} = \mathcal{P}_{[\mathbf{L},\mathbf{R}]}$, $\tilde{\mathcal{P}} = \mathcal{P}_{[\tilde{\mathbf{L}},\tilde{\mathbf{R}}]}$. Similarly one defines partition $\mathcal{P}_{\mathbf{\Delta} \oplus \tilde{\mathbf{\Delta}}}$ for witness $(\mathbf{a}, \mathbf{Q} \oplus \tilde{\mathbf{Q}})$.

First, suppose $\ell \leq \tilde{\ell}$, and let $\tilde{P}_{max}$ a maximal part of $\tilde{\mathcal{P}}$. Consider the refinement $\mathcal{P}_{max}^{\cap} = \tilde{P}_{max} \cap \mathcal{P}$. If j_1, j_2 lie in the same part of $\mathcal{P}_{max}^{\cap}$ then $j_1 \mathcal{R}_{\mathbf{\Delta} \oplus \tilde{\mathbf{\Delta}}} j_2$ iff $j_1 \mathcal{R}_{\mathbf{\Delta}} j_2$. This follows from the fact that if j_1 and j_2 are both in $\tilde{P}_{max}$, then $\tilde{\delta}^{j_1} = \tilde{\delta}^{j_2}$. In particular, each part of $\mathcal{P}_{max}^{\cap}$ lies in a different part of $\mathcal{P}_{\mathbf{\Delta} \oplus \tilde{\mathbf{\Delta}}}$.

Now, look at the auxiliar partition $\{[n]\backslash \tilde{P}_{max}, \mathcal{P}_{max}^{\cap}\}$. The maximum size we can hope for a part in $\mathcal{P}_{\mathbf{\Delta} \oplus \tilde{\mathbf{\Delta}}}$ occurs when $[n]\backslash \tilde{P}_{max}$ collapses with a single maximal part of $\mathcal{P}_{max}^{\cap}$. Even in this case, the size of a maximal part of $\mathcal{P}_{\mathbf{\Delta} \oplus \tilde{\mathbf{\Delta}}}$ is upper bounded by

$$n(1 - \tilde{\ell}) + n\ell = n(1 - (\ell + \epsilon) + \ell) = n(1 - \epsilon).$$

This follows from observing that $\tilde{P}_{max}$ is of size $n\tilde{\ell}$, and $\mathcal{P}_{max}^{\cap}$ have parts upper bounded by $n\ell$. The case $\tilde{\ell} \leq \ell$ is analogous (using auxiliar partition $\{[n]\backslash P_{max}, P_{max} \cap \tilde{\mathcal{P}}\}$).

Next, we estimate the acceptance probability of $\pi_{m,n,t}$ on any possible input of Bob. Note that the first witness obtained in the commit phase is the output of $\pi_{m,n,t}$.

Proposition 1. *Let $W = (\mathbf{a}, \mathbf{Q})$ the first witness that Alice obtains in the commit phase of $\pi_{m,n,t}$. Then, if W has thickness $\ell \leq t$, Alice accepts any adversarial proof with probability $p \leq 2^{-n(1-t)/2+2}$. In that case, Bob knows at least $n(1 - t)/2$ bits of $\mathbf{a}$.*

Proof. Recall that in the protocol $r = \lceil \frac{1-t}{2} n \rceil$, and let $E = (E_1, \ldots, E_r)$ be the random variable (uniformly distributed over $\mathbb{F}_2^r$) that Alice uses to challenge Bob*. For $i \in [r]$, let $B_i^* = (L_i^*, R_i^*)$ be the adversarial random variables that Bob* uses to sample the r matrices in the commit phase of π. Let $[\mathbf{L}_i, \mathbf{R}_i] = \mathbf{B}_i \leftarrow B_i^*$ the actual matrices. Denote with $\mathbf{\Delta}_i$ their correspondent underlying matrices. Each $\mathbf{\Delta}_i$ defines a unique partition $\mathcal{P}_i$ of $[n]$, with thickness $\ell_i \in [\frac{1}{n}, 1]$.

We want to upper bound the probability of Alice accepting in $\pi_{m,n,t}$ with $\ell \leq t$. Denote with Accept this event. Consider the r.v. $E^* = (E_1^*, \ldots, E_r^*)$, given by:

$$E_i^* = \begin{cases} 0 & i \text{ if } \ell_i > t' + \ell \\ 1 & \text{if } \ell_i \leq t' + \ell \end{cases}$$

where $t' = \frac{1-t}{2}$ is positive if $t \in [\frac{1}{n}, 1)$. We first look at the probability of Alice accepting the ith proof,

$$P[\mathsf{Accept}_i] = \frac{1}{2}(P[\mathsf{Accept}_i \mid E_i \to 0] + P[\mathsf{Accept}_i \mid E_i \to 1])$$

$$\leq \frac{1}{2}(P[\mathsf{Accept}_i \mid E_i \to 0, E_i^* \to 0] + (P[\mathsf{Accept}_i \mid E_i \to 0, E_i^* \to 1]$$

$$+ P[\mathsf{Accept}_i \mid E_i \to 1, E_i^* \to 0] + P[\mathsf{Accept}_i \mid E_i \to 1, E_i^* \to 1])$$

$$= p_{0,0} + p_{0,1} + p_{1,0} + p_{1,1}.$$

Consider the cases:

$(e_i, e_i^*) = (0,1)$. If Alice uses $\tilde{W}_i = W_i$ and $\ell_i \leq t' + \ell$ (i.e. $1 - \ell_i \geq 1 - t' - \ell$), by Lemma 1 we bound $p_{0,1} \leq 2^{-n(1-\ell_i)} \leq 2^{-n(1-t'-\ell)}$.

$(e_i, e_i^*) = (1,0)$. If Alice uses $\tilde{W}_i = W_+ = W_i + W$ and $\ell_i \geq t' + \ell$ (i.e. $\ell_i - \ell \geq t'$, with $t' \geq 0$ is equivalent to $|\ell_i - \ell| \geq t'$), by Lemma 2 we have $\ell_+ \leq 1 - |\ell_i - \ell| \leq 1 - t'$, and Lemma 1 bounds $p_{1,0} \leq 2^{-n(1-\ell_+)} \leq 2^{-n(1-(1-t'))} = 2^{-nt'}$.

Now, observe that by hypothesis $\ell \leq t$, and therefore $\frac{1-t}{2} = t' = min\{1 - t' - \ell, t'\}$. From the above we deduce, (1) if Bob* does not guess Alice's coin E_i with his own coins E_i^* then he has to guess at least $n(1-t)/2$ bits of $\mathbf{a}$, (2) in that case we bound $p_{b,b+1} \leq 2^{-nt'} = 2^{-n(1-t)/2}$.

We have to give up in bounding $p_{0,0}$ and $p_{1,1}$ as Bob* can always choose ℓ_i appropriately to pass the test with high probability (e.g. $\ell_i = 1$, $\ell_i = \ell$ respectively). As observed, in these cases Bob* is guessing Alice's coin e_i with his own coin e_i^*. It is now easy to finish the proof as follows:

Let Guess be the event $\{\mathbf{e} \leftarrow E\} \cap \{\mathbf{e} \leftarrow E^*\}$, is clear that if $\neg$Guess, then exist i_0 s.t. $e_{i_0} \leftarrow E_{i_0}$ and $e_{i_0} \oplus 1 \leftarrow E_{i_0}^*$, we can write

$$P[\mathsf{Accept}] = P[\cap_{i=1}^r \mathsf{Accept}_i]$$

$$\leq P[(\cap_i^r \mathsf{Accept}_i) \cap \mathsf{Guess}] + P[(\cap_i^r \mathsf{Accept}_i) \mid \neg\mathsf{Guess}]$$

$$\leq P[\mathsf{Guess}] + P[\mathsf{Accept}_{i_0} \mid E_{i_0} \to e_i, E_{i_0}^* \to e_i + 1]$$

$$\leq 2^{-r} + 2^{-n\frac{1-t}{2}+1}$$

Therefore Alice accepts and Bob* knows $n(1-t)/2$ bits of $\mathbf{a}$ with probability at most $2^{-n(1-t)/2+2}$.

Remark on Proposition 1. The above result ensures two things: First, if Bob inputs a matrix whose columns do not add to the same constant value *he is forced to take a guess on some bits of* $\mathbf{a}$. As we saw in Sect. 3.3 this is enough to implement the extended transfers securely. Second, setting the thick parameter t appropriately we can rule out a wide range of adversarial inputs with overwhelming probability in n. For example, the adversarial input $\mathbf{I}_{IKNP} = [\mathbf{L}, \mathbf{R}]$ of the attack in the IKNP extension has all its columns adding up to distinct elements, i.e. its underlying partition is the *thinnest* possible partition of $[n]$, $\mathcal{P}_{IKNP} = \{\{1\}, \ldots, \{n\}\}$. Since $t \geq \frac{1}{n}$, this input is rejected with overwhelming probability.

Simulating a malicious Alice* $\mathcal{S}$ externally sends (Alice, corrupt) to $c\mathcal{PAD}_{m,n}$. Next, it runs an internal execution of π. In step 1 $\mathcal{S}$ does nothing (acting as π_{B}). In step 2, $\mathcal{S}$ internally gets adversarial $\tilde{\mathbf{a}}$ as input to the inner $\mathcal{OT}$. $\mathcal{S}$ externally sends (corruptAlice, sid, $\tilde{\mathbf{a}}$) to $c\mathcal{PAD}_{m,n}$, obtaining matrix $\tilde{\mathbf{Q}}_0$. It samples at random r vectors $\boldsymbol{\sigma}_i \in \mathbb{F}_2^m$, and for each it sets n sharings $[\mathbf{L}_i, \mathbf{R}_i]$ (i.e. $\mathbf{l}_i^j \oplus \mathbf{r}_i^j = \boldsymbol{\sigma}_i$ for $j \in [n]$). Let $\mathbf{Q}_i$ a $m \times n$ matrix such that $\mathbf{q}_i^j = \mathbf{l}_i^j \oplus \tilde{\mathbf{a}}^j \cdot \boldsymbol{\sigma}_i$, $\mathcal{S}$ internally gives $[\tilde{\mathbf{Q}}_0, \mathbf{Q}_1, \ldots, \mathbf{Q}_r]$ to $\mathcal{A}$ in step 2.
Let $\tilde{\mathbf{e}} \in \mathbb{F}_2^r$ the adversarial challenge that $\mathcal{S}$ internally gets from $\mathcal{A}$ in step 3. If $e_i = 0$, $\mathcal{S}$ prepares proof $(\boldsymbol{\sigma}_i, [\mathbf{L}_i, \mathbf{R}_i])$. If $e_i = 1$, $\mathcal{S}$ prepares proof $(\boldsymbol{\sigma}_{i,+}, [\mathbf{L}_{i,+}, \mathbf{R}_{i,+}])$, where $\boldsymbol{\sigma}_{i,+}$ is sampled at random, and $[\mathbf{L}_{i,+}, \mathbf{R}_{i,+}]$ defines n sharings of $\boldsymbol{\sigma}_{i,+}$. Then $\mathcal{S}$ internally sends the r proofs to $\mathcal{A}$ in step 4. If π_{A} aborts in step 5, $\mathcal{S}$ externally sends to $c\mathcal{PAD}_{m,n}$ message (corruptAlice, sid, $\perp$). Lastly, $\mathcal{S}$ outputs whatever π_{A} outputs and halts.

Simulating a malicious Bob* $\mathcal{S}$ externally sends (Bob, corrupt) to $c\mathcal{PAD}_{m,n}$ and as a response obtains input $\mathbf{B}_{\mathcal{E}} = [\mathbf{L}_{\mathcal{E}}, \mathbf{R}_{\mathcal{E}}]$. It then sets π_{B}'s input to $\mathbf{B}_{\mathcal{E}}$, and runs an internal execution of π up to step 5 (π_{B} is controlled by $\mathcal{A}$). In step 2, $\mathcal{A}$ specifies an $m(r+1) \times 2n$ matrix $[[\mathbf{L}_0, \ldots \mathbf{L}_r], [\mathbf{R}_0, \ldots \mathbf{R}_r]]$ as input to $\mathcal{OT}^n_{m(r+1)}$, and in step 4 $\mathcal{A}$ specifies r proofs $(\tilde{\boldsymbol{\sigma}}_i, [\tilde{\mathbf{L}}_i, \tilde{\mathbf{R}}_i])_{i \in [r]}$.
Next, $\mathcal{S}$ runs step 5 of its internal copy of $\pi_{m,n,t}$, it sets flag Rabort to true iff it resulted in abort, but it does not tell $\mathcal{A}$ whether or not she passed. If Rabort is true $\mathcal{S}$ externally sends (corruptBob, sid, $\perp$) to $c\mathcal{PAD}_{m,n}$, outputs what π_{B} outputs and halts. Otherwise, it computes the $r + 1$ associated $m \times n$ matrices $\boldsymbol{\Delta}_i$ of the (adversarial) input [a] given to $\mathcal{OT}^n_{m(r+1)}$. For each $i \in [r]$, $\mathcal{S}$ finds the indices $j \in [n]$ such that $\tilde{\boldsymbol{\sigma}}_i \neq \boldsymbol{\delta}_i^j \oplus e_i \cdot \boldsymbol{\delta}_0^j$ (if any). Denote this subset of $[n]$ as G. Now, for those $j \in G$, $\mathcal{S}$ finds the first $k \in [m]$ such that $\tilde{\sigma}_{i,k} \neq \delta_{i,k}^j$, then it sets $\tilde{a}^j = l_{i,k}^j \oplus \tilde{l}_{i,k}^j$. Lastly, if G is empty, $\mathcal{S}$ externally sends (deliver, sid) to $c\mathcal{PAD}_{m,n}$. Otherwise it sends (corruptBob, sid, $[\mathbf{L}_0, \mathbf{R}_0]$, $\tilde{\mathbf{a}}$, G) to $c\mathcal{PAD}_{m,n}$. $\mathcal{S}$ tells to abort to $\mathcal{A}$ iff $c\mathcal{PAD}_{m,n}$ says so, outputs what π_{B} outputs and halts.

Simulating an honest execution $\mathcal{S}$ gets (sid) from $c\mathcal{PAD}_{m,n}$, runs an internal execution of π and halts.

---

[a] Recall how they are defined, i.e. $\boldsymbol{\Delta}_i$ has columns $\boldsymbol{\delta}_i^j = \mathbf{l}_i^j \oplus \mathbf{r}_i^j$ for $i \in [r] \cup \{0\}$, $j \in [n]$.

Fig. 6. The ideal adversary for $c\mathcal{PAD}$

Putting the Pieces in the UC Framework. We have not yet captured the notion of having blocks of $\mathbf{a}$ randomly distributed in Bob's view, it is resolved with a simulation argument. More concretely, we show a reduction to $\mathcal{OT}^n_{m(r+1)}$ with *perfect* security against Alice*, and *statistical* security against Bob*.

Theorem 1. *In the* $\mathcal{OT}^n_{m(r+1)}$*-hybrid, in the presence of static active adversaries, the output of protocol* $\pi_{m,n,t}$ *and the output of the ideal process involving* $c\mathcal{PAD}_{m,n}$ *are* $2^{-n(1-t)/2+2}$ *close.*

Proof. Let $\mathcal{E}$ denote the environment, and $\mathcal{S}$ be the ideal world adversary. $\mathcal{S}$ starts invoking an internal copy of $\mathcal{A}$ and setting dummy parties π_{A} and π_{B}. It then runs an internal execution of π between $\mathcal{A}$, π_{A}, π_{B}, where every incoming communication from $\mathcal{E}$ is forwarded to $\mathcal{A}$ as if it were coming from $\mathcal{A}$'s environment, and any outgoing communication from $\mathcal{A}$ is forwarded to $\mathcal{E}$. The description of $\mathcal{S}$ is in Fig. 6.

We now argue indistinguishability. Let Dist be the event of having $\mathcal{E}$ distinguishing between the ideal and real process. We bound the probability of Dist occurring conditioned on corrupting at most one of the parties.

Perfect security for Bob ($\mathsf{EXEC}_{\pi,\mathcal{E},\mathcal{A}} \equiv \mathsf{EXEC}_{\phi,\mathcal{E},\mathcal{S}}$). If Alice is malicious, then what $\mathcal{E}$ gets to see from Bob's ideal transcript is $(\mathbf{B}, [\tilde{\mathbf{Q}}_0, \mathbf{Q}_1 \ldots, \mathbf{Q}_r])$, $(\tilde{\boldsymbol{\sigma}}_i$,

$[\tilde{\mathbf{L}}_i, \tilde{\mathbf{R}}_i])_{i\in[r]})$, where $\mathbf{B} = [\mathbf{L}, \mathbf{R}]$ is the input, i.e. n sharings of, say, $\boldsymbol{\sigma}_0$. Matrix $\tilde{\mathbf{Q}}_0$ is consistent with $\mathbf{B}$ and with adversarial choice $\tilde{\mathbf{a}}$ (see Fig. 1), hence by definition of $\mathcal{S}$ and the robustness of $\mathcal{OT}^n_{m(r+1)}$, matrix $[\tilde{\mathbf{Q}}_0, \mathbf{Q}_1, \ldots, \mathbf{Q}_r]$ is exactly distributed as in the real process. Furthermore, if $\tilde{e}_i = 1$, then $\tilde{\boldsymbol{\sigma}}_i = \boldsymbol{\sigma}_{i,+}$ is randomly sampled, whereas in the real process $\tilde{\boldsymbol{\sigma}}_i = \boldsymbol{\sigma}_0 \oplus \boldsymbol{\sigma}_i$, with $\boldsymbol{\sigma}_i$ being in the *private* part of Bob's transcript. Therefore the proofs of the ideal and real process are identically distributed. We conclude that real and ideal transcripts are identically distributed, and therefore $Pr[\text{Dist}|\text{corruptAlice}] = 0$.

Statistical security for Alice ($\text{EXEC}_{\pi,\mathcal{E},\mathcal{A}} \overset{s}{\approx} \text{EXEC}_{\phi,\mathcal{E},\mathcal{S}}$). For the case Bob is corrupted, we first note that up to step 5, both processes are *identically* distributed because $\mathcal{S}$ runs an internal copy of $\pi_{m,n,t}$ using input $[\mathbf{L}_\mathcal{E}, \mathbf{R}_\mathcal{E}]$ specified by $\mathcal{E}$. Next, say $[\mathbf{L}_0, \mathbf{R}_0]$ is ℓ-thick. Then, if $\ell \leq t$, by Proposition 1, the size of G is at least $n(1-t)/2$ with overwhelming probability (in n), thus $c\mathcal{PAD}_{m,n}$ does not abort with probability $p \leq 2^{-n(1-t)/2}$. By Proposition 1 again the ideal and the real processes abort on thin inputs except with probability $p \leq 2^{-n(1-t)/2+2}$ (i.e. we do not care if $\mathcal{E}$ distinguishes in this case). On the other hand, if $\ell > t$ and the internal copy of $\pi_{m,n,t}$ did not abort (if aborts, so does $c\mathcal{PAD}_{m,n}$ by definition of $\mathcal{S}$), then we claim that the output of both processes are identically distributed. This follows from (1) the output matrix $[\mathbf{L}_0, \mathbf{R}_0]$ is extracted by $\mathcal{S}$, and looking closely at the proof of Lemma 1, we deduce (2) if $j \in G$, then for some $i \in [r]$, Bob* "decommits" to $\tilde{\boldsymbol{\sigma}}_i \neq \boldsymbol{\delta}_i^j \oplus e_i \cdot \boldsymbol{\delta}_0^j$, the real bit a^j is exactly as the one extracted by $\mathcal{S}$; (3) if $j \notin G$, then j is such that for each $i \in [r]$, Bob* is decommitting to $\tilde{\boldsymbol{\sigma}}_i = \boldsymbol{\delta}_i^j \oplus e_i \cdot \boldsymbol{\delta}_0^j$. In this case, the system of equations given in the proof of Lemma 1 collapses to $\mathbf{l}^j = \tilde{\mathbf{l}}^j$; $\mathbf{r}^j = \tilde{\mathbf{r}}^j$. One sees that if $\mathcal{E}$ could tell anything from $\mathbf{a}_{|[n]\backslash G}$, he could equally tell the same *before* the prove phase, contradicting the security of the underlying $\mathcal{OT}^n_{m(r+1)}$.

We have argued $Pr[\text{Dist}|\text{corruptBob}] \leq 2^{-n(1-t)/2+2}$.

Completeness. For the case none of the parties are corrupted, indistinguishability follows from the security of the underlying $\mathcal{OT}^n_{m(r+1)}$.

Adding up, $\mathcal{E}$ distinguishes with probability $Pr[\text{Dist}] \leq 2^{-n(1-t)/2+2}$. This concludes the proof.

4.4   Another Look at the Outer Reduction

Here we take a different perspective for the IKNP reduction that fits better with our partition point of view (as defined in Sect. 4.3). We aim to give some intuition of the underlying ideas, and the reader should by no means take the following discussion as formal arguments.

For an illustrative example let us first look at the attack of the protocol in the IKNP extension. A malicious Bob* was giving input matrix $\mathbf{B}$ with all the columns adding up to distinct elements. Consequently its underlying partition is $\mathcal{P}_{IKNP} = \{\{1\}, \ldots, \{n\}\}$. This structure on $\mathbf{B}$ is such that all but one of the bits of both pads are known to Bob*. One can see this as splitting the query space $\mathbb{F}_2^n$ as n copies of $\mathbb{F}_2$, namely $Q = \bigoplus_{i=1}^n \mathbb{F}_2$. To search for the secret vector $\mathbf{a}$, one just have to brute-force each summand separately and use the CRF to compare. After $n \cdot |\mathbb{F}_2| = 2n$ calls the query space is exhausted, i.e. even computationally bounded environments would distinguish between the ideal and the real process.

We want to assign to each possible matrix input $\mathbf{B} = [\mathbf{L}, \mathbf{R}]]$ a unique structure of the query space that the environment is forced to use towards distinguishing. In other words, we want to establish a correspondence between the partition implicitly defined in $\mathbf{B}$, and the possible ways to split the query space $Q = \mathbb{F}_2^n$.

Let $\mathcal{P}$ be any partition of $[n]$, express it as $\mathcal{P} = \{P_{1,1}, \ldots, P_{q_1,1}, \ldots, P_{1,n}, \ldots, P_{q_n,n}\}$ where for $i \in [n]$, $j \in [q_i]$, part $P_{j,i}$ is either empty or is of size i (i.e. there are q_i parts of size i in $\mathcal{P}$). The *type* of $\mathcal{P}$ is the vector $\mathbf{q} = (q_1, \ldots, q_n) \in \{[n] \cup \{0\}\}^n$. The $\mathbf{q}$-type query space, is the vectorial space $Q_{\mathbf{q}} = \bigoplus_{i=1}^n Q_{\mathbf{q},i}$, where $Q_{\mathbf{q},i}$ is the ith *block of* $Q_{\mathbf{q}}$, and stands for q_i copies of an $\mathbb{F}_2$-vectorial space of dimension i.

Thus, the type of $\mathcal{P}_{IKNP}$ corresponds to vector $\mathbf{q} = n \cdot \mathbf{e}_1$, and the query space the environment was using to brute-force Alice's secret $\mathbf{a}$ is precisely $Q_{n \cdot \mathbf{e}_1}$. On the other hand, honest inputs always define the trivial partition $\mathcal{P}_H = \{[n]\}$ with type $\mathbf{q} = \mathbf{e}_n$, the reduction against a semi-honest receiver in [IKNP03], based security arguing that the environment would have to brute-force $\mathbb{F}_2^n$, which is the query space $Q_{\mathbf{e}_n}$.

Now, the map $f : \mathbf{B} \mapsto Q_{\mathbf{q}}$, where $\mathcal{P}_{\mathbf{B}}$ is $\mathbf{q}$-type, is well defined. To see this, just observe that the relation in $\mathcal{P}^{[n]}$ defined as $\mathcal{P} \sim \mathcal{P}'$ iff "both partitions are of same type" is an equivalence relation, and look at the chain

$$\mathcal{M}_{m \times n} \xrightarrow{g_1} \mathcal{P}^{[n]} \xrightarrow{g_2} (\mathcal{P}^{[n]}/\sim) \xrightarrow{g_3} \mathcal{V}$$
$$\Delta \mapsto \mathcal{P}_{\Delta} \mapsto [\mathcal{P}_{\Delta}]_{\sim} = \mathbf{q} \mapsto Q_{\mathbf{q}}$$

We see that $f = g_3 \circ g_2 \circ g_1$ is well defined.

From this one can imagine how the reduction would work. $c\mathcal{PAD}$ could check the thickness of the adversarial $\mathbf{B}$, and reject if is less than a fixed parameter t. This ensures that the structure of the query space contains at least one block of size big enough, wasting the chances of the environment to search through it in reasonable time. Unfortunately, with this reduction the composition of the inner and outer protocols renders worst choices of parameters.

5   Concluding the Construction

In this section we prove the main result of the paper. For a given security parameter n recall that t is a parameter lying in interval $[\frac{1}{n}, 1)$, and $r = \lceil \frac{1-t}{2} n \rceil$. Observe that the results of Sect. 4 break down for $t = 1$. This corresponds to a superfluous $\pi_{m,n,1}$ (no checks at all). In other words, a malicious Bob* can input *any* possible bad-formed matrix $\mathbf{B}$ to the IKNP extension, in which case there is no security.

Corollary 1. *In the* $\mathcal{OT}^n_{m(r+1)}$*-hybrid, for any* $t \in [\frac{1}{n}, 1)$ *protocol* $\rho^{\pi_{m,n,t}/c\mathcal{PAD}_{m,n}}$ *UC-realizes* $\mathcal{OT}^m_n$ *in the presence of static active adversaries, provided the environment is given access to at most* $2^{o(n)}$ *queries to a* CRF.

Proof. The result follows applying the Composition Theorem of [Can01]. By Claim the error simulation for ρ is $e_\rho = 2^{-n/2+o(n)+2}$, and by Theorem 1 the error simulation for $\pi_{m,n,t}$ is $e_\pi = 2^{-n(1-t)/2+2}$. Using that $(1-t)/2 < 1/2$ if $t > 0$, and the transitivity of the composition operation, the error simulation for $\rho^{\pi_{m,n,t}/c\mathcal{P}\mathcal{A}\mathcal{D}_{m,n}}$ is $e = e_\rho + e_\pi \leq 2^{-n(1-t)/2+o(n)+3}$.

5.1   Complexity and Choice of Parameters

For the computational overhead, we emphasize that a cryptographic primitive is still needed to implement the actual extended transfers (we are using the IKNP extension). To implement $m = poly(n)$ transfers, in the test Alice and Bob have to XOR $rm(2n+1)$ bits. Thus, per extended OT each participant needs to XOR $O((1-t)n^2)$ bits. The communication complexity (number of bits transferred per OT) turns out to be equivalent. The test adds a *constant* number of rounds to the overall construction, concretely 2 extra rounds.

In terms of the seed expansion we can do it better. For a security level of s bits in the reduction, one need roughly $n \approx \frac{2}{1-t}(s + o(n) + 3)$ OT seeds. One can measure the quality of the reduction looking at the seed expansion factor $exp(t) = \frac{2}{1-t}$. It is clear that $exp(t)$ tends to 2, when $t \to \frac{1}{n}$ and $n \to \infty$. One only need to halve the security parameter of the IKNP reduction (asymptotically).

Practical choice of parameters are also very efficient. For example, to implement about $1,000,000$ transfers, with security of $s = 64$ bits, setting $t = \frac{1}{16}$, one needs roughly $n \approx 186$ OT seeds. For security level $s = 128$, one would need roughly 323 OT seeds.

5.2   Open Problems

In the reductions for ρ and π the security parameter suffers an expansion factor of 2. We ask whether one can remove this overhead whilst still maintaining security against computational unbounded receivers in the inner protocol.

In the area of secure function evaluation, recently OT has been used to boost the efficiency of two-party protocols [NNOB12] and their counterparts in the multiparty case [LOS14]. A key part on the design of such protocols was the generation of authenticated bits, which in turn borrows techniques from the IKNP extension. It would be interesting to see whether (a suitable modification of) our protocol π can be used to generate such authenticated bits. This would immediately give unconditional security (currently both constructions need a random oracle), in terms of efficiency we do not know if this replacement would bring any improvement at all.

Acknowledgments. This work has been supported in part by EPSRC via grant EP/I03126X.

References

BCR86a. Brassard, G., Crépeau, C., Robert, J.M.: All-or-nothing disclosure of secrets. In: Odlyzko, A.M. (ed.) CRYPTO 1986. LNCS, vol. 263, pp. 234–238. Springer, Heidelberg (1987)

BCR86b. Brassard, G., Crépeau, C., Robert, J.-M.: Information theoretic reductions among disclosure problems. In: FOCS, pp. 168–173 (1986)

Bea96. Beaver, D.: Correlated pseudorandomness and the complexity of private computations. In: STOC, pp. 479–488 (1996)

Can01. Canetti, R.: Universally composable security: a new paradigm for cryptographic protocols. In: FOCS, pp. 136–145 (2001)

CK88. Crépeau, C., Kilian, J.: Weakening security assumptions and oblivious transfer. In: Goldwasser, S. (ed.) CRYPTO 1988. LNCS, vol. 403, pp. 2–7. Springer, Heidelberg (1990)

CLOS02. Canetti, R., Lindell, Y., Ostrovsky, R., Sahai, A.: Universally composable two-party and multi-party secure computation. In: STOC, pp. 494–503 (2002)

Cré87. Crépeau, C.: Equivalence between two flavours of oblivious transfers. In: Pomerance, C. (ed.) CRYPTO 1987. LNCS, vol. 293, pp. 350–354. Springer, Heidelberg (1988)

Cré89. Crépeau, C.: Verifiable disclose for secrets and applications. In: Quisquater, J.-J., Vandewalle, J. (eds.) EUROCRYPT 1989. LNCS, vol. 434, pp. 150–154. Springer, Heidelberg (1990)

EGL85. Even, S., Goldreich, O., Lempel, A.: A randomized protocol for signing contracts. Commun. ACM **28**(6), 637–647 (1985)

GMW86. Goldreich, O., Micali, S., Wigderson, A.: Proofs that yield nothing but their validity and a methodology of cryptographic protocol design (extended abstract). In: FOCS, pp. 174–187 (1986)

GV87. Goldreich, O., Vainish, R.: How to solve any protocol problem - an efficiency improvement. In: CRYPTO, pp. 73–86 (1987)

HIKN08. Harnik, D., Ishai, Y., Kushilevitz, E., Nielsen, J.B.: OT-combiners via secure computation. In: Canetti, R. (ed.) TCC 2008. LNCS, vol. 4948, pp. 393–411. Springer, Heidelberg (2008)

IKNP03. Ishai, Y., Kilian, J., Nissim, K., Petrank, E.: Extending oblivious transfers efficiently. In: Boneh, D. (ed.) CRYPTO 2003. LNCS, vol. 2729, pp. 145–161. Springer, Heidelberg (2003)

IPS08. Ishai, Y., Prabhakaran, M., Sahai, A.: Founding cryptography on oblivious transfer – efficiently. In: Wagner, D. (ed.) CRYPTO 2008. LNCS, vol. 5157, pp. 572–591. Springer, Heidelberg (2008)

IR89. Impagliazzo, R., Rudich, S.: Limits on the provable consequences of one-way permutations. In: STOC, pp. 44–61 (1989)

Kil88. Kilian, J.: Founding cryptography on oblivious transfer. In: STOC, pp. 20–31 (1988)

LOS14. Larraia, E., Orsini, E., Smart, N.P.: Dishonest majority multi-party computation for binary circuits. In: Garay, J.A., Gennaro, R. (eds.) CRYPTO 2014, Part II. LNCS, vol. 8617, pp. 495–512. Springer, Heidelberg (2014)

LZ13. Lindell, Y., Zarosim, H.: On the feasibility of extending oblivious transfer. In: Sahai, A. (ed.) TCC 2013. LNCS, vol. 7785, pp. 519–538. Springer, Heidelberg (2013)

Nie07.   Nielsen, J.B.: Extending oblivious transfers efficiently - how to get robust-
          ness almost for free. IACR Cryptology ePrint Arch. **2007**, 215 (2007)
NNOB12.   Nielsen, J.B., Nordholt, P.S., Orlandi, C., Burra, S.S.: A new approach to
          practical active-secure two-party computation. In: Safavi-Naini, R., Canetti,
          R. (eds.) CRYPTO 2012. LNCS, vol. 7417, pp. 681–700. Springer, Heidel-
          berg (2012)
PVW08.   Peikert, C., Vaikuntanathan, V., Waters, B.: A framework for efficient and
          composable oblivious transfer. In: Wagner, D. (ed.) CRYPTO 2008. LNCS,
          vol. 5157, pp. 554–571. Springer, Heidelberg (2008)
Rab81.   Rabin, M.O.: How to exchange secrets with oblivious transfer. IACR Cryp-
          tology ePrint Arch. 187 (1981)
Wie83.   Wiesner, S.: Conjugate coding. SIGACT News **15**, 78–88 (1983)
Yao82.   Yao, A.C.-C.; Protocols for secure computations (extended abstract). In:
          FOCS, pp. 160–164 (1982)

Author Index

Printed in the United States
By Bookmasters